Degradation of Implant Materials

Noam Eliaz
Editor

Degradation of Implant Materials

Springer

Editor
Noam Eliaz
Faculty of Engineering
School of Mechanical Engineering
Tel Aviv University
Tel-Aviv, Israel

ISBN 978-1-4614-3941-7 ISBN 978-1-4614-3942-4 (eBook)
DOI 10.1007/978-1-4614-3942-4
Springer New York Heidelberg Dordrecht London

Library of Congress Control Number: 2012945197

Printed on acid-free paper

Springer is part of Springer Science+Business Media (www.springer.com)

metal ion release, biofilm formation, and contamination are discussed briefly. Materials such as titanium and its alloys, calcium phosphates, and zirconia are analyzed. In Chap. 4, *T. Eliades*, *S. Zinelis* and *W.A. Brantley* extend the discussion on dental implants—their aging, corrosion, and failure.

Chapter 5, by *F. Witte* and *A. Eliezer*, starts a series of chapters which describe positive effects and applications of controlled degradation. This chapter covers different aspects of biodegradable implants, with focus on magnesium- and iron-based alloys, and absorbable stents. In Chap. 6, *A.R. Boccaccini et al.* discuss bioactive and biodegradable scaffolds for bone tissue engineering, which are made of a synthetic biodegradable polymer matrix, such as poly(D,L-lactide) (PDLLA), and an inorganic reinforcement such as Bioglass®. In Chap. 7, *J.A. Jansen et al.* discuss different factors that influence the biodegradation of calcium phosphate cements. These cements are good candidates as bone grafts in dental, orthopaedic, and reconstructive surgery. In Chap. 8, *N. Lotan*, *R. Azhari*, and *T. Gold* present the design and performance evaluation of an implantable, degradable, drug delivery device, the function of which is controlled by the concentration and activity of a given enzyme present at the site of implantation. In Chap. 9, *J. Chevalier et al.* comprehensively review the degradation mechanisms of ceramic implants, both bioinert and bioactive, and the interactions between them and their environment. Crack propagation mechanisms are discussed, along with a variety of materials such as hydroxyapatite and other calcium phosphates, calcium sulfate, bioactive glasses, and composites of them.

Chapter 10, by *N. Eliaz* and *K. Hakshur*, starts a series of chapters which focus on the outcomes of wear and fatigue of biomaterials and implants. This chapter first gives brief introduction to the three elements of tribology—friction, lubrication, and wear. Subsequently, the principles and use of Ferrography and Bio-Ferrography in isolating wear debris and monitoring either the degradation of bone and cartilage during osteoarthritis or the degradation of hip and knee artificial joints are reviewed comprehensively. The concept of soft bearing materials such as polycarbonate-urethane (PCU) is presented. In Chap. 11, *S.H. Teoh*, *Y.L. Teo*, and *S.K. Chong* discuss fatigue failures of medical devices. First, basic fundamentals and terminologies in fatigue mechanics are presented. Subsequently, case studies of failures of hip and knee prostheses as well as of dental restoratives are discussed. In Chap. 12, *N.J. Hallab* reviews the topic of hypersensitivity to implant debris. Metal sensitivity, related tests, and case studies are presented.

Chapter 13, by *C.R. Arciola*, *D. Campoccia*, and *L. Montanaro*, discusses implant infections, historical approach to their prevention, and infection-resistant materials. In Chap. 14, *N. Vyavahare*, *F. Schoen*, and *A. Munnelly* discuss mechanisms and prevention strategies of calcification in the two classes of implants: biologically derived and synthetic. Calcification of bioprosthetic heart valves, collagen and elastin tissue engineering scaffolds, polyurethane, silicone breast implants, hydrogels and ophthalmic implants, and intrauterine devices (IUDs) is reviewed.

Chapter 15, by *L.C. Jones*, *A.K. Tsao*, and *L.D.T. Topoleski*, reviews the significance of retrieved orthopaedic implants and failure analysis to their long-term

Preface

The degradation of biomaterials is one of the major considerations in their design, processing, and use. It might be undesirable (e.g., corrosion of metallic dental implants, wear of polyethylene in artificial hip joints, or calcification of stents), but in other cases it may be desirable (e.g., absorbable sutures, biodegradable polymers for drug delivery, or tissue engineering). In both cases, the kinetics of the degradation is of paramount importance. The undesirable degradation might lower the structural integrity of the implant or release metal ions and debris that elicit an adverse biological reaction (possibly causing synovitis, periprosthetic bone loss, cytotoxicity, allergy, or even cancer).

This book summarizes the current understanding of the mechanical, chemical, and biological processes that are responsible for the degradation of a variety of implant materials. The 18 chapters were written by internationally renowned experts and address both fundamental and practical aspects. Different failure mechanisms such as corrosion, fatigue, and wear are reviewed, together with experimental techniques for monitoring them, either *in vitro* or *in vivo*. Procedures for implant retrieval and analysis are presented. A variety of biomaterials (stainless steels, titanium and its alloys, magnesium alloys, polyethylene, polycarbonate-urethane, biodegradable polymers, calcium phosphates, etc.) and medical devices (orthopaedic and dental implants, stents, heart valves, etc.) are analyzed in detail.

In Chap. 1, *J.L. Gilbert* and *S. Mali* present the complex interactions that occur in the human body during corrosion of metallic implants. The concepts associated with oxide films and their interaction with the biological, mechanical, and electrochemical environments are discussed, as is mechanically assisted corrosion. Specific examples of biotribocorrosion are discussed. In Chap. 2, *S. Virtanen* summarizes the state-of-the-art knowledge on the degradation modes of titanium and biomedically relevant Ti-based alloys. Issues such as passivity, the effect of alloying elements, and tribocorrosion are discussed comprehensively. A short discussion on some relevant implant design-related aspects of degradation is also provided.

In Chap. 3, *T. Hanawa* reviews different aspects of the degradation of dental implants. Issues such as wear and fracture, the biological environment, corrosion, survival. Issues such as the musculoskeletal system, orthopaedic implants, failure, preclinical and clinical testing, implant retrieval programs, the role of materials selection and implant design, and biological responses to implants are discussed comprehensively. In Chap. 16, *M. Wu* and *P. Briant* summarize the use of finite element analysis (FEA) in design, life prediction, and failure analysis of biomaterials and medical devices. Nitinol wire frame based inferior vena cava filter (IVCF) is given as a case study.

Chapters 17 and 18 discuss the biological responses to and toxicity of polymers and nanomaterials, respectively. First, in Chap. 17, *J.C. Park* and *B.J. Park* review the biological response following implantation of biodegradable polymers and some methods for biological safety evaluation of biodegradable materials recommended by the International Organization for Standardization (ISO) and the US Food and Drug Administration (FDA). Nanomaterial safety and toxicity are of great importance for nanomaterial-based medical implants. Then, in Chap.18, *T.J. Webster* and *L. Yang* introduce the host responses to implant materials and properties of nanomaterials pertinent to their altered biological responses. Next, the advances and progression of biological responses (especially concerning the toxicity of nanoscale implant materials, either after production or implantation) are summarized.

I hope that this book will become a reference source for undergraduate and graduate students, college and university professors, scientists, engineers, implant manufacturers, venture capitalists, regulatory entities, and research professionals working both in academia and industry. It may be of interest to materials, mechanical, biomedical, and corrosion engineers; biologists and medical doctors; chemists and electrochemists; surface scientists; failure analysts; etc.

I dedicate this book to my wife Billie and our three children—Ofri, Shahaf, and Shalev—for their infinite love and support.

Tel-Aviv, Israel Noam Eliaz

Contents

Biography

Professor Noam Eliaz

Noam Eliaz is an Associate Professor at Tel-Aviv University, Israel, where he serves as the Head of the Biomaterials and Corrosion Laboratory. He also serves as a Chief Editor of the journal *Corrosion Reviews* (jointly with Ron Latanision). He received his B.Sc. and Ph.D. (direct track) in Materials Engineering, both cum laude, from Ben-Gurion University. Next, he became the first ever materials scientist to receive, simultaneously, a Fulbright postdoctoral award and a Rothschild postdoctoral fellowship and worked for 2 years in the H.H. Uhlig Corrosion Laboratory at M.I.T. To-date, he has contributed more than 230 journal and conference publications, including 31 plenary and invited talks, as well as 5 book chapters. In addition to editing this *Degradation of Implant Materials* book, he has edited a double volume entitled "Applications of Electrochemistry and Nanotechnology in Biology and Medicine" for the reputed book series *Modern Aspects of Electrochemistry* (Springer). He has garnered numerous accolades, including the T.P. Hoar Award for the best paper published in *Corrosion Science* during 2001 (on corrosion of Ti–Ag-based alloys processed by three-dimensional printing for biomedical applications) and NACE International's Herbert H. Uhlig Award (2010) and Fellow Award (2012). His main research interests include environment-induced degradation of materials, failure analysis, Bio-Ferrography, biomaterials (with focus on electro-crystallization of hydroxyapatite and other calcium phosphates), and electrochemical processing (namely, electrodeposition, electroless deposition, and electropolishing) of materials.

Chapter 1
Medical Implant Corrosion: Electrochemistry at Metallic Biomaterial Surfaces

Jeremy L. Gilbert and Sachin A. Mali

Abstract Metallic biomaterials represent the class of materials with the largest use in medical devices in humans today. This fact will likely continue for decades to come because of the unique combination of strength, wear resistance, and corrosion resistance. However, metallic biomaterials also pose unique and specific concerns related to electrochemical behavior in the body. This chapter will focus on the elements of most importance in understanding the complex interactions present in the human body during corrosion of metallic implants. The concepts associated with oxide films and their interaction with the biological, mechanical, and electrochemical environments are discussed to provide insight into why corrosion is a critically important factor in the long-term performance of devices. Mechanically assisted corrosion in the biological system is discussed in terms of the structural, electrochemical, and biological interactions, and the idea of electrochemical history is presented to explain why such severe evidence of corrosion is observed *in vivo*. Finally, specific examples of mechanically assisted corrosion *in vivo* (or biotribocorrosion) are presented, and recent observations concerning the important role the reduction half-cell plays in the biological response to corrosion are discussed.

1 Introduction

1.1 A Short History on Metallic Biomaterials and Corrosion

Corrosion of metallic biomaterials has been an issue for as long as metals were considered for surgical repair. Early efforts at bone fracture repair, dental restoration, sutures, and other surgical applications dating back to the early twentieth

J.L. Gilbert (✉) • S.A. Mali
Department of Biomedical and Chemical Engineering, Syracuse Biomaterials Institute, Syracuse University, Syracuse, NY 13244, USA
e-mail: gilbert@syr.edu

N. Eliaz (ed.), *Degradation of Implant Materials*, DOI 10.1007/978-1-4614-3942-4_1,

century and earlier [1, 2] showed significant advancements in utilizing metals and alloys in the human body. With the attempts to introduce metallic biomaterials into the body came the inevitable problems associated with their use. Typically, metals were used in high-repetitive-load situations (i.e., fatigue environments), where surface wear processes take place, and where they are exposed to body fluids that led to electrochemical attack.

Early use of gold and silver alloys in applications like dental fillings and skull plates pointed to the early recognition of corrosion as a significant problem in the application of metals to the body. Silver–mercury dental amalgam usage dates back about 180 years (1833) [3], while there are early references to the assessment of steels and magnesium alloys dating back to the early 1900s [2, 4]. Dentistry pioneered the early use of metals in the body in terms of dental restorations, bridges, and crowns. One of the first uses of cobalt was in the development of base-metal alloy partial dentures. Indeed, Haynes Stellite 21, also referred to as Vitallium, was utilized as a medical/dental alloy in the late 1920s, well before it was adapted for use in high-temperature aircraft parts in 1941 [4].

Development of the currently used medical alloys accelerated in the 1920s with the application of both Co–Cr–Mo alloys and stainless steels (18% Cr–8% Ni) in the 1920s. Stainless steel was further improved by the addition of 2–3% molybdenum (Mo) to enhance pitting resistance—ultimately leading to the development of AISI 316L stainless steel and some more recent alloys. Titanium alloys were not significantly applied to biomaterials applications until the 1960s and were adapted from the aircraft industry. The work-horse alloy, Ti–6Al–4V, continues to be a primary alloy for use in orthopedic and spinal applications where high strength and high corrosion resistance are required; however, other alloys including commercially pure titanium, β titanium alloys, and alloys that eliminate vanadium are in use as well. More recently, other alloys are seeing increased use in biomaterials applications. This includes zirconium alloys, tantalum, niobium, and others. Shape memory alloys, in particular NiTi, are also seeing increased use in medical applications.

In most of the efforts to develop and use metallic biomaterials, the goal has been to have a material with suitable mechanical properties for the application while also having appropriate biocompatibility. Early studies by Laing et al. [5] looked at as many metal coupons as they could identify and implanted them into the paravertebral pocket of rabbits, allowed the tissue to heal, and then evaluated the fibrous capsule thickness. The basic finding from this work was that the more corrosion resistant the metal was, the more biocompatible the interaction (smallest fibrous capsule).

1.2 The More Corrosion Resistant, the More Biocompatible

This is the central paradigm of metallic biomaterials biocompatibility today. *The more corrosion resistant, the more biocompatible* has been the view of metals in the body for decades. Thus, the major implant alloys in use today are those that have been shown to be highly corrosion resistant. However, corrosion resistance can result from several factors, including no or small driving force for corrosion (e.g., noble alloys: Au, Ag, Pt) and a barrier to corrosion in the form of a passive metal-oxide thin film that spontaneously forms on the surface. Alloys based on steel, Co–Cr, Ti, Zr, Nb, Ta, all owe their corrosion resistance to the development of this metal oxide thin film which acts as a kinetic barrier to corrosion. It needs to be clearly stated here that these alloys have high driving forces to corrode. Thus, if the oxide films on their surfaces are breached or disrupted, then highly energetic oxidation reactions of the underlying metal will occur, releasing ions into the environment until the oxide film reforms, which will occur within milliseconds. Thus, any discussion of corrosion of medical alloys needs to include a discussion of conjoint effects between mechanical processes that can disrupt the oxide films and the biological and electrochemical processes that arise during the repassivation processes that lead to reestablishment of the oxide film.

The oxide films on these surfaces, while referred to as passive films, are not, in fact, passive but are highly dynamic in their response and interaction with the biological environment [6–9]. The oxides are responsive to the prior electrochemical history of the surface and can change their structure and properties (e.g., impedance [10]) with exposure to different solutions and voltage conditions on their surface. That is, the electrochemical history of the passive oxide film is very important in determining subsequent behavior. Electrochemical history can be thought of as the time course of solution and voltage conditions to which the surface is exposed. The behavior of a surface will depend on this prior history, sometimes in very profoundly different ways than is typically expected for an alloy surface. This is important since one consequence of mechanical disruption of the oxide films and their repassivation is the large transient shift in voltage across the interface that results from the burst repassivation and slow reduction-coupled reactions that result. Prolonged exposure to cathodic potentials that results from triboelectrochemical interactions of the surface may result in dramatically altered corrosion resistance of the entire oxide-film covered surface, even in locations not exposed to mechanical abrasion.

The consequences of corrosion in the biological system have, for the most part, focused on the effects of the ions and particles released from the metal surface on the adjacent biological system. Ion toxicity and local particulate burden have significant effects on the body and need to be carefully considered [11]; however, these are not the only factors and effects worth noting. Recently, the metallic biomaterials biocompatibility paradigm has come under some pressure to be modified. For example, recent studies of Mg alloys *in vivo* have shown that highly corroding alloys can induce high rates of bone formation [2, 12]. Interestingly, this effect was noted as early as 1910 [1]. Second, other work has shown that titanium and Co–Cr–Mo can be induced to cause cell death *in vitro* with relatively small impressed cathodic voltages [13–15]. Also, published work has shown that protein adsorption and conformation on metallic surfaces can be affected by substrate electrochemical state [16]. With these recent developments, it is clear that metallic biomaterials can have profound effects on the biological system, beyond the effects of ions and particles, and the biological system can also profoundly influence the electrochemical behavior.

Indeed, corrosion of metallic biomaterials is mostly a story of conjoint effects. That is, corrosion, in and of itself, in most of the alloys used is not a significant problem when no other effects are at play. However, when combined with mechanics, restricted crevice-like geometries, inflammation, or any combination thereof, significantly increased corrosion rates can occur that may lead to adverse biological process or failure of the device.

This chapter will discuss metallic biomaterials corrosion from the materials science perspective and will also provide insight into some of the biological consequences and interactions that are possible. This chapter will not focus on the consequences of metal ion release or particle generation on the local and systemic tissues of the body. There are several good reviews of this in the literature [11].

2 Basics of Corrosion

2.1 Oxidation and Reduction

Corrosion processes are electrochemical reactions that take place at electrode surfaces which include charge transfer across the electrode–electrolyte interface [17]. Two basic types of electrode reactions can occur: oxidation and reduction [see (1.1) and (1.2)]. In oxidation, a chemical species reacts to raise its valence state by releasing electrons, while reduction processes reduce the valence state of other species by taking on an electron. The convention used here is that when positive charge moves from electrode to electrolyte (i.e., metal atom in the metal surface going to metal ion in solution), this represents a positive (or anodic, or oxidation) current. Similarly, when positive charge moves from electrolyte to electrode, it is a negative (or cathodic, or reduction) current. In all electrochemical systems that are closed systems (i.e., no external source or sink for charge), a fundamental rule of electrode reactions is that the sum of all anodic currents must balance out (averaged over time) with the sum of all cathodic currents. That is, the electrode–electrolyte interface will adjust its voltage to arrive at the conditions where both anodic and cathodic currents are equal. Or, put another way, there can be no long-term net accumulation of charge. This is called the "Mixed Potential Theory" [17], and a clear understanding of this concept is essential for understanding how corrosion processes work.

$$\begin{array}{c} Co \rightarrow Co^{2+} + 2e^- \\ 2Cr + 3H_2O \rightarrow Cr_2O_3 + 6H^+ + 6e^- \end{array} \quad \text{(Oxidation reactions)} \qquad (1.1)$$

$$O_2 + H_2O + e^- \rightarrow HO_2^{\bullet} + OH^-$$
$$HO_2^{\bullet} + H_2O \rightarrow H_2O_2 + OH^{\bullet} \text{ (Reduction reactions involving oxygen)} \quad (1.2)$$
$$H_2O_2 + 2e^- \rightarrow 2OH^-$$

$$O_2 + 2H_2O + 4e^- \rightarrow 4OH^-$$
$$O_2 + 2H_2O + 2e^- \rightarrow H_2O_2 + 2OH^-$$

Note the two oxidation reactions (1.1) represent ionic dissolution and oxide formation reactions. Both liberate electrons into the metal which will act to lower the potential of the implant (i.e., make more negative) until the reduction reactions can remove these electrons. The reduction reactions listed are a limited subset of possible reduction reactions that may be present at implant surfaces. They involve mostly oxygen, water, and their reactive oxygen by-products, including hydroxide radicals and hydrogen peroxide. These are listed to emphasize that there are intermediate species that may arise from reduction processes which are known to have significant effects on the biological system and can induce oxidative stress in cells [18]. These oxidation and reduction reactions are not meant to be an exhaustive listing, but rather a representative list of possible reactions. In particular, there are numerous other potential reduction reactions possible in the biological system, including reduction of disulfide bonds [19] and other protein-like molecules. The effect of redox reactions at implant surfaces on the biological system has received very little in-depth study and much is still unknown about the interactions that may arise. Indeed, the redox state of cells is highly dependent on numerous redox processes across the cell membrane, within the cell and outside of the cell. The redox state of the cell is known, for example, to influence adhesion sites, actin cytoskeletal formation, membrane potential, permeability, etc. [20–24]. Thus, the redox state of cells is likely affected by the local redox environment presented by a metallic implant alloy which can serve as a source or sink for oxidation or reduction equivalents.

Recent studies have focused on the consequences of reduction of disulfide bonds in terms of protein structure and properties, membrane transport, and intracellular behavior (e.g., redox sensitive chaperones, NAD/H and NADPH behavior, ROS and oxidative stress, etc.). Indeed, some of these studies have shown the importance of disulfide bonds, present primarily in cysteine and methionine amino acids, which are highly susceptible to redox processes [20], and which are involved in the oxidative stress response and in intracellular chaperoning. Thiol and disulfide redox status also affects the ease of uptake of cell-penetrating peptides (CPPs) [21] including transmembrane transport of prions and other short peptides [22]. The redox status of a cell is critical to its oxidative stress state. It seems very likely that when cells are adjacent to a half-space (metal surface), actively reducing species available at their surface, including disulfide-containing molecules, will affect their redox balance.

Additional work in the redox biology of cells has involved reactive oxygen species (primarily, NO and H_2O_2) and their role on a variety of biological systems.

Oxidative stress and the generation of H_2O_2 by the cell have been shown to be able to depolymerize actin [23] and alter cellular adhesion [24].

All half-cell reactions (i.e., those reactions that are either oxidation or reduction reactions) establish a potential at the electrode interface, called the equilibrium potential, to balance the forward and reverse reactions for that half-cell. Note the forward reaction for a half-cell can be defined as the reduction reaction. Running this reaction in reverse results in an oxidation reaction. That is, a half-cell reaction is a reduction reaction in one direction and an oxidation reaction in the other. Which reaction occurs depends on the potential of the electrode relative to the equilibrium potential for the half-cell. When multiple half-cell reactions are present on the same surface with different half-cell equilibrium potentials, then the electrode cannot satisfy the equilibrium potential for all simultaneously. Therefore, the electrode potential will change until all oxidation processes balance with all of the reduction half-cell reactions. This potential is called the Open-Circuit Potential (OCP) and it is the potential that defines where all half-cell reactions balance out to net zero current. If the OCP is above a particular half-cell equilibrium potential, then that half-cell reaction will be an oxidation reaction. Whereas if the OCP is below the half-cell potential, the reaction will be a reduction reaction. The rate of the specific reaction will be governed by the magnitude of the voltage difference between OCP and the equilibrium potential, the availability of the species, and the voltage–current characteristic behavior governing the kinetics of the reaction. For nonoxide film covered electrodes, the Butler–Volmer equation [25] governs this behavior and shows that currents rise exponentially with voltage away from the half-cell potential. However, with passive films, other kinetic factors alter the electrode behavior and give more complex behavior.

It is important to note that on real metallic biomaterial surfaces, the OCP of the surface can undergo significant changes depending on what reactive species are present and how they affect the overall balance of reactions. For example, increasing concentrations of hydrogen peroxide (a by-product of inflammation) typically results in a more positive OCP for a particular alloy surface, while additions of proteins to the electrolyte will typically decrease the OCP of the surface. Because most of these surfaces are oxide film covered, and these oxide films are sensitive to voltage and solution conditions [6, 9], OCP shifts can occur not only due to the presence of a new half-cell reaction, but also because of alterations in the film's kinetic barrier behavior. That is, the oxide film's charge transfer characteristics and how they are affected by the surrounding species is an important element in the overall balance that is present.

2.2 *Oxide Film Kinetic Effects and Disruption*

Oxide film charge transfer behavior is highly complex and is comprised of both the mechanisms of ionic transport across the film and the electronic charge transfer [26]. Ionic transport is a function of the process of migration where the high electric field that develops across this oxide will drive cations (i.e., metal ions) to move in one direction (usually from metal to electrolyte), while also driving anions (e.g., oxygen ions) from the electrolyte to the metal. The process of migration involves movement due to the presence of an electric field and the atomistic mechanisms by which ions jump through these thin film oxides (e.g., vacancy motion, boundary transport, etc.). Also important in the movement of charge through the oxide is the transport of electrons and electronic holes (regions of positive charge) through the oxide. This involves adoption of the concepts from the semiconducting field, including electron tunneling, Schottky barriers, charge carrier densities, Fermi energy levels, and flat band potentials. Basically, electron (or electronic charge) movement through oxide films can be dramatically altered by the electronic nature of the oxide film present. For a more detailed discussion, please look at the following reference [27].

Thus, the nature of the oxide thin film on these surfaces is highly influential in terms of the severity and extent of corrosion reactions that can occur. For example, when the oxide film is mechanically removed from the surface such that bare metal is exposed directly to electrolyte, several events transpire at very high rates. Within the first nanosecond or two, the surface metal atoms begin to oxidize and either leave the metal surface as ions, or remain as oxidized ions on the surface. Electrons are liberated from the atoms and attach themselves, typically to oxygen atoms at the surface, and an intense electric field forms that can reach the order of 10^7 V/cm. Within one to two milliseconds this electric field will provide the driving force for repassivation as metal oxidation results in oxide film formation by interaction with water molecules [i.e., metal plus water goes to metal oxide plus hydrogen ion plus electrons, see (1.1)].

This burst reaction will reform the oxide film within milliseconds, according to the low-temperature, high-field growth models first proposed by Cabrerra and Mott, and Gunthershultze and Betz [28, 29], but will also accumulate a large number of electrons in the metal that will result in an overall shift in the surface potential to more negative values. That is, an oxide film covered surface that is abraded will see its OCP shift to more cathodic potentials. As the potential shifts to more negative values, the corresponding reduction reactions will increase until the excess electrons are removed and the potential slowly moves more positive back to its starting OCP. These shifts in OCP are important to understand because as mentioned, they (1) change the structure of the oxide film, (2) alter the reactions that are taking place, and (3) can result in propagating electric fields that may affect biological processes in the adjacent tissue [30].

3 The Electrochemical, Biological, and Mechanical Environment

3.1 Conjoint Effects in Biomaterials Corrosion

To understand medical alloy corrosion and its impact on the use of alloys in the human body, it is useful to have a clear understanding of the environments in which these alloys operate. This includes the chemical/electrochemical environment, the biological environment, and the mechanical environment. Additionally, a detailed understanding of the nature of the surface of alloys and the interfacial zone of the biological system is crucial for an understanding of the complex processes related to corrosion. The role of the biological system on the corrosion process, as well as the effect of the corrosion processes on the biological system, is also necessary to be able to design and utilize metals in the body.

Also, virtually all corrosion testing takes place in *in vitro* tests that try to simulate, in a simplified fashion, the electrochemical conditions of the body. Moving to simplified environments may make control and interpretation of experiments more feasible, but they also remove important potential interactions that may dominate particular metal–body behavior. As a result of the fact that most corrosion testing is done in simplified physiological solutions (e.g., pH 7.4 buffered isotonic saline solutions), very little systematic assessment of how complex "real" biological systems affect corrosion behavior has been explored. This includes assessment of restricted crevice-like geometries associated with, for example, bone cement–metal interfaces [31], modular tapers, and deep implant locations like diaphyseal intramedullary canals.

3.2 The Electrochemical Environment

In most *in vitro* tests assessing the corrosion performance of metallic biomaterials, the solutions used are highly approximate substitutes for the real biological environment. To first-order approximation, the body is a pH 7.4, 0.9% (0.154 M) NaCl solution. This is the most simplified solution used to evaluate corrosion. More complex solutions used *in vitro* include phosphate buffered saline (PBS), Ringer's solution, Hank's solution, simulated biological fluid (SBF), and several others. All of these are inorganic salt solutions that approximate the inorganic constituents in body solutions. In some cases additional constituents are included that are typically organic components like amino acids, single protein solutions (e.g., albumin), cell culture medium, and medium with serum additions. These combined solutions again attempt to get more representative of the actual body environment, yet do not come close. They also raise the complexity of the possible electrochemical reactions present and make interpretation of experiments more difficult, yet they also give better insight into how complex biological systems may interact with metal surfaces.

Some general comments can be made about *in vitro* electrochemical environments. These are typically aqueous salt solutions that provide the electrolyte to complete the electrochemical circuit (to allow both oxidation and reduction processes). Typical redox-active species in these solutions are dissolved oxygen and water. The ionic salts are not thought to participate in reduction reactions. When additional species are added, depending on the redox potential and the ease of the reaction, these additional species may participate in the processes. Examples of additions include hydrogen peroxide. If the added species do not participate in redox processes, but do adsorb to surfaces, that may (or may not) significantly alter the overall redox balance of the surface.

3.3 The Biological Environment

The biological environment into which metallic biomaterials are placed is highly complex, time and spatially varying, and can result in very severe corrosion conditions. It is important to understand, in any evaluation of metallic biomaterials corrosion, that metal implants are placed into a wound site. That is, the surgical intervention creates a zone of damage around the metal where the wound healing process must occur. Thus, the environment adjacent to the metal, at least initially (during acute inflammation) or chronically, can be dramatically altered in terms of its electrochemical effects. Many of the cellular and biochemical constituents in a wound site can produce redox processes. As mentioned, species like hydrogen peroxide, hypochlorous acid, peroxynitrite, and other enzymes are redox active and can create conditions locally that are very different than the *in vitro* solutions used in corrosion testing [32].

Researchers have shown that hydrogen peroxide can, for example, alter the oxide film on titanium alloys [33]. Indeed, significant alterations in corrosion behavior of titanium alloys can be induced by a combination of hydrogen peroxide and voltage history. Chandrasekaran et al. showed, for example, that cathodic polarization of Ti–6Al–4V in the presence of 0.03 M hydrogen peroxide in PBS to -1 V for 10 min followed by anodic polarization can selectively dissolve the β-phase regions of the surface [34]. Selective dissolution did not occur when the potential was not held cathodically prior to anodic polarization. This implies that cathodic voltages result in alterations of the oxide film, lowering its resistance to electrochemical transport processes and ultimately allowing the oxide film on the β-phase to lose its passive character, resulting in β-phase dissolution. Ehrensberger et al. [10] followed up on these studies with electrochemical impedance spectroscopy (EIS) studies on the voltage-dependent oxide film resistance and showed that titanium oxides can undergo three orders of magnitude decrease in oxide resistance over the voltage range of 0 V to -1 V vs. Ag/AgCl.

This combination of increased inflammatory oxidizing agents and negative voltage excursions that can arise from mechanical factors (oxide film abrasion) may lead to the kinds of severe corrosion attack seen in the selective dissolution processes discussed earlier. Indeed, in a recent retrieval study, Rodrigues et al. showed severe corrosion attack of Ti–6Al–4V modular body hip replacement junctions where the microstructure appeared to be attacked preferentially at the β-phase regions (i.e., selective dissolution) [35], and showed extensive pitting corrosion attack and hydrogen embrittlement of the Ti–6Al–4V alloy in the human body.

Additionally, it is important to remember that metal surfaces in the body may be adjacent to highly restricted crevice-like solution conditions. This is particularly true in metal-on-metal (MoM) interfaces including modular tapers and MoM hip replacements. When such restricted crevices arise, the local solution conditions can be dramatically altered from the commonly assumed pH 7.4, 0.154 M saline environment [26]. Indeed, early work on the corrosion of modular taper interfaces showed that the pH and Cl^- concentration within the modular taper deviate significantly from the physiological level. There are indications from retrieval analyses and other experiments that the pH within taper crevices can, under certain conditions, reach below one [26]. Additionally, the cationic concentrations within tapers can be orders of magnitude higher than those seen in the peri-implant tissues. The role of cation concentrations in solutions adjacent to corroding implant surfaces remains a mostly unexplored area of study.

3.4 The Mechanical Environment

The mechanical environment for metallic biomaterials typically consists of high cyclic loads associated with the activities of daily living and can also include significant surface interactions that include wear against a counter face. The general processes of wear can be further classified into large cyclic motion (wear) and small cyclic motion (fretting) processes. Typically, wear processes are thought to occur at occlusal surfaces (in dentistry) and joint surfaces for total joint prostheses. Fretting, on the other hand, is thought to primarily occur at screw–countersink junctions in fracture fixation devices [36], dental implants, and modular taper interfaces of total joint prostheses [37, 38]. While these are important areas where fretting occurs, others are also present and likely to be influencing the corrosion behavior of devices. These additional locations include back-side contacts in joint prostheses, bone-implant contact regions, overlapping cardiovascular stent contact points [39], and other locations where surface contact and small cyclic motion takes place.

4 Biotribocorrosion Fundamentals

4.1 Wear and Corrosion in the Biological System

For wear of, for example, MoM implants, much of the research on the longevity of these devices has focused entirely on wear (ignoring the corrosion aspects). However, recently, researchers [40] have realized the importance of corrosion processes during wear in these and other devices. It is important to state that in the biological system one cannot have wear of a metal device and NOT have corrosion reactions taking place as well. That is, wear of metals in the body is a tribocorrosion (wear and corrosion) process. The phrase tribocorrosion does not fully and adequately describe the full range of what is more appropriately called mechanically assisted corrosion, where other mechanical mechanisms like elastic straining, plastic deformation, etc., which can affect the oxide film covered surfaces, are included.

Tribocorrosion (i.e., mechanically assisted corrosion) is a conjoint failure mechanism that is highly complex and comprised of surface mechanical processes and surface electrochemical processes. Tribocorrosion in the biological system (biotribocorrosion) has the additional factors of biological processes at play. Thus, biotribocorrosion is the process of corrosion, modified and assisted by the processes of wear (wear and fretting) occurring in and influenced by the adjacent biological environment.

There are several research programs studying the fretting corrosion behavior of medical alloys. Mischler et al. [40–43] have developed several interesting methods for systematic evaluation of fretting corrosion of Ti alloys (amongst others) in physiological solutions and have investigated the effects of potential and proteins (amongst other factors).

To understand biotribocorrosion, the following sections will lay out the basic factors, processes, and mechanisms at play in each aspect of the process. There are three basic aspects (mechanical, electrochemical, biological) at play; however, in each of these primary areas highly complex processes are ongoing. Additionally, the materials in contact and engaged in the processes play an important role. Thus, the materials science of the surface of the metallic biomaterials is critical.

Below is a short, and incomplete, listing of the materials factors, the mechanical factors, and the electrochemical factors that are important in biotribocorrosion. Within each of these broader categories, more specific information of the structure, properties, and processes present during biotribocorrosion is provided. This is meant to help the reader understand the wide range of factors that interact to give rise to mechanical-assisted corrosion in the biological system.

4.2 Materials Factors

4.2.1 Structural Factors

Surface Oxide

Thickness, chemistry (e.g., chemistry and oxidation states of cations), structure, e.g., suboxides present, defect density (vacancy distribution, pipe-diffusion paths, amorphous vs. crystalline), residual stresses (volumetric mismatch with substrate), growth mechanism (high-field growth with electromigration transport), electronic states, semiconducting behavior, ionic transport behavior.

Substrate Alloy

Crystal structure and orientation to the surface, volumetric mismatch with oxide, chemistry of near-surface alloy, transport processes (diffusion) within metal, dislocation structure (cold-working state), precipitate and defect inclusions.

Biological Environment

Proteins adsorbed and their conformation, pH and ion levels (in crevices and restricted geometry environments), oxidizing agents present (ROS, enzymes, proteins, cations, etc.), electrical double layer, surface water structure.

4.2.2 Properties

Surface oxide

Semiconducting properties (flatband potential, Fermi level, work function, defect density), ionic transport properties (vacancies, oxide boundaries for pipe diffusion and migration, etc.), oxide modulus, oxide fracture strain/stress, coefficient of friction with opposing surface, interfacial fracture strain (stress), electric field generated.

Substrate alloy

Yield stress, work hardening sensitivity, hardness, diffusion coefficients, redox sensitivity of alloy constituents.

4.3 Mechanical Factors

Contact asperity geometry, contact stresses, coefficient of friction, fracture strain, interfacial strain to failure, modulus, yielding, fracture mechanics of thin films, relative sliding distances, elastic displacements, structural rigidity of constructs.

4.4 Electrochemical Factors

Thermodynamic condition (potential), active redox processes, surface oxide and electric field, relative anodic and cathodic areas, impedance characteristics (kinetic barriers to oxidation and reduction charge transport processes), solubility of cations, oxide forming tendencies and stability, solution chemistry.

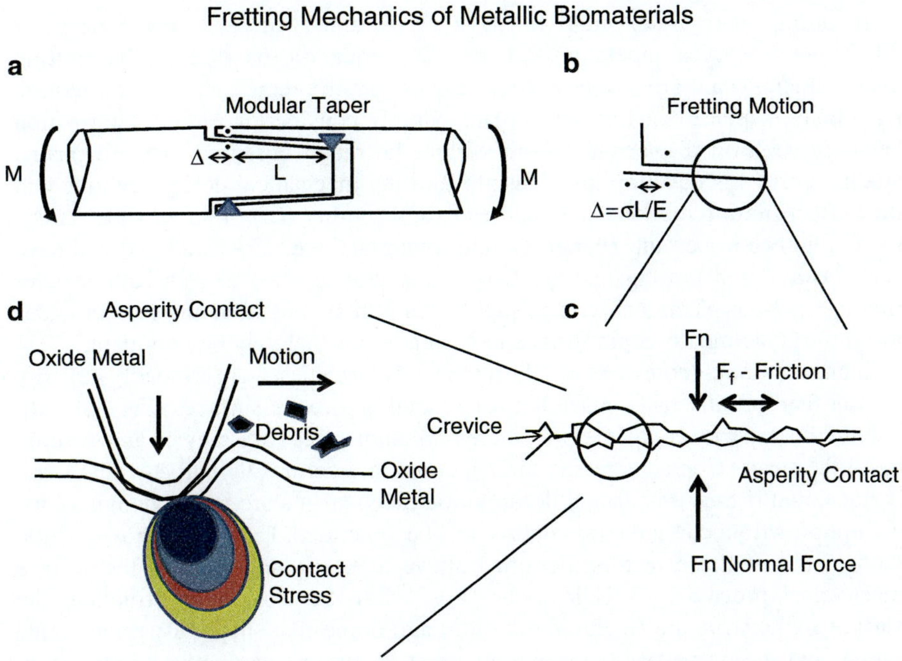

Fig. 1.1 Schematic representations across spatial scale of the mechanics of fretting. (**a**) Elastic bending strains, in conjunction with rigidly connected contact point (*triangles*) can give rise to elastically based displacements, (**d**), in the taper. This does not require rigid body sliding to occur. (**b**) Zoom-in of elastic fretting strains with the displacement depends on the bending stress, modulus, and distance from rigid contact. (**c**) Schematic of a zoom-in on the contact region within a modular taper. Crevice solution can fit within the contacts and the interface will consist of asperity–asperity contact and normal and frictional stresses. (**d**) Close up of metal–oxide surfaces in asperity contact and causing contact stresses, local surface deformation, and oxide debris

One can see from the above listing of factors, that these interfaces and the processes associated with mechanically assisted corrosion *in vivo* are highly complex and interactive and span several different physical, chemical, and biological disciplines.

To understand mechanically assisted corrosion, a series of figures are presented to provide insight into the mechanical and electrochemical elements. Figure 1.1 is a series of schematics of a modular taper interface. In Fig. 1.1a, there is a representative modular taper where an applied bending moment, M, is shown. If one assumes that there are two points within the taper (triangle regions) that can be considered rigidly connected to the opposing surface, then the elastic strain associated with cyclic bending will give rise to a displacement, Δ, within the taper interface (see Fig. 1.1b), which results from taking the flexure formula for stress, assuming this stress is constant along the length, L, of the taper from the fixed point to the point of interest. One can see from this that the displacement, Δ, will be given by:

$$\Delta = \varepsilon L = \frac{\sigma L}{E} = \frac{MyL}{EI}$$

Inputting appropriate values for *M*, *L*, *y*, *E*, and *I* yields Δ in the range of 20–40 μm for typical tapers in use today. This equation shows, first, that fretting displacements can arise solely from elastic deformation and do not require rigid body displacements of the implant. Simply considering elastic deformation behavior, fretting displacements are within the range observed in these tapers. Additionally, this equation gives insight into the mechanical design features that may affect fretting motion. That is, the flexural rigidity (EI) of the modular cross-section will be important. Higher modulus materials (e.g., Co–Cr alloys) will have less elastic-based fretting motion. Larger diameter tapers will also have smaller fretting motions. This analysis is highly simplified, but none-the-less provides insight into the macroscopic stress and design issues that may be important.

If one now looks more closely at the interfacial mechanics, there is asperity–asperity contact that initially takes place between metal and oxide surfaces (Fig. 1.1c, d). Within this asperity contact region, there is enough space for a crevice and solution to fill the space. Contact stresses and frictional interactions will arise within these regions, and if the stresses and deformation mechanisms are large enough, oxide disruption will occur and oxide debris will be generated. Figure 1.2 shows a little more detail of those factors identified above as being important to the surface mechanical processes. This includes (Fig. 1.2a) the dislocation structure, the surface oxide structure (including the film and dome-like shape of passive oxide films), and the properties (e.g., moduli, fracture strains, hardness, etc.) that will impact on the mechanical disruption of the oxide. The nominal contact stresses needed to induce oxide disruption sufficient to cause enhanced corrosion is an important property and is likely to be very low for Co–Cr alloy and even lower for Ti alloys. The local asperity–asperity contact stresses are likely much higher; however, it is also likely that only relatively low contact stresses are necessary for oxide abrasion processes.

Also shown in Fig. 1.2b is what transpires electrochemically just behind a moving asperity that is scraping oxide film from the metal surface. Oxide debris is released and two basic oxidation reactions occur just behind the moving asperity: ion dissolution and oxide repassivation. The first of these releases cations into the crevice solution, while the second regenerates the oxide and releases hydrogen ions into solution. Both oxidation reactions shown result in the accumulation of positive charge in the crevice solution and electrons accumulating in the metal (causing a drop in potential of the surface). The electrons will move to any other location on the metal surface where reduction reactions are present (and still, in electrical contact with the solution) and will be consumed in those reactions. Additionally, anions (Cl^-, PO_4^{3-}, etc.) will be drawn into the crevice in order to maintain charge neutrality and balance the cation release. This combination results in lowering of pH and phosphate and chloride ion levels (among others) in the crevice. These reactions, if they continue within the crevice and are slow to clear from there, will result in a highly aggressive acidic solution that may result in continued corrosion processes after the fretting motion ceases. Indeed, in *in vitro* testing of modular tapers, one can detect this continuing process by the increased currents that remain after loading is stopped [44].

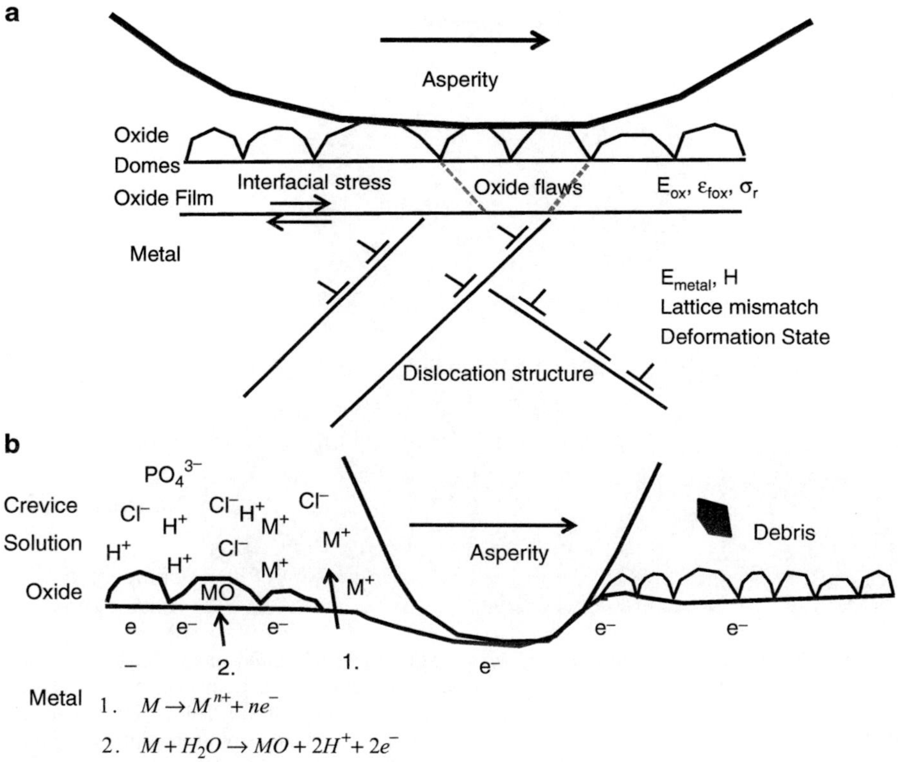

Fig. 1.2 Schematic views of asperity–oxide–metal interfaces showing some of the factors associated with fretting corrosion processes. (**a**) The surface oxide film in contact with an opposing asperity has a structure that typically consists of an oxide film–oxide dome morphology with specific properties of modulus, fracture strain, and residual stresses present in the oxide. A lattice mismatch between oxide and metal gives rise to residual stresses and interfacial stresses as does the asperity interaction. Metal substrate structure and chemistry is also important including the dislocation structure (amount of cold working, hardness, etc.). (**b**) A second schematic of an asperity moving across a metal–oxide surface where contact stresses and motion scrape oxide from the metal surface, creating oxide debris while behind the moving asperity are electrochemical reactions reforming oxide and dissolving metal ions into the crevice solution. Hydrogen ions are generated as well and anions (Cl^-, PO_4^{3-}, etc.) migrate into the crevice to maintain charge neutrality and result in pH drops

Figure 1.3 schematically represents the atomic-scale view of the metal–oxide–solution crevice interface (Fig. 1.3a). Also shown are two scanning electron micrographs taken of the cross-section of a retrieved Ti–6Al–4V alloy (Fig. 1.3b) [35]. These images are from a pitted region of the interface where severe pitting attack of the alloy had taken place while inside the human body. This is the first documented case of pitting corrosion attack of Ti alloys in the human body. What is interesting to note in this figure is that there are a network of tunnels and microsized crevices between the alloy and the oxide where the highly aggressive solution

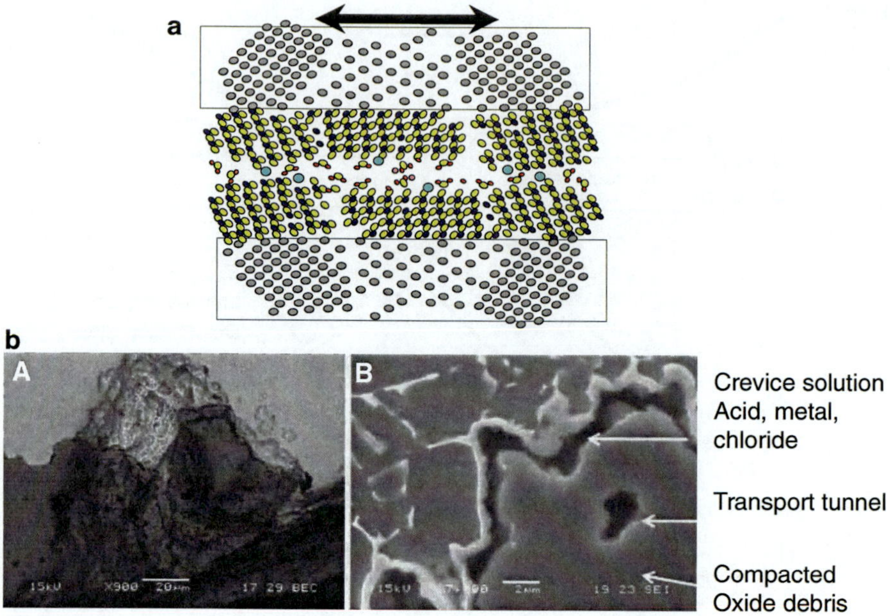

Fig. 1.3 (**a**) Atomic scale schematic of metal–oxide–crevice solution interface. (**b**) Scanning electron micrograph images of Ti–6Al–4V cross-sections from modular body implant tapers. *Gray regions* (*lower parts*) are oxide regions within a pit on the Ti alloy surface. This pitting penetrated up to 200 μm in to the taper surface. The high magnification image shows some interesting and important structural details at the interface between the metal and the oxide. There are small (1–2 mm) channels and tunnels for crevice solution transport. It appears that dissolution of the alloy occurs with subsequent precipitation of the oxide and enrichment of the Ti alloy surface of a precipitated layer (*light line* along the interface). This oxide debris structure likely affects solution chemistry and helps drive the pitting attack (images in (**b**) are reprinted with permission from Wiley [35], all rights reserved)

accumulates and proceeds to corrode the metal. It appears, based on these images, that there is a dissolution process of the metal into the crevice solution and then a precipitation back out of the debris on the oxide side. Also, there appears to be about a 1 μm thick zone on the alloy surface which has precipitated from the reactions taking place (bright line on alloy surface). Thus, the alloy dissolves while the oxide front grows and this process continues, leading to a penetrating pitting-like corrosion attack. The exact details of this process are not known; however, it is clear that because the pits are up to 300 μm deep, fretting is not likely a part of the process ongoing in these pits.

Electrochemically, abrasion of surface oxides dramatically raises the current density and alters the overall polarization behavior, as can be seen in Fig. 1.4 [45]. In this work, a Co–Cr–Mo surface was subject to potentiostatic conditions (from −1 V to +1 V vs. Ag/AgCl) and abraded with 600 grit emery paper for 15 s, followed by a recovery period. The abrasion current density and recovered current

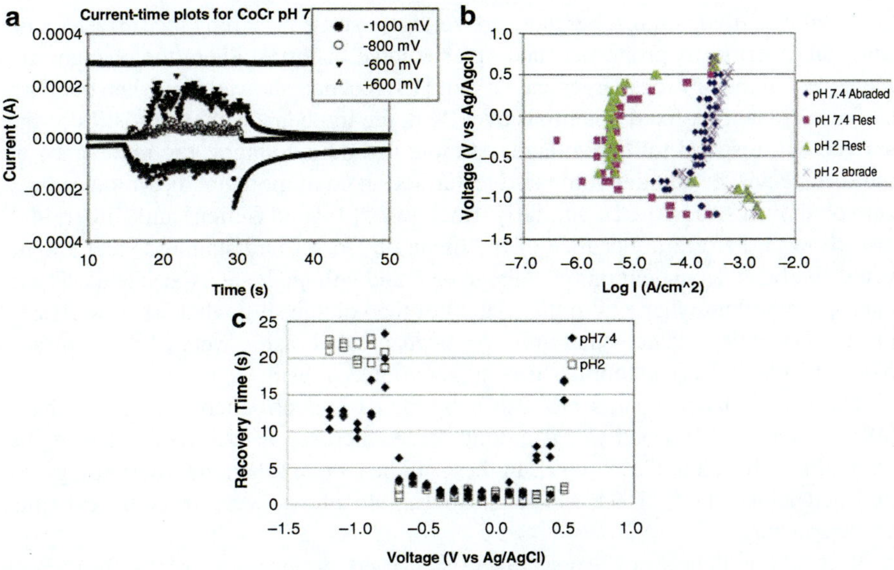

Fig. 1.4 Results of polarization testing of Co–Cr–Mo alloy in pH 7.4 and pH 2 PBS solutions. Alloy surfaces were potentiostatically held and abraded with 600 grit emery paper and then allowed to recover. (**a**) Current versus time plots for abrasion results at four different potentials (−1000, −800, −600, and 600 mV vs. Ag/AgCl). (**b**) Polarization plot of the resulting data over the range of potentials from −1,000 mV to +700 mV. Note that the current density is about 50 times larger during abrasion and that pH did not have a significant effect over most of the range. (**c**) Time constant for recovery of the current after abrasion is stopped. Note the voltage dependence of the recovery that reflects both the presence of the oxide film and the predominant electrochemical reaction being affected. (Images reprinted by permission of ASM International, all rights reserved [42])

levels were tested at pH 7 and pH 2 in PBS, and the time constant for the recovery curves was determined (Fig. 1.4b, c respectively). Clearly (Fig. 1.4b), the abrasion currents are about 50 times greater than the recovered current densities, and there are systematic changes in the time constant for recovery with voltage. These abrasion currents are unlikely to be the maximum possible since the abrasion was not uniform and complete at any time point.

5 Electrochemical History Effects

An element of corrosion behavior that is often poorly understood, especially when it comes to oxide-film covered alloys, is the concept of electrochemical history. That is, how the prior voltage–time solution conditions affect present and future behavior of the surface. For example, oxide films (e.g., TiO_2) are known to have an anodization rate of about 2 nm/V [46], which is not fully irreversible. That is, oxide films can grow with positive voltage, but there are voltage histories whereby the surface oxide film can revert in its properties, and other voltage histories where the changes are more irreversible. Ehrensberger and Gilbert [10] recently showed that when titanium surfaces are held at fixed potentials over 24 h, the impedance characteristics of the surface will respond to that voltage. At more cathodic voltages, the impedance of the surface will be smaller compared to surfaces held at more anodic voltages (i.e., zero or above vs. Ag/AgCl). Similarly, Haeri et al. [47] and Gettens and Gilbert [48] also showed that the impedance of the oxide on Co–Cr and stainless steel alloys, respectively, is dependent on the voltage path and voltage level experienced. These data are important when put in the context of the voltage shifts that are now clearly present in medical devices that undergo mechanical abrasion/wear of their surface oxides and have large cathodic shifts in potential as a result.

These impedance results are also supported by electrochemical scratch tests [49–51] and by electrochemical atomic force microscope (AFM) studies of the oxide films on Ti and CoCr [6–9]. In these studies, the oxide dome morphology on polished and etched Ti–6Al–4V and Co–Cr–Mo alloys were responsive to the voltage history.

It is likely that during mechanically assisted crevice corrosion, there is a combination of large negative voltage shifts (up to $-1{,}000$ mV) that may serve to activate the oxide and large changes in solution chemistry. Solution changes include large changes in pH, recruitment of anions to cause formation of hydrochloric and phosphoric acids (among others), and the accumulation of high concentrations of multivalent cations that may alter the oxide film and corrosion behavior, etc. all present in crevices. This solution, while still in the body, can become highly deviated from the normal body solution. Besides affecting the ongoing corrosion reactions, this solution may also leak out and exchange with the adjacent body solution, and these elements may affect local biological processes. Additionally, this stimulus may bring a stronger inflammatory response from the chronic inflammatory system which will bring to bear larger numbers of activated macrophages that may release significant amounts of additional oxidizing agents (e.g., ROS) that will accelerate the processes.

Taken together, this complex interplay between mechanics, electrochemistry, crevice environments, and biological response has resulted in some extreme examples of severe corrosion in the most corrosion resistant alloys we have today for metallic biomaterials applications. We are also seeing more extreme examples of biological response to these processes, including osteolysis due to corrosion [52] and pseudotumor formation [53, 54].

6 *In Vivo* Fretting Corrosion of Metallic Orthopedic Implants

In the following sections, examples of fretting corrosion of orthopedic implants *in vivo* will be summarized.

6.1 Spinal Devices

In the past 3 decades, a wide variety of spinal instrumentation have been developed for the treatment of spinal disorders. In spinal devices, there has been a progressive increase in the use of multisegmental spinal rods, with screw fixation systems for various spinal conditions like advanced scoliosis or deformity, spondyloarthropathies, spinal stenosis syndrome, spondylolisthesis, and posttraumatic spine instability.

Spinal fixation devices consist of several components (pedicle screws, hooks, and rods) that connect together to form a spinal implant assembly. Spinal fixation systems provide a means of gripping the spinal segment and providing mechanical stability to the affected vertebrae. Within these systems, designs have developed that contain recessed, crevice-like regions where screw head–countersink contact points occur. These can also be locations where different alloy combinations come into contact (e.g., Co–Cr rods with Ti alloy screws, etc.). The three main alloy systems, stainless steel, titanium, and Co–Cr–Mo, can be present in these spinal systems, and the interaction between them within loaded crevice regions is not well understood at this time.

Corrosion is a serious problem in spinal instrumentation, causing local and systematic complications. In a study done by Aulisa et al. [55], fretting corrosion was reported in 20 Harrington's spinal systems (stainless steel rod fitted with hooks on both ends and a rachet) which were removed after 2 years of implantation and recommended removal of the systems after consolidation of bone grafts. The study demonstrated that fretting occurred at the rod–hook junctions (due to relative motion of rotation about the axis of the rod). This resulted in removal of the metal and produced wear debris and tracks on the interior surfaces of inferior hooks. As a result of this, severe biological reactions were observed in the host tissues (brownish color with high concentration of ferrous material) surrounding the rod–hook junction.

Akazawa et al. [56], in a study of corrosion of spinal implants retrieved from 11 patients suffering from scoliosis, reported macroscopic evidence of corrosion in 66% of rod junctions (rod–hook, rod–crosslink junction) after long-term implantation. The study demonstrated that intergranular corrosion and fretting (repeated surface motion) had contributed to corrosion of the implants. Based on these findings, removal of implant is recommended after solid bony union.

Vieweg et al. [57] reported on 13 spinal fixators and showed corrosion of pedicle screws and telescopic rods as a result of fretting and crevice corrosion, respectively, after a mean length of implantation of 10 months.

Rio et al. [58] observed radiological signs of corrosion at the proximal end of the spinal rods in 11 patients. Progressive decrease in metal density of rods was observed radiographically, which was later confirmed on implant retrieval.

6.2 Fracture Fixation Devices

In 1968, Cohen et al. [59] reported on fretting corrosion in a retrieval study on 25 retrieved hip-nail plate assemblies made of type 316 stainless steel and vitallium. Fretting corrosion was observed on the plates, on the surface of countersinks around the holes, on the screws, and on the underhead chamfer surface. It was stated that crevice formation was inevitable feature of the geometry of the screw–plate interface, while fretting was the result of micromotion of the plate and the screw. Thus, corrosion was enhanced in the presence of fretting and due to differences in electrochemical potential between passivated and nonpassivated regions on the surface of the metal. It was hypothesized that fretting corrosion is a combination of abrasive, corrosive, and adhesive wear, and corrosive component plays a significant role in all situations where metal-to-metal contact is made.

In retrieval study of 17 orthopedic internal fixation devices (plates and screw) made of 316L stainless steel, Weinstein et al. [60] reported evidence of fretting and crevice corrosion in all the countersink holes in which screws were used as implants for few months.

Brown et al. [61] found significant amount of corrosion in a series of tests conducted on osteosynthesis stainless steel plates and screws, which was a result of fretting corrosion accelerating crevice corrosion. They reported that patient ambulation causes a relative motion between screw head and plate hole, which leads to fretting corrosion with continuous disruption of passive oxide layer and subsequent depletion of oxygen in crevices, accelerating the initiation of crevice corrosion.

6.3 Modular Joint Replacement Retrieval Analyzes

Much of what we know about severe *in vivo* corrosion of metallic biomaterials comes from retrieval studies of modular taper interfaces in total hip replacements. These tapers are typically conical in shape and can consist of Co–Cr–Mo/Ti alloy/ Stainless Steel couples in any combination, or indeed could interface alumina or zirconia ceramic heads. In some designs, thimble-like connectors are added to convert from metal–metal tapers to metal–ceramic tapers. Most modular taper corrosion studies have focused on the head–neck taper interface; however, more recent studies have shown both modular body and neck–stem tapers as being susceptible to severe corrosion attack [35, 62, 63]. Neck–stem modular tapers are a relatively new approach to hip stems that allows greater flexibility in construction of the prosthesis during the operation. However, these tapers are typically more elliptical in cross-section and are subject to higher bending moments than head–neck tapers. Recent work has shown that these tapers are susceptible to severe corrosion and fracture [62, 64].

Jacobs et al. [65] characterized the local and distant distribution of solid and soluble products of corrosion from the taper junction of modular femoral total hip prostheses. Particles of metal oxides, metal chlorides, and chromium phosphate (chromium orthophosphate hydrate) with elevated levels of serum cobalt and urine chromium concentrations were observed in the implant and within the periprosthetic tissues. It was suggested that fretting corrosion may be a dominant mechanism of metal release in patients with modular hip implants, and that solid corrosion products from modular femoral stems may accelerate wear and fretting corrosion via three-body wear mechanism. These corrosion products raise the issue of systemic toxicity.

A retrieval study by Urban et al. [66] on the effects of migration of solid corrosion products from the modular head/neck junction in 15 total hip replacements to the periprosthetic tissues reported that corrosion products migrate along the bone implant interface and to the prosthetic bearing surfaces and indirectly contribute to periprosthetic bone loss, osteolysis, and aseptic loosening of the implant.

Another retrieval study by Urban et al. [67] on 16 modular body titanium alloy stems used in cementless hip replacements showed evidence of corrosion. The surface damage was characterized as fretting scars, pitting, and etching.

Cook et al. [68] analyzed 108 porous coated uncemented modular femoral stems with modular femoral heads retrieved after 25 months of implantation. Wear and corrosion were observed in 7% of the similar-metal and 34.5% of the mixed-metal implants due to combine effects of fretting, crevice, and galvanic corrosion. It was hypothesized that fretting led to breakdown of the passive oxide film on the alloy surfaces, followed by a combined attack due to crevice and galvanic corrosion.

In a retrieval study of 148 modular hip prostheses, Gilbert et al. [38] reported evidence of fretting and crevice corrosion (moderate to severe in 16–35% of 148 retrieved implants) at head–neck tapers in both mixed (Ti–6Al–4V/Co–Cr) and similar-metal (Co–Cr/Co–Cr) combinations. Restricted crevice condition at taper geometry and mechanical factors like high cyclic loading, fretting, or stresses related corrosion create an unstable environment at crevice, leading to fracture and abrasion of the passive oxide layer covering these passive metal surfaces. This causes the metal surface potential to go more negative, resulting in deaeration (loss of oxygen) in crevice solution with decrease of the pH in the crevice, leading to increased corrosion rates [69, 70]. These ultimately impact the local biological processes and/or the mechanical integrity of the implant.

Another retrieval study by Gilbert et al. [70] on two modular hip implants with cobalt–alloy (CoCr) head and stem showed that fracture occurred in the neck of femoral components (close to head/neck junction). It was hypothesized that the fracture of implant was due to combination of mechanical factors like porous grain boundaries, intergranular corrosive attack initiated at the head–neck taper surface, and cyclic fatigue loading of the stem.

In a study by Brown et al. [71], 79 retrieved modular hip tapers were examined with stereo and scanning electron microscopy to evaluate the incidence and nature of corrosion. The study showed that fretting corrosion played a major role in the initiation of modular interface corrosion, and the correlation between corrosion and length of neck indicated that long head neck is more prone to fretting corrosion due to instability at the interface. The problem of fretting corrosion can be overcome by making design changes so as to increase the stability at the interface.

Lieberman et al. [72], in a retrieval study, reported that fretting and crevice corrosion is evident in 2 out of 10 mixed-metal implants, whereas no corrosion was reported in 26 similar-metal implants. It was hypothesized that corrosion damage was due to fretting (micromotion at interfaces) and trapped fluid at the taper interfaces.

Collier et al. [73], in a retrieval study of 139 modular femoral components of hip prostheses, reported corrosion in the tapered interface between head and the neck of 25 out of 48 mixed-metal (CoCr head/Titanium alloy stem) prostheses, whereas no corrosion was observed in 91 same-alloy combination. Based on this study and other studies by Collier et al., it was hypothesized that galvanically accelerated crevice corrosion plays major role in failure of modular prostheses [74, 75]. Galvanism was proposed because only Ti–6Al–4V/Co–Cr–Mo couples were observed to suffer corrosion; however, corrosion of all combinations of metals has been demonstrated, thus reducing the likelihood that galvanism is playing any significant role in the process.

6.4 Stents

Coronary stents are small metallic wire mesh devices implanted in heart arteries to prevent the arteries from closing up. Endovascular stents have been placed in human vessels for more than a decade as a treatment of atherosclerosis, yet, information regarding their *in vivo* electrochemical and mechanical stability remains limited.

"Late problems with stents can be divided into two broad categories: mechanical failure due to material fatigue resulting from the considerable stress imposed to stents by cardiac contractions, and chemical failure where corrosion or depolymerization can release potential toxic substances such as nickel, degradation products, or contaminants" [76].

In a retrieval study, Halwani et al. [39] analyzed nine excised human vascular segments with implanted stents ($n = 16$) made of stainless steel, nickel–titanium, tantalum, and cobalt-based alloys. Fretting accelerated crevice corrosion and pitting corrosion were observed as a result of electrochemical and mechanical factors. Mass spectroscopy determined the metal debris in surrounding tissues, which is a potential cause for contributing biological mechanism of in-stent restenosis.

7 Recent Biological Assessment

While the role of ions and particles has been well studied by the cell-culture community, there has been relatively little direct work on the effects of reduction reactions on biocompatibility. Gilbert et al. [13] and Ehrensberger et al. [14] have demonstrated that cathodic potentials give rise to conditions that can kill cells in culture when they are cultured directly on polarized titanium surfaces. That titanium is NOT "the most biocompatible metal" under cathodic conditions, where reduction reactions predominate, is a fundamentally important finding that may provide insight into why implants subjected to significant corrosion processes may be developing severe adverse biological response. Very recent work by Sivan et al. [76] shows that cathodic preconditioning of titanium surfaces in the presence of cell culture medium results in the formation of a biological film (likely, a reaction product of amino acids), that subsequently renders the titanium surfaces lethal to cultured MC3T3-E1 cells. That is, if cells are cultured on surfaces that have been preconditioned at −1,000 mV for 24 h in cell culture medium (AMEM), and then placed into fresh solution and cells cultured on them, these cells die. The reaction product appears to be both particulate (likely, a calcium phosphate precipitate) and film like, where the film has characteristics of an organic polymeric-like structure and may derive from the amino acids, vitamins, or other constituents of the medium. No serum additions were needed for the killing effects to be seen, and the conditioned medium, when transferred into fresh surfaces for culture, had no effect on the cells. The mechanisms at play in this reduction-based toxicity are still poorly understood but may play out in other systems where large negative potentials are developed (e.g., Mg alloys).

8 Summary

In this chapter, the elements of corrosion related to metallic biomaterials were described. Several unique aspects of the mechanical, electrochemical, and biological environments were presented to show the complex interplay giving rise to severe corrosion attack of some of the world's most corrosion resistant alloys. The human body, in conjunction with implants designed with crevices and high cyclic stress, can induce pitting, hydrogen embrittlement, selective dissolution, and intergranular corrosion of Co–Cr–Mo and Ti alloys in ways unimaginable just a few decades ago. The complexity and adjustability of the biological system has profound effects on corrosion, and the reduction half-cells present in biological systems are poorly understood and in need of significantly more study. Overall, as long as metallic biomaterials experience stresses, abrasion, crevices, and biological fluids, corrosion processes will continue to be a significant issue. Advances in coating technologies and clear-eyed understanding of means to mitigate corrosion processes are needed to further advance medical device technology and the longevity that is still lacking in most of these devices.

In today's world, patients are forced to modify their behavior to fit the needs of the implant. What is needed are implants that fit the needs of the patient. A major limitation in this regard is the susceptibility of these metals to mechanically assisted corrosion in the human body (biotribocorrosion). Until we can advance beyond these limitations, joint replacements, spinal devices, dental implants, cardiovascular stents, and all other medical devices in which metals play a role will continue to be susceptible to corrosion.

Acknowledgements JLG would like to acknowledge all of those colleagues and students who have worked with him over the 20 plus years of study of these issues.

References

1. Bechtol CO, Fergeson AB, Laing PG (1959) Metals in engineering in bone and joint surgery. William and Wilkins, Baltimore
2. Witte F (2010) The history of biodegradable magnesium implants: a review. Acta Biomater 6 (5):1680–1692
3. Craig RG (1989) Restorative dental materials, 8th edn. Mosby, St. Louis
4. Gilbert JL (2006) Basic science: metals. In: Callaghan J, Rubash H, Rosenberg A (eds) The adult hip, 2nd edn. Lippencott-Raven, New York
5. Laing PG, Furgeson AB, Hodge ES (1967) Tissue reaction in rabbit muscle exposed to metallic implants. J Biomed Mater Res 1:135–149
6. Bearinger JP, Orme CA, Gilbert JL (2001) Direct observation of hydration of TiO_2 on Ti using AFM: freely corroding versus potentiostatically held. Surf Sci 491:370–387
7. Bearinger JP, Orme CA, Gilbert JL (2003) In situ imaging and impedance measurements of titanium surfaces using AFM and SPIS. Biomaterials 24(11):1837–1852
8. Bearinger JP, Orme CA, Gilbert JL (2003) Effect of hydrogen peroxide on titanium surfaces: in situ imaging and step polarization impedance spectroscopy of commercially pure and Ti-6Al-4V. J Biomed Mater Res 67A(3):702–712
9. Gilbert JL, Bai Z, Bearinger J, Megremis S (2004) Dynamics of oxide films on metallic biomaterials. In: Shrivastava S (ed) Proceedings of the ASM conference on medical device materials, Anaheim, CA, September, pp 139–143
10. Ehrensberger MT, Gilbert JL (2010) The effect of static applied potential on the 24 hour impedance behavior of commercially pure titanium in simulated biological conditions. J Biomed Mater Res 93B(1):106–112
11. Jacobs JJ, Gilbert JL, Urban RM (1998) Current concepts review: corrosion of metal orthopaedic implants. J Bone Joint Surg 80A(2):268–282
12. Witte F, Eliezer A (2012) Biodegradable metals. In: Eliaz N (ed) Degradation of implant materials. Springer, New York
13. Gilbert JL, Zarka L, Chang E, Thomas C (1998) The reduction half-cell in biomaterials corrosion: oxygen concentration profiles near and cell response to polarized titanium. J Biomed Mater Res 42:321–330
14. Ehrensberger M, Sivan S, Gilbert JL (2010) Titanium is NOT "The Most Biocompatible Metal" under cathodic potentials: the relationship between voltage and MC3T3 pre-osteoblast behavior on electrically polarized cpTi surfaces. J Biomed Mater Res 93A(4):1500–1509
15. Haeri M, Gilbert JL (2010) Cellular response to anodic and cathodic surface voltage, and metal ion release in polarized CoCr biomedical alloy. In: Gilbert JL (ed) Proceedings of 2009 conference on materials and processes for medical devices. ASM International, Metal Park, OH
16. Gettens RT, Gilbert JL (2008) Fibrinogen adsorption onto 316L stainless steel under polarized conditions. J Biomed Mater Res 85(1):176–187
17. Jones DA (1992) Principles and prevention of corrosion. Macmillan, New York
18. Kalbacova M, Roessler S, Hempel U, Tsaryk R, Peters K, Scharnweber D, Kirkpatrick J, Deiter P (2007) The effect of electrochemically simulated titanium cathodic corrosion

products on ROS production and metabolic activity of osteoblasts and monocytes/macrophages. Biomaterials 28:3263–3272

19. Stankovich MT, Bard AJ (1978) The electrochemistry of proteins and related substances, part III: bovine serum albumin. J Electroanal Chem 86:189
20. Kumsata C, Jakob U (2009) Redox-regulated chaperones. Biochemistry 48:4666–4676
21. Aubrey S, Burlina F, Dupont E, Delaroche D, Joliot A, Lavielle S, Chassaing G, Sagan S (2009) Cell-surface thiols affect cell entry of disulfide-conjugated peptides. FASEB J 23:2956–2967
22. Shin JY, Shin JI, Kim JS, Yang YS, Shin YK, Kim KK, Lee S, Kweon DH (2009) Disulfide bond as a structure determinant of prior protein membrane insertion. Mol Cells 27(6):673–680
23. Lassing I, Schmitzberger F, Bjornstedt M, Holmgren A, Norland P, Schutt CA, Lingberg U (2007) Molecular and structural basis for redox regulation of b-actin. J Mol Biol 370:331–348
24. Chiarugi P, Pani G, Giannoni E, Taddei L, Colavitti R, Raugei G, Symnos M, Borrello S, Galeotti T, Ramponi G (2003) Reactive oxygen species as essential mediators of cell adhesion: the oxidative inhibition of a FAK tyrosine phosphatase is required for cell adhesion. J Cell Biol 161(5):933–944
25. Bard AJ, Faulkner LR (1980) Electrochemical methods: fundamentals and applications. Wiley, New York
26. Gilbert JL (2006) Mechanically assisted corrosion of metallic biomaterials. In: ASM handbook, vol 13C, Corrosion. ASM International, Metals Park, OH, pp 826–836
27. Gilbert JL (2011) Electrochemical behaviour of metals in the biological milieu, chapter 13. In: Ducheyne P, Healy KE, Kirkpatrick J (eds) Comprehensive biomaterials. Elsevier, London
28. Cabrera N, Mott NF (1948) Theory of the oxidation of metals. Rep Prog Phys 12:163
29. Gunthershultze A, Betz H (1932) Cold temperature glow Kaltes Temperatur leuchten Zeitschrift fur Physik 74(9–10):681–691
30. Goldberg S, Gilbert JL (2001) Transient electric fields induced by mechanically assisted corrosion of Ti-6Al-4V. J Biomed Mater Res 56:184–194
31. Willert HG, Broback LG, Buchhorn GH et al (1996) Crevice corrosion of cemented titanium alloy stems in total hip replacements. Clin Orthop Relat Res (333):51–57
32. Ratner B, Hoffman A, Schoen F, Lemons J (1996) Biomaterials science, 2nd edn. Academic, New York, pp 249–252
33. Pan J, Thierrry D, Leygraf C (1996) Hydrogen peroxide toward enhanced oxide growth on titanium in PBS solution: blue coloration and clinical relevance. J Biomed Mater Res 30:393–402
34. Chandrasekaran N, Bai Z, Gilbert JL (2006) Titanium electrochemistry in the presence of the inflammatory species H2O2. Transactions of the Society for Biomaterials Annual Meeting, Pittsburgh, PA
35. Rodrigues DC, Urban RM, Jacobs JJ, Gilbert JL (2009) Severe corrosion and hydrogen embrittlement in vivo in Ti-6Al-4V modular body hip stems. J Biomed Mater Res 88B (1):206–219
36. Piehler HR, Portnoff MA, Sloter LE, Vegdahl EJ, Gilbert JL, Weber MJ (1985) Corrosion-fatigue performance of hip nails: the influence of materials and design. In: Fraker AC, Griffin CD (eds) ASTM special technical publication 859. American Society for Testing and Materials, Philadelphia, pp 93–104
37. Flemming CAC, Brown SA, Payer JH (1994) Mechanical testing for fretting corrosion of modular total hip tapers. In: Kambic HE, Yokobori AT (eds) Biomaterials mechanical properties. ASTM special technical publication 1173. American Society for Testing and Materials, Philadelphia, pp 156–166
38. Gilbert JL, Buckley CA, Jacobs JJ (1993) In-vivo corrosion of modular hip prosthesis components in mixed and similar metal combinations: the effect of crevice, stress, motion and alloy coupling. J Biomed Mater Res 27(12):1533–1544

39. Halwani DO, Anderson PG, Brott BC, Anayiotos AS, Lemons JE (2010) Clinical device-related article surface characterization of explanted endovascular stents: evidence of in vivo corrosion. J Biomed Mater Res B Appl Biomater 95B(1):225–238
40. Landholt D, Mischler S, Stemp M (2001) Electrochemical methods in tribocorrosion: a critical review. Electrochem Acta 46:3913–3929
41. Barril S, Mischler S, Landolt D (2005) Electrochemical effects on the fretting corrosion of Ti6Al4V in 0.9% sodium chloride solution. Wear 259:282–291
42. Hiromoto S, Mischler S (2006) The influence of proteins on the fretting-corrosion behavior of a Ti6Al4V alloy. Wear 261:1002–1011
43. Mischler S (2008) Triboelectrochemical techniques and interpretation methods in tribocorrosion: a comparative evaluation. Tribol Int 41:573–583
44. Gilbert JL, Mehta M, Pinder B (2009) In-vitro fretting crevice corrosion of stainless steel-cobalt chrome modular hip stems: effect of material, assembly and offset. J Biomed Mater Res 88B(1):162–173
45. Gilbert JL, Lam B (2009) Polarization behavior of Ti-6Al-4V, CoCr, and 316L stainless steel during manual abrasion in pH 7.4 and pH 2 PBS. In: Gilbert JL (ed) Proceedings of 2009 MPMD conference, Minneapolis, MN. ASM International, Metals Park, OH
46. Shams El Din AM, Hammond AA (1988) Oxide film formation and thickness on titaniumin water. Thin Solid Films 167:269–280
47. Haeri M, Gilbert JL (2011) The voltage-dependent electrochemical impedance spectroscopy of CoCrMo medical alloy using time-domain techniques: generalized Cauchy-Lorentz, and KWW-Randles functions describing non-ideal interfacial behavior. Corros Sci 53(2):582–588
48. Gettens RT, Gilbert JL (2009) The electrochemical impedance of polarized 316L stainless steel: structure–property-adsorption correlation. J Biomed Mater Res 90A(1):121–132
49. Gilbert JL, Buckley CA, Lautenschlager EP (1996) Titanium oxide fracture and repassivation: the effect of potential, pH and aeration. In: Brown SA, Lemons J (eds) Medical applications of titanium and its alloys. The materials and biological issues, ASTM special technical publication 1272. American Society for Testing and Materials, Philadelphia, PA, pp 199–215
50. Goldberg JR, Lautenschlager EP, Gilbert JL (1997) Electrochemical response of CoCrMo to high speed fracture of its metal oxide using an electrochemical scratch test method. J Biomed Mater Res 37(2):421–433
51. Goldberg JR, Gilbert JL (2004) The electrochemical and mechanical behavior of passivated and TiN/AlN-coated CoCrMo and Ti6Al4V alloys. Biomaterials 25(5):851–864
52. Jones D, Marsh JL, Nepola JV, Jacobs JJ, Skipor AK, Urban R, Gilbert JL, Buckwalter J (2001) Focal osteolysis at the junctions of a modular stainless steel femoral intramedulary nail. J Bone Joint Surg 83-A(4):537–548
53. Kwon YM, Glyn-Jones S, Simpson DJ, Kamali A, McLardy-Smith P, Gill HS, Murray DW (2010) In vivo wear analysis of metal-on-metal hip resurfacing implants revised due to pseudotumours. J Bone Joint Surg 92(4):356–361
54. Kwon YM, Ostlere SJ, McLardy-Smith J, Athanasou NA, Gill HS, Murray DW (2011) Assymptomatic pseudotumors after metal-on-metal hip resurfacing arthroplasty. J Arthroplasty 26(4):511–518
55. Aulisa L, Di Benedetto A, Vinciguerra A, Lorini G, Tranquilli-Leali P (1982) Corrosion of the harrington's instrumentation and biological behaviour of the rod-human spine system. Biomaterials 3(4):246–248
56. Akazawa T, Minami S, Takahashi K, Kotani T, Hanawa T, Moriya H (2005) Corrosion of spinal implants retrieved from patients with scoliosis. J Orthop Sci 10(2):200–205
57. Vieweg U, van Roost D, Wolf HK, Schyma CA, Schramm J (1999) Corrosion on an internal spinal fixator system. Spine 24(10):946
58. Del Rio J, Beguiristain J, Duart J (2007) Metal levels in corrosion of spinal implants. Eur Spine J 16(7):1055–1061
59. Cohen J, Lindenbaum B (1968) Fretting corrosion in orthopaedic implants. Clin Orthop Relat Res (61):167–178

60. Weinstein AM, Spires WP, Klawitter JJ, Clemow AJT, Edmunds JO (1979) Orthopaedic implant retrieval and analysis study. In: Syrett BC, Acharya A (eds) Corrosion and degradation of implant materials, ASTM STP 684. American Society for Testing and Materials, Philadelphia, PA, pp 212–228
61. Brown SA, Merritt K (1983) Fretting corrosion of plates and screws: an in vitro test method. In: Fraker AC, Griffin CD (eds) Corrosion and degradation of implant materials. ASTM STP 859. American Society for Testing and Materials, Philadelphia, PA, pp 105–116
62. Wright G, Sporer S, Urban RM, Jacobs JJ (2010) Fracture of a modular femoral neck after total hip arthroplasty: a case report. J Bone Joint Surg A 92:1518–1521
63. Urban RM, Gilbert JL, Hall DJ, Levin BR, Sporer SM, Galante JO, Jacobs JJ (2010) Fretting corrosion and fracture of modular neck-body junctions in hip replacement femoral components. J Bone Joint Surg Am 92(6):1518–1521
64. Lackstein D, Eliaz N, Levi O, Backstein D, Kosashvili Y, Safir O, Gross AE (2011) Fracture of cementless femoral stems at the mid-stem junction in modular revision hip arthroplasty systems. J Bone Joint Surg Am 93(1):57–65
65. Jacobs JJ, Urban RM, Gilbert JL, Skipor AK, Black J, Jasty M, Galante JO (1995) Local and distant products from modularity. Clin Orthop Relat Res (319):94–105
66. Urban RM, Jacobs JJ, Gilbert JL, Galante JO (1994) Migration of corrosion products from modular hip prostheses. Particle microanalysis and histopathological findings. J Bone Joint Surg Am 76(9):1345–1359
67. Urban RM, Gilbert JL, Jacobs JJ (2005) Corrosion of modular titanium alloy stems in cementless hip replacement. J ASTM Int 2(10):215–224
68. Cook SD, Barrack RL, Clemow AJ (1994) Corrosion and wear at the modular interface of uncemented femoral stems. J Bone Joint Surg Br 76(1):68–72
69. Jacobs JJ, Gilbert JL, Urban RM (1998) Current concepts review-corrosion of metal orthopaedic implants. J Bone Joint Surg 80(2):268
70. Gilbert J, Buckley C, Jacobs J, Bertin K, Zernich M (1994) Intergranular corrosion-fatigue failure of cobalt-alloy femoral stems. A failure analysis of two implants. J Bone Joint Surg 76 (1):110
71. Brown SA, Flemming CAC, Kawalec JS, Placko HE, Vassaux C, Merritt K, Payer JH, Kraay MJ (1995) Fretting corrosion accelerates crevice corrosion of modular hip tapers. J Appl Biomater 6(1):19–26
72. Lieberman JAYR, Rimnac CM, Garvin KL, Klein RW, Salvati EA (1994) An analysis of the head-neck taper interface in retrieved hip prostheses. Clin Orthop Relat Res (300):162
73. Collier JP, Surprenant VA, Jensen RE, Mayor MB, Surprenant HP (1992) Corrosion between the components of modular femoral hip prostheses. J Bone Joint Surg Br 74(4):511–517
74. Collier JP, Surprenant VA, Jensen RE, Mayor MB (1991) Corrosion at the interface of cobalt-alloy heads on titanium-alloy stems. Clin Orthop Relat Res (271):305–312
75. Collier JP, Mayor MB, Jensen RE, Surprenant VA, Surprenant HP, McNamar JL, Belec L (1992) Mechanisms of failure of modular prostheses. Clin Orthop Relat Res (285):129–139
76. Bertrand OF, Sipehia R, Mongrain R, Rodes J, Tardif J.C, Bilodeav L, Cote G, Bourassa MG (1998) Biocompatibility aspects of new stent technology. J Am Coll Cardiol 32(3):562–571
77. Sivan S, Oeullette E, Gilbert JL (2010) Cathodic voltage preconditioning of Ti-6Al-4V in media affects MC3T3 preosteoblast cell viability. Transactions of the Society for Biomaterials, Seattle, WA

Biography

Dr. Gilbert is Professor of Biomaterials in the Department of Biomedical and Chemical Engineering in the L.C. Smith College of Engineering and Computer Science at Syracuse University. He is currently Editor-in-Chief of the Journal of Biomedical Materials Research—Part B: Applied Biomaterials, and President of the Society for Biomaterials. Gilbert was elected as a Fellow of the American Institute for Medical and Biological Engineers in 2004 and is a fellow of the International Union of Societies of Biomaterials Science and Engineering (inducted in 2012 at the World Congress of Biomaterials, Chengdu, China). He is the Founder of the Syracuse Biomaterials Institute at Syracuse University.

He received his PhD in Metallurgical Engineering and Materials Science, and Biomedical Engineering at Carnegie Mellon in 1987. Gilbert was an Assistant and Associate Professor in the Department of Biological Materials at Northwestern University Dental School and Department of Biomedical Engineering in the Engineering School for 11 years before moving to Syracuse University as Professor in the Department of Bioengineering and Neuroscience. He became Chair of the Department in 2002, and subsequently was named Associate Dean for Research and Doctoral Programs in the College of Engineering from 2004 to 2008.

Chapter 2
Degradation of Titanium and Its Alloys

Sannakaisa Virtanen

Abstract This chapter summarizes the state-of-the-art knowledge on the degradation modes of Ti and biomedically relevant Ti-based alloys. First, general aspects of passivity of Ti as well as special corrosion modes of passive Ti are shortly described. Then, the influence of alloying on the electrochemical dissolution modes is summarized, emphasizing the specific corrosion modes relevant for the biomedical application. Degradation of materials in biomedical applications can, in addition to purely chemical or electrochemical processes, be strongly influenced by mechanical/tribological processes. Therefore, tribocorrosion of Ti and Ti-based alloys is described. In addition, the role of the biological environment in the degradation process of Ti alloys is discussed. Moreover, a short discussion on some relevant implant design-related aspects of degradation is provided.

1 Introduction

Titanium and Ti-based alloys belong to the most important metallic implant materials, due to their generally good biocompatibility as well as their high chemical, electrochemical, and mechanical stability [1–3]. The generally excellent corrosion resistance of Ti-based alloys is due to the spontaneous formation of a stable TiO_2 passive film on the material surface in most environments. This passive film is an efficient barrier against further dissolution, and hence the dissolution rate is orders of magnitude lower than for metals and alloys actively corroding (i.e., for metals without the self-protection of a passivating oxide layer).

S. Virtanen (✉)
Institute for Surface Science and Corrosion, Department of Materials Science and Engineering (WW-4), University of Erlangen-Nuremberg, Martensstrasse 7, 91058 Erlangen, Germany
e-mail: Virtanen@ww.uni-erlangen.de

N. Eliaz (ed.), *Degradation of Implant Materials*, DOI 10.1007/978-1-4614-3942-4_2,

However, even for passive metals and alloys, a certain amount of material degradation by electrochemical dissolution mechanisms always takes place, as resistance is not given by thermodynamic inertness but by the kinetic barrier of the passive film. In addition, passive metals and alloys can be susceptible to specific corrosion modes, such as pitting or crevice corrosion. Depending on the type of application, materials in implants can experience not only chemical attack, but also mechanical stresses, and wear processes can significantly contribute to materials degradation.

Different aspects of degradation of Ti-based implant alloys have been widely studied, due to the great significance of these materials in the field. Corrosion behavior of Ti and Ti alloys (including alloys used in biomedical applications) has been reviewed, for instance in [4–6]. Information on the mechanical properties and wear behavior of different classes of Ti-based implant alloys (α/β-alloys and β-alloys), including considerations on mechanical biocompatibility, can be found in [7, 8]. Many aspects of corrosion, fatigue, and corrosion fatigue of implant alloys, with an emphasis on Ti-based alloys, are summarized in [9]. Moreover, fretting corrosion in biomedical implants (including Ti alloys) has been reviewed in [10]. Many international conferences and symposia include presentations on this topic, to mention are the proceedings in [11–13].

This chapter aims to summarize state-of-the-art knowledge on the degradation modes of Ti and biomedically relevant Ti-based alloys. In view of the very large amount of work published, it is not possible to discuss all specific findings in detail. Instead, the most important mechanisms and principles of degradation are considered. First, general aspects of passivity of Ti are shortly described, as this is the basis for understanding the good corrosion resistance of Ti-based materials. Then, the influence of alloying on the electrochemical dissolution modes is summarized, emphasizing the specific corrosion modes relevant for the biomedical application. Degradation of materials in the biomedical applications can, in addition to purely chemical or electrochemical processes, be strongly influenced by mechanical/tribological processes. Therefore, tribocorrosion of Ti and Ti-based alloys is described. The chapter also discusses the role of the biological environment in the degradation process of Ti alloys. It should be pointed out that this research field is still strongly evolving, therefore many open questions remain, especially concerning the detailed understanding of the complex interplay between the materials surface and the *in vitro* or *in vivo* biological environment. As degradation processes are influenced not only by the material and environment, but also by the design of a device, a short discussion on some relevant implant design-related aspects of degradation is provided, such as possible galvanic corrosion effects in modular systems. However, this discussion is only based on basic principles and does not cover a review of failure case studies or explant analysis. An overview of implant retrieval and failure analysis is provided by another chapter in this book.

2 Passivity of Ti

2.1 Active and Passive Dissolution, Depassivation

Pure titanium is a base metal with a low standard potential (−1.63 V vs. NHE for the couple Ti/Ti^{2+}). This indicates a high reactivity of a bare Ti surface in aqueous environments. In most natural environments, however, titanium shows an excellent corrosion resistance due to the spontaneous formation of a thin but highly protective TiO_2 passive layer on the surface. Thermodynamically, TiO_2 is stable in a large pH range, as illustrated by the potential–pH diagram of the Ti/H_2O system (Fig. 2.1) [14]. Complexing species, such as HF and H_2O_2, can, however, lead to significant dissolution [14]. The effect of fluorides on Ti corrosion can be especially relevant for Ti in dental applications (considering the use of prophylactic fluoride-containing products in dental treatments), as has therefore been studied. For instance, [15, 16] demonstrate the activating effect of fluoride, which becomes especially drastic at lower pH values. General information on the corrosion behavior of metallic alloys used in dentistry is summarized in [17]. Also, H_2O_2 may be of special relevance generally in biomedical applications, as it can be generated in biological reactions—this case is further discussed later in this chapter.

As the passivation potential of Ti is more negative than the equilibrium potential of hydrogen evolution, and the critical current density for passivation is relatively low, spontaneous passivation of a bare Ti surface, even in deaerated acidic solutions, is expected. The passivation reaction can, however, be hampered by the slow kinetics of the cathodic hydrogen evolution reaction on the Ti surface (high overpotential of the reduction reaction). This explains findings of active dissolution of Ti under reducing conditions in acidic solutions (depending on the acid type and concentration) [18]. In oxidizing acids, and in most neutral solutions, Ti is spontaneously passive.

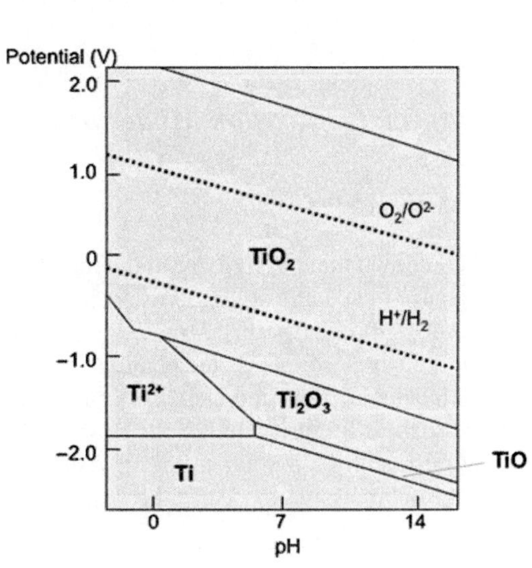

Fig. 2.1 Potential–pH (Pourbaix) diagram for the system Ti/H_2O

The corrosion behavior at equilibrium (corresponding to immersion of samples without polarization) can be studied by monitoring the open-circuit potential or by electrochemical impedance spectroscopy (EIS) measured at the corrosion potential. In the case of stable passivity, the open-circuit potential values remain in the passive region as a function of time; however, if depassivation takes place, typically a sudden decay of the potential to low potentials corresponding to the active dissolution region of the material takes place (examples are shown later in this chapter in Fig. 2.3). The value of the open-circuit potential does not give direct information on the corrosion rate, but it is clear that the active dissolution rate will be orders of magnitude higher than dissolution in the passive state. Hence, depassivation would be the "worst" case in degradation. Depassivation of previously passivated Ti surfaces (Ti with a native, air-formed passive film or anodically formed oxide film), leading to uniform corrosion of Ti, has been observed in deoxygenated acidic media (depending on the acid concentration) [19–23].

The native passive film on titanium is a thin, some nm thick layer, mostly consisting of TiO_2. The passive film on titanium shows an *n*-type semiconducting nature, with a band gap of about 3.1 eV. Anodic polarization can be used to thicken the oxide layer; the oxide growth rate has been reported to be in the range of 2–3 nm/V. Anodic oxidation of titanium and its alloys was reviewed in 1973 [24]. Since then, many more details on electrochemical growth and properties of anodic oxide layers on Ti have been resolved, involving intense activities in the field of electrochemical formation of self-organized anodic TiO_2 nanotube surface layers [25]. However, it is beyond the scope of this chapter to give an extensive overview on anodization processes of titanium. The passive TiO_2 layer—also without specific anodization treatments—is an efficient barrier against metal dissolution; in electrochemical terms, dissolution rates $<10^{-6}$ $\mu A/cm^2$ are measured in the passive range.

More information on passivity of metals and alloys can be found in review articles [26] and [27].

2.2 *Special Corrosion Modes of Passive Titanium*

2.2.1 Pitting Corrosion

It is well known that passivity of many metals and alloys can be locally destroyed in the presence of specific aggressive species in the environment, most typically halide ions. Alloys prone to this type of attack include many technically important materials such as stainless steels and aluminum alloys. Pitting corrosion describes a very localized breakdown of passivity, often related to heterogeneities on the surface such as impurity inclusions or second-phase particles—for an overview of pitting corrosion of metals see [28]. After pit initiation, pits can either continue growing, or repassivation can take place. An alloy with a high pitting corrosion resistance should in best case combine a low susceptibility to pit initiation, high tendency and rate of repassivation, and a slow pit propagation rate. It should be mentioned, however, that for metals with electron-conducting passive films (e.g., stainless steels), the number of pits typically inversely correlates with the average depth of the pits, as the cathodic current provided by the large passive area drives the anodic dissolution in the pits—then depending on the number of active pits, the pits grow with different rates. Due to the great significance of pitting corrosion in practical applications in limiting the life time of many technically important materials, mechanisms of localized breakdown of passivity have been widely studied, and for the most important alloys the major parameters influencing the pitting corrosion resistance have been determined.

Considering the biomedical application, in which the materials are exposed to a Cl^--containing environment, it is interesting to note that titanium is susceptible to passivity breakdown in Cl^--containing solutions only at very high anodic potentials, which are not encountered in the practical case. For instance, for CP-Ti, pitting potentials >10 V have been reported in chloride-containing solutions. In fact, Ti is an interesting exception to other metals and alloys prone to pitting corrosion in halide solutions, in that for Ti-based materials the chloride ion is the least pit-triggering anion of the halides [29–31]. One can therefore conclude that classical pitting corrosion should not be expected as a failure mode for titanium in biomedical applications.

It may be of interest to note, however, that metastable pitting corrosion has been observed for CP-Ti (and for the Ti–6Al–4V implant alloy) in NaCl and in simulated body solutions far below the pitting potential, even at the open-circuit potential [32–34]. This means that local passivity breakdown events can take place in the passive region of titanium, but instead of pit propagation, the small pits do not remain active, and fast repassivation takes place. This type of metastable pitting events can be experimentally observed as current transients in potentiodynamic or potentiostatic experiments in the passive region, provided that the experiments are carried out with a sufficient current- and time-resolution; for Ti in chloride solutions these events typically only can be resolved using microelectrodes. Even though metastable pitting does not lead to a complete deterioration of the system, it nevertheless indicates that passivity of the metal is not fully stable in the environment, and the small activation/repassivation events contribute to passive dissolution modes. It may be noteworthy—as pointed out by authors reporting on metastable pitting events on Ti [33]—that in contrast to the metastable pitting behavior of stainless steels, in which case the number of pit initiation events typically decreases to almost zero as a function of time, for titanium the metastable pitting activity remained high over the duration of the experiment (some hours). This behavior may have some relevance to the mode of metal ion release from Ti-based implants.

Since pitting corrosion for Ti is much more prominent in bromide solutions than in presence of chloride, mechanisms of pitting have been far more studied for the bromide case (see, e.g., [35, 36]). However, as this case is not related to implant failure, it is not further discussed in this chapter.

2.2.2 Crevice Corrosion

Another mode of localized corrosion often encountered with passive metals and alloys—and closely related to pitting corrosion in its mechanism—is crevice corrosion. This type of attack occurs preferentially in regions on the metal surface, where mass transport is limited (e.g., in narrow crevices or under deposits), and hence concentration of aggressive species (halides), a pH decrease, and depletion of oxygen, can rapidly lead to activation of the surface in the crevice area. Metals susceptible to pitting corrosion also suffer from crevice corrosion. The presence of crevices on the surface often triggers localized corrosion already under conditions where stable pitting would not take place (e.g., with lower concentration of aggressive halides).

Typically, crevice corrosion for Ti has been reported to take place only at elevated temperatures, e.g., for grade-2 titanium a temperature threshold of 65°C in 0.27 mol/dm^3 NaCl solution was reported for crevice corrosion initiation [37]. However, since Ti can undergo depassivation in deoxygenated acidic media also at room temperature, as described previously, such processes could also take place in crevices with a locally aggressive environment. Even if complete depassivation would not take place, the local passive dissolution rate could be increased, due to the low pH values and high chloride concentration prevailing in the crevice area. In concentrated HCl and H_2SO_4 solutions, it has been shown that the passive current density for CP-Ti increases with the acid concentration [38, 39]. In LiCl solutions of varying concentrations, the passive current density of CP-Ti, as measured in potentiodynamic polarization curves, increases upon the increase of the Cl^- concentration in the solution (Fig. 2.2). Therefore, even though no complete activation of the Ti surface took place in crevice-type situations, significant

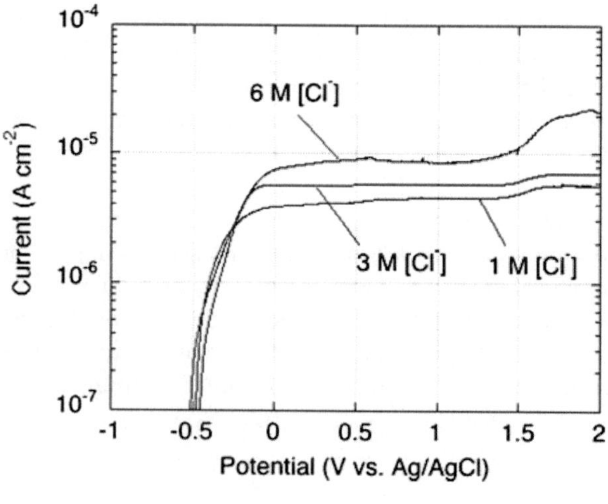

Fig. 2.2 Open-circuit potential as a function of time for CP-Ti and Ti–6Al–4V in 0.5 M H_2SO_4

increase of the passive dissolution rate may be expected by the occluded cell chemistry. In biomedical applications, failures in cemented Ti-based implants have been associated with crevice corrosion, but mostly related to Ti alloys [40, 41]—this case is discussed in more detail later in this chapter. In this context already, one may question, if crevice corrosion of Ti can be initiated in laboratory experiments under similar conditions than prevailing *in vivo* (especially, at 37°C). As the crevice corrosion susceptibility of Ti is influenced by many factors, including the solution chemistry and pH, crevice geometry, surface condition of the metal, metallurgical features (alloying, impurities) [37], and these factors can vary for real implant/cement interfaces, an exact simulation of the real case in the laboratory is difficult. In how far only electrochemical activation effects in the crevice are responsible for the observations in implant retrieval analysis, and the possible role of the biological environment in enhancing the degradation remain unresolved issues. In any case, it is important to note that CP-Ti is significantly less susceptible to crevice corrosion than, for instance, typical grades of stainless steel, and significantly more aggressive conditions (higher temperature, lower pH) together with very narrow, deep crevices are required for crevice corrosion initiation of Ti [5].

2.2.3 Fretting Corrosion/Tribocorrosion

For orthopedic implants, fretting corrosion (or generally, wear-assisted corrosion/tribocorrosion) is often discussed as the dominant failure mode, especially in case of high-corrosion resistance passive alloys, for which the tribological load leads to local mechanical destruction of the passive film. Even though a subsequent spontaneous repassivation reaction takes place in many cases, continuous activation/repassivation cycles lead to an increased material loss. Very recently, fretting-corrosion map in Ringer's solution has been developed for CP-Ti, with the aim to identify the predominant mechanisms (e.g., wear, corrosion, or fretting corrosion) for a given set of parameters [42]. For this, the fretting-corrosion behavior and restoration ability were studied under various combinations of load, frequency, and number of fretting cycles. The authors concluded that fretting-corrosion occurrence can be a serious concern for a safe application of CP-Ti, for instance in hip or knee implants. The fretting corrosion resistance of CP-Ti can, however, be significantly improved by surface treatments, such as anodizing on thermal oxidation [43]. In addition to increasing the fretting-corrosion resistance, surface modification was also found to change the wear mechanism of CP-Ti against alumina: adhesive galling was predominant for untreated CP-Ti, adhesive wear and delamination were found to be operative for anodized samples, whereas for thermally oxidized CP-Ti mainly an abrasive wear mechanism was operative. These findings indicate that the thickness and nature of oxide layers on Ti strongly influence the mechanism and the intensity of tribocorrosion.

3 Effect of Alloying on Degradation of Ti

3.1 Ti–Al–(V, Nb) Alloys

Until now, the most common orthopedic implant alloys are α/β alloys with the composition Ti–6Al–4V and Ti–6Al–7Nb (and some minor variations). These alloys provide good mechanical and corrosion behavior. The corrosion and degradation behavior of these alloys has been studied in a very large number of publications; the aim of this chapter is to summarize the most important findings and to evaluate the role of the alloying elements in the corrosion behavior.

Similar to commercially pure Ti, all Ti-based implant alloys show a spontaneously passive behavior in neutral and near-neutral solutions and, at first hand, one may expect no large differences in the degradation behavior of the different alloys. However, a more detailed look on the electrochemical and corrosion behavior of different alloys illustrates significant alloying effects. It is obvious that the presence of alloying elements Al, V, or Nb in the material not only raises questions on the general corrosion behavior, but especially on metal ion release in specific corrosion processes, as Al and V ions are of concern in biocompatibility considerations.

Experiments under aggressive conditions (e.g., low pH values) can reveal drastic differences between pure Ti and Ti alloys, even at open circuit without polarization to high anodic potentials. An example is shown in Fig. 2.3, comparing the open-circuit potential measured as a function of time for Ti and the implant alloy Ti–6Al–4V in 0.5 M H_2SO_4. Even though this environment is not relevant for the biomedical application, the example simply illustrates that significant differences between the stability of passive films on pure Ti and the Ti alloy can be seen: the Al-containing alloy shows a fast depassivation, whereas CP-Ti remains stable in

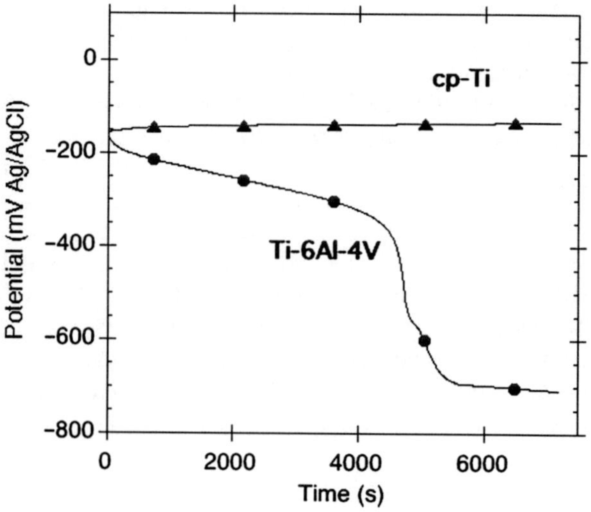

Fig. 2.3 Polarization resistance (R_p) values for CP-Ti and Ti–6Al–4V determined from EIS measured at the corrosion potential in 0.14 M NaCl, 37 °C, at different pH values

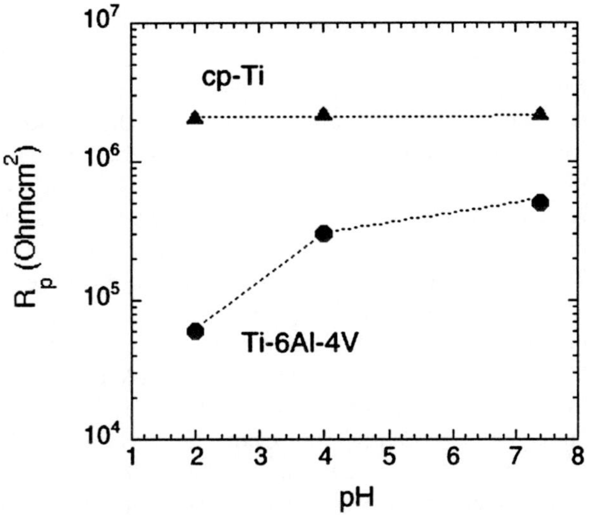

Fig. 2.4 Polarization resistance (R_p) values for CP-Ti and Ti–6Al–4V as a function of solution pH, determined from EIS measurements at the corrosion potential in 0.14 M NaCl solution, T = 37°C

the passive state, at least for the duration of the 2-h experiment. Similarly, the Ti–6Al–7Nb also has been shown to depassivate very fast in H_2SO_4 solutions [44]. However, the behavior was drastically changed when the Ti–Al alloys were passivated in H_2SO_4 instead of studying the native, air-formed passive films [44]. After 0.5 h passivation in H_2SO_4, the alloys also remained stable in the passive state for the 2-h duration of the experiment at open circuit. This behavior can be explained by studying the composition of the different type of passive films. In the air-formed passive films on Ti–Al alloys, significant amounts of Al oxides are incorporated, in addition to TiO_2. The passive films formed in the acidic solution, however, contain a much smaller amount of Al oxide. Hence, the stability of the passive films on the Al-containing alloys seems to be correlated to the presence of Al oxides in the passive layers, as Al oxides are not stable in the acidic pH range. Also, in HCl, the Ti–6Al–4V alloy has been shown to activate at the open-circuit potential at lower acid concentrations than CP-Ti [38]. In agreement with this, the critical current density for passivation in HCl solutions was higher for the Ti–Al–V alloy than for CP-Ti; the aggressive effect of HCl was significantly influenced by a temperature increase from 22 to 37°C.

In less aggressive solutions, in which complete depassivation does not take place (and, hence, such strong shifts of the open-circuit potential cannot be observed), differences in the dissolution behavior between alloys can be assessed, for instance by EIS measurements at the corrosion potential. For measurements carried out in 0.14 M NaCl at different pH values (Fig. 2.4), for CP-Ti the polarization resistance R_p values determined from the impedance spectra are almost independent of the pH value, in the range of pH = 2–7.4. In contrast to this, for the Ti–6Al–4V alloy, the R_p value decreases with the pH, indicating a larger dissolution rate in the acidic solution. Clearly, such effects could be relevant in crevices or any kind of occluded cells on the surface, where local acidification may take place.

Even in neutral solutions, such as different types of simulated body fluids, differences in the passive state between CP-Ti and Ti–Al–(Nb,V) alloys can be detected. These different behaviors are mostly observed at elevated potentials, which however are not directly relevant to the biomedical application. For instance, stable pitting corrosion (assessed by the measurement of the pitting potential) takes place at somewhat lower potentials for the Ti–Al–(V,Nb) alloys than for CP-Ti. As an example, $E_{pit} \approx 4$ V was reported for the Ti–6Al–4V alloy in Hank's Balanced Salt Solution, whereas, in the same study, for CP-Ti and for the Ti–6Al–7Nb alloy, pitting corrosion was not observed up to 10 V [45]. Other studies in Hank's physiological solution reported the stable range of passivity of the Ti–6Al–4V to be limited by pitting corrosion at potentials >3.5 V [46], whereas for the Ti–6Al–7Nb alloy the range of stable passivity was extended up to potential of ca. 6 V [47]. However, also for the alloys, the potential range of stable pitting corrosion is beyond the redox potentials encountered in the body. A more detailed look on the passive current density (with a high time- and current-resolution) reveals that current transients, corresponding to metastable pitting events, can be detected in the passive region far below the pitting potential—examples from magnification of potentiodynamic polarization curves in the passive region are shown for CP-Ti and the Ti–6Al–4V alloy in acidified simulated body fluid, pH 4, at 37°C (Fig. 2.5). With the current resolution of the experiment, current transients were only detected for the Ti alloy, but not for CP-Ti. Pit nucleation/ repassivation events of CP-Ti and the Ti–6Al–4V alloy have also been studied with microelectrodes, enabling the detection of smaller current transients, and the number of these pit nucleation events was higher for the alloy [33]. Systematic studies on the role of the alloying elements in different Ti-based alloys in the metastable pitting process (e.g., influence of the alloying elements on pit initiation, repassivation, and propagation) have not been carried out. The influence of the alloying elements on the passive behavior can be a direct follow-up of the different compositions of the passive films on the alloys vs. CP-Ti, as discussed in [45–47]. Incorporation of very stable Nb oxides may be beneficial for the passivity of Ti–Al–Nb, whereas incorporation of oxidized Al and V (which do not show stable passivity in a large potential region in chloride-containing solutions) may weaken the passive films.

As already mentioned earlier, fretting corrosion—or, in more general terms, tribocorrosion—has been recognized as a dominant failure mode for many types of implants, and therefore has been widely studied. However, astonishingly, few studies can be found comparing different alloys in the same setup. As tribocorrosion is a complex phenomenon, influenced by the material, the environment, and the type of tribological load, comparison between findings in different studies is hardly possible [48, 49]. Williams and Khan studied a number of Ti alloys in regard to corrosion, wear, and wear-accelerated corrosion behavior; the α/β alloys were found to show the best combination of the corrosion and wear resistance, but the wear-accelerated effects on corrosion were less profound for the Ti–6Al–7Nb alloy than for the Ti–6Al–4V alloy and CP-Ti [50]. In other studies on Ti–6Al–4V alloy in contact with alumina, the tribocorrosion behavior was reported to critically

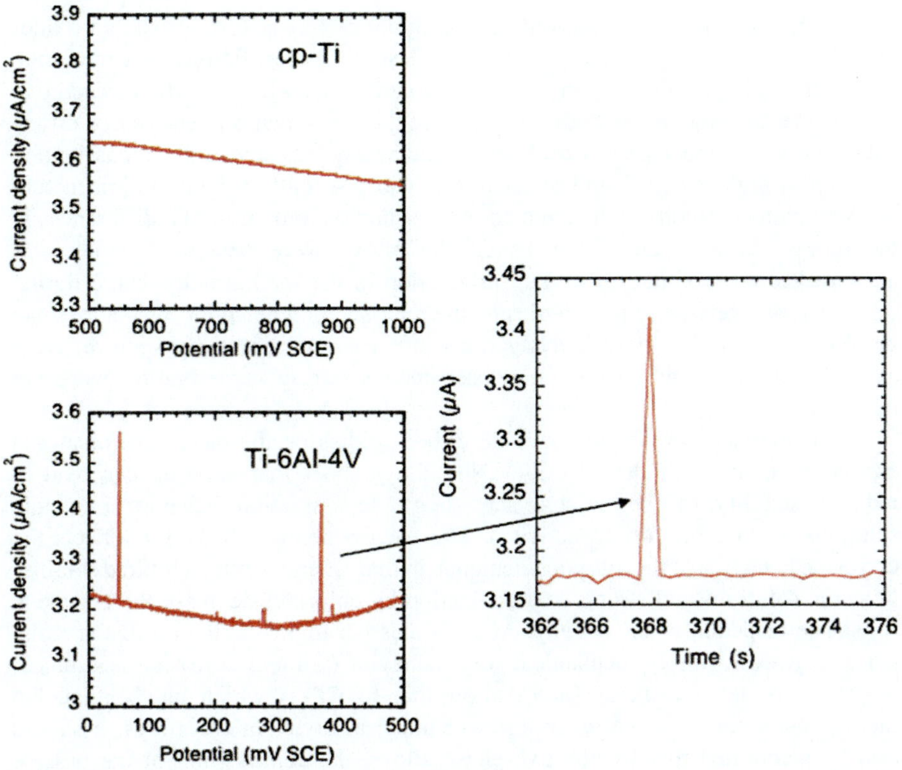

Fig. 2.5 Magnification of potentiodynamic polarization curves in the passive range measured for CP-Ti and Ti–6Al–4V in simulated body fluid, pH 4, 37 °C. The right figure shows a magnified single transient for Ti–6Al–4V

depend on the fretting regime and on the prevailing electrochemical conditions [51, 52]. It is also interesting that no mechanical damage of the Ti–6Al–4V alloy surface could be detected when the contact partner was poly-methyl-methacrylate (PMMA). In this case, also the alloy corrosion was only slightly enhanced by fretting; however, the wear of the PMMA was large [53].

In considerations of wear-assisted corrosion, not only the total damage is of relevance for the biomedical applications, also the influence of the tribological loads on the metal ion release is of interest. For CP-Ti, only Ti cations may be released; for alloys, of course, also oxidized species of the alloying elements. For metals and alloys that are spontaneously passive in the environment in which wear takes place, the metal ion release will be affected by the repassivation behavior of the alloys. In best case, repassivation would take place almost as a solid-state surface reaction, with as little as possible metal ion dissolution. The repassivation rate of different alloys can vary, but another aspect is that for alloys the metal ion release during the activation/repassivation cycles need not be stoichiometric; instead, it might be expected that the alloying elements less stable in the environment will be dissolved, and alloying elements forming stable oxides under these conditions would dissolve in a lesser manner. Preferential metal ion release from alloys under fretting corrosion conditions has been shortly discussed in [54]. However, systematic studies comparing the repassivation behavior of different Ti-based materials are scarce. Kolman and Scully [55] studied the repassivation behavior of high-purity Ti and selected α, β, and β + α alloys by thin film fracture and scratching methods in different chloride solutions: only minimal differences in the repassivation parameters between the alloys were observed, which was attributed to the dominance of Ti^{3+} production in the total anodic charge during repassivation. Relatively few studies on the repassivation behavior of Ti alloys can be found [56–61]. In these investigations, often a comparison between different alloys is missing, and metal ion release modes during repassivation were not reported.

In summary, many of the above-described studies on the electrochemical and corrosion behavior of the Ti–Al–(V,Nb) alloys show that alloying can lead to reduced stability of the passive state (e.g., depassivation behavior in acidic solutions, metastable pitting activity). The electrochemical behavior reflects the corrosion behavior of the alloying elements, in that Al and V with a limited stability in acidic solutions and strong de-stabilization by chlorides decrease the corrosion resistance (especially, as concerns Al), whereas Nb improves the corrosion resistance. However, purely mechanical wear behavior (without corrosion assistance) has been reported to be better for α/β alloys than for CP-Ti, which may be related to the microstructure and the resulting mechanical behavior of the alloys. It should also be mentioned that for the two-phase alloys, the composition of the passive film shows lateral variations related to the composition of the underlying phases [62, 63]. Therefore, also a nonuniform (phase-dependent) corrosion behavior could be expected, but very little information on phase-dependent dissolution can be found in the literature. For Ti–6Al–4V and Ti–6Al–7b alloys, some indications of selective dissolution of the higher Al-containing α-phase in 0.5 M H_2SO_4 solution have been observed [44].

3.2 New Ti-Based Implant Alloys

Research is exploring possibilities to improve the compatibility of Ti alloys in implants, e.g., by tailoring the elastic modulus to reach better mechanical biocompatibility, or by reducing the presence of "critical" alloying elements such as aluminum to decrease the risk of release of toxic elements by corrosion. Recent overviews on some of these developments are given in [8, 64, 65]. Interesting from corrosion point of view are alloys, which contain only valve metals as alloying elements: as all valve metals form very stable oxide layers, stable passivity with very little metal ion release is expected for such alloys. In this chapter, as an example we discuss the behavior of two alloys: Ti–13Nb–13Zr [66] and Ti–29Nb–13Ta–4.6Zr [67]. These alloys show a homogeneous β-phase structure, which leads to a lower elastic modulus, nearer to the modulus of cortical bone than, for instance, in the case of the alloy Ti–6Al–4V; hence, these materials are of interest especially for orthopedic applications. Moreover, the alloying elements are not only chosen by corrosion considerations (metals forming stable oxides), but they are also considered to be biocompatible, similarly to Ti.

The corrosion behavior of the Ti–13Nb–13Zr alloy in different heat treatments, in comparison to CP-Ti and other Ti-based biomedical implant alloys, has been studied in Ringer's solution and in 5 M HCl simulating occluded cell environments [68]. In Ringer's solution, all alloys studied were spontaneously passive, showing similar low passive current densities. However, in deaerated 5 M HCl, significant differences between the alloys were observed, and only the Ti–13Nb–13Zr alloy in its diffusion-hardened state showed spontaneous passivity; this was attributed to the presence of a thick thermal oxide layer on the diffusion-hardened material, and therefore is not only a consequence of the alloy composition. Generally, the addition of Nb and Zr (and lack of Al) in Ti was discussed as beneficial for a high corrosion resistance and improved passivity of the Ti–13Nb–13Zr alloy in comparison to other alloys studied. The effect of the microstructure, adjusted by heat treating from single β-phase to α/β-phase alloys, depended on the alloy composition: the corrosion behavior was inferior with alloying containing Al where the partitioning of the alloying elements led to Al-rich α-phase (this phase being less stable).

The excellent corrosion behavior of the Ti–13Nb–13Zr alloys in different types of simulated body fluids has been confirmed by a number of authors (see, e.g., [69] and references therein). It is hardly surprising that an alloy with such a composition is highly passive in many environments, considering that all alloying elements show a large range of stable passivity. Only few authors have studied the tribocorrosion behavior of this alloy in comparison to other alloys; the overall degradation behavior when both corrosion and wear processes take place was found to be lowest for the Ti–13Nb–13Zr alloy and highest for the Ti–6Al–4V alloy (Ti–6Al–7Nb was somewhat better) [70]. Regarding pure corrosion and pure mechanical wear, the Ti–13Nb–13Zr alloy showed the best performance from the three alloys studied.

Also, for the β-alloy Ti–29Nb–13Ta–4.6Zr, an excellent corrosion behavior, both in Ringer's solution and in 0.5% HCl solution, has been reported [71]. The corrosion and fatigue behavior is influenced by heat treatments: the corrosion behavior in different conditions was always superior to the Ti–6Al–4V alloy, but the fatigue strength of this β-alloy can be significantly lower than that of Ti–6Al–4V. However, with specific heat treatments, the fatigue limit almost reaches the values of Ti–6Al–4V. Even though a significant amount of studies have been conducted on the mechanical behavior of this novel alloy, the corrosion behavior has not been studied in great detail at all.

In summary, novel alloys containing only valve metals as alloying elements show a very promising combination of excellent corrosion performance and optimized mechanical properties. Here we only described the behavior of two alloys; other alloys containing purely valve metals have also been found to exhibit excellent behavior. For instance, for Ti–Ta alloys (0–70% Ta) it has been demonstrated that the corrosion resistance increases with additions of Ta, the wear resistance of the Ti–Ta alloys was better than that of the alloy Ti–6Al–4V, and all Ti–Ta were noncytotoxic, similar to pure Ti [72]. The corrosion behavior of the novel alloys has, however, been studied in less detail than for the classic implant alloys. The corrosion behavior in general is higher than for the classic Ti–Al–(V,Nb) alloys, which can be attributed to the absence of alloying elements that show limited stability of the passive film (e.g., Al). In addition, even in wear-assisted corrosion processes, which typically lead to increased metal ion release from passive metals, less concerns arise as the alloys only contain noncytotoxic elements.

4 Influence of the Biological Environment on the Degradation Behavior of Ti-Based Alloys

The biomedical application brings new challenges for the degradation behavior of materials, in view of wear, corrosion, and tribocorrosion. Apart from implant design-related aspects (to be shortly discussed later in this chapter), the slightly increased temperature (37°C), the specific chemical environment of the body fluids, biomolecules, and living cells can all have important effects on the degradation modes and rates. Due to the very complex nature of the interactions between material surfaces and the biological environment (also due to the high dynamics of the "living" interface), many open questions exist in this field. In this chapter, only selected significant findings are summarized. For more information the reader is referred to the review articles in [38, 73–78].

It was already mentioned previously in this chapter that the environment temperature plays a critical role for occurrence of crevice corrosion of Ti and Ti alloys. The critical temperature for crevice corrosion is mostly assessed in the range of 70°C. However, even the physiological temperature of 37°C can change the degradation behavior as compared with room temperature. For instance, for the Ti–6Al–4V alloy, in contrast to CP-Ti, an increase of the temperature from room temperature to 37°C has been shown to lead to slightly increased passive dissolution rate in neutral physiological saline solutions (as evaluated from EIS measurements) [38, 79]. For more aggressive acidic solutions, the temperature effect is even more crucial in that at higher temperature less acidic solutions can lead to surface activation; also, in this case the effect is more significant for the Ti–6Al–4V alloy than for CP-Ti [38]. Temperature increase has also been shown to have a very significant effect on the metastable pitting activity of CP-Ti and the Ti–6Al–4V alloy [33]. It is noteworthy that at 37°C, the frequency of pit nucleation did not decay to zero with time, and therefore in a long-term exposure, even such small activation/repassivation events could cumulatively contribute significantly to metal ion release. Apart from influencing the passive dissolution behavior, the increase of the temperature to 37°C has been found to influence the adsorption of Ca^{2+} and PO_4^{3-} ions on the surface of passive Ti or Ti alloys; EIS monitoring of the electrochemical interface indicated fast changes in the electrochemical interface (observed as changes in the interfacial capacitance) in the presence of these ions at 37°C, but such effects were not observed at 23 or 25°C [79, 80]. Clearly, in order to evaluate the chemical, electrochemical, and degradation behavior of Ti alloys properly, experiments should be carried out at the physiological temperature of 37°C.

The specific chemistry of inorganic simulated body fluids, in comparison to simple NaCl solutions of a physiological concentration, can significantly change the corrosion behavior of Ti and Ti alloys. As mentioned earlier, Ca and phosphate ions present in the (simulated) body fluids modify the surface of Ti and Ti-based alloys [81–88]. As the solubility of the various calcium phosphates depends on the concentration of the ions in solution, as well as on the alkalinity and the temperature [89], different types of surface layers have been reported (i.e., variations in thickness, surface coverage, Ca/P ratio, etc.). By a prolonged exposure to the biological or simulated body environment, typical observations report an increase in the Ca/P ratio in the surface layer as a function of time, leading to the presence of hydroxyapatite on the surface. A combination of real-time quartz crystal microbalance measurements with X-ray photoelectron spectroscopy on nucleation and growth of calcium phosphate on Ti indicates that hydroxyapatite deposits most likely form via transformation of precursor phases, such as octacalcium phosphate [90]. Different types of surface pretreatments can very significantly influence the calcium phosphate precipitation; a review of CaP-based coatings on titanium is given in [91].

In spite of the many observations that a spontaneous modification of Ti alloy surfaces takes place in the presence of calcium and phosphate ions, the influence of the inorganic ions in simulated body solution on the corrosion behavior has been less widely studied; a reason for this could be that a common consensus is that Ti-based alloys are "highly corrosion resistant," and therefore small changes in the passivity seem not highly relevant. It may be nevertheless interesting to approach this spontaneous surface modification from point of view of influencing the passive behavior. It has been reported that the presence of calcium phosphate on titanium surfaces decreases the corrosion resistance as compared with the behavior in pure saline solutions, both in terms of the general passive behavior and the passivity breakdown at high anodic potentials [82]. In contrast to this, formation of a calcium phosphate deposit has been shown to decrease the cathodic current densities and increase the corrosion resistance of Ti at the open-circuit potential [92]. In the same study, at high anodic potentials (above 2.6 V), the deposit layer led to decreased anodic current densities; however, the passive current densities were not significantly influenced by the presence of the deposit layer on the surface. The origin of the opposite findings is difficult to deduce, as the experimental approaches are quite different in those studies. Another study reports that in simulated body fluids the spontaneous formation of the CaP layer on top of the passive film hinders the native aging of the passive film, which takes place in pure NaCl solutions. In NaCl, a significant increase of the protectiveness of the passive film can be observed as a function of exposure time (determined by EIS) [93]. Even though the protectiveness of the passive film is very high also in the presence of calcium and phosphate ions in the solution, such experiments nevertheless indicate higher passive dissolution rates (i.e., higher metal ion release) for Ti exposed to simulated body fluids as compared with a simple saline solution. Furthermore, the influence of P and Ca ions on the repassivation behavior has been investigated, and retardation of repassivation by their presence was observed [94]. These findings indicate that the formation of a P- and Ca-containing precipitate layer on the passive Ti surface is not only of relevance to the biological performance, but it can influence the passive behavior of titanium as well. Even though the picture on the exact mechanisms is not quite clear (partially due to strong variations in the experimental approaches used, e.g., the type of simulated body solution), it therefore seems important to study the corrosion behavior of Ti in the presence of the inorganic ions of the body fluids, instead of in simple saline solutions. In contrast to thin CaP adsorbate layers formed on Ti spontaneously upon exposure to (simulated) body solutions, thick hydroxyapatite coatings can be prepared on Ti surface by many methods. Such thick coatings, of course, act as an additional barrier and reduce the degradation rate. This can be of relevance for metal ion release by dissolution, but such coatings can strongly influence the tribocorrosion behavior as well.

Body fluids not only contain inorganic components, but in addition a variety of biomolecules and living cells. Also, the effects of proteins and cells on the degradation behavior have been investigated; however, in this field (due to its complex nature), many open questions exist on the exact mechanisms. Also, there is some controversy between the findings of different studies.

The influence of proteins on the corrosion behavior has not been unambiguously clarified, especially as concerns detailed mechanisms. In principle, a protein adsorption layer could act as an additional barrier between the metal surface and the surroundings, hence inhibiting corrosion. On the other hand, proteins may complex metal cations and thus accelerate dissolution. In the case of wear or wear-assisted corrosion reactions, proteins can act as lubricants on the surface. The approaches used to explore protein effects on metals vary largely, in that either serum addition (containing many proteins) or addition of single proteins (most frequently, albumin, being the most abundant protein in the blood) has been studied. Generalized statements on the effects of proteins on biometal surfaces are not possible, as the behavior depends on the protein and metal in question. For instance, it has been shown that for Al and Ti (being highly passive), proteins (albumin and fibrinogen) showed a negligible effect on corrosion, whereas for transition metals a notable increase of corrosion was observed, the effect being dependent on the metal [95]. The findings were explained by the tendency of transition metals to form stable complexes with potential ligands, such as proteins. In contrast to this early study which showed no influence of proteins on the corrosion rate of Ti, another study reported an increase of the Ti corrosion rate by addition of calf serum to saline solution [96]. However, no effect was found for the Ti–6Al–4V alloy. The serum effect on corrosion was found to depend on the surface condition (e.g., ground samples or HNO_3 passivated surfaces). Therefore, the different observations in different studies could stem not only from different approaches to study protein effects (serum vs. single proteins), but also from differences in the surface chemistry of the Ti samples. It has also been reported that proteins have a beneficial effect on the corrosion behavior of Ti, as long as the surface is in the passive state, but in case of film breakdown proteins were reported to have an adverse effect on the corrosion behavior [82]. Both albumin and fibrinogen have been shown to significantly inhibit the cathodic reduction reaction on Ti, but the corrosion rate at the open-circuit potential was not significantly affected, which was explained by the chelating effect of proteins on the anodic dissolution rate [92]. The inhibition of the cathodic reactions by a protein adsorption layer was explained to be due to blocking of the cathodically active reactions sites and/or blocking of the transport of dissolved oxygen to the electrode surface. This is similar to the discussion in a very recent study, indicating that BSA and fetal bovine serum (FBS) adsorption to Ti surface act as a diffusive barrier to oxygen [97]. The modification of the electrochemical properties of Ti surface by protein adsorption at cathodic potentials, but not under anodic polarization, has been reported by other authors too [98]. The barrier effects were greater for serum than for BSA. A comparison of protein effects on corrosion of different Ti-based implant alloys indicated an improvement of the corrosion resistance of the Ti–6Al–4V by addition of serum to PBS, whereas serum addition reduced the corrosion resistance of Ti–13Nb–13Zr and Ti–6Al–7Nb [99].

This short overview on the influence of proteins on the corrosion behavior of Ti alloys demonstrates that the behavior has not been unambiguously elucidated. Even more complex are findings in studies on the effects of inorganic ions of simulated body fluids, amino acids, and proteins on the repassivation behavior of activated Ti surfaces [100]. It could be demonstrated that proteins accelerated repassivation, but some amino acids retarded it. However, the dominant factors in a complex electrolyte combining the different single parameters could not be determined. The influence of proteins on the fretting-corrosion behavior of Ti–6Al–4V has also been explored [101]; the fretting wear was not affected by the presence of collagen or albumin in buffered saline solutions. A small lubricating effect of collagen was observed at cathodic potentials. Moreover, in agreement with previous studies described earlier, the authors observed the proteins to act as cathodic inhibitors. In summary, protein adsorption on Ti-based alloy surfaces can modify the passive corrosion behavior, has mostly been reported to influence the electrochemical properties only at cathodic potentials, and has also been reported to influence formation of CaP deposits on Ti surfaces [88]. However, protein effects on dissolution of Ti-based materials seem to be less prominent than the effects of proteins on the dissolution behavior of some other metals and alloys.

The corrosion behavior of implant metals in the presence of cells has been reviewed in [74]. Similar to what was discussed earlier on protein adsorption effects on metals corrosion, different types of effects for the corrosion behavior could be expected when cells adhere on surfaces: the cell layer could act as a physical barrier, blocking the surface and hence bringing an additional resistance to the surface. Moreover, cell metabolism products could influence the surface reactions. For instance, macrophages can generate active oxygen species (e.g., O_2^-), and these active oxygen species can lead to increased metal release from titanium, in the absence of wear and fretting [102]. Another study reports a reduction of metal ion release from Ti–6Al–4V in the presence of mouse macrophages [103]. In that case, reactive chemical species generated by the macrophages were found to lead to enhancement of the surface oxides [104], hence reducing corrosion. Electrochemical experiments in the presence of cell adhesion layers on Ti surfaces have been carried out [105–107], indicating that cell adhesion layers retard diffusion through the biomolecule adsorption layer due to the generation of the extracellular matrix [105, 106]. A decrease in the cathodic current densities and, consequently, a negative shift of the open-circuit potential, was also observed. The resistivity of the Ti surface/cell layer interface (measured by EIS) was found to increase with time, together with the increase of cell coverage of the surface [107]. Additionally, changes of the surface composition of Ti upon culturing murine fibroblasts have been observed [108]. The CaP precipitation layer was influenced by the cells and, in addition, precipitation of sulfur as S° or S^{2-} was detected. The latter finding indicates that the presence of the cell layer on the surface renders reducing conditions at the Ti surface/cell layer interface, leading to the reduction of sulfates from the surroundings. Further indications of such conditions on passivity of Ti were not discussed in that publication, but clearly such a situation could be of high relevance for metal ion release modes. In addition, pH decrease in the intracell layer on stainless steel and Ti specimens has been measured and found to be caused by corrosion reactions as well as chemicals generated by cells [109]. Cell adhesion layers on the surface, similarly to protein adsorption layers, therefore form an occluded cell type of a situation on the alloy surface, and hence limit mass transport between the metal surface and its surroundings. Therefore, "cell-induced crevice corrosion effects" may be an additional factor to be considered when regarding the behavior of materials in a biological environment.

As already mentioned earlier, reactive chemical species may be generated in cell metabolism reactions. For Ti alloys in the biological environment, special concern has been given to the role of H_2O_2 generated in biological reactions in enhanced metal ion release [110–113]. Figure 2.6 shows an example of H_2O_2-induced dissolution of Ti. In this experiment, *in situ* X-ray absorption near edge spectroscopy (XANES) measurements were carried out on thin-film Ti samples under anodic polarization, and dissolution was monitored by determining the edge height of the spectra as a function of time (explanation on XANES studies can be found in [114]). The results of this and other experiments indicate that dissolution of Ti in the presence of H_2O_2 in the solution does not depend on pH or the anion type studied (as long as the potential is below the pitting potential of Ti), but the dissolution rate strongly increases with increasing potential. In addition to effects on Ti dissolution, the modification of the passive film by hydrogen peroxide by formation of an outer porous oxide layer has been brought into connection to the well-known high osseointegration property of Ti surfaces [112]. The effects of hydrogen peroxide were found to be significantly more dominant on CP-Ti than on the Ti–6Al–4V alloy [113]. Moreover, reactive oxygen species may form in the cathodic oxygen reduction reaction, and these species may influence the viability of

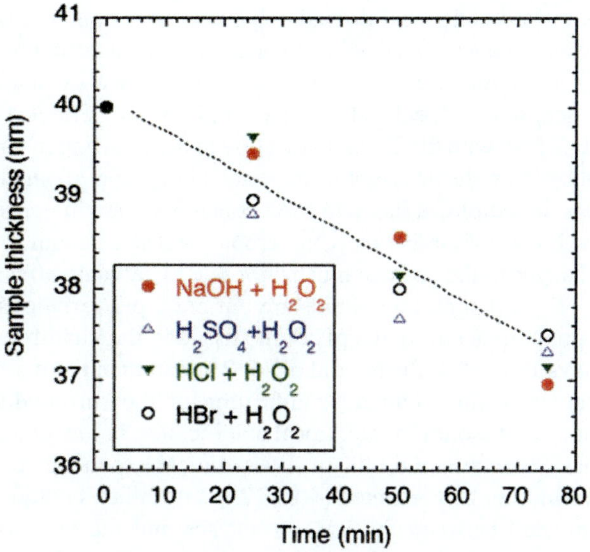

Fig. 2.6 Dissolution of thin-film Ti samples at 1 V SCE in different 0.1 M solutions containing 0.5 M H_2O_2

the surrounding cells [115]. Therefore, a cascade of interactions between the Ti surface and the biological environment is onset. How far peroxide formation in the cathodic reduction reaction is part of the origin of the recently observed reduction in spreading and viability of cells cultured on cathodically polarized Ti surfaces still remains a question to be elucidated [116].

5 Implant Design-Related Degradation Issues

Biomedical implants often are of a modular design, combining different materials in the design. Therefore, possible galvanic corrosion effects resulting from different nobility of the materials involved should be considered. In most cases, however, the metallic materials used are passive. Hence, galvanic coupling will not significantly change the dissolution rate of the less noble metal, as the dissolution rate is controlled by the passive film. Also, passive surfaces in many cases are less efficient cathodes than oxide-free metals, due to slower kinetics of the cathodic reactions on many oxide-covered surfaces. Coupling of more noble passive metals (e.g., Ti alloys) together with low-grade stainless steels could lead to polarization of the stainless steel surface above the pitting potential and, hence, be deleterious for the performance. However, examples can be found in the literature where coupling of stainless steel together with a Ti alloy in spinal implants led to insignificant galvanic corrosion [117], as the pitting corrosion resistance of the specific steel was sufficiently high in the body environment. Actually, in most reported cases, Ti is the more noble partner of a dissimilar metal couple. For instance, coupling of CP-Ti with carbon fiber reinforced polyetheretherketone (CFRP) (as a candidate for bone fixation material) led neither to galvanic corrosion nor to crevice corrosion effects on Ti in simulated body fluid. This is in contrast to stainless steel, which was more susceptible to localized corrosion and coupling effects [118]. Similarly, coupling of Zr-2.5Nb with the Ti–6Al–4V alloy in neutral Ringer's solution indicated that the Zr alloy was the less noble material [119]. Based on thermodynamic and kinetic considerations, it has also been concluded that, in general, coupling of Ti–6Al–4V with Co-28Cr-6Mo or with carbon electrodes should not lead to any detrimental changes in the corrosion behavior due to galvanic effects [120].

Even though body fluids are buffered, pH variations are possible. For instance, upon implantation, the pH of the tissue in the vicinity of the implant may decrease to values below the natural pH of 7.4, and then recover within weeks. As discussed earlier in this chapter, acidification of the surroundings typically increases the passive dissolution rate. Local acidification in the vicinity of Ti-based hip implants has been measured [40, 121]. As clinical experience demonstrates, Ti-based stems should not be cemented [40, 122, 123]; failures could be due to crevice corrosion initiated between the implant surface and the bone cement. On the other hand, cementation also affects micromotions at the implant surface/bone interface, and hence plays a role in the occurrence of fretting corrosion. It should be pointed out that *in vivo* failures on titanium alloy hip stems have also been reported for uncemented implants (see, e.g., [124]). Generally, design of load-bearing implants strongly affects the biomechanics of the system, and hence all degradation modes involving wear or stresses. Moreover, it should be considered that, *in vivo*, the implant surface is experiencing different hydrodynamic conditions (e.g., blood flow) than often used for laboratory testing; hence, mass transport controlled reactions may take place with considerably different rates than expected.

6 Concluding Remarks

As discussed in this chapter, the degradation of biomaterials in the biological environment is a complex process that depends on all materials parameters (e.g., chemical composition, microstructure), environment parameters (e.g., chemistry, temperature), as well as the construction (e.g., presence of crevices). For materials in biomedical applications, the situation is often more complex than in "classical" engineering applications, in that the biological environment of the host is a highly dynamic system.

In the field of surface engineering of biomaterials, a very impressive amount of research is being carried out. Many different types of surface modification approaches on Ti-based materials have been, or are being developed, to optimize the surface compatibility of implants. These approaches range from morphological, to chemical, to biological functionalization of the surface. Of course, all surface modifications may change not only the biological performance (e.g., cell adhesion or tissue integration), but can also influence the degradation behavior of the materials.

Finally, even though today's Ti-based alloys generally show very good performance, novel alloys are being developed. These include novel compositions of alloys as well as alloys with modified microstructures. For instance, interesting from the point of view of changing the mechanical properties of alloys without changing the chemical composition, are ultrafine grained or nanocrystalline alloys, achieved with advanced processing such as severe plastic deformation. However, it is not sufficient to exploit the enhanced mechanical strength of small grain size materials, but instead the effects of the small grain size on the degradation behavior are of high significance for the safe application. Details of the corrosion behavior of nanograined or ultrafine grained materials, however, have been far less studied than effects of the small grain size on the mechanical properties.

References

1. Brunette DM, Tengvall P, Textor M, Thomsen P (eds) (2001) Titanium in medicine. Springer, New York
2. Tengvall P, Lundström I (1992) Physico-chemical considerations of titanium as a biomaterial. ClinMater 9:115–134
3. Geetha M, Singh AK, Asokami R, Gogia AK (2009) Ti based biomaterials, the ultimate choice for orthopaedic implants—a review. Prog Mater Sci 54:397–425
4. Schenk R (2001) The corrosion properties of titanium and titanium alloys. In: Brunette DM, Tengvall P, Textor M, Thomsen P (eds) Titanium in medicine. Springer, New York
5. Been J, Grauman JS (2000) Titanium and titanium alloys. In: Revie RW (ed) Uhlig's corrosion handbook. Wiley, New York
6. Shoesmith DW, Noel JJ (2010) Corrosion of titanium and its alloys. In: Richardson TJA, Cottis R, Lindsay R, Stuart L, Scantlebury DJD, Graham MJ (eds) Shreir's corrosion, vol 3. Elsevier, Oxford, UK
7. Long M, Rack HJ (1998) Titanium alloys in total joint replacement—a materials science perspective. Biomaterials 19:1621–1639
8. Niinomi M (2008) Mechanical biocompatibilities of titanium alloys for biomedical applications. J Mech Behav Biomed Mater 1:30–42
9. Fleck C, Eifler D (2010) Corrosion, fatigue and corrosion-fatigue behavior of metal implant materials, especially of titanium alloys. Int J Fatigue 32:929–935
10. Hallab NJ, Jacobs JJ (2003) Orthopedic implant fretting corrosion. Corros Rev 21:183–213
11. Syrett BC, Acharya A (eds) (1978) Corrosion and degradation of implant materials, ASTM STP 684. ASTM, Philadelphia
12. Fraker AC, Griffin CD (1985) Corrosion and degradation of implant materials, ASTM STP 859. ASTM, Philadelphia
13. Kovacs P, Istephanous NS (eds) (1994) Compatibility of biomedical implants, proceedings volume 94-15. The Electrochemical Society, Pennington
14. Pourbaix M (1963) Atlas d'Equilibres Electrochimique. Gautiers-Billars & Vie, Paris
15. Fovet Y, Gal J-Y, Toumelin-Chemla F (2001) Influence of pH and fluoride concentration on titanium passivating layer: stability of titanium oxide. Talanta 53:1053–1063
16. Robin A, Meirelis JP (2007) Influence of fluoride concentration and pH on corrosion behavior of titanium in artificial saliva. J Appl Electrochem 37:511–517
17. Upadhyay D, Panchal MA, Dubey RS, Srivastava VK (2006) Corrosion of alloys used in dentistry—a review. Mater Sci Eng A 432:1–11

18. Kelly EJ (1982) Electrochemical behavior of titanium. In: Bockris JOM, Conway BE, White RE (eds) Modern aspects of electrochemistry, No. 14, chapter 5. Plenum Press, New York
19. El-Taib Heakal F, Mokoda AS, Mazhar AA, El-Basiouny MS (1987) Kinetic studies on the dissolution of the anodic oxide film on titanium in phosphoric acid solutions. Corros Sci 27:453–482
20. Blackwood DJ, Peter LM, Williams DE (1988) Stability and open circuit breakdown of the passive oxide film on titanium. Electrochim Acta 33:1143–1149
21. Blackwood DJ, Greef R, Peter LM (1989) An ellipsometric study of the growth and open-circuit dissolution of the anodic oxide film on titanium. Electrochim Acta 34:875–880
22. Yu SY, Brodwick CW, Ryan MR, Scully JR (1999) Effects of Nb and Zr alloying additions on the activation behavior of Ti in hydrochloric acid. J Electrochem Soc 146:4429–4438
23. Azumi K, Nakajima M, Okamoto K, Seo M (2007) Dissolution of Ti wires in sulphuric acid and hydrochloric acid solutions. Corros Sci 49:469–480
24. Aladjem A (1973) Review: anodic oxidation of titanium and its alloys. J Mater Sci 8:688–704
25. Ghicov A, Schmuki P (2009) Self-ordering electrochemistry: a review on growth and functionality of TiO_2 nanotubes and other self-aligned MO_x structures. Chem Commun 20:2791–2808
26. Schultze JW, Lohrengel MM (2000) Stability, reactivity and breakdown of passive films. Problems of recent and future research. Electrochim Acta 45:2499–2513
27. Schmuki P (2002) From Bacon to barriers: a review on the passivity of metals and alloys. J Solid State Electrochem 6:145–164
28. Zsklarska-Smialowska Z (1986) Pitting corrosion of metals. National Association of Corrosion Engineers, Houston, TX
29. Dugdale I, Cotton JB (1964) The anodic polarization of titanium in halide solutions. Corros Sci 4:397–400
30. Beck TR (1973) Pitting of titanium. J Electrochem Soc 120:1310–1324
31. Rätzer-Scheibe HJ (1978) Relationship between repassivation behavior and pitting corrosion. Corrosion 34:437–442
32. Burstein GT, Souto RM (1995) Observations of localized instability of passive titanium in chloride solutions. Electrochim Acta 40:1881–1888
33. Burstein GT, Liu C, Souto RM (2005) The effect of temperature on the nucleation of corrosion pits on titanium in Ringer's solution. Biomaterials 26:245–256
34. Burstein GT, Liu C (2008) Depassivation current transients measured between identical twin microelectrodes in open circuit. Corros Sci 50:2–7
35. Casillar N, Charlebois S, Smyrl WH (1994) Pitting corrosion of titanium. J Electrochem Soc 141:636–642
36. Basame SB, White HS (2000) Pitting corrosion of titanium the relationship between pitting potential and competitive anion adsorption at the oxide film/electrolyte interface. J Electrochem Soc 147:1376–1381
37. He X, Noel JJ, Shoesmith DW (2002) Temperature dependence of crevice corrosion initiation on titanium grade-2. J Electrochem Soc 149:B440–B449
38. Virtanen S (2008) Corrosion of biomedical implant materials. Corros Rev 26:147–171
39. Mogoda AS, Ahmad YH, Badawy WA (2004) Corrosion behavior of Ti-6Al-4V alloy in concentrated hydrochloric and sulfuric acids. J Appl Electrochem 34:873–878
40. Willert H-G, Broback L-G, Buchhorn GH, Jensen PH, Köster G, Ochsner P, Schenk R (1996) Crevice corrosion of cemented titanium alloy stems in total hip replacements. Clin Orthop Relat Res (333):51–75
41. Thomas SR, Shukla D, Latham PD (2004) Corrosion of cemented titanium femoral stems. J Bone Joint Surg 86-B:974–978
42. Kumar S, Sivakumar S, Sankara Narayanan TSN, Ganesh Sundara Raman S, Seshadri SK (2010) Fretting-corrosion mapping of CP-Ti in Ringer's solution. Wear 268:1537–1541
43. Kumar S, Narayanan TSNS, Ganesh Sundara Raman S, Seshadri SK (2010) Surface modification of CP-Ti to improve the fretting-corrosion resistance: thermal oxidation vs. anodizing. Mater Sci Eng C 30:921–927

44. Ruzickova M, Hildebrand H, Virtanen S (2005) On the stability of passivity of Ti-Al alloys in acidic environment. Z Phys Chem 219:1447–1459
45. Metikos-Hukovic M, Kwokal A, Pilkac J (2003) The influence of niobium and vanadium passivity of titanium-based implants in physiological solution. Biomaterials 24:3765–3775
46. Milosev I, Metikos-Hukovic M, Strehblow H-H (2000) Passive film on orthopaedic TiAlV alloy formed in physiological solution investigated by X-ray photoelectron spectroscopy. Biomaterials 21:2103–2113
47. Milosev I, Kosec T, Strehblow H-H (2008) XPS and EIS study of the passive film formed on orthopaedic Ti-6Al-7Nb alloy in Hank's physiological solution. Electrochim Acta 52:3547–3558
48. Landolt D, Mischler S, Stemp M (2001) Electrochemical methods in tribocorrosion: a critical appraisal. Electrochim Acta 46:3913–3929
49. Mischler S (2008) Triboelectrochemical techniques and interpretation methods in tribocorrosion: a comparative evaluation. Tribol Int 41:573–583
50. Khan MA, Williams RL, Williams DF (1996) In-vitro corrosion and wear of titanium alloys in the biological environment. Biomaterials 17:2117–2126
51. Barril S, Mischler S, Landolt D (2004) Influence of fretting regimes on the tribocorrosion behaviour of Ti6Al4V in 0.9 wt-% sodium chloride solution. Wear 256:963–972
52. Barril S, Mischler S, Landolt D (2005) Electrochemical effects of the fretting corrosion behavior of Ti6Al4V in 0.9 wt-% sodium chloride solution. Wear 259:282–291
53. Mischler S, Barril S, Landolt D (2009) Fretting corrosion behaviour of Ti-6Al-4V/PMMA contact in simulated body fluid. Tribol Mater Surf Interfaces 3:16–23
54. Hanawa T (2004) Metal ion release from metal implants. Mater Sci Eng C 24:745–752
55. Kolman DG, Scully JR (1996) On the repassivation behavior of high-purity titanium and selected α, β, and $\beta + \alpha$ alloys in aqueous chloride solutions. J Electrochem Soc 143:1847–1860
56. Beck TR (1973) Electrochemistry of freshly-generated titanium surface—I. Scraped-rotating-disk experiments. Electrochim Acta 18:807–814
57. Beck TR (1973) Electrochemistry of freshly-generated titanium surface—II. Rapid fracture experiments. Electrochim Acta 18:815–827
58. Buhl H (1973) Repassivation behavior of the titanium alloy TiAl6V4 in aqueous sodium halides. Corros Sci 13:639–646
59. Raetzer-Scheibe H-J, Buhl H (1979) Zum Repassivierungsverhalten metallischer Werkstoffe am Beispiels einer Titanlegierung—Contribution to the understanding of repassivation behavior for a titanium alloy. Werkst Korros 30:846–853
60. Gilbert JL, Buckley CA, Lautenschlager EP (1996) Titanium oxide film fracture and repassivation: the effect of potential, pH and aeration. In: Brown SA, Lemons JE (eds) Medical applications of titanium and its alloys: the material and biological issues, ASTM STP 1272. American Society for Testing and Materials, Philadelphia, pp 199–215
61. Goldberg JR, Gilbert JL (2004) The electrochemical and mechanical behavior of passivated and TiN/AlN-coated CoCrMo and Ti6Al4V alloys. Biomaterials 25:851–864
62. Lausmaa J (1996) Surface spectroscopic characterization of titanium implant materials. J Electron Spectrosc Rel Phen 31:343–361
63. Sittig C, Hähner G, Marti A, Textor M, Spencer ND, Hauert R (1999) The implant material, Ti6Al7Nb: surface microstructure, composition and properties. J Mater Sci Mater Med 10:191–198
64. Schuh A, Bigoney J, Hönle W, Zeiler G, Holzwarth U, Forst R (2007) Second generation (low modulus) titanium alloys in total hip arthoplasty. Materialwiss Werkst 38:1003–1007
65. Guillemot F (2005) Recent advances in the design of titanium alloys for orthopedic applications. Exp Rev Med Dev 2:741–748
66. Davidson JA, Mishra AK, Kovacs P, Poggie RA (1994) New surface-hardened, low-modulus, corrosion-resistant Ti13Nb13Zr alloy for total hip arthoplasty. Biomed Mater Eng 4:231–243
67. Niinomi M (2003) Fatigue performance and cyto-toxicity of low rigidity titanium alloys Ti-29Nb-13Ta-4.6Zr. Biomaterials 24:2673–2683

68. Yu SY, Scully JR (1997) Corrosion and passivity of Ti-13%Nb-13%Zr in comparison to other biomedical implant alloys. Corrosion 53:965–976
69. Robin A, Carvalho OAS, Schneider SG, Schneider S (2008) Corrosion behavior of Ti-xNb-13Zr alloys in Ringer's solution. Mater Corros 59:929–933
70. Khan MA, Williams RL, Williams DF (1999) Conjoint corrosion and wear in titanium alloys. Biomaterials 20:765–772
71. Akahori T, Niinomi M, Fukui H, Suzuki A (2004) Fatigue, fretting fatigue and corrosion characteristics of biocompatible beta type titanium alloy conducted with various thermo-mechanical treatments. Mater Trans 45:1540–1548
72. Zhoua YL, Niinomi M, Akahori T, Fukui H, Toda H (2005) Corrosion resistance and biocompatibility of Ti-Ta alloys for biomedical applications. Mater Sci Eng A 398:28–36
73. Virtanen S, Milosev I, Gomez-Barrena E, Trebse R, Salo J, Konttinen YT (2008) Special modes of corrosion under physiological and simulated physiological conditions. Acta Biomater 4:468–476
74. Hiromoto S, Hanawa T (2006) Corrosion of implant metals in the presence of cells. Corros Rev 24:323–352
75. Galante JO, Lemons J, Spector M, Wilson PD Jr, Wright TM (1991) Review. The biologic effect of implant materials. J Orthop Res 9:760–775
76. Bundy KJ (1994) Corrosion and other electrochemical aspects of biomaterials. Crit Rev Biomed Eng 22:139–251
77. Mathew MT, Srinisiva Pai P, Pourzal R, Fischer A, Wimmer MA (2009) Significance of tribocorrosion in biomedical applications: overview and current status. Adv Tribol; article no. 250986
78. Cadosch D, Chan E, Gautschi OP, Filgueira L (2009) Metal is not inert: role of metal ions released by biocorrosion in aseptic loosening—current concepts. J Biomed Mater Res 91A:1252–1262
79. Alves VA, Reis RQ, Santos ICB, Souza DG, de F Goncalves T, Pereira-da-Silva MA, Rossi A, da Silva LA (2009) In situ impedance spectroscopy study of the electrochemical corrosion of Ti and Ti-6Al-4V in simulated body fluid at 25°C and 37°C. Corros Sci 51:2473–2482
80. Hodgson AWE, Mueller Y, Forster D, Virtanen S (2002) Electrochemical characterization of passive film on Ti alloys under simulated biological conditions. Electrochim Acta 47:1913–1923
81. Hanawa T, Ota M (1991) Calcium phosphate naturally formed on titanium in electrolyte solution. Biomaterials 12:767–774
82. Sousa SR, Barbosa MA (1993) Corrosion resistance of titanium cp in saline physiological solution with calcium phosphate and proteins. Clin Mater 14:287–294
83. Wu W, Nancollas GH (1998) Kinetics of heterogeneous nucleation of calcium phosphates on anatase and rutile. J Colloid Interface Sci 199:206–211
84. Li P, Ducheyne P (1998) Quasi-biological apatite film induced by titanium in a simulated body fluid. J Biomed Mater Res 41:341–348
85. Frauchiger L, Taborelly M, Aronsson BO, Descouts P (1999) Ion adsorption on titanium surfaces exposed to a physiological solution. Appl Surf Sci 143:57–77
86. Healy KE, Ducheyne P (1992) Hydration and preferential molecular adsorption on titanium in vitro. Biomaterials 13:553–561
87. Sundgren JE, Bodo P, Lundstrom I (1986) Auger electron spectroscopic study of the interface between human tissue and implants of titanium and stainless steel. J Colloid Interface Sci 110:9–20
88. Lima J, Sousa SR, Ferreira A, Barbosa MA (2001) Interactions between calcium, phosphate, and albumin on the surface of titanium. J Biomed Mater Res 55:45–53
89. Ban S, Maruno S (1995) Effect of temperature on electrochemical deposition of calcium phosphate coatings in a simulated body fluid. Biomaterials 16:977–981
90. Eliaz N, Kopelovitch W, Burstein L, Kobayashi E, Hanawa T (2008) Electrochemical processes of nucleation and growth of calcium phosphate on titanium supported by real-time quartz crystal microbalance measurements and X-ray photoelectron spectroscopy analysis. J Biomed Mater Res 89A:270–280

91. Narayanan R, Seshadri SK, Kwon TY, Kim KH (2008) Calcium phosphate-based coatings on titanium and its alloys. J Biomed Mater Res B 85:279–299
92. Cheng R, Roscoe SG (2005) Corrosion behavior of titanium in the presence of calcium phosphate and serum proteins. Biomaterials 26:7350–7356
93. Alkhateeb E, Virtanen S (2005) Influence of surface self-modification in Ringer's solution on the passive behavior of titanium. J Biomed Mater Res A 75:934–940
94. Hanawa T, Asami K, Asaoka K (1998) Repassivation of titanium and surface oxide film regenerated in simulated bioliquid. J Biomed Mater Res 40:530–538
95. Clark GCF, Williams DF (1982) The effects of proteins on metallic corrosion. J Biomed Mater Res 16:125–134
96. Williams RL, Brown SA, Merritt K (1988) Electrochemical studies on the influence of proteins on the corrosion of implant alloys. Biomaterials 9:181–186
97. Burgos-Asperilla L, Garcia-Alonso MC, Escudero ML, Alonso C (2010) Study of the interaction of inorganic and organic compounds of cell culture medium with a Ti surface. Acta Biomater 6:652–661
98. Ehrenberger MT, Gilbert JL (2010) The effect of scanning electrochemical potential on the short-term impedance of commercially pure titanium in simulated biological conditions. J Biomed Mater Res 94A:781–789
99. Khan MA, Williams RL, Williams DF (1999) The corrosion behavior of Ti-6Al-4V, Ti-6Al-7Nb and Ti-13Nb-13Zr in protein solutions. Biomaterials 20:631–637
100. Hanawa T, Kohayama Y, Hiromoto S, Yamamoto A (2004) Effects of biological factors on the repassivation current of titanium. Mater Trans 45:1635–1639
101. Hiromoto S, Mischler S (2006) The influence of proteins on the fretting-corrosion behavior of a Ti6Al4V alloy. Wear 261:1002–1011
102. Mu Y, Kobayashi T, Sumita M, Yamamoto Y, Hanawa T (2000) Metal ion release from titanium with active oxygen species generated by rat macrophages *in vitro*. J Biomed Mater Res 49:238–243
103. Lin H-Y, Bumgardner JL (2004) In vitro biocorrosion of Ti-6Al-4V implant alloy by a mouse macrophage cell line. J Biomed Mater Res 68A:717–724
104. Lin H-Y, Bumgardner JL (2004) Changes in the surface composition of the Ti-6Al-4V implant alloy by cultured macrophage cells. Appl Surf Sci 225:21–28
105. Hiromoto S, Noda K, Hanawa T (2002) Development of electrolytic cell with cell-culture for metallic biomaterials. Corros Sci 44:955–965
106. Hiromoto S, Noda K, Hanawa T (2002) Electrochemical properties of an interface between titanium and fibroblasts L929. Electrochim Acta 48:387–396
107. Garcia-Alonso MC, Saldana L, Alonso C, Barranco V, Munoz-Morris MA, Escudero ML (2009) In situ cell culture monitoring on a Ti-6Al-4V surface by electrochemical techniques. Acta Biomater 5:1374–1384
108. Hiromoto S, Hanawa T, Asami K (2004) Composition of surface oxide films of titanium with culturing murine fibroblasts L929. Biomaterials 25:979–986
109. Hiromoto S, Hanawa T (2004) pH near cells on stainless steel and titanium. Electrochem Solid State Lett 7:B9–B11
110. Tengvall P, Elwing H, Sjöqvist L, Lundström I (1989) Interaction between hydrogen peroxide and titanium: a possible role in the biocompatibility of titanium. Biomaterials 10:118–120
111. Pan J, Thierry D, Leygraf C (1996) Hydrogen peroxide toward enhanced oxide growth on titanium in PBS solution: blue coloration and clinical relevance. J Biomed Mater Res 30:393–402
112. Pan J, Liao H, Leygraf C, Thierry D, Li J (1998) Variation of oxide films on titanium induced by osteoblast-like cell culture and the influence of an H_2O_2 pretreatment. J Biomed Mater Res 40:244–256
113. Bearinger JP, Orme CA, Gilbert JL (2003) Effect of hydrogen peroxide on titanium surface: in situ imaging and step-polarization impedance spectroscopy of commercially pure titanium and titanium, 6-aluminum,4-vanadium. J Biomed Mater Res 67A:702–712

114. Virtanen S, Isaacs HS, Schmuki P (2002) In situ X-ray absorption near edge studies of mechanisms of passivity. Electrochim Acta 47:3117–3125
115. Tsaryk R, Kalbacova M, Hempel U, Scharnweber D, Unger RE, Dieter P, Kirkpatrick CJ, Peters K (2007) Response of human endothelial cells to oxidative stress on Ti6Al4V alloy. Biomaterials 28:806–813
116. Ehrensberger MT, Sivan S, Gilbert JL (2009) Titanium is not "the most biocompatible metal" under cathodic potential: the relationship between voltage and MC3T3 preosteoblast behavior on electrically polarized cpTi surfaces. J Biomed Mater Res 93A:1500–1509
117. Serhan H, Slivka M, Albert T, Kwak SD (2004) Is galvanic corrosion between titanium alloy and stainless steel spinal implant a clinical concern? Spine J 4:379–387
118. Mueller Y, Tognini R, Mayer J, Virtanen S (2007) Anodized titanium and stainless steel in contact with CFRP: an electrochemical approach considering galvanic corrosion. J Biomed Mater Res 82A:936–946
119. Marek M, Pawar V, Tsai S, Thomas R, Sprague J, Salehi A, Hunter G (2006) Galvanic corrosion evaluation of Zr-25Nb coupled with orthopaedic alloys. In: Medical device materials III, vol 2006., pp 195–201
120. Griffin CD, Buchanan RA, Lemons JE (1983) In vitro electrochemical corrosion study of coupled surgical implant materials. J Biomed Mater Res 17:489–500
121. Konttinen YT, Takagi M, Mandelin J, Lassus J, Salo J, Ainola M (2001) Acid attack and cathepsin K in bone resorption around total hip replacements. J Bone Miner Res 16:1780–1786
122. Schöll E, Eggli S, Ganz R (2000) Osteolysis in cemented titanium alloy hip prosthesis. J Arthoplasty 15:570–575
123. Kovac S, Trebse R, Milosev I, Pavlovcic V, Pisot V (2006) Long-term survival of a cemented titanium-aluminum-vanadium alloy straight-stem femoral component. J Bone Joint Surg 88:1567–1573
124. Paliwal M, Gordon Allan D, Filip P (2010) Failure analysis of three uncemented titanium-alloy modular total hip stems. Eng Fail Anal 17:1230–1238

Biography

Sannakaisa Virtanen studied materials science at the Helsinki University of Technology, Finland (M.Sc. in 1984) and received her PhD degree from the Swiss Federal Institute of Technology, ETH-Zurich, in 1989. From 1989 to 1998 she worked as a senior scientist at the ETH-Zurich (Institute for Materials Chemistry and Corrosion). During this time, she was a visiting scientist at Brookhaven National Laboratory, USA (1994–1995) and at McMaster University, Canada (1996). In 1997 she was elected Assistant Professor at the Department of Materials of the ETH-Zurich. Since 2003 she is professor at the Department of Materials Science (Institute for Surface Science and Corrosion) of the University of Erlangen-Nuremberg, Germany. Her research activities are related to electrochemistry and surface science, and range from studies on corrosion, passivity, and localized breakdown of advanced metals and alloys to surface functionalization of metallic materials. Of a special interest are corrosion behavior and surface reactivity of metallic materials used in biomedical applications.

Chapter 3
Degradation of Dental Implants

Takao Hanawa

Abstract In this chapter, the degradation of dental implants is discussed. First, the dental implant system and the biological environment surrounding it are explained. Hydroxyapatite coating layer is sometimes fractured, causing loosening and infection, as discussed too. To understand biodegradation of dental implants, general properties and electrochemical properties must be understood. The passive surface oxide film on titanium and the reconstruction of the passive film are significant to understand an electrochemical degradation or corrosion of titanium. On the other hand, the release of metallic ions from titanium and the behavior of the released ions govern the biological reaction or toxicity. Of course, the contamination of implant surface is important to determine the tissue compatibility. The fracture of dental implants is also discussed. Finally, biofilm formation on dental implant, inducing infection, is explained.

1 Introduction

A dental implant is an artificial tooth root used in dentistry to support restorations that resemble a tooth or group of teeth: dental prostheses, including crowns, implant-supported bridges, or dentures. Virtually all dental implants placed today are root-form endosseous implants. The majority of dental implants currently available are shaped like small screws, with either tapered or parallel sides, as shown in Fig. 3.1. A typical implant consists of a titanium (Ti) screw with a roughened or smooth surface. The majority of dental implants are made of commercially pure titanium (CP-Ti), which is available in four grades, depending upon the amount of carbon, oxygen, nitrogen, and iron contained. Ti alloy with 6 wt%

T. Hanawa (✉)
Institute of Biomaterials and Bioengineering, Tokyo Medical and Dental University,
2-3-10 Kanda-surugadai, Chiyoda-ku, Tokyo 101-0062, Japan
e-mail: hanawa.met@tmd.ac.jp

N. Eliaz (ed.), *Degradation of Implant Materials*, DOI 10.1007/978-1-4614-3942-4_3,

Fig. 3.1 Dental implant

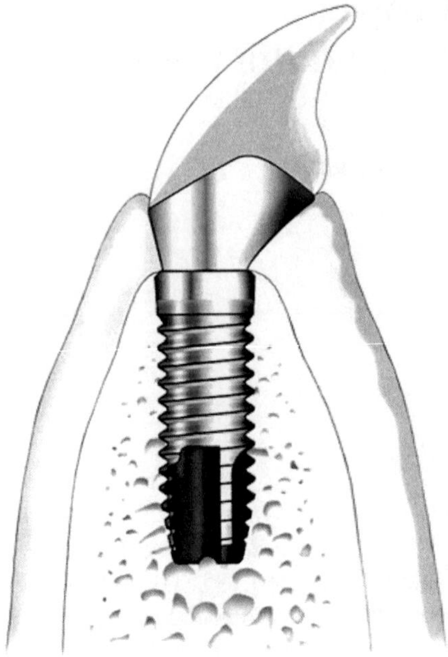

aluminum (Al) and 4 wt% vanadium (V), Ti–6Al–4V alloy, is believed to offer similar osseointegration levels as CP-Ti. Today, most implants are still made out of CP-Ti, but some implant systems are fabricated from the Ti–6Al–4V alloy. The difference in periodontal tissues around a natural tooth and around a dental implant is illustrated in Fig. 3.2. The implant directly contacts alveolar bone and junctional epithelium. Therefore, hard tissue compatibility (bone formation or bone bonding) for the fixation, and soft tissue compatibility (sealing cell adhesion) for the inhibition of bacterial invasion, are required. Implant surfaces may be modified by plasma spraying of hydroxyapatite (HA), anodizing (micro-arc oxidation), etching, or blasting to increase the surface area and the integration potential of the implant [1]. Currently, HA is also used for fixture base. Some current research in dental implantology is focusing on the use of ceramic materials such as zirconia (ZrO_2). Zirconia is the dioxide of zirconium, a metal close to titanium in the periodic table. However, the surface properties of zirconium are different from those of Ti [2], as described later. Although generally having the same shape as Ti implants, zirconia, which has been used successfully for orthopedic surgery for a number of years, has the advantage of being more cosmetically esthetic owing to its bright tooth-like color [3]. However, long-term clinical data are necessary before one-piece zirconia implants can be recommended for daily practice [4].

As expected from the above dental implant structure, the "degradation of dental implants" contains comprehensive meanings. From the viewpoint of mechanical behavior, a design based on the mechanical properties and manufacturing processes

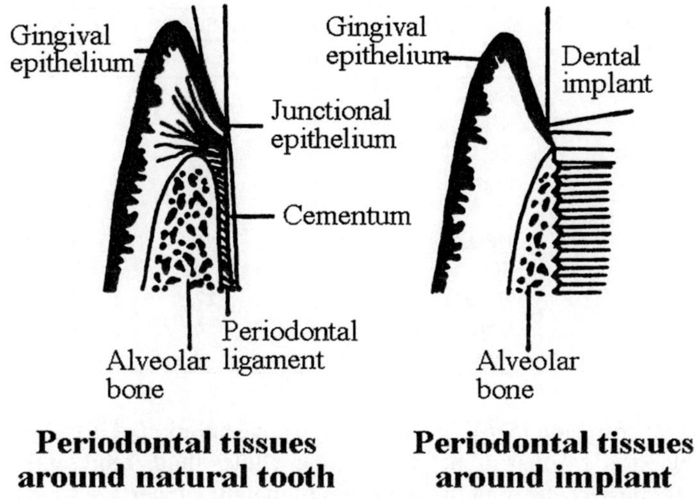

Fig. 3.2 Difference in periodontal tissue between natural tooth and dental implant

allows to prevent fatigue and fracture; thus fracture of Ti implants hardly occurs *in vivo*, with the exception of the recent thin implants. In dental implants, wear is also little during the implantation. On the other hand, HA coating on Ti is sometimes not perfect and fails *in vivo*. From electrochemistry viewpoint, the surface oxide film on Ti is microscopically unstable and changes its composition during implantation, which influences the tissue compatibility. The electrochemical properties, especially passivity, should be understood when discussing ion release from titanium materials in the human body. Yet, the durability of most surface-modified implants is unclear, while the success cases are actively demonstrated. In addition, little information on the fracture of zirconia implants has been accumulated. Therefore, degradation of the surface-modified surface and zirconia implants could not be discussed at present.

Implant failure has been associated with factors such as poor bone quality, insufficient bone volume, implant instability, and unfavorable implant loading. Infection and host responses may also be important factors in dental implant failure. Therefore, the fracture and infection inducing implant failure are briefly explained.

2 The Biological Environment

As summarized in Table 3.1, the concentrations of chloride ions in serum and interstitial fluid are 113 and 117 mEq L^{-1}, respectively, which is 1/3 of the concentration in brine. Thus, a severe corrosive environment is established for metallic materials. Body fluids contain various amino acids and proteins that

Table 3.1 Electrolyte concentrations per 1 L water (mEq)

	Extracellular fluid		
Species	Serum	Interstitial	Intracellular fluid
Na^+	152	143	14
K^+	5	4	157
Ca^{2+}	5	5	–
Mg^{2+}	3	3	26
Cl^-	133	117	–
HPO_4^{2-}	2	2	113 (PO_4^{3-})
SO_4^{2-}	1	1	–
HCO_3^-	27	27	10
Organic acid	6	6	–
Protein	16	2	74

influence metal corrosion [5, 6]. In addition, the concentration of dissolved oxygen in venous blood is 1/4 that in air, and in intercellular fluids represents 1/80–1/4 that in air [7], thus accelerating the corrosion of metallic materials. Changes in the pH of body fluids are small because the fluids are buffered solutions; hence the pH usually remains between 7.0 and 7.35 [7]. The pH of the hard tissue into which a material is implanted decreases to approximately 5.2 and then recovers to 7.4 within 2 weeks [8].

3 Failure of Hydroxyapatite Coating Layer

The most important reason for the surface modification of Ti is the improvement of hard-tissue compatibility through the formation of a HA film. Currently, plasma spraying of HA on metallic materials is widely used to form the HA layer—which is the nucleus for active bone formation and conductivity. In the case of plasma-sprayed HA, however, the HA–Ti interface or HA itself may fracture under relatively low stress because of the low interface bonding strength and low toughness of the sprayed layer itself [9]. The solubility of ceramics increases as their crystallinity decreases. The crystallinity of coated HA is an important factor because crystallinity governs solubility in the human body. The crystallinity of a thin film formed with ion beam is low, thus its solubility is high. A poorly crystalline film on Ti dissolves rapidly when the Ti is implanted into a human body. Thus, heat treatment of HA films is necessary to increase their crystallinity and reduce their solubility [9, 10]. Calcium ions are implanted during the mixing process to induce strong bonding between the HA film and the Ti substrate, with implanted calcium ions serving as binders [11]. HA and calcium phosphate coatings with radio frequency (RF) magnetron sputtering have been applied [12, 13].

Electrochemical treatments have been evaluated in an attempt to form HA layers on Ti [14, 15]. For example, carbonate-containing HA with a desirable morphology such as plate, needle, or particle can be precipitated on a Ti substrate, which is sometimes heated to obtain a better coating layer [16, 17]. β-TCP has also been coated cathodically on Ti, for immobilization of collagen [18]. Low-voltage alternating current has been reported to be effective in precipitating calcium phosphate on Ti [19]. This technique can be useful for the treatment of thin wires and fibers without the dissolution of Ti. HA can also be electrodeposited with pulse current [20]. Nano-grained calcium phosphate can be electrochemically deposited on Ti using acidic electrolytes [21]. The coating layer contains dicalcium phosphate dihydrate (55–85 nm in grain size), with a small amount of HA (20–25 nm); the content of HA increases with the increase of the current density [22]. An electrochemical method for producing nanocrystalline HA coatings on Ti surface has been reported [23, 24]. Also, HA has been coated by dynamic voltage during electrophoretic deposition [25]. Electrocrystallization of octacalcium phosphate (OCP) and HA on CP-Ti and Ti–6Al–4V alloy has also been applied with much success [26, 27]. The goal of these coatings is to achieve rapid bone conduction. Unfortunately, the durability of the coated HA is often questionable.

4 General Properties of Titanium and Its Alloys

4.1 CP-Ti

Pure Ti is composed of hcp crystals (α phase) at ambient temperatures, but becomes composed of bcc crystals (β phase) at temperatures higher than 882°C. Real pure Ti does not exist, because Ti easily dissolves oxygen, carbon, and nitrogen, and contains them as impurities. Ti containing these impurities is described as CP-Ti. CP-Ti is classified into four grades, according to its impurity content and mechanical properties (Table 3.2). The higher the grade number, the higher the impurity content, tensile strength, and proof strength (offset yield stress), and the lower the elongation (Fig. 3.3). In other words, CP-Ti is a type of alloy, from the viewpoint of the definition of an alloy and the general effects of alloying. CP-Ti is used for maxillofacial prosthetic plates, sternal wire, dental implants, dental restoratives, and dental denture bases. Sometimes, CP-Ti containing more impurities than grade 4 displays almost the same strength as Ti alloys, and since the tensile strength greatly increases with cold working, bone screws that are exposed to high loads are sometimes made of CP-Ti.

The attractive properties of Ti as an implant material are related to the thermodynamically active titanium element, as shown in Fig. 3.4. The biocompatibility of titanium is associated with the formation of an insoluble oxide film, which imparts sufficiently high corrosion resistance in biological systems [28]. As widely known, the corrosion resistance is usually obtained by a surface oxide. A passive film forms rapidly on titanium in water. However, the biocompatibility, e.g., in terms of toxicity and allergy, has become of great interest in recent years due to the rapidly growing interest in implants and restorative materials. In the case of allergy, it is proposed that the allergy by titanium could not be detected so far because of the

Table 3.2 Compositions and mechanical properties of commercially pure titanium (CP-Ti)

	Grade 1	Grade 2	Grade 3	Grade 4
Element	Composition (wt.%)			
Fe	<0.15	<0.2	<0.25	<0.3
O	<0.18	<0.25	<0.35	<0.45
N	<0.03	<0.03	<0.05	<0.05
H	<0.0125	<0.0125	<0.0125	<0.0125
C	<0.1	<0.1	<0.1	<0.1
Ti	Balance	Balance	Balance	Balance
	Mechanical properties			
Tensile strength (MPa)	275–412	343–510	481–618	>550
0.2 % proof strength (MPa)	170	275	380	>440
Elongation (%)	>27	>23	>18	>15
Young's modulus (GPa)	114			

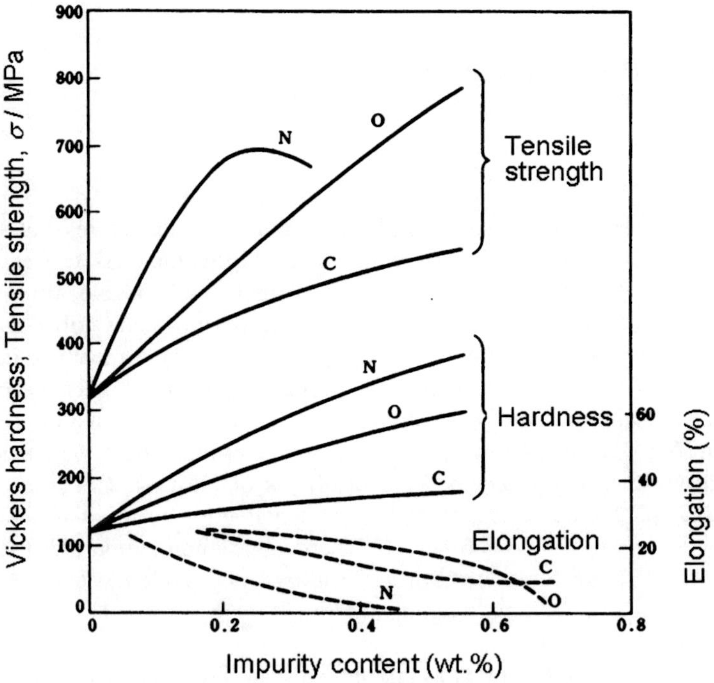

Fig. 3.3 Effect of impurity content on mechanical properties of titanium

short history of titanium as a biomaterial. In contrast, it is proposed that titanium is inert in biological systems and is an allergy-free material. One way or another, it is necessary that the titanium ions released from materials combine with biomolecules in order to demonstrate toxicity and allergy. Discussion on the basis of the physicochemical (thermodynamic and electrochemical) theory is needed to

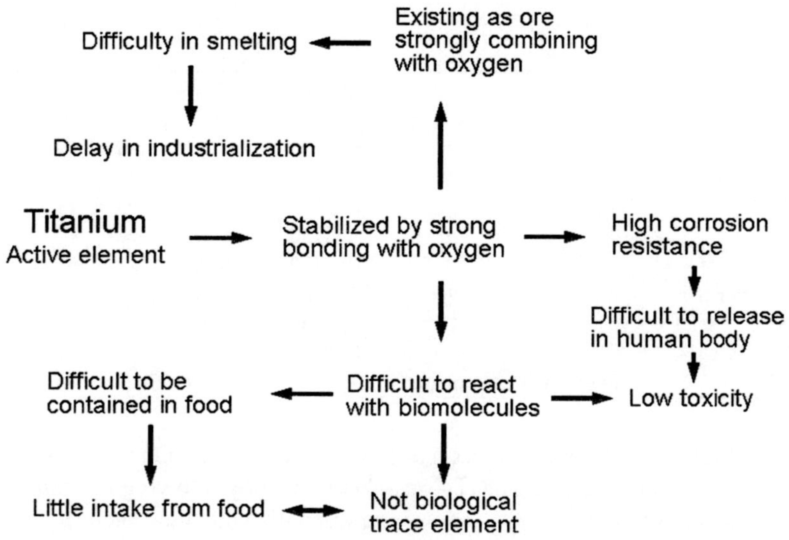

Fig. 3.4 Properties of titanium induced by thermodynamic activity of titanium

determine whether or not the release of titanium ions, or corrosion of titanium in its narrow sense, occurs in biological systems.

In spite of the passivity of titanium, several researches have revealed that a significantly larger amount of titanium is detected in biological systems when titanium materials are implanted without abrasion, compared to controls in water [29–34]. Therefore, the passivity of titanium in biological systems may occur under a specific mechanism, which is different from that in water. What is this specific mechanism? The possible dissolution mechanisms of metal ions will be considered later in this chapter.

In addition, the behavior of titanium ions released into a bio-liquid is also important. As mentioned earlier, titanium ions cause various physiological hindrances if the ions combine with biomolecules. The chemical state of titanium ions in bio-liquid and the possibility of combination with biomolecules will be discussed from the electrochemistry viewpoint.

4.2 Ti Alloys

Ti alloys at ambient temperatures are categorized as α, α+β, or β type alloys, according to the quantities and types of their alloying elements. Many kinds of alloys have been developed (Table 3.3), but Ti–6Al–4V (α+β alloy) is still the most common one in medical use. This alloy shows good workability, heat treatment ability, and weldability, as well as high corrosion resistance, high strength, and good biocompatibility. The extra-low interstitials (ELI) grade, containing small

Table 3.3 Standard titanium alloys

Composition (wt.%)	Type	UNS	ASTM	ISO
Ti–3Al–2.5V	α+β	R56320	ASTM B 348	–
Ti–5Al–2.5Fe	α+β	–	–	ISO 5832-10
Ti–6Al–4V	α+β	R56400	ASTM F 1472	ISO 5832-3
Ti–6Al–4V ELI	α+β	R56401	ASTM F 136	ISO 5832-3
Ti–6Al–7Nb	α+β	R56700	ASTM F 1295	ISO 5832-11
Ti–15Mo	β	R58150	ASTM F 2066	–
Ti–13Nb–13Zr	β	R58130	ASTM F 1713	–
Ti–12Mo–6Zr–2Fe	β	R58120	ASTM F 1813	–
Ti–45Nb	β	R58450	AMS 4982	–
Ti–35Nb–7Zr–5Ta	β	R58350	–	–
Ti–55.8Ni	Intermetallic compound	–	ASTM F 2063	–

amounts of interstitial impurities (oxygen, carbon, nitrogen, and hydrogen) is used for biomedical applications. The Ti–6Al–4V ELI alloy shows great toughness, because the impurities decrease the fatigue strength with the notch effect. ELI alloy is used for bone fixation plates and the stems of artificial hip joints. Ti–6Al–4V alloy has an extremely high 0.2% offset yield strength of 895 MPa, which is much larger than that of stainless steel and Co–Cr–Mo alloys, making plastic deformation difficult even under large load.

The Ti–6Al–7Nb (niobium) alloy [35] has been developed as a replacement for the Ti–6Al–4V alloy. It is mainly used in Europe, because the V contained in the Ti–6Al–4V shows strong toxicity, although there are no reports of accidents when using implants made of Ti–6Al–4V. The Ti–6Al–7Nb alloy has been formed by substituting the V with Nb of the same atomic concentration. Its corrosion resistance and safety are improved compared to Ti–6Al–4V.

5 Electrochemical Properties of Titanium

Ti (clean surface without surface oxide) readily reacts with water and is passivated even at ambient temperature when contacting water [36, 37]. The partial pressure of oxygen in biological systems is about one fourth that in air, thus the dissolved oxygen is deficient in bio-liquids. In the case of metallic titanium, the passivation in water is independent of dissolved oxygen because the reaction with H_2O is preferential [38]. Titanium is passivated in water at ambient conditions due to oxidation with water, as follows [39]:

$$\mathrm{Ti} + 2\mathrm{H_2O} \leftrightarrow \mathrm{TiO_2} + 4\mathrm{H^+} + 4\mathrm{e^-} \tag{3.1}$$

The Nernst equation for this reaction at standard conditions yields

$$E_{\mathrm{rev}} = -1.072 - 0.0591\,\mathrm{pH} \tag{3.2}$$

Fig. 3.5 Potential–pH diagram of titanium in water

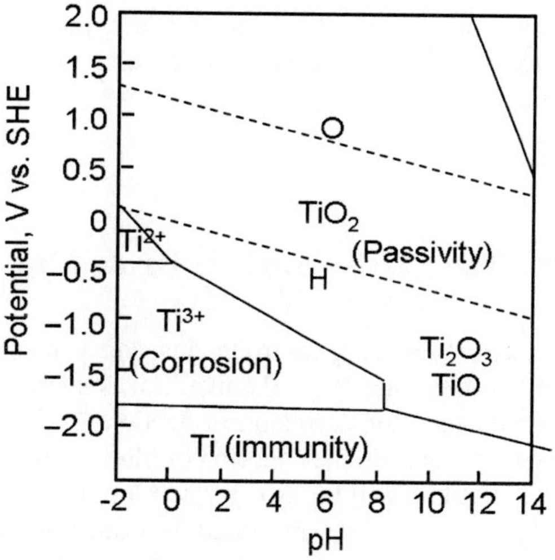

In acidic solutions, a possible cathodic reaction in water is

$$2H^+ + 2e^- \leftrightarrow H_2 \tag{3.3}$$

For which the Nernst equation yields

$$E = -0.0591\ \text{pH} - 0.0295 \log P_{H2} \tag{3.4}$$

Therefore, titanium is passive in aqueous solution, i.e., it is in the "passivity" area in the Pourbaix diagram (pH-potential diagram), Fig. 3.5. The surface oxide on titanium is electrochemically stable. It is necessary that the potential shifts to the corrosion area on the higher potential side or to the Ti^{2+} or Ti^{3+} stable areas on the lower potential side in Fig. 3.5 in order to destroy the surface oxide. The dashed line H in this figure shows the equilibrium potential of the hydrogen evolution reaction. The surface oxide film on Ti could not be removed electrochemically, by cathodic sweep of potential, because hydrogen evolution occurs preferentially. On the other hand, a relatively high potential must be applied, entering the corrosion area, in order for corrosion to occur. However, it is unlikely to achieve such a high potential in biological systems.

The pH should remain below 3 for the destruction of the surface film and the release of titanium ions. Assuming that the concentration of titanium ion is $[Ti^{2+}] = 1 \times 10^{-6}\ \text{mol L}^{-1}$, the stable limit of pH is given as follows [40]:

$$\text{pH} = -8.4E_{\text{rev}} - 2.75 \tag{3.5}$$

Equation (3.5) indicates that titanium is passive when the redox potential is more noble than −0.3 V at pH = 0. Titanium is corroded in alkaline solutions at pH 13–14, which hardly occurs in biological systems. The surface oxide film will be regenerated unless the material is exposed to pH and potential values that are sufficient for the destruction of passivity, no matter how the surface oxide film was temporarily destroyed.

6 The Passive Surface Oxide Film on Titanium

Except in reducing environments, corrosion processes cause a reaction film to form on metallic materials. Passive film is such a reaction film, which is of particular significance for corrosion protection. When the solubility is extremely low and pores are absent, the adhesion of film—which is formed in an aqueous solution—to the substrate will be strong. The film then becomes a corrosion resistant or passive film. Passive films are about 1–5 nm thick and are transparent. Due to the tremendously fast rate at which they are formed, passive films readily become amorphous. For example, the film on a titanium metal substrate can be formed in 30 ms. This can be estimated from the current transient of titanium at 1 V vs. SCE after exposing the metal surface, as shown in Fig. 3.6. Since amorphous films hardly contain grain boundaries and structural defects, they are usually corrosion resistant.

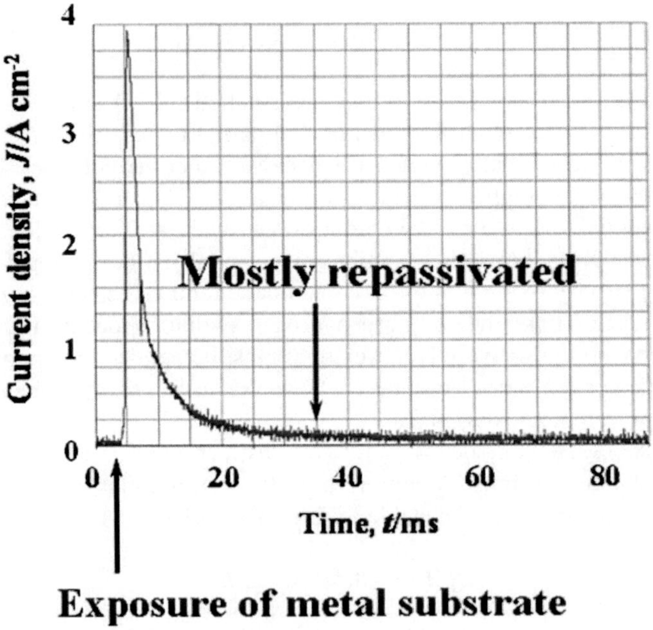

Fig. 3.6 Time transient of anodic current of titanium in Hanks' solution under the 1 V charge versus a saturated calomel electrode (SCE). Anodic current is generated with the dissolution and repassivation of titanium

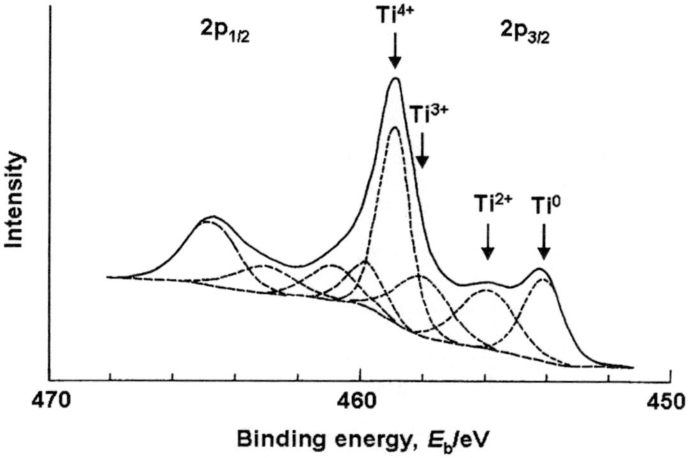

Fig. 3.7 Decomposition of Ti 2p XPS spectrum obtained from titanium abraded and immersed for 300 s in water into eight peaks ($2p_{3/2}$ and $2p_{1/2}$ electron peaks in four valences). Numbers with *arrows* are valence numbers

However, corrosion resistance decreases with crystallization. Fortunately, passive films contain water molecules that promote and maintain amorphousness.

When titanium is polished in de-ionized water and analyzed using X-ray photoelectron spectroscopy (XPS), the Ti 2p spectrum obtained from the titanium gives four doublets, which correspond to the valences Ti^0, Ti^{2+}, Ti^{3+}, and Ti^{4+}, respectively, as shown in Fig. 3.7. A distinct Ti^0 peak for the metallic state is observed, which accounts for a very thin surface oxide film, i.e., thinner than a few nanometers. Ti^{4+} (TiO_2), Ti^{3+} (Ti_2O_3), and Ti^{2+} (TiO) are also detected. Though Ti^{2+} oxide exists in the surface oxide film, Ti^{2+} formation is always thermodynamically less favorable than Ti^{3+} formation at the surface. In the same specimen, O 1s spectrum is observed, as shown in Fig. 3.8. The surface oxide contains hydroxide or hydroxyl group (OH^-) and water. The surface film on titanium consists mainly of amorphous, or low-crystalline and non-stoichiometric TiO_2, and the film stands firm against chloride ions.

Since a considerable portion of oxidized titanium remains as Ti^{2+} and Ti^{3+} in the surface film, the oxidation process might proceed to the end just at the uppermost part of the surface film. As shown in Fig. 3.9, the proportion of Ti^{4+} among titanium cations in the film decreases with the increase of photoelectron take-off angle [38], indicating that more Ti^{4+} exists near the outer layer in the film. Based on the take-off angle dependence of $[OH^-]/[O^{2-}]$ in Fig. 3.9, oxygen atoms in the hydroxyl group are mainly located in the outer part of the surface film. This means that dehydration proceeds inside the surface film.

The thickness of the film is about 2 nm just after polishing and about 5 nm at 1 week after polishing. The thickness of the surface oxide film increases according to a logarithmic rule, which is common to the initial growth of oxide films on metallic materials.

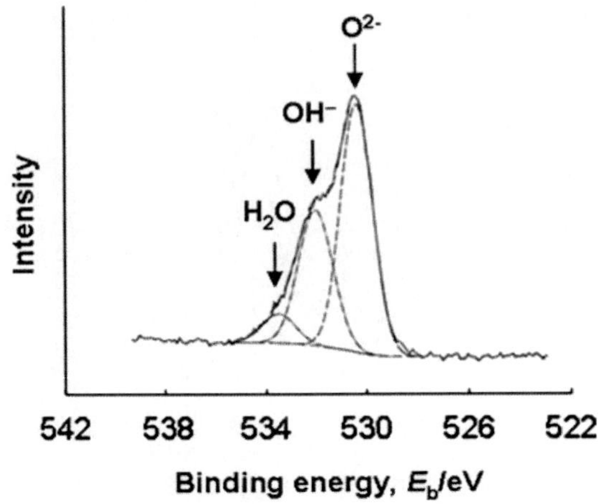

Fig. 3.8 Decomposition of O 1s XPS spectrum obtained from titanium abraded and immersed for 300 s in water into three peaks

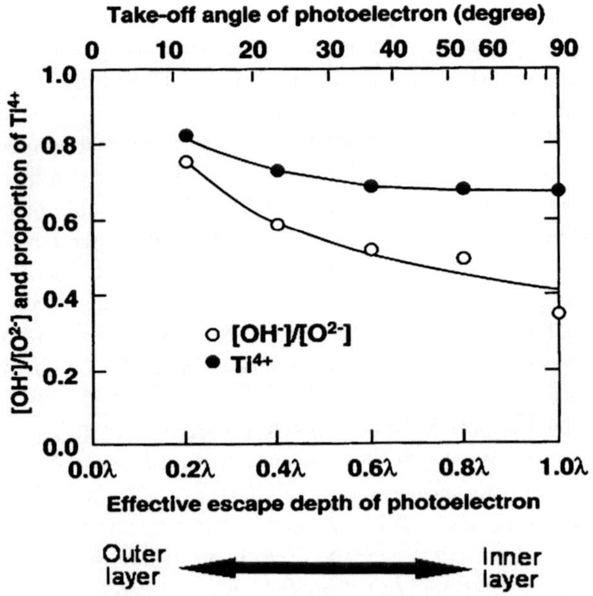

Fig. 3.9 The ratio of the proportion of the concentration of OH^- to that of O^{2-}, $[OH^-]/[O^{2-}]$ and proportion of cationic fraction of Ti^{4+} among titanium species, in surface oxide film on titanium polished in water plotted against the average effective escape depth of photoelectrons for angle-resolved XPS measurements. Lambda (λ) is the average escape depth of O 1s and Ti $2p_{3/2}$ photoelectrons, and the effective escape depth is the escape depth times sin (take-off-angle). The values at small take-off angle of photoelectron or effective escape depth of photoelectron represent outer region information and those larger ones represent inner region information

The film on Ti–6Al–4V alloy is almost the same as that on pure titanium, but contains a small amount of aluminum oxide [41, 42]. In other words, the surface oxide film on Ti–6Al–4V is TiO_2 containing small amounts of Al_2O_3, hydroxyl groups, and bound water. Vanadium contained in the alloy is not detected in the oxide film after the alloy is polished.

The Ti–56Ni shape-memory alloy is covered by TiO_2-based oxide, with minimal amounts of nickel in both the oxide and metallic states [41, 42]. The film on Ti–56Ni is TiO_2 containing small amounts of metallic nickel, NiO, hydroxyl groups, and bound water.

In the case of Ti–Zr alloy, the surface oxide film consists of titanium and zirconium oxides [43]. The relative concentration ratio of Ti-to-Zr in the film is almost the same as that in the alloy. In the surface oxide film, Ti and Zr are uniformly distributed along the depth. The thickness of the film increases with the increase of Zr content. The chemical state of Zr is more stable than that of Ti in the film.

7 Reconstruction of Passive Films

Reactions between the surfaces of metallic materials and living tissues are the initial events, which occur when the materials are implanted into the human body. Tissue compatibility is governed by the reactions in the initial stage. In this regard, the surface properties of materials are important. The composition of the surface oxide film changes even though the film is macroscopically stable. Passive surfaces exist simultaneously in contact with electrolytes, undergoing a continuous process of partial dissolution and re-precipitation from the microscopic viewpoint [36]. Therefore, the surface composition should change according to the environment. Due to abrasion against bone and other materials, the surface oxide film might be scratched and destroyed during insertion and implantation into living tissues. Fretting corrosion also leads to film destruction. Fortunately, the surface oxide is immediately regenerated in a biological environment where biofluid surrounds the metallic material. However, the composition and properties of the oxide film regenerated in a biological environment may be different from those in water.

Calcium, phosphorus, and sulfur are incorporated into the surface film of titanium after it is surgically implanted into the human jaw [44]. Calcium phosphates are formed on titanium and its alloys by immersion in Hanks' solution and other solutions [41–43, 45, 46], resulting in the preferential adsorption of phosphate ions [47, 48]. Hydrated phosphate ions are adsorbed by a hydrated titanium oxide surface during the release of protons. Calcium ions are adsorbed by phosphate ions being adsorbed on a titanium surface, with calcium phosphate being formed eventually. In this regard, when titanium is immersed in Hanks' solution containing albumin, a nonuniform and porous apatite-containing albumin is formed [49]. The above phenomena are characteristic of titanium and its alloys [42].

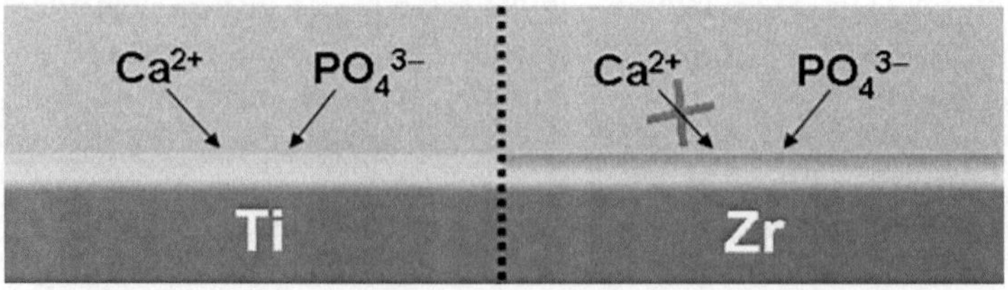

Fig. 3.10 Both calcium and phosphate ions attract to Ti surface, while only phosphate ion can be incorporated by Zr surface

The surface oxide film on Ti is not completely oxidized and is relatively reactive, while that on Zr is stably oxidized; that on Zr is more passive and protective than that on Ti. Neither calcium nor phosphate stably exists alone on Ti; stable, protective calcium phosphate is formed on Ti in biological environments. On the other hand, calcium is never incorporated on Zr, while zirconium phosphate formed on Zr is highly stable and establishes a protective layer; therefore, no calcium reacts with the layer, as shown in Fig. 3.10. Surface oxide films as passive films on Ti and Zr are nearly amorphous and are different from crystallized titanium oxide and zirconium oxide bulk ceramics with regard to their chemical properties [2]. Recently, zirconia has been used for dental implants, and its hard tissue compatibility has been claimed to be the same as that of Ti. This, however, is not the case. Zirconia is a bioinert ceramic, thus no chemical reaction in living tissues is expected.

8 Metallic Ion Release from Titanium

Trace metal elements in human body are sometimes originated from biomaterials implanted in the body, although they are usually originated from food and water. Metal ions are released from metallic biomaterials, such as dental implants in the body. The released metal ions might cause various phenomena: transportation, metabolism, accumulation in organs, allergy, and carcinoma. If a large amount of metal ions are released, it could be generally harmful for human health. Release of metal ions from metallic materials *in vivo* should be understood when discussing the safety and biocompatibility of the materials. Since the release of metal ions depends on electrochemical rules, many efforts have been made to electrochemically analyze biomaterials in simulated body fluids.

According to the theory of passivity, metallic biomaterials in aqueous solutions are systems in which active and passive surfaces exist simultaneously in contact with an electrolyte [36]. Therefore, it is now thought that the surface oxide film on the materials undergoes a process of partial dissolution and re-precipitation in aqueous solutions. If the dissolution rate is larger than the re-precipitation rate, metal ions are gradually released. This process is "anodic dissolution" in a narrow sense. If the potential of a material changes in the anodic direction, the anodic dissolution rate increases. Metal ions release from the remaining surface oxide film is relatively slow *in vivo* because the change in the potential of a material *in vivo* is usually small.

The dissolution rate of the film is accelerated by amino acids and proteins [5, 6]. When pure metal powders are immersed in saline, with or without serum albumin or fibrinogen, molybdenum, copper, cobalt, and nickel ions are released, but titanium is not released and is not influenced by proteins [50]. In the case of Ti–6Al–4V, titanium and aluminum ions are released in Hanks' solution with 2% EDTA, while titanium, aluminum, and vanadium ions are released in Hanks' solution with 0.05 M sodium citrate [51]. Ti–Ni alloy initially releases more nickel than stainless steel immersed in medium with osteoblast or fibrinogen, but after 2 days the amount released decreases [52]. Fretting corrosion depends on (the charge of) proteins; preferential dissolution of nickel increases in the presence of proteins [5]. Consequently, metal ions are released in rabbit in the absence of wear and are detected in tissues around materials, as well as in serum and urine [34]. The cause of metal ion release is possibly biomolecules containing proteins. Although the mechanism of acceleration of metal ions release in the presence of amino acids and proteins is not elucidated, imbalance between partial dissolution and re-precipitation in oxide films may accelerate the ions release.

When a material is implanted, it is recognized as a foreign body by immunological processes and macrophages (Mϕs) adhered to the surface of the material [53]. Mϕ generates active oxygen species without response to particles that can be phagocytosed. An intracellular dismutation of O_2^- catalyzed by superoxide dismutase (SOD) produces H_2O_2, which has a much longer lifetime and higher permeability against cell membrane than O_2^-: H_2O_2 reaches the surface where Mϕ has adhered. The titanium surface is hyperoxidized by H_2O_2 [54, 55] that may induce the release of titanium ions. H_2O_2 reacts with the surface oxide of titanium according to the following equation [55]:

$$Ti^{4+} + H_2O_2 \rightarrow Ti^{5+} + OH^- + OH^* \tag{3.6}$$

where * represents radical. Dissolution of Ti with active oxygen generated by Mϕ is elucidated [56]. In other words, when Ti implants are inserted and macrophage accumulates generate active oxygen, Ti implant always dissolve by inflammation after operation. Active oxygen species produced by Mϕ that phagocytose HDPE particles induce ion release from titanium *in vitro*, as shown in Fig. 3.11.

On the other hand, major causes of titanium release were surgical handling in implantation and wear and/or fretting during experiment for 48 weeks [57]. Titanium was also released in the absence of wear. No morphological abnormality was observed in the tissue around the implant by biopsy at postoperation week 48.

Fig. 3.11 Titanium ion concentration in culture medium in which titanium disk and macrophage are incubated. (**a**) No addition, (**b**) addition of high density polyethylene powder (HDP), (**c**) addition of HDP + antioxidant enzyme

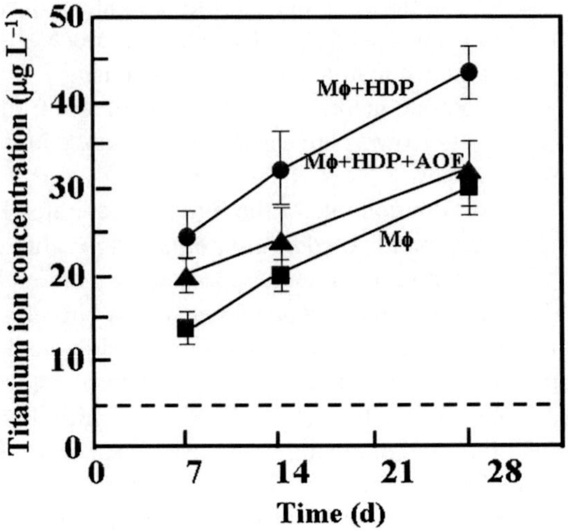

9 Behavior of Released Metal Ions

Metal ions released into the human body do not always damage it. The partner for combination with metal ion is very important (Fig. 3.12). Every molecule has a chance to combine with the ion. An ion is active and immediately reacts with water molecules or inorganic anions depending on the number and size (weight) of the molecules. For example, titanium ion is very active and readily reacts with hydroxyl radicals and anions, forming oxide and salt in body fluid, indicating that the possibility of combination with biomolecules is low. However, the reaction of titanium ion with biomolecules may occur, even if it has a slight chance. Zirconium, niobium, and tantalum ions belong to the titanium-type ions. These elements are expected to be promising components to improve the safety of titanium alloys. On the other hand, inactive ions, e.g., of nickel and copper, do not immediately combine with water molecules and inorganic anions, and survive in the ionic state for relatively long time. Therefore, these ions have higher chance to combine with biomolecules and reveal toxicity. In the case of titanium alloy, other component elements in the alloy, such as nickel, exist as ions for long time. Therefore, when toxicity of metallic biomaterials is discussed, the following factors must be considered: (1) corrosion resistance of the material, (2) ions released by corrosion and wear, (3) activity of the released ion, and (4) toxicity of the ion itself.

No biomolecule contains Ti, Zr, Nb, and tantalum (Ta) atoms, to the best of the author's knowledge. On the process of evolution of life, these atoms have not been incorporated into biomolecules. In other words, these elements could not survive in primitive ocean as ions. This indicates the difficulty of the combination of these ions with biomolecules.

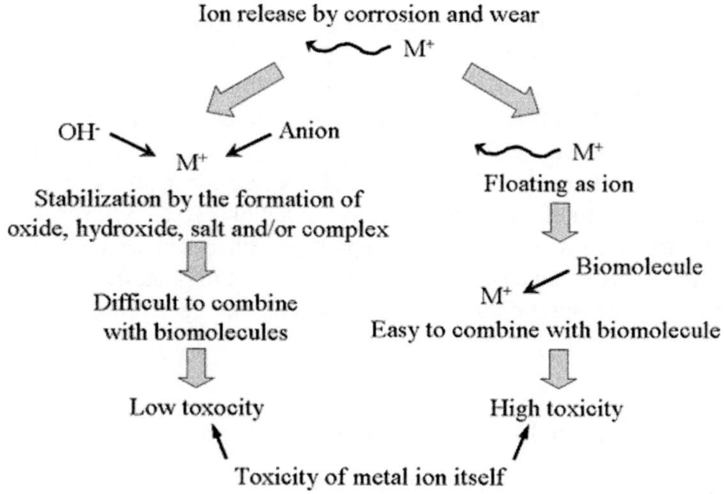

Fig. 3.12 Behavior of metallic ions released in body fluids and their toxicity

10 Contamination

The shelf life of implantable materials has rarely been addressed. The stability of osteoconductivity of titanium has been examined [58]. Rat bone marrow-derived osteoblasts were cultured on new titanium disks (immediately after acid etching), 3-day-old disks (stored after acid etching for 3 days in dark ambient conditions), 2-week-old disks, and 4-week-old disks. Protein adsorption capacity and osteoblast migration, attachment, spreading, proliferation, and mineralization decreased substantially on old titanium surfaces, in an age-dependent manner. When the 4-week-old implants were placed into rat femurs, the biomechanical strength of the bone–titanium integration was less than half that for newly processed implants at the early healing stage. More than 90% of the new implant surface was covered by newly generated bone, compared to 58% for 4-week–old implants. This time-dependent biological degradation was also found for machined and sandblasted titanium surfaces and was associated with progressive accumulation of hydrocarbon on titanium surfaces. The new surface could attract osteoblasts even under a protein-free condition, but its high bioactivity was abrogated by masking the surface with anions. These results uncover an aging-like, time-dependent biological degradation of titanium surfaces, from bioactive to bioinert.

11 Fracture

Osseointegrated threaded Ti screw-type implants rarely lose integration after the first year of clinical function. Implant failure can occur for other reasons, with implant fracture being one of the major reasons for late failure. This has been determined by a prospective evaluation of 4,937 implants [59]. All observed fractures occurred with CP-Ti, 3.75 mm in diameter, threaded implants, and prosthetic or abutment screw loosening preceded implant fracture for the majority of the implants. On the other hand, metal fatigue and biomechanical overload [60] and hydrogen embrittlement following hydrogen absorption [61] cause fracture of implants. The identification of only one main failure mechanism is not possible since the failure of dental implants can be attributed to several factors related to the material, the fixation process, and the acceptor quality, as well as to the operational conditions [62]. As explained earlier, the mechanism of implant fracture is often not certain, and only few systematic investigations have been performed in this regard.

12 Biofilm Formation and Infection

Bacterial factor is one of the causes of implant failures. Three major etiologies might have been implicated in the failure processes: impaired healing ability of the host bone site, disruption of a weak bone-to-implant interface after abutment connection, and infection in situations with complicated surgery [63]. An overview of the literature associated with common peri-implant microbiology and an assessment as to whether bacteria associated with periodontitis exert a possible risk for peri-implant tissue breakdown has been provided [64]. A critical examination of the literature on the colonization of the peri-implant environment and its impact on clinically significant infections in the context of a comparison with the microbial etiology of periodontitis has also been provided [65]. Some implants fail as a result of infection, but the frequency of failure by infection or any other cause is unknown. The main histologic features of peri-implantists consisted of the presence of a bone sequestrum near the implant, many bacteria present on the implant surface, and the inflammatory infiltrate (macrophages, lymphocytes, and plasma cell) nearby [66]. There are neither standardized antibiotic prophylactic regimens for dental implant placement nor universally accepted treatment for peri-implantists. The treatment of infected implants is difficult and usually requires removal [67].

References

1. Hanawa T (2010) Biofunctionalization of titanium for dental implant. Jpn Dent Sci Rev 46:93–101
2. Tsutsumi Y, Nishimura D, Doi H, Nomura N, Hanawa T (2009) Difference in surface reactions between titanium and zirconium in Hanks' solution to elucidate mechanism of calcium phosphate formation on titanium using XPS and cathodic polarization. Mater Sci Eng C 29:1702–1708
3. Andreiotelli M, Kohal RJ (2009) Fracture strength of zirconia implants after artificial aging. Clin Implant Dent Relat Res 11:158–166
4. Natali AN (ed) (2003) Dental biomechanics. Taylor & Francis, London, pp 69–87

5. Merritt K, Brown SA (1988) Effect of proteins and pH on fretting corrosion and metal ion release. J Biomed Mater Res 22:111–120
6. Williams RL, Brown SA, Merritt K (1988) Electrochemical studies on the influence of proteins on the corrosion of implant alloys. Biomaterials 9:181–186
7. Black J (1984) Biological performance of materials. Plenum, New York
8. Hench LL, Ethridge EC (1975) Biomaterials – the interfacial problem. Adv Biomed Eng 5:35–150
9. Ong L, Lucas LC (1994) Post-deposition heat treatment for ion beam sputter deposited calcium phosphate coatings. Biomaterials 15:337–341
10. Yoshinari M, Watanabe Y, Ohtsuka Y, Dérand T (1997) Solubility control of thin calcium-phosphate coatings with rapid heating. J Dent Res 76:1485–1494
11. Yoshinari M, Ohtsuki Y, Derand T (1994) Thin hydroxyapatite coating produced by the ion beam dynamic mixing method. Biomaterials 15:529–535
12. Yoshinari M, Hayakawa T, Wolke JCG, Nemoto K, Jansen JA (1998) Influence of rapid heating with infrared radiation on RF magnetron-sputtered calcium phosphate coatings. J Biomed Mater Res A 37:60–67
13. Ozeki K, Yuhta T, Fukui Y, Aoki H, Nishimura I (2002) A functionally graded Ti/hydroxyapatite film obtained by sputtering. J Mater Sci Mater Med 13:253–258
14. Ban S, Maruno S, Arimoto N, Harada A, Hasegawa J (1997) Effect of electrochemically deposited apatite coating on bonding of bone to the HA-G-Ti composite and Ti. J Biomed Mater Res 36:9–15
15. Ban S, Maruno S (1998) Morphology and microstructure of electrochemically deposited calcium phosphates in a modified simulated body fluid. Biomaterials 19:1245–1253
16. Yuda A, Ban S, Izumi Y (2005) Biocompatibility of apatite-coated titanium mesh prepared by hydrothermal-electrochemical method. Dent Mater J 24(4):588–595
17. Kuroda K, Moriyama M, Ichino R, Okido M, Seki A (2008) Formation and in vivo evaluation of carbonate apatite and carbonate apatite/$CaCO_3$ composite films using the thermal substrate method in aqueous solution. Mater Trans 49:1434–1440
18. Hosaka M, Shibata Y, Miyazaki T (2006) Preliminary beta-tricalcium phosphate coating prepared by discharging in a modified body fluid enhances collagen immobilization onto titanium. J Biomed Mater Res B 78B:237–242
19. Tanaka Y, Kobayashi E, Hiromoto S, Asami K, Imai H, Hanawa T (2007) Calcium phosphate formation on Ti by low-voltage electrolytic treatments. J Mater Sci Mater Med 18:797–806
20. Hayakawa T, Kawasaki M, Takaoka GH (2008) Coating of hydroxyapatite films on titanium substrates by electrodeposition under pulse current. J Ceram Soc Jpn 116:68–73
21. Narayanan R, Seshadri SK, Kwon TY, Kim KH (2007) Electrochemical nano-grained calcium phosphate coatings on Ti-6Al-4 V for biomedical applications. Scr Mater 56:229–232
22. Narayanan R, Kwon TY, Kim KH (2008) Preparation and characteristics of nano-grained calcium phosphate coatings on titanium from ultrasonated bath at acidic pH. J Biomed Mater Res B 85:231–239
23. Narayanan R, Kwon TY, Kim KH (2008) Direct nanocrystalline hydroxyapatite formation on titanium from ultrasonated electrochemical bath at physiological pH. Mater Sci Eng C 28:1265–1270
24. Narayanan R, Kim SY, Kwon TY, Kim KH (2008) Nanocrystalline hydroxyapatite coatings from ultrasonated electrolyte: preparation, characterization, and osteoblast response. J Biomed Mater Res A 87:1053–1060
25. Meng XW, Kwon TY, Kim KH (2008) Hydroxyapatite coating by electrophoretic deposition at dynamic voltage. Dent Mater J 27:666–671
26. Eliaz N, Kopelovitch W, Burstein L, Kobayashi E, Hanawa T (2009) Electrochemical processes of nucleation and growth of calcium phosphate on titanium supported by real-time quartz crystal microbalance measurements and X-ray photoelectron spectroscopy analysis. J Biomed Mater Res A 89:270–280

27. Lakstein D, Kopelovitch W, Barkay Z, Bahaa M, Hendel D, Eliaz N (2009) Enhanced osseointegration of grit-blasted, NaOH-treated and electrochemically hydroxyapatite-coated Ti-6Al-4V implants in rabbits. Acta Biomater 5:2258–2269
28. Williams DF (1981) Titanium and titanium alloys. In: Williams DF (ed) Biocompatibility of clinical implant materials, vol 1, CRC. Boca Raton, FL, pp 9–44
29. Meachin G, Williams DF (1973) Change in non-osseous tissue adjacent to titanium implants. J Biomed Mater Res 7:555–572
30. Woodman JL, Jacobs JJ, Galante JO, Urban RM (1984) Metal ion release from titanium-based prosthetic segmental replacements of long bones in baboons: a long-term study. J Orthop Res 1:421–430
31. Bessho K, Fujimura K, Iizuka T (1995) Experimental long-term study of titanium ions eluted from pure titanium miniplates. J Biomed Mater Res 29:901–904
32. Ektessabi AM, Otsuka T, Tsuboi Y, Yokoyama K, Albrektsson T, Sennerby L, Johansson C (1994) Application of micro beam PIXE to detection of titanium ion release from dental and orthopaedic implants. Int J PIXE 4:81–91
33. Ektessabi AM, Otsuka T, Tsuboi Y, Horino Y, Mokuno Y, Fujii K, Albrektsson T, Sennerby L, Johansson C (1996) Preliminary experimental results on mapping of the elemental distribution of organic tissues surrounding titanium-alloy implants. Nucl Instrum Methods Phys Res B 109 (110):278–283
34. Bianco PD, Ducheyne P, Cuckler JM (1996) Local accumulation of titanium released from a titanium implant in the absence of wear. J Biomed Mater Res 31:227–234
35. Semlitsch M, Staub F, Weber H (1985) Titanium-aluminium-niobium alloy, development for biocompatible, high strength surgical implants. Biomed Eng 30:334–339
36. Kelly EJ (1982) Electrochemical behaviour of titanium. Mod Aspect Electrochem 14:319–424
37. Uhlig HH, Revie RW (1985) Corrosion and corrosion control, 3rd edn. Wiley, New York, NY
38. Hanawa T, Asami K, Asaoka K (1998) Repassivation of titanium and surface oxide film regenerated in simulated bioliquid. J Biomed Mater Res 40:530–538
39. Pourbaix M (1974) Atlas of electrochemical equilibria in aqueous solution. National Association of Chemical Engineers, Houston, TX
40. Asami K, Chen SC, Habazaki H, Hashimoto K (1993) The surface characterization of titanium and titanium-nickel alloys in sulfuric acid. Corros Sci 35:43–49
41. Hanawa T, Ota M (1991) Calcium phosphate naturally formed on titanium in electrolyte solution. Biomaterials 12:767–774
42. Hanawa T (1991) Titanium and its oxide film: a substrate for formation of apatite. In: Davies JE (ed) The bone-biomaterial interface. University of Toronto Press, Toronto, pp 49–61
43. Hanawa T, Okuno O, Hamanaka H (1992) Compositional change in surface of Ti-Zr alloys in artificial bioliquid. J Jpn Inst Met 56:1168–1173
44. Sundgren JE, Bodö P, Lundström I (1986) Auger electron spectroscopic studies of the interface between human tissue and implants of titanium and stainless steel. J Colloid Interface Sci 110:9–20
45. Wever DJ, Veldhuizen AG, de Vries J, Busscher HJ, Uges DRA, van Horn JR (1998) Electrochemical and surface characterisation of a nickel-titanium alloy. Biomaterials 19:761–769
46. Li P, Ducheyne P (1998) Quasi-biological apatite film induced by titanium in a simulated body fluid. J Biomed Mater Res 41:341–348
47. Healy KE, Ducheyne P (1992) Hydration and preferential molecular adsorption on titanium in vitro. Biomaterials 13:553–561
48. Healy KE, Ducheyne P (1992) The mechanisms of passive dissolution of titanium in a model physiological environment. J Biomed Mater Res 26:319–338
49. Serro AP, Fernandes AC, Saramago B, Lima J, Barbosa MA (1997) Apatite desorption on titanium surfaces – The role of albumin adsorption. Biomaterials 18:963–968
50. Clark GCF, Williams DF (1982) The effects of proteins on metallic corrosion. J Biomed Mater Res 16:125–134

51. Bruneel N, Helsen JA (1988) In vitro simulation of biocompatibility of Ti-Al-V. J Biomed Mater Res 22:203–214
52. Ryhanen J, Niemi E, Serlo W, Niemela E, Sandvik P, Pernu H, Salo T (1997) Biocompatibility of nickel-titanium shape memory metal and its corrosion behaviour in human cell cultures. J Biomed Mater Res 35:451–457
53. Tang L, Eaton JW (1993) Fibrinogen mediates acute inflammatory responses to biomaterials. J Exp Med 178:2147–2156
54. Tengvall P, Lundström I, Sjöqvist L, Elwing H, Bjursten LM (1989) Titanium-hydrogen peroxide interaction: model studies of the influence of the inflammatory response on titanium implants. Biomaterials 10:166–175
55. Pan J, Liao H, Leygraf C, Thierry D, Li J (1998) Variation of oxide films on titanium induced by osteoblast-like cell culture and the influence of an H_2O_2 pretreatment. J Biomed Mater Res 40:244–256
56. Mu Y, Kobayashi T, Sumita M, Yamamoto A, Hanawa T (2000) Metal ion release from titanium with active oxygen species generated by rat macrophages in vitro. J Biomed Mater Res 49:238–243
57. Mu Y, Kobayashi T, Tsuji K, Sumita M, Hanawa T (2002) Causes of titanium release from plate and screws implanted in rabbits. J Mater Sci Mater Med 13:583–588
58. Att W, Hori N, Takeuchi M, Ouyang J, Yang Y, Anpo M, Ogawa T (2009) Time-dependent degradation of titanium osteoconductivity: an implication of biological aging of implant materials. Biomaterials 30:5352–5363
59. Eckert SE, Meraw SJ, Cal E, Ow RK (2000) Analysis of incidence and associated factors with fractured implants: a retrospective study. Int J Oral Maxillofac Implants 15:662–667
60. Piattelli A, Scarano A, Piattelli M, Vaia E, Matarasso S (1998) Hollow implants retrieved for fracture: a light and scanning electron microscope analysis of 4 cases. J Periodontol 69:185–189
61. Yokoyama K, Ichikawa T, Murakami H, Miyamoto Y, Asaoka K (2002) Fracture mechanisms of retrieved titanium screw thread in dental implant. Biomaterials 23:2459–2465
62. Manda MG, Psyllaki PP, Tsipas DN, Koidis PT (2009) Observations on an in-vivo failure of a titanium dental implant/abutment screw system: a case report. J Biomed Mater Res B Appl Biomater 89:264–273
63. Esposito M, Thomsen P, Ericson LE, Lekholm P (1999) Histopathologic observations on early oral implant failures. Int J Oral Maxillof Implants 14:798–810
64. Heydenrijk K, Meijer HJA, van der Reijden WA, Raghoebar GM, Vissink A, Stegenga B (2002) Microbiota around root-form endosseous implants: a review of the literature. Int J Oral Maxillofac Implants 17:829–838
65. Ellen RP (1998) Microbial colonization of the peri-implant environment and its relevance to long-term success of osseointegrated implants. Int J Prosthodont 11:433–441
66. Piattelli A, Scarano A, Piattelli M (1998) Histologic observations on 230 retrieved dental implants: 8 years' experience (1989–1996). J Periodontol 69:178–184
67. Pye AD, Lockhart DE, Dawson MP, Murray CA, Smith AJ (2009) A review of dental implants and infection. J Hosp Infect 72:104–110

Biography

Dr. Takao Hanawa is a full professor at the Department of Metallic Biomaterials, Institute of Biomaterials and Bioengineering, Tokyo Medical and Dental University (since July 2004) as well as at the Graduate School of Engineering, The University of Tokyo (since April 2009). He received a B.Sc. in Metallurgical Engineering from Hokkaido University in 1981, a D.D.Sc. from Hokkaido University in 1989, and a Ph.D. from Tohoku University in 1998. His professional experience includes a Deputy-Director-General position at the Biomaterials Research Center of the National Institute for Materials Science. His research interests include metallic biomaterials, dental materials, bio-functionalization of metals, and metal–tissue interactions. Hanawa has contributed nearly 250 journal papers and 37 book chapters. He has won many awards, including The Young Investigator award of the Japanese Society for Biomaterials (1995), The Minister Award of the Japanese Science and Technology Agency (1999), Award for Distinguished Service by Japan Institute of Metals (2001), Award for Developmental Engineering by Japan Institute of Metals (2004), and an award by the Japanese Society for Biomaterials (2009).

Chapter 4
In Vivo Aging and Corrosion Aspects of Dental Implants

Spiros Zinelis, Theodore Eliades, and William A. Brantley

Abstract The modern endosseous dental implants are made by Ti thanks to its biocompatibility, corrosion resistance, and adequate mechanical properties of this material. Additionally Ti is characterized by osseoinductive properties providing increased fixation with the adjacent human bone. Despite the beneficial properties of Ti over the previously used stainless steel dental implants, the Ti implants are not free of problems and failure of dental implants is a great concern in everyday clinical practice. This chapter offers fundamental information for the current status of Ti implants in dental surgery providing also a picture for the failure mechanisms based on current clinical data. The text also deals with the corrosion aspect of Ti under clinical conditions and galvanic phenomena that may be triggered by the presence of implant-retained superstructures made of precious and base alloys or other metallic materials which are used in restorative dentistry.

1 Introduction

Modern endosseous dental implants are increasingly used in restorative dentistry for a variety of clinical treatments. Implants function as transmucosal structures to support crowns [1], fixed partial dentures [2], complete removable dentures [3], and complete arch superstructures [4]. Implant dentistry significantly broaden the

S. Zinelis
Department of Biomaterials, School of Dentistry, University of Athens, Athens, Greece

T. Eliades (✉)
Department of Orthodontics and Paediatric Dentistry, Center of Dental Medicine, University of Zurich, Platternstrasse 11, CH-8032 Zurich
e-mail: teliades@ath.forthnet.gr

W.A. Brantley
Section of Restorative and Prosthetic Dentistry, College of Dentistry, The Ohio State University, Columbus, OH, USA

N. Eliaz (ed.), *Degradation of Implant Materials*, DOI 10.1007/978-1-4614-3942-4_4,

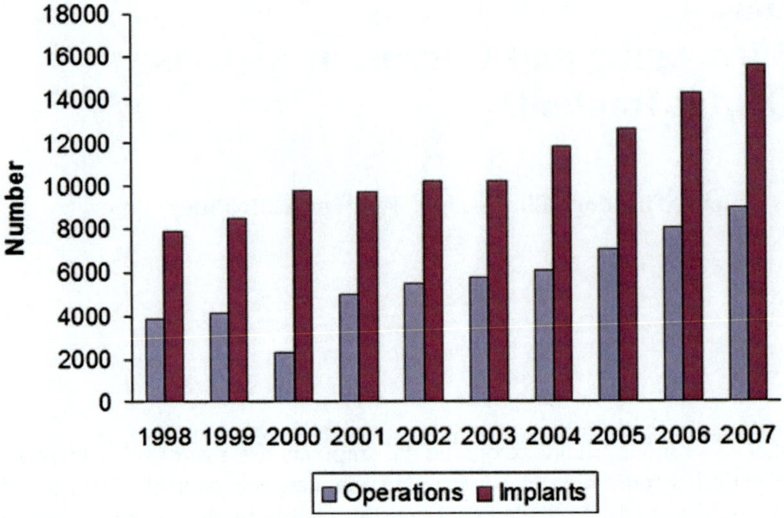

Fig. 4.1 The number of operations and implants placed from 1998 till 2007 in Finland, as provided by the Finland National Agency for Medicine [6]. Image is redrawn from the 2007 report

treatment options of proving alternative clinical treatment plans as well as new solutions to edentulous patients. In the past 2 decades, the number of dental implant procedures has increased steadily worldwide. It is estimated that this number is above one million dental implantations per annum [5]. Figure 4.1 shows this trend according to the database of the Finland National Agency for Medicine [6].

Although the modern implants are solely made of Ti and its alloys, in the past a variety of materials were used for their fabrication, including different alloys, ceramics, and polymers [7]. Although a variety of polymers have been tested in the past, they have been finally abandoned due to adverse immunologic reactions, inferior mechanical properties, and lack of adhesion with living tissues [8, 9]. Ceramic materials are still represented in the dental market with a few commercially available products made of stabilized tetragonal ZrO_2 [10].

Dental implants are currently manufactured by Ti and its alloys due to their superior biocompatibility [11], enhanced corrosion resistance, and lower modulus of elasticity when compared to conventional cobalt-based alloys and stainless steels. Figure 4.2 shows a dental implant made of commercially pure (CP) Ti. In dental implants, the roughened surface resembling the root portion in contact with bone is readily distinguished from the well-polished area of the cervical collar where gingival tissue is attached. Ti is also favorable for these applications due to its osseoinductive properties [12–14]. Bone response is related to direct bone-implant contact without an intervening connective tissue layer. This is what is known as osseointegration [5, 15, 16], which was discovered accidentally by Branemark and his associates in 1969 when they noticed that a metallic piece of Ti embedded in a rabbit bone became firmly connected and hard to remove. After 1 year, the peri-implant bone remained without any indication of inflammation while soft tissue had

Fig. 4.2 *Left*: Dental implant. The treated region shows a more dull appearance due to surface roughening. *Right*: Backscattered electron image taken in topographic mode of the treated region after surface roughening

developed an attachment to the metallic surface, and bone to the Ti [17]. The first dental implant system was introduced in the market in 1971, and since then it is estimated that about 80 different manufactures provide 220 implant brands [18, 19].

The osseointegration property of Ti and its alloys has been appended to the surface layer of Ti oxide. Starting from this point, a priority in the research field is surface modification to positively modulate the host-implant tissue response, accelerating the osseointegration of dental implants and decreasing the healing time. To achieve this goal, numerous surface engineering methods such as grit blasting, acid etching, anodization, and others have been applied to modify dental implant topography and chemistry [5, 20]. It is worth noting that many manufactures perform a combination of grit blasting and acid etching for surface roughening. Figure 4.3 shows secondary electron images from commercially available products prepared by grit blasting, acid etching, and anodization; a great difference is evident in surface morphology.

Despite the intense research in the field, the precise role of surface topography and chemistry on the initial stages in dental implants osseointegration remains uncovered, leaving biomaterials scientists without rationale guidelines for implant surface design [5, 20]. On the other hand, it is widely accepted among scientists and clinicians that the current scientific literature does not provide any clear evidence for the superiority of one surface treatment over the others [15, 21].

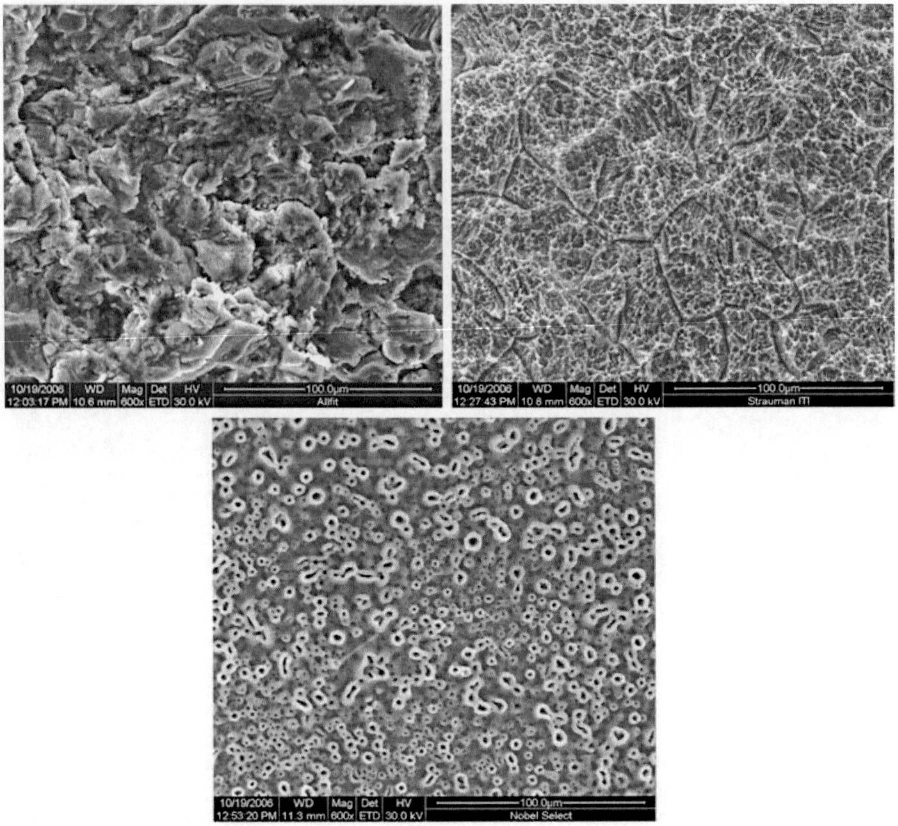

Fig. 4.3 Surface morphology of Ti dental implants prepared by three different surface roughening techniques. *Left*: grit blasting, *center*: acid etching, *right*: anodization. Scale bar = 100 μm

2 *In Vivo* Aging and Failure Rate of Dental Implants

Criteria for successful integration of dental implants were initially described in 1978 by the NIH, and later by Albrektsson et al. [22] and Smith and Zarb [23]. Currently, a lack of mobility is of crucial importance, as loosening is the most often cited factor for implant removal [6]. The rest criteria are the radiographic evidence of minimal bone loss, absence of pain, gingival inflammation, damage to surrounding tissues, and nerve damage, as well as non-satisfactory prosthetic rehabilitation. Following the records of failure causes of dental implants, it is profound that implants fail mainly by two causes. The first one is loss of implant fixation (loosening), and the other is infection. All other causes (e.g., implants position, external trauma, fracture of implant, and other uncategorized reasons) have limiting effect, as shown in Fig. 4.4.

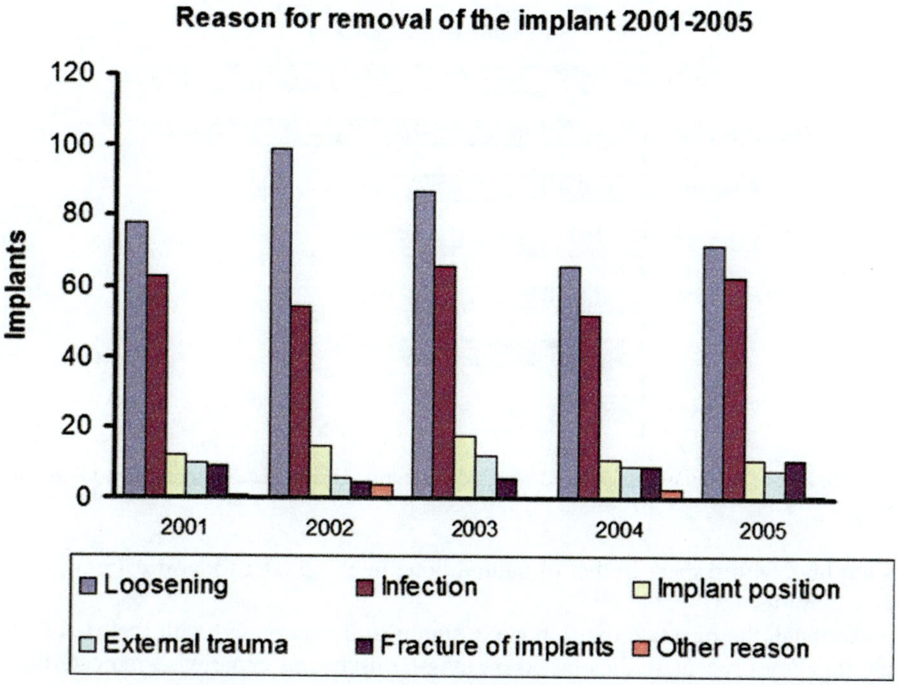

Fig. 4.4 Causes of implant removal from 2003 to 2007, according to the database of the Finland National Agency for Medicine. Loosening and infection are the predominant factors for implant failure under clinical conditions [6]

Bone surrounding an osseointegrated dental implant responds to strain by remodeling. This process of bone resorption and formation at the same time allows replacement of existing bone, which ideally will be firmly anchored through osseointegration on dental implant. In general, a crucial factor for the clinical success of an implant is the manner in which stresses are transferred to the surrounding bone. Although the precise mechanism is not fully understood, it is widely accepted that there is an adaptive remodeling response of the adjacent bone [24]. Overloading can cause bone resorption while underloading of the bone may lead to atrophy and bone loss. This phenomenon, commonly known as "*stress shielding*," has been related to the elastic moduli and is dependent on flexibility differences between the implant material and natural bone [25]. Based on the above approach, any reduction in the modulus of elasticity (stiffness) of the metallic implant is expected to enhance stress distribution to the adjacent bone tissues, thus minimizing the effect of stress shielding, and eventually increasing the longevity of the prosthetic device. The currently used α-phase commercially pure (CP) Ti and $\alpha + \beta$ dual phase alloys such as Ti–6Al–4V, Ti–6Al–7Nb, and Ti–5Al–2.5Fe have almost half the Young's modulus compared to conventional cobalt–chromium (Co–Cr) and stainless steel (SS) alloys. The latter are now obsolete in dental implantology. However, the Young's modulus of these materials

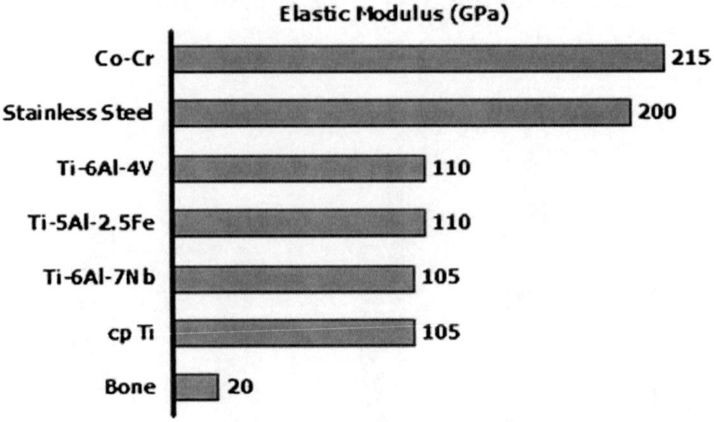

Fig. 4.5 Elastic modulus values of alloys used for implant manufacturing compared to crestal bone

is too high with respect to that of natural bone (e.g., 20 GPa for cortical bone), as shown in Fig. 4.5.

Although the mismatch with bone is significantly decreased with the use of Ti and its alloys, recent studies based on long-term clinical experience indicate that extensive load transfer from the artificial implant to the adjacent remodeling bone results in bone resorption, and eventually—loosening of the prosthetic device [26–30]. Previous clinical studies have shown a mean marginal bone loss around prosthetic implants, up to 2 mm (Fig. 4.6, left) in the first 6 months after prosthetic restoration [31–36].

The aforementioned crestal bone loss has been appended to the overloading of adjacent hard tissues [37–39], resulting in bone resorption and, finally, to implant loosening, which is the more important reason for fixation loss and implant failure [6]. Based on current clinical data acquired from the Finland National Agency for Medicine [6], the failure rate of dental implants due to loosening is estimated at about 50% (Fig. 4.4).

Infection is the second important factor contributing to the failure of dental implants, although to date no single microorganism has been associated with infection of any commercially available implant system [40]. In general, peri-implantitis is considered an inflammatory process which adversely affects the surrounding tissues of an osseointegrated implant, resulting in bone loss [41]. Signs of a failing dental implant are identified during clinical and radiographic inspection, and the diagnosis made resembles periodontitis [42]. To avoid infection during placement of dental implants, a few published guidelines are available, advising that the surgical field should be isolated and free of contamination [43]. Although such a requirement is difficult to be achieved in oral cavity during operation, it has been elucidated that contamination from patient's saliva does not prevent success [44]. Furthermore, the implant placement under controlled

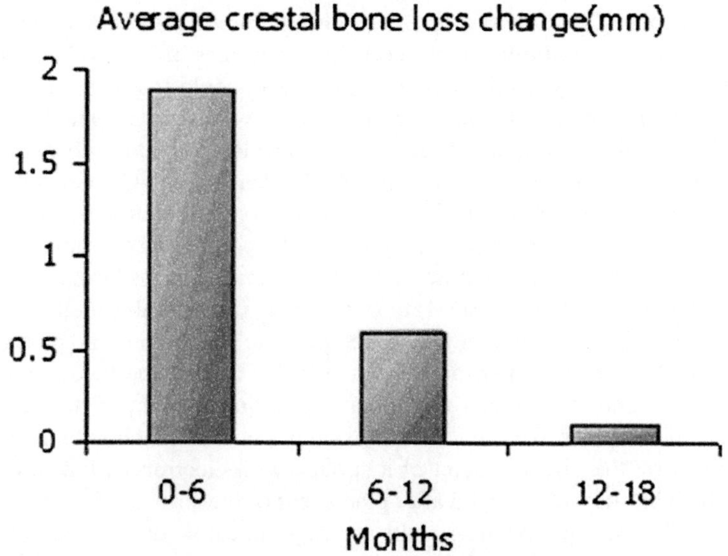

Fig. 4.6 Average crestal bone change versus time after implantation [36]

operating theater conditions did not illustrate significant differences in osseointegration success rates compared to those placed to less environmentally controlled dental school clinic [45] implying that the operative environment has not such a critical effect on the success of dental implants compared to implant placement in other body sites.

3 The Corrosion Aspect

Metallic structures in the mouth are constantly exposed to the corrosive attack of the oral environment and, thus, high corrosion resistance is very important for biomedical application of metallic materials. *In vivo* corrosion and ionic release might have adverse biological consequences due to toxicity of certain metallic elements that are contained in metallic biomaterials. Ti and its alloys (Ti–6Al–4V, Ti–6Al–7Nb, etc.) demonstrate high corrosion resistance and till today have been used for a variety of dental applications such as cast dental prosthetic restorations [46], orthodontic brackets and archwires [47], and endodontic files [48]. Although the aforementioned materials have been extensively used in biomedical implants production, they are not free of problems. Ti alloys containing Al and/or V have been lately accused of causing health problems, because of the cytotoxicity of Al and V. Previous studies have pointed out that V causes cytotoxic effects and adverse tissue reactions, while Al is associated with potential neurological disorders [49–54]. In light of these concerns, the manufacturers have adopted commercially pure Ti (grade II or grade IV) for the production of dental implants. Although the biocompatibility and the corrosion resistance of Ti is well accepted by the scientific community, its placement in the mouth could trigger galvanic corrosion due to coupling with other metallic structures previously placed in the patient's mouth. Namely, the Ti implant could be galvanically coupled with a variety of different metallic materials such as Au–Pt, Pd–Ag, Ni–Cr, Co–Cr, Ni–Ti, and different grades of stainless steels used for orthodontic purposes. Especially, Ni–Cr and Co–Cr are the materials of choice for the production of implant-retained superstructures. Some superstructures are currently made of cast Ti and prepared by Electro Discharge Machining (EDM) to improve press fit, but this method enriches the Ti surface with Cu [55] and might trigger galvanic phenomena and possible ions release [56]. Despite the experimental work on this aspect, the laboratory results unfortunately cannot be directly extrapolated in clinical practice due to various limitations. In first place, the intraoral corrosion behavior of metallic biomaterials is not described by the direct contact of a surface with electrolyte, but through the biofilm which has been developed after placement of the structure in the biological environment. Thus, experimental results are only indicative of the electrochemical behavior of metallic structures. Fortunately, previous studies on this matter, despite using different electrolytes (e.g., artificial saliva), yielded similar ranking of metallic biomaterials with the following order of decreasing corrosion resistance [57–59]: high gold alloys > low gold alloys > Ag-based alloys > Pb-based alloys > Ti > Ni–Cr alloys > Co–Cr alloys > amalgams. This means that Ti undergoes accelerated corrosion in the presence of precious alloys, and in contrast causes accelerated corrosion of base metals and amalgams. Certainly, to minimize galvanic effects Ti should be coupled with alloys as close as possible to it in the galvanic series, for example Ag-based alloys and Ni–Cr alloys, while coupling with amalgam or high gold alloys should be avoided from corrosion standpoint.

Implant dentistry has attracted much research, as many fundamental aspects remain puzzling to both clinicians and scientists. Loosening of implants remains the predominant failure cause. This phenomenon is controlled by biomechanical factors. Hence, the next breakthrough may be associated not with surface modification for rapid osseointegration, but with the design, selection, and production of the material itself.

References

1. Sabri R (1998) Four single-tooth implants as supernumerary premolars in the treatment of diastemas and microdontia: report of a case. Int J Oral Maxillofac Implants 13:706–709
2. Knabe C, Hoffmeister B (1998) The use of implant-supported ceramometal titanium prostheses following sinus lift and augmentation procedures: a clinical report. Int J Oral Maxillofac Implants 13:102–108
3. Mericske-Stern R (1998) Treatment outcomes with implant-supported overdentures: clinical considerations. J Prosthet Dent 79:66–73

4. Balshi TJ, Wolfinger GJ, Balshi SF (1999) Analysis of 356 pterygomaxillary implants in edentulous arches for fixed prosthesis anchorage. Int J Oral Maxillofac Implants 14:398–406
5. Le Guehennec L, Soueidan A, Layrolle P et al (2007) Surface treatments of titanium dental implants for rapid osseointegration. Dent Mater 23:844–854
6. National Agency for Medicines (2006) The 2006 Dental implant Yearbook. Publication of the National Agency for Medicines, Helsinki
7. Sykaras N, Iacopino AM, Marker VA et al (2000) Implant materials, designs, and surface topographies: their effect on osseointegration. A literature review. Int J Oral Maxillofac Implants 15:675–690
8. Lemons JE (1990) Dental implant biomaterials. J Am Dent Assoc 121:716–719
9. Kawahara H (1983) Cellular responses to implant materials: biological, physical and chemical factors. Int Dent J 33:350–375
10. Zinelis S, Thomas A, Syres K et al (2010) Surface characterization of zirconia dental implants. Dent Mater 26:295–305
11. Lautenschlager EP, Monaghan P (1993) Titanium and titanium alloys as dental materials. Int Dent J 43:245–253
12. Javier Gil F, Planell JA, Padros A et al (2007) The effect of shot blasting and heat treatment on the fatigue behavior of titanium for dental implant applications. Dent Mater 23:486–491
13. Long M, Rack HJ (1998) Titanium alloys in total joint replacement—a materials science perspective. Biomaterials 19:1621–1639
14. Yokoyama K, Ichikawa T, Murakami H et al (2002) Fracture mechanisms of retrieved titanium screw thread in dental implant. Biomaterials 23:2459–2465
15. Albrektsson T, Wennerberg A (2004) Oral implant surfaces: part 2—review focusing on clinical knowledge of different surfaces. Int J Prosthodont 17:544–564
16. Albrektsson T, Wennerberg A (2004) Oral implant surfaces: part 1—review focusing on topographic and chemical properties of different surfaces and in vivo responses to them. Int J Prosthodont 17:536–543
17. Branemark PI (1983) Osseointegration and its experimental background. J Prosthet Dent 50:399–410
18. Brunski JB (1999) Mechanical properties of trabecular bone in the human mandible: implications for dental implant treatment planning and surgical placement. J Oral Maxillofac Surg 57:706–708
19. Jokstad A, Braegger U, Brunski JB et al (2004) Quality of dental implants. Int J Prosthodont 17:607–641
20. Coelho PG, Granjeiro JM, Romanos GE et al (2009) Basic research methods and current trends of dental implant surfaces. J Biomed Mater Res B Appl Biomater 88B:579–596
21. Gross KA (2002) Preparation of thin histological sections for bone attachment and implant material analysis. Bioceramics 14 218–220:315–318
22. Albrektsson T, Zarb G, Worthington P et al (1986) The long-term efficacy of currently used dental implants: a review and proposed criteria of success. Int J Oral Maxillofac Implants 1:11–25
23. Smith DE, Zarb GA (1989) Criteria for success of osseointegrated endosseous implants. J Prosthet Dent 62:567–572
24. Geng JP, Tan KB, Liu GR (2001) Application of finite element analysis in implant dentistry: a review of the literature. J Prosthet Dent 85:585–598
25. Sumner DR, Galante JO (1992) Determinants of stress shielding: design versus materials versus interface. Clin Orthop Relat Res (274):202–212
26. Pilliar RM, Deporter DA, Watson PA et al (1991) Dental implant design—effect on bone remodeling. J Biomed Mater Res 25:467–483
27. Dujovne AR, Bobyn JD, Krygier JJ et al (1993) Mechanical compatibility of noncemented hip prostheses with the human femur. J Arthroplasty 8:7–22
28. Vaillancourt H, Pilliar RM, McCammond D (1995) Finite element analysis of crestal bone loss around porous-coated dental implants. J Appl Biomater 6:267–282

29. Vaillancourt H, Pilliar RM, McCammond D (1996) Factors affecting crestal bone loss with dental implants partially covered with a porous coating: a finite element analysis. Int J Oral Maxillofac Implants 11:351–359
30. Lin CL, Lin YH, Chang SH (2010) Multi-factorial analysis of variables influencing the bone loss of an implant placed in the maxilla: prediction using FEA and SED bone remodeling algorithm. J Biomech 43:644–651
31. Hurzeler M, Fickl S, Zuhr O et al (2007) Peri-implant bone level around implants with platform-switched abutments: preliminary data from a prospective study. J Oral Maxillofac Surg 65:33–39
32. Canullo L, Fedele GR, Iannello G et al (2010) Platform switching and marginal bone-level alterations: the results of a randomized-controlled trial. Clin Oral Implants Res 21:115–121
33. Canullo L, Goglia G, Iurlaro G et al (2009) Short-term bone level observations associated with platform switching in immediately placed and restored single maxillary implants: a preliminary report. Int J Prosthodont 22:277–282
34. Canullo L, Iurlaro G, Iannello G (2009) Double-blind randomized controlled trial study on post-extraction immediately restored implants using the switching platform concept: soft tissue response. Preliminary report. Clin Oral Implants Res 20:414–420
35. Cappiello M, Luongo R, Di Iorio D et al (2008) Evaluation of peri-implant bone loss around platform-switched implants. Int J Periodontics Restorative Dent 28:347–355
36. Nowzari H, Chee W, Yi K et al (2006) Scalloped dental implants: a retrospective analysis of radiographic and clinical outcomes of 17 NobelPerfect (TM) implants in 6 patients. Clin Implant Dent Relat Res 8:1–10
37. Brunski JB, Puleo DA, Nanci A (2000) Biomaterials and biomechanics of oral and maxillofacial implants: current status and future developments. Int J Oral Maxillofac Implants 15:15–46
38. Sahin S, Cehreli MC, Yalcin E (2002) The influence of functional forces on the biomechanics of implant-supported prostheses—a review. J Dent 30:271–282
39. Cehreli M, Sahin S, Akca K (2004) Role of mechanical environment and implant design on bone tissue differentiation: current knowledge and future contexts. J Dent 32:123–132
40. Lee KH, Maiden MF, Tanner AC et al (1999) Microbiota of successful osseointegrated dental implants. J Periodontol 70:131–138
41. Mombelli A, Lang NP (1998) The diagnosis and treatment of peri-implantitis. Periodontol 2000 17:63–76
42. Bragger U, Hugel-Pisoni C, Burgin W et al (1996) Correlations between radiographic, clinical and mobility parameters after loading of oral implants with fixed partial dentures. A 2-year longitudinal study. Clin Oral Implants Res 7:230–239
43. Garg AK, Reddi SN, Chacon GE (1994) The importance of asepsis in dental implantology. Implant Soc 5:8–11
44. Ivanoff CJ, Sennerby L, Lekholm U (1996) Influence of soft tissue contamination on the integration of titanium implants. An experimental study in rabbits. Clin Oral Implants Res 7:128–132
45. Scharf DR, Tarnow DP (1993) Success rates of osseointegration for implants placed under sterile versus clean conditions. J Periodontol 64:954–956
46. Zinelis S (2000) Effect of pressure of helium, argon, krypton, and xenon on the porosity, microstructure, and mechanical properties of commercially pure titanium castings. J Prosthet Dent 84:575–582
47. Gioka C, Bourauel C, Zinelis S et al (2004) Titanium orthodontic brackets: structure, composition, hardness and ionic release. Dent Mater 20:693–700
48. Zinelis S, Eliades T, Eliades G (2010) A metallurgical characterization of ten endodontic Ni-Ti instruments: assessing the clinical relevance of shape memory and superelastic properties of Ni-Ti endodontic instruments. Int Endod J 43:125–134
49. Steinemann SG (1996) Metal implants and surface reactions. Injury 27(Suppl 3):SC16–SC22
50. Steinemann SG (1998) Titanium—the material of choice? Periodontol 2000 17:7–21

51. Buser D, Broggini N, Wieland M et al (2004) Enhanced bone apposition to a chemically modified SLA titanium surface. J Dent Res 83:529–533
52. Zreiqat H, Valenzuela SM, Nissan BB et al (2005) The effect of surface chemistry modification of titanium alloy on signalling pathways in human osteoblasts. Biomaterials 26:7579–7586
53. Okazaki Y, Gotoh E, Manabe T et al (2004) Comparison of metal concentrations in rat tibia tissues with various metallic implants. Biomaterials 25:5913–5920
54. Okazaki Y, Gotoh E (2005) Comparison of metal release from various metallic biomaterials in vitro. Biomaterials 26:11–21
55. Zinelis S (2007) Surface and elemental alterations of dental alloys induced by electro discharge machining (EDM). Dent Mater 23:601–607
56. Cortada M, Giner L, Costa S et al (2000) Galvanic corrosion behavior of titanium implants coupled to dental alloys. J Mater Sci Mater Med 11:287–293
57. Reclaru L, Meyer JM (1994) Study of corrosion between a titanium implant and dental alloys. J Dent 22:159–168
58. Grosgogeat B, Reclaru L, Lissac M et al (1999) Measurement and evaluation of galvanic corrosion between titanium/Ti6A14V implants and dental alloys by electrochemical techniques and auger spectrometry. Biomaterials 20:933–941
59. Taher NM, Al Jabab AS (2003) Galvanic corrosion behavior of implant suprastructure dental alloys. Dent Mater 19:54–59

Biography

Dr. Spiros Zinelis graduated from the Department of Mining Engineering and Metallurgy at the National Technical University of Athens, Greece, in 1994. In 2001 he received his PhD. His dissertation title: "Development of new Ti alloys for biomedical applications." He worked as a research fellow in the Department of Biomaterials at the School of Dentistry, University of Athens, from 2001 to 2004, when he became a faculty member (Lecturer position). Since 2009 he is Assistant Professor of Dental Biomaterials. He has been awarded with two scholarships from the Greek State Scholarship Foundations. He is the recipient of the first award in 2004 from the European Federation of Orthodontics (FEO), the second award from the European Academy of Esthetic Dentistry in 2008, and 4 research grants from Greek scientific societies. He is a fellowship member in the Academy of Dental Materials and a member in five scientific societies (ASM, ASTM, IADR, HSB, and TCG). He is a member of the editorial boards of Dental Materials, Journal of Dental Biomechanics, and Recent Patents on Materials Science, and a reviewer for 12 international journals. He has contributed more than 100 presentations at domestic and international conferences, 13 publications in Greek scientific journals, and 52 publications in peer reviewed journals.

Theodore Eliades is Professor and Director of Department of Orthodontics and Paediatric Dentistry at University of Zurich. He is also affiliated with Texas, Marquette, Manchester, and Bonn Universities. He graduated from the University of Athens and the postgraduate orthodontic program of the Ohio State University and holds 3 degrees in biomaterials: a Master's from Ohio State, a doctorate from the University of Athens, School of Medicine, and a PhD from the University of Manchester. Eliades has published 120 articles and 20 book chapters in 8 books, which have been cited more than 1,200 times; he has also edited 6 textbooks published by major houses, some translated into 4 languages. The diffusion of this work into fields associated with natural and engineering sciences led to his election as a Fellow in the Royal Society of Chemistry, and the Institute of Materials, Minerals and Mining, as well as a Member in the Institute of Physics. He is the founding Editor of the Journal of Dental Biomechanics (Sage), Guest Editor of Seminars in Orthodontics and Mini Reviews in Organic Chemistry, Associate Editor of the European Journal of Orthodontics, and the American Journal of Orthodontics and Dentofacial Orthopedics, Editorial Board member in 5 other journals, and reviewer for 30 journals.

William A. Brantley is Professor at the Division of Restorative and Prosthetic Dentistry, College of Dentistry, Ohio State University, and Director of the Graduate

Program in Dental Materials Science. He also has a Courtesy Appointment in the Department of Biomedical Engineering, College of Engineering. Professor Brantley has been at Ohio State University since 1989, and was previously Professor and Chair, Department of Dental Materials, Marquette University School of Dentistry, where he was a faculty member from 1974 to 1989. He received his PhD from Carnegie Mellon University in Metallurgy and Materials Science, and was a member of the Ceramics Laboratory of the US Army Materials and Mechanics Research Center, and then member of the technical staff, Compound Semiconductor Materials Department, Bell Laboratories, prior to joining the faculty at Marquette University. Brantley has research interests in dental alloys and dental materials science, focusing on alloys for prosthodontics, orthodontics and endodontics. He is coeditor of two textbooks: Orthodontic Materials: Scientific and Clinical Aspects (Thieme), and Dental Materials *in vivo*: Aging and Related Phenomena (Quintessence). Brantley is author or coauthor of nearly 170 published articles, 30 book chapters, 270 IADR/AADR abstracts, and 20 published abstracts for other professional societies. He has mentored 7 postdoctoral researchers, 9 PhD students, and 65 MS students.

Chapter 5
Biodegradable Metals

Frank Witte and Amir Eliezer

Abstract Temporary implants based on controlled corrosion of metals are currently investigated as biodegradable metals. Intended corrosion of implants is a paradigm-changing approach in biomaterial sciences. This chapter will provide a comprehensive overview on historic approaches to biodegradable metals and will discuss current research and challenges in this novel field of biomaterial sciences.

1 Introduction to Biodegradable Metals

In the history of metals in surgical fixation of fractured bones and wound closure, most metals have been used as permanent implant materials and were selected on a trial and error basis. The basic idea of the temporary use of metal implants is reported by the Romans who used metal clips for the adaptations of skin [1]. Various metals have been used in the beginning of osteosynthesis, such as gold, silver, platinum, copper, lead, and iron [1]. Since then, however, these metals have been rejected for surgical use due to diverse reasons. Gold and platinum were desirable from a corrosion resistance standpoint, but were very expensive and suffered from poor mechanical properties, while lead was rejected due to its toxicity. Silver and iron were generally considered suitable biomaterials [1]. However, pure silver was mechanically insufficient for osteosynthesis, but its antibacterial effect has been

F. Witte (✉)
Laboratory for Biomechanics and Biomaterials, Hannover Medical School, Anna-von-Borries- Street, 1-7 30625, Hannover, Germany

Department of Biotechnology of Bioengineering, Ben-Gurion University of the Negev, P.O. Box 653, 84105, Beer-Sheva, Israel
e-mail: witte.frank@mh-hannover.de

A. Eliezer
Corrosion Research Center Nano-Bio and Advanced Materials, Sami Shamoon College of Engineering, P.O. Box 653, 84105, Beer-Sheva, Israel

N. Eliaz (ed.), *Degradation of Implant Materials*, DOI 10.1007/978-1-4614-3942-4_5,

already appreciated. There have been some controversial discussions on the biocompatibility of iron since metallosis has been observed after implantation of iron implants. Metallosis is a local destruction of soft and hard tissue based on mechanical–biological, electro-energetic, and chemical-toxic effects of metal implants [1]. Metallosis is also occasionally observed as an infiltration of periprosthetic soft tissues and bone by metallic debris resulting from wear of joint arthroplasties. Metallosis is often associated with significant osteolysis. Therefore, the identification of metallosis is an indication for revision surgery [2]. These clinical observations have led to the paradigm that metal implants should be generally corrosion resistant.

However, some early and also recent findings have proven that implants made of magnesium-based and iron-based materials may be suitable as temporary biomaterials which degrade *in vivo* by corrosion [3–6]. In fact, this is a paradigm-breaking approach to a new class of temporary implants — biodegradable metals. The basic concept of biodegradable metals is to compose the metals of elements which can be cleared from the body by physiological pathways and which do not exceed the toxicity limits during the corrosion process.

Research interest is rapidly growing in magnesium-based alloys for medical applications. However, in the words of the author of Ecclesiastes, "there is nothing new under the sun" [7]. The first time that magnesium was mentioned as a biodegradable implant material was in 1878, in a report about absorbable ligature for the closure of bleeding vessels [8], while pure iron had been used as implants even before. The first studies on degradable metals in the musculoskeletal field were published at the beginning of the last century based on the experiences with osteosynthesis implants made of magnesium alloys [6]. At that time, no attention was paid to metal allergies because of more severe surgical complications at that time, such as infection or implant failure. Skin sensitizing reactions to metal implants were reported more often after the introduction of aseptic surgery and less corrosive osteosynthesis materials, such as stainless steel, and are clinically still apparent in 10–15% of all implanted metals [9]. However, biodegradable magnesium alloys have shown no skin sensitizing potential in animal experiments [9]. The advantage of biodegradable metals over existing biodegradable materials such as polymers, ceramics, or bioactive glasses in load-bearing applications is a higher tensile strength and a Young's modulus that is closer to bone (Table 5.1). These facts provide enough evidence to investigate selected metallic biomaterials as temporary implant materials.

1.1 Brief History

Today, pure metals are used as biomaterials in some selected applications such as silver coatings for antibacterial surface properties, copper for contraceptive coils, CP-Ti for dental applications, and Pt electrodes, e.g., in cardiovascular applications. Recently, magnesium, iron, and tungsten have been investigated as biomaterials for

Table 5.1 Mechanical properties of implant materials

Material	Density (g/cm^3)	Young's modulus (GPa)	Tensile strength (MPa)	Elongation (%)
Surgical steel (X2CrNiMo18164)	8.0	193	585	55
Surgical titanium (TiAl6V4)	4.43	100–110	930–1,140	8–15
Cortical bone	1.7–2.0	3–30	80–150	3–4
DL-PLA (DL-polylactide)	1.24	1.9	29	5.0
Magnesium alloy AZ9E, casted	1.81	45	240	3.0
Magnesium alloy Mg10Gd, casted	N/A	44	131	2.5
Magnesium alloy RS66, rapid-solidified, extruded	N/A	45	400	23.5

temporary implants [3–5, 10]. Magnesium is the most extensively investigated element, especially for cardiovascular stent applications and musculoskeletal fixative devices. Tungsten has been investigated as a cardiovascular coil to close persistent postnatal shunts in animal experiments [10]. More promising, but still in the experimental stage, are iron-based cardiovascular devices, especially stents.

After magnesium was first mentioned as an absorbable implant material for the closure of bleeding vessels in 1878 [8], it was used as fast absorbable wound closure and fixatures of fractured bone in the early twentieth century [6, 11]. Payr, Chlumpsky, and Lespinasse used magnesium for anastomosis of blood vessels and intestines. Payr used magnesium tubes for vessel and nerve suture, as well as for tiny arrows to treat hemangioma [6]. Andrews and Seelig used magnesium for ligature of blood vessels, and as a surgical suture material. Even though Seelig obtained large amounts of hydrogen cavities while using pure magnesium implants, he could not detect any negative systemic effects [11]. In musculoskeletal surgery, pure magnesium was first used by Lambotte in 1906 as fracture fixation plates [12]. He observed a fast degradation of the magnesium plate within 8 days when he combined the magnesium plate with steel screws. This was the first indication that magnesium implants degrade *in vivo* by a corrosion process.

1.2 Basic Concepts of Biodegradation

The control and adaptation of the implant degradation rate are crucial, since the resorption capacity of the tissue is limited. Moreover, the local physiology of the implant environment determines the maximal degradation rate of a temporary implant.

There are many reasons that contribute to the corrosion of metals when implants are placed inside the human body. For magnesium corrosion, the most critical factor seems to be the local pH, while iron corrosion seems to be mainly dependent on the local oxygen concentration. After surgery, the pH surrounding the implant is reduced to a value between 5.3 and 5.6, typically due to the trauma of surgery. This process may accelerate initial magnesium corrosion, while infectious microorganisms and crevices formed between components can reduce the local oxygen concentration. The main challenge in iron implants is to accelerate corrosion, while magnesium corrosion is generally very fast *in vivo*. An option to design appropriate magnesium alloys is to slow down the initial implant corrosion by appropriate alloying, microstructure design, processing, and/or an additional coating. However, it has to be kept in mind that the corrosion rate will be further reduced *in vivo* after implantation due to adherent proteins and inorganic salts such as calcium phosphates, which stabilize the corrosion layer [3, 13]. Based on this theory, a magnesium implant with an initially reduced corrosion rate could finally lead to an arresting corrosion process *in vivo*. Thus, the right balance of a reduced corrosion rate and an assured complete corrosion *in vivo* will create a useful biodegradable magnesium implant; otherwise, parts of the implant may persist locally and will act as long-term biomaterials.

1.2.1 Corrosion *In Vitro* and *In Vivo*

The interdependence of corrosion and the human body is of most interest in this field. However, direct electrochemical testing in humans ranges from difficult (in the mouth) to virtually impossible (e.g., for orthopedic devices) because of ethical, safety, legal, and regulatory considerations [14]. Consequently, much effort has been devoted to identify alternative environments which simulate the corrosion conditions *in vivo*. A possible alternative is testing under cell culture conditions, but they are less related to real-life conditions than tests in laboratory animals. The experimental parameters that simulate the corrosion *in vitro* are not fully known. Thus, the existing *in vitro* corrosion test system always needs to be adapted to the corresponding *in vivo* application.

The physiological temperature of 37°C is an important factor influencing *in vivo* electrochemical behavior [13]. Dissolved salts, particularly chlorides, are probably the most influential components for implant corrosion *in vivo* [14]. Also gases dissolved in body fluids play an important role in implant alloy corrosion. Oxygen is one of the most important physiological gases; its partial pressure varies widely within the body from about 2.67×10^2 to 1.33×10^4 Pa [14]. Thus, an implant surface can be in contact with anatomical environments of widely different pO_2, creating the possibility to develop different aeration cells [14]. Carbon dioxide is another gas that can be important for *in vivo* corrosion, because of its influence on the pH value [14]. While the pH is usually homeostatically regulated at about pH 7.4, it may fall below 4.5 at the sites of inflammation, for a period of hours, or longer if acute inflammation processes convert into chronic inflammation [14]. These initial pH changes after surgery or during inflammation are especially critical for the corrosion resistance of magnesium implants.

The role played by proteins regarding corrosion *in vivo* is one of the most important aspects of the unique environment within the body, and can cause differences between the corrosion behavior in laboratory, chloride-containing environments, and *in vivo* [13–15]. Especially for magnesium corrosion, the influence of albumin and other serum proteins on the formation of a corrosion protective film has been investigated [16–19]. In general, from the moment an implant is inserted into the body, it becomes covered with a layer of adsorbed proteins [20]. The properties of this layer change with time, because (1) both thermodynamic and kinetic factors are involved in protein adsorption [21] and (2) the adjacent cells are actively involved in protein synthesis [22]. The effects of proteins on metal corrosion are complex and may increase or decrease the corrosion rate [22]. The *in vivo* environment is characterized by dynamic, constantly changing chemical and physiological processes, mechanical loading patterns, and bioelectric potentials [14]. There is certainly more than one standardized *in vivo* environment. However, in current biodegradable research, no considerable attention has been devoted to the interaction between mechanical loading and corrosion phenomena yet. However, corrosion fatigue and stress corrosion cracking (SCC) are actual force patterns *in vivo* that may increase the corrosion rate, while the mechanical integrity of the biomaterial is temporarily maintained [14]. Special attention has to be given to the implant design, which can alter the corrosion performance of alloys *in vivo*. Localized corrosion could occur, for example, between screws and plates for crevice corrosion, or with parts in relative motion for fretting corrosion, or both [14, 23].

1.2.2 Methods to Measure Corrosion *In Vivo*

The elemental components of biodegradable magnesium alloys (Mg, Al, Li, Zn, rare earth elements) have been investigated by various analytical methods in histological sections, bone, tissue, and body fluids [24]. However, the application of these methods for trace and ultra trace analysis in small sample volumes is typically hampered by the detection limits of the elemental concentration in physiological liquids and tissues. Current detection limits are given in serum (<1 μg/L to about 1 mg/L), and in liver and bone (<1 mg/kg up to about 500 mg/kg) [24]. Further limitations are caused by time-consuming sample preparation (AES, OES, ICP-MS), access to appropriate methods (NAA, synchrotron-based methods), lack of sufficient lateral resolution for solid sample analysis (GD-OES), or challenging interferences during the measuring process (AAS, AES, ICP-MS, XRF) (Table 5.2) [24].

Magnesium corrosion can be determined *in vivo* using microtomography. In particular, synchrotron-based microtomography (SRμCT) is a non-destructive method with a high density and high spatial resolution. Furthermore, element-specific SRμCT can determine the spatial distribution of the alloying elements during *in vivo* corrosion [25]. The remaining non-corroded metal volume as well as the surface morphology can be determined in three dimensions non-destructively on a micrometer scale using SRμCT. The reduction of the metallic implant volume can be converted into a corrosion rate by using a modification of the ASTM G31-72, 2004 eq. (5.1) for weight loss measurements:

$$\mathrm{CR} = \frac{W}{At\rho}, \tag{5.1}$$

Table 5.2 Explanation of abbreviations

AAS	Atomic absorption spectroscopy
AES	Atomic emission spectroscopy
GD-OES	Glow-discharge optical emission spectroscopy
ICP-MS	Mass spectrometry with inductively coupled plasma
NAA	Neutron activation analysis
XRD	X-ray diffraction
XRF	X-ray fluorescence analysis

where CR (mm per year) is the corrosion rate, W (g) is the weight loss of the metal or alloy, A (mm^2) is the initial surface area exposed to corrosion, ρ (g/mm^3) is the standard density of the metal or alloy, and t (365 days = 1 year) is the time of immersion. Herein, the weight loss will be substituted by the reduction in volume (ΔV) multiplied by the standard density (ρ) resulting in eq. (5.2):

$$\mathrm{CR} = \frac{\Delta V}{At}, \tag{5.2}$$

where ΔV (mm^3) is the reduction in implant volume that is equal to the remaining metal volume subtracted from the initial metal implant volume. This method provides a general corrosion rate of the implanted metal. A better analysis of the local corrosion can be obtained if the corrosion rates are calculated based on the pitting depth [26].

2 Magnesium-Based Biodegradable Metals

In general, magnesium and its alloys are light metals and provide a density of about 1.74 g/cm^3, which is 1.6 times less dense than aluminum, and 4.5 times less dense than steel [24]. The fracture toughness of magnesium is higher than ceramic biomaterials such as hydroxyapatite, while the elastic modulus and compressive yield strength of magnesium are closer to those of natural bone than other commonly used metallic osteosynthesis materials (Table 5.1). Furthermore, magnesium has a low corrosion resistance, which is due to its electrochemical standard potential of $E_o = -2.363$ V vs. SHE, or alternatively $E_o = -2.159$ V vs. SHE if magnesium oxides or magnesium hydroxides are formed. According to Pourbaix, these potentials may be shifted if the Mg ions are complexed in the corrosion layer. It is obvious that magnesium carbonates and phosphates are formed during Mg corrosion under cell culture conditions and has been detected in a complex corrosion layer *in vivo*. Thus, complexed Mg is truly changing the local potential and promotes local pitting *in vivo*. However, detailed Mg corrosion processes remain currently uncovered *in vivo*. If exposed to air, the surface of magnesium is passivated by a growing thin gray layer of magnesium hydroxide, which reduces further by chemical reactions. Magnesium hydroxides are slightly soluble in water;

however, severe corrosion occurs in saline media, as well as in the human body, where high chloride ion concentrations of about 150 mM are present. Magnesium hydroxide accumulates on the underlying magnesium matrix as a corrosion protective layer in water, but when the chloride concentration in the corrosive environment rises above 30 mM [27], magnesium hydroxide starts to convert into highly soluble magnesium chloride. Therefore, severe pitting corrosion can be observed on magnesium alloys *in vivo* [3, 15, 28]. Importantly, magnesium corrosion is relatively insensitive to various oxygen concentrations around implants in different anatomical locations. The overall corrosion reaction of magnesium in aqueous environments is given as follows:

$$Mg_{(s)} + 2H_2O_{(aq)} \rightleftharpoons Mg(OH)_{2(s)} + H_{2(g)} \tag{5.3}$$

This overall reaction may include the following partial reactions:

$$Mg_{(s)} \rightleftharpoons Mg^{2+}{}_{(aq)} + 2e^- \quad \text{(anodic reaction)} \tag{5.4}$$

$$2\,H_2O_{(aq)} + 2\,e^- \rightleftharpoons H_{2(g)} + 2OH^-{}_{(aq)} \quad \text{(cathodic reaction)} \tag{5.5}$$

$$Mg^{2+}{}_{(aq)} + 2OH^-{}_{(aq)} \rightleftharpoons Mg(OH)_{2(s)} \quad \text{(product formation)} \tag{5.6}$$

2.1 *Physiological and Toxicological Aspects*

During the corrosion process, a mass of magnesium ions is released from the magnesium implant, which can be eliminated from the body very rapidly via the blood serum and the kidneys [24]. Magnesium can also be stored in muscle (39% of total Mg) or bone (60%), which are the natural storages of the 21–35 g of elemental magnesium of an average adult person who weighs about 70 kg [24]. The level of magnesium in the extracellular fluid is kept constant at levels between 0.7 and 1.05 mM. While serum magnesium levels exceeding 1.05 mM can lead to muscular paralysis, hypotension, and respiratory distress, cardiac arrest occurs only for severely high serum levels of 6–7 mM. However, magnesium is essential to human metabolism and is the fourth most abundant cation in the human body. Furthermore, magnesium is also a co-factor for many enzymes, and stabilizes the structures of DNA and RNA.

2.2 *Evolution of Hydrogen During* In Vivo *Corrosion*

In magnesium and its alloys, impurities and cathode sites with a low hydrogen overpotential facilitate hydrogen evolution [29], thus causing substantial galvanic corrosion and potential local gas cavities *in vivo*. However, there are contradicting reports on the occurrence of gas cavities after magnesium implantation. Gas cavities have been observed after subcutaneous implantation while intravasal application showed no local gas accumulation. An explanation for this observation may be based on the diffusion and solubility coefficient of hydrogen in biological tissues, which has been widely reviewed [30]. The solubility of hydrogen in tissues is influenced by the content of lipids, proteins, and salinity, but in fat and oils, the solubility seems to be approximately independent of temperature in the physiological range [30, 31]. Not only viscosity but also different tissue components and structures like lipids, proteins, and glycosaminoglycans influence the numeric value of the hydrogen diffusion coefficient [30, 32]. Depending on the experimental configuration, the diffusion coefficient may be underestimated in both stagnant and flowing media due to a boundary layer formation, which increases the effective diffusion distance [30]. This finding may be important for intravascular magnesium applications. Correlating the hydrogen diffusion coefficients from various biological media having fractional water contents from about 68% to 100%, it is found that the diffusion coefficient of hydrogen increases exponentially with the increasing water fraction of the tissue [32]. The tissue water content increases from adipose tissue to skin, and from bone to muscles, in animals and humans [24]. This can explain why different corrosion rates and gas cavities were observed for magnesium alloys in different anatomical implantation sites [3, 28, 33, 34]. In an animal study with rats, it was shown that the adsorption of hydrogen gas from subcutaneous gas pockets was limited by the diffusion coefficient of hydrogen in the tissue; the overall hydrogen adsorption rate was determined as 0.954 mL per hour [31]. Thus, the local blood flow and the water content of the tissue surrounding the implant are important parameters, which need to be considered in designing biodegradable magnesium alloys with an appropriate corrosion rate. Concomitantly, it can be assumed that local hydrogen cavities occur when more hydrogen is produced per time interval than can be dissolved in the surrounding tissue, or diffuse from the implant surface into the extracellular medium, which is renewed depending on the local blood flow. This means that magnesium alloys are corroding *in vivo* with an appropriate corrosion rate when no local gas cavities are observed during the implantation period in a specific anatomical site.

2.3 Recent Research and Product Development

Magnesium has been intensively investigated as a biodegradable material in cardiovascular application for some years. The German company Biotronik is producing an absorbable magnesium stent (AMS-1) that has been already used in several clinical trials covering peripheral and coronary indications [35, 36]. However, despite the good clinical results, Biotronik is still improving its magnesium stent (AMS-2) by prolonged mechanical stability, improved stent design, modified alloy, and enhanced surface passivity (Fig. 5.1). A further aim of Biotronik is the reduction of neointima hyperplasia by creating a drug-eluting biodegradable

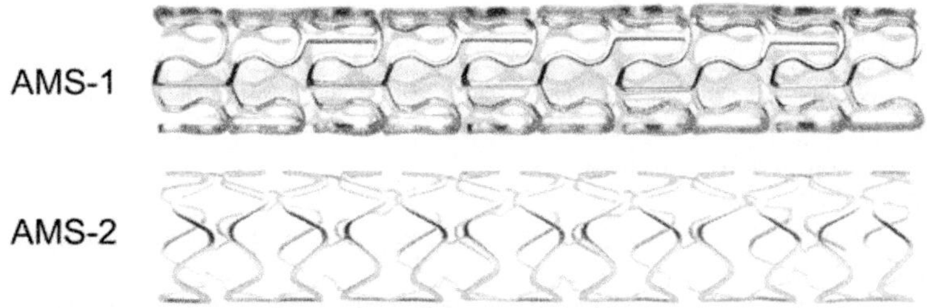

Fig. 5.1 Following preliminary clinical trials, the absorbable metal stent (AMS-1) has been improved in regard to alloy composition, mechanical stability, and design, resulting in the AMS-2 stent. With kind permission from Biotronik AG

Fig. 5.2 The first magnesium-based implants, such as this prototype of an interference screw, have been produced for musculoskeletal applications. With kind permission from AAP Biomaterials GmbH

magnesium stent (AMS-3, DREAMS). Following the first successful attempts to control the corrosion rate of magnesium implants *in vivo*, the industrial attention to biodegradable magnesium in musculoskeletal applications is rising. The first prototypes of magnesium implants have already been produced (Fig. 5.2). However, although the first biodegradable prototypes have been developed and some basic knowledge has been summarized in reviews [6, 24, 37–41], there are still several open questions about the corrosion process *in vivo* and the local tissue–material interaction.

1. choose the implant application and clinlical indication

↓

2. get information about the biological environment

↓

3. define the property profile of the implant

↓

4. know the effectiveness of the alloying elements on intermetallic phases and microstructure

↓

5. determine the alloy composition with regard to requirements on effectiveness of the alloying elements

↓

6. define the processing route to achieve a proper microstructure - property relationship

↓

7. use testing routines which are appropriate for the material and the specific application

↓

8. reconsider step 5 to 7 based on results obtained from step 7

Fig. 5.3 Following the steps in this flowchart may help to select the appropriate magnesium alloy for the intended implant application. Modified diagram from [24]

2.3.1 How to Control Corrosion *In Vivo*?

Magnesium corrosion *in vivo* can be controlled by alloying, processing, and coating. The first empirical approach in biodegradable stent development has been successfully leading to magnesium alloys containing rare earth elements. However, a more systematic approach is needed. For the use in humans, the authors of this chapter recommend to use aluminum-free magnesium alloy systems, while aluminum-containing Mg alloys may be used for research purposes, e.g., to investigate the corrosion protective effect of specific coatings by measuring the release of aluminum. As indicated in Fig. 5.3, it seems to be of major importance that an interdisciplinary team of researchers designs the magnesium alloy and the production process according to the intended application, reviews available data from the literature, and performs a critical analysis of the methods (Fig. 5.3).

3 Iron-Based Biodegradable Metals

The degradation of iron–manganese based implants has been lately described by research groups [42–44]. These groups suggested that immediately after immersion in the test solution, the iron alloy will be oxidized to metal ions in anodic spots, following eqs. (5.7) and (5.8). The electrons from the anodic reaction will be consumed by a corresponding cathodic reaction and the reduction of dissolved oxygen, following eq. (5.9). These reactions should occur randomly over the entire surface where potential differences exist, e.g., at grain boundaries and at the interface between different phases.

$$Fe \rightarrow Fe^{2+} + 2e^{-} \tag{5.7}$$

$$Mn \rightarrow Mn^{2+} + 2e^{-} \tag{5.8}$$

$$2H_2O + O_2 + 4e^{-} \rightarrow 4OH^{-} \tag{5.9}$$

The released metal ions should then react with the hydroxyl ion (OH^{-}) released from the cathodic reaction, forming insoluble hydroxides according to eqs. (5.10) and (5.11). Since iron is usually the main component of the implant alloy, the equations hereafter are written for iron only.

$$2Fe^{2+} + 4OH^{-} \rightarrow 2Fe(OH)_2 \text{ or } 2FeO \times 2H_2O \tag{5.10}$$

$$4Fe(OH)_2 + O_2 + 2H_2O \rightarrow 4Fe(OH)_3 \text{ or } 2Fe_2O_3 \times 6H_2O \tag{5.11}$$

The corroded surface layer of iron alloys normally consists of $FeO{\cdot}nH_2O$ (black) at the bottom, $Fe_3O_4{\cdot}nH_2O$ (black) in the middle, and $Fe_2O_3{\cdot}nH_2O$ (red-brown) on the top [45]. Continuing corrosion is driven by chloride ions from the solution which can penetrate the inconsistent hydroxide layers on the samples. The formed metal chloride can then be hydrolyzed by water, forming hydroxide, and free acid (eq. 5.12) which lowers the pH in the pits while the bulk solution remains neutral. This autocatalytic reaction leads to the formation of wide and deep pits [46].

$$Fe^{2+} + 2Cl^{-} \rightarrow FeCl_2 + H_2O \rightarrow Fe(OH)_2 + HCl \tag{5.12}$$

As the degradation process continues, a new layer enriched with calcium and phosphorus is agglomerating in physiological environments over the previously formed flat degradation layer and slows down the underlying corrosion process.

3.1 Physiological and Toxicological Aspects

Iron is an essential trace element in the human body and, due to its high toxic potential, its pathways are heavily regulated in mammals. Iron in its Fe^{2+} state can only be absorbed in the intestine. Iron in its Fe^{3+} state needs to be reduced before it can be absorbed. There are several other metals and ligands (e.g., oxalic acid), which form less soluble iron complexes, and thus inhibit the iron absorption. Iron in its Fe^{3+} state is transported by the protein transferrin. In human blood, approximately 70% of iron is bound to hemoglobin; the rest is bound to myoglobin, or is stored as ferritin or hemosiderin. Most of the iron released from the degradation of hemoglobin and myoglobin will be reused, and only a small portion of ~10% will be excreted via feces, urine, or sweat. Hemoglobin contains Fe^{2+} and forms an unstable, reversible bond with oxygen in the erythrocytes (oxyhemoglobin) and appears bright red, while the oxygen-unloaded form appears purple-blue and is called desoxyhemoglobin. If the iron in the hemoglobin is oxidized (Fe^{3+}), ferrihemoglobin (methemoglobin) is formed, which cannot bind oxygen. In erythrocytes, the methemoglobin reductase reactivates hemoglobin ($Fe^{3+} \rightarrow Fe^{2+}$). Otherwise, methemoglobin will be enzymatically degraded and eliminated from the body.

Since iron is usually bound to proteins and highly regulated in mammals, there are several known iron storage diseases. Most of them are based on genetic defects and result in the pathological storage of iron in liver, heart, skin, endocrine organs, and the reticuloendothelial system. If the natural storage capacity of free iron is exceeded and transferrin cannot bind any more iron, then iron causes oxidative stress, which leads to cell and tissue damage, and the excess iron will be pathologically stored in heart, liver, and endocrine organs. In this case, free iron ions harm cells by forming reactive free radicals, e.g., hydroxyl radicals, which have a very strong oxidation capacity. The hydroxyl radical reacts with organic molecules on the cell membrane and starts a lipid-peroxidation process, which finally leads to cell death.

Thus, for biodegradable iron implants it is most important that the local storage capacity and the amount of iron that is released per time interval do not exceed the local and general transport and storage capacities. Moreover, iron implants may need to be applied only in individuals without any iron-related disease. However, the slow degradation of an iron stent that weighs ~40 mg will unlikely cause system toxicity, taking into account the high iron load of normal blood (about 447 mg/L) [47].

3.2 Recent Research and Product Development

Iron-based implants have been developed mainly for cardiovascular stent applications. Initial implantation of iron stents into the descending aorta of New Zealand White Rabbits was performed by Peuster et al. [5]. The long-term biocompatibility of iron stents was evaluated in the descending aorta of minipigs and showed similarity in the amount of neointimal proliferation between iron and 321L stainless steel stents, without any sign of iron-related toxicology, e.g., due to corrosion products [48]. Iron stents assessed in coronary arteries of juvenile domestic pigs performed similar to Co–Cr stents [49]. The ions released by iron stent corrosion have been shown to reduce vascular smooth muscle cell proliferation [50]. These studies seem to suggest the safety of iron stents, although most authors are demanding a faster degradation rate. Schinhammer et al. [44, 51] provide design strategies to enhance the corrosion rates compared to pure iron while keeping suitable strength and ductility.

3.2.1 How to Accelerate Corrosion *In Vivo*?

Manganese was used as an alloying element to enhance the degradation of Fe–Mn alloys. It was shown that the formed corrosion product consisted of biocompatible metal hydroxides and calcium phosphates. Furthermore, the austenite (γ) phase formed in these Fe–Mn alloys reduced the magnetic susceptibility compared to 316L stainless steel, which provided an enhanced compatibility with the magnetic resonance imaging (MRI) [47, 52]. Other design strategies to achieve an enhanced corrosion rate using the (Fe,Mn)–Pd system have been introduced by Schinhammer et al. [44]. Moreover, electroforming is a novel technique for fabrication of near-net shaped Fe-based stents *in situ* [52]. Pure Fe-foils produced by electroforming showed higher corrosion rates and exhibited a fine microstructure with small grains and a high volume fraction of the grain boundaries. Another strategy is to create microstructural defects and strong preferred texture, which initiates corrosion. Both strategies can be produced by electroforming [52]. The mechanical properties of these Fe-foils include 18% elongation after annealing of 550°C for 1 h, and a decreased UTS of 292 MPa [52]. Thus, the corrosion rate of future iron-based implants may be enhanced by further alloy design and the innovative production routes. However, the biocompatibility and efficiency of larger iron-based implants still needs to be proven *in vivo*.

References

1. Schuster J (1975) The metallosis. Ferdinand Enke, Stuttgart
2. Heffernan EJ, Hayes MM, Alkubaidan FO, Clarkson PW, Munk PL (2008) Aggressive angiomyxoma of the thigh. Skeletal Radiol 37:673–678
3. Witte F, Kaese V, Haferkamp H, Switzer E, Meyer-Lindenberg A, Wirth CJ, Windhagen H (2005) In vivo corrosion of four magnesium alloys and the associated bone response. Biomaterials 26:3557–3563
4. Heublein B, Rohde R, Kaese V, Niemeyer M, Hartung W, Haverich A (2003) Biocorrosion of magnesium alloys: a new principle in cardiovascular implant technology? Heart 89:651–656
5. Peuster M, Wohlsein P, Brugmann M, Ehlerding M, Seidler K, Fink C, Brauer H, Fischer A, Hausdorf G (2001) A novel approach to temporary stenting: degradable cardiovascular stents produced from corrodible metal—results 6–18 months after implantation into New Zealand white rabbits. Heart 86:563–569
6. Witte F (2010) The history of biodegradable magnesium implants: a review. Acta Biomater 6:1680–1692
7. Mantovani D, Witte F (2010) Editorial. Acta Biomaterialia 6:1679
8. Huse EC (1878) A new ligature? Chicago Med J Exam 172:2
9. Witte F, Abeln I, Switzer E, Kaese V, Meyer-Lindenberg A, Windhagen H (2008) Evaluation of the skin sensitizing potential of biodegradable magnesium alloys. J Biomed Mater Res A 86A:1041–1047
10. Peuster M, Fink C, Wohlsein P, Bruegmann M, Gunther A, Kaese V, Niemeyer M, Haferkamp H, Schnakenburg C (2003) Degradation of tungsten coils implanted into the subclavian artery of New Zealand white rabbits is not associated with local or systemic toxicity. Biomaterials 24:393–399

11. Seelig MG (1924) A study of magnesium wire as an absorbable suture and ligature material. Arch Surg 8:669–680
12. Lambotte A (1932) L'utilisation du magnésium comme matériel perdu dans l'ostéosynthèse. Bull Mém Soc Nat Cir 28
13. Eliezer A, Witte F (2010) The role of biological environments on magnesium alloys as biomaterials. In: Kainer KU (ed) 7th International conference on magnesium alloys and their applications. Wiley-VCH, Dresden, pp 822–827
14. Bundy K (1995) In vivo. In: Baboian R (ed) Corrosion tests and standards. ASTM, Fredericksburg, VA, pp 411–419
15. Witte F, Fischer J, Nellesen J, Crostack HA, Kaese V, Pisch A, Beckmann F, Windhagen H (2006) In vitro and in vivo corrosion measurements of magnesium alloys. Biomaterials 27:1013–1018
16. Mueller W-D, Mele MFLd, Nascimento ML, Zeddies M (2009) Degradation of magnesium and its alloys: dependence on the composition of the synthetic biological media. J Biomed Mater Res A 90A:487–495
17. Mueller WD, Lucia Nascimento M, Lorenzo de Mele MF (2010) Critical discussion of the results from different corrosion studies of Mg and Mg alloys for biomaterial applications. Acta Biomater 6:1749–1755
18. Rettig R, Virtanen S (2009) Composition of corrosion layers on a magnesium rare-earth alloy in simulated body fluids. J Biomed Mater Res A 88A:359–369
19. Eliezer A, Witte F (2010) Corrosion behavior of magnesium alloys in biomedical environments. J Adv Mater Res 95:17–22
20. Rudee ML, Price TM (1985) The initial stages of adsorption of plasma derived proteins on artificial surfaces in a controlled flow environment. J Biomed Mater Res 19:57–66
21. Neumann AW, Hum OS, Francis DW, Zingg W, van Oss CJ (1980) Kinetic and thermodynamic aspects of platelet adhesion from suspension to various substrates. J Biomed Mater Res 14:499–509
22. Clark GC, Williams DF (1982) The effects of proteins on metallic corrosion. J Biomed Mater Res 16:125–134
23. Eliaz N (2008) Biomaterials and corrosion. In: Raj UKMaB (ed) Corrosion science and technology: mechanism, mitigation and monitoring. Narosa Publishing House, New Delhi, pp 356–397
24. Witte F, Hort N, Vogt C, Cohen S, Kainer KU, Willumeit R, Feyerabend F (2008) Degradable biomaterials based on magnesium corrosion. Curr Opin Solid State Mater Sci 12:63–72
25. Witte F, Fischer J, Nellesen J, Beckmann F (2006) Microtomography of magnesium implants in bone and their degradation. Progress in Biomedical Optics and Imaging - Proceedings of SPIE 2006:6318, art. no. 631806
26. Witte F, Fischer J, Nellesen J, Vogt C, Vogt J, Donath T, Beckmann F (2009) In vivo corrosion and corrosion protection of magnesium alloy LAE442. Acta Biomater 6(5):1792–1799
27. Shaw BA (2003) Corrosion resistance of magnesium alloys. In: ASM handbook, vol 13a: Corrosion: fundamentals, testing and protection. ASM International, London
28. Xu L, Yu G, Zhang E, Pan F, Yang K (2007) In vivo corrosion behavior of Mg-Mn-Zn alloy for bone implant application. J Biomed Mater Res A 83:703–711
29. Song GL, Atrens A (1999) Corrosion mechanisms of magnesium alloys. Adv Eng Mater 1:11–33
30. Lango T, Morland T, Brubakk AO (1996) Diffusion coefficients and solubility coefficients for gases in biological fluids and tissues: a review. Undersea Hyperb Med 23:247–272
31. Piiper J, Canfield RE, Rahn H (1962) Absorption of various inert gases from subcutaneous gas pockets in rats. J Appl Physiol 17:268
32. Vaupel P (1976) Effect of percentual water-content in tissues and liquids on diffusion-coefficients of O_2, Co_2, N_2, and H_2. Pflugers Arch Eur J Physiol 361:201–204
33. Wen CE, Yamada Y, Shimojima K, Chino Y, Hosokawa H, Mabuchi M (2004) Compressibility of porous magnesium foam: dependency on porosity and pore size. Mater Lett 58:357–360

34. Witte F, Ulrich H, Rudert M, Willbold E (2007) Biodegradable magnesium scaffolds. Part I: Appropriate inflammatory response. J Biomed Mater Res A 81:748–756
35. Bosiers M, Deloose K, Verbist ü, Peeters P (2005) First clinical application of absorbable metal stents in the treatment of critical limb ischemia: 12-month results. Vasc Dis Manag 2(4):86–91
36. Erbel R, Di Mario C, Bartunek J, Bonnier J, de Bruyne B, Eberli FR, Erne P, Haude M, Heublein B, Horrigan M, Ilsley C, Bose D, Koolen J, Luscher TF, Weissman N, Waksman R (2007) Temporary scaffolding of coronary arteries with bioabsorbable magnesium stents: a prospective, non-randomised multicentre trial. Lancet 369:1869–1875
37. Zhu YY, Wu GM, Zhao Q (2010) Research progress of magnesium-based alloy in biomedical application. Chin J Biomed Eng 29:932–938
38. Zeng R, Dietzel W, Witte F, Hort N, Blawert C (2008) Progress and challenge for magnesium alloys as biomaterials. Adv Eng Mater 10:B3–B14
39. Virtanen S (2011) Biodegradable Mg and Mg alloys: corrosion and biocompatibility. Mater Sci Eng B 176(20):1600
40. Xin Y, Hu T, Chu PK (2011) In vitro studies of biomedical magnesium alloys in a simulated physiological environment: a review. Acta Biomater 7:1452–1459
41. Chen H, Zhao XX (2011) Biodegradable magnesium-alloy stent: current situation in research. J Interv Radiol 20:62–64
42. Moravej M, Purnama A, Fiset M, Couet J, Mantovani D (2010) Electroformed pure iron as a new biomaterial for degradable stents: in vitro degradation and preliminary cell viability studies. Acta Biomater 6:1843–1851
43. Hermawan H, Purnama A, Dube D, Couet J, Mantovani D (2010) Fe-Mn alloys for metallic biodegradable stents: degradation and cell viability studies. Acta Biomater 6:1852–1860
44. Schinhammer M, Hanzi AC, Loffler JF, Uggowitzer PJ (2010) Design strategy for biodegradable Fe-based alloys for medical applications. Acta Biomater 6:1705–1713
45. Roberge P (2000) Handbook of corrosion engineering. McGraw-Hill, Toronto
46. Shreir L, Jarman R, Burstein G (2000) Corrosion, meal/environment reactions. Butterworth-Heinemann, Oxford
47. Hermawan H, Dube D, Mantovani D (2010) Developments in metallic biodegradable stents. Acta Biomater 6:1693–1697
48. Peuster M, Hesse C, Schloo T, Fink C, Beerbaum P, von Schnakenburg C (2006) Long-term biocompatibility of a corrodible peripheral iron stent in the porcine descending aorta. Biomaterials 27:4955–4962
49. Waksman R, Pakala R, Baffour R, Seabron R, Hellinga D, Tio FO (2008) Short-term effects of biocorrodible iron stents in porcine coronary arteries. J Interv Cardiol 21:15–20
50. Mueller PP, May T, Perz A, Hauser H, Peuster M (2006) Control of smooth muscle cell proliferation by ferrous iron. Biomaterials 27:2193–2200
51. Moszner F, Sologubenko AS, Schinhammer M, Lerchbacher C, Hanzi AC, Leitner H, Uggowitzer PJ, Loffler JF (2011) Precipitation hardening of biodegradable Fe-Mn-Pd alloys. Acta Mater 59:981–991
52. Moravej M, Prima F, Fiset M, Mantovani D (2010) Electroformed iron as new biomaterial for degradable stents: development process and structure-properties relationship. Acta Biomater 6:1726–1735

Biography

Dr. Frank Witte is an Associate Professor for Experimental Medicine (Orthopedics) at Hannover Medical School. He also keeps an Adjunct Professorship for Bioengineering at University of Pittsburgh and University of North Carolina A&T State. Dr. Witte is the Director of Biomaterial Research in the Laboratory for Biomechanics and Biomaterials and is also leading the research group on implant immunology at CrossBIT at the Hannover Medical School. Dr. Witte graduated from Hannover Medical School in 2000 and completed his PhD thesis in 2001. Since then he has published extensively in the area of biodegradable metals, especially on implant performance *in vivo*. Moreover, he is the co-founded of the International Symposium on Biodegradable Metals (www.biodegradablemetals.org).

Dr. Amir Eliezer is the Director of the Corrosion Research Center Nano-Bio & Advanced Materials as well as a faculty member of both Civil Engineering and Mechanical Engineering at the Sami Shamoon College of Engineering. He is active as the Vice President of the World Corrosion Organization (WCO), CAMPI Chair, NACE International Europe Board Member and faculty advisor of the Israel Negev student section.

Chapter 6
Degradable and Bioactive Synthetic Composite Scaffolds for Bone Tissue Engineering

A.R. Boccaccini, X. Chatzistavrou, J.J. Blaker, and S.N. Nazhat

Abstract There is significant information and knowledge acquired in the last years on the fabrication, characterisation, and application of bioactive composite scaffolds based on combinations of biodegradable polymers and inorganic fillers intended for bone tissue engineering. Of particular importance is the complete understanding of the degradation behaviour of these scaffolds in order to assess the effects of pore structure, scaffold geometry, permeability, and the influence of bioactive fillers on the scaffold mechanical properties and biological performance. The present chapter examines the development of such bioactive and biodegradable scaffolds discussing their mechanical properties and degradation behaviour. A general background on biodegradable polymers with focus on the fabrication and properties of scaffolds made from polyesters is included, followed by a complete overview of the development of scaffolds based on biodegradable polymer/bioactive glass composites. The general degradation behaviour of composite polymer/inorganic phase scaffolds is presented and a specific example is discussed, e.g. poly(D,L lactide) (PDLLA)/bioactive glass composite, based on recent results on a long-term (600 days) degradation study in simulated body fluid. Remaining areas of research in this field are indicated, highlighting the need for appropriate characterisation techniques coupled with predictive analytical models

A.R. Boccaccini (✉) • X. Chatzistavrou
Department of Materials Science and Engineering, Institute of Biomaterials,
University of Erlangen-Nuremberg, 91058 Erlangen, Germany
e-mail: aldo.boccaccini@ww.uni-erlangen.de

J.J. Blaker
Department of Chemical Engineering, Polymer and Composite Engineering (PaCE) Group,
Imperial College London, London SW7 2AZ, UK

S.N. Nazhat
Department of Mining and Materials Engineering, McGill University, Montreal,
Quebec, Canada H3A 2B2

N. Eliaz (ed.), *Degradation of Implant Materials*, DOI 10.1007/978-1-4614-3942-4_6,

and the requirement for accurate characterisation of the interface between the polymer matrix and the inorganic fillers in order to ascertain the scaffold degradation behaviour *in vitro* and *in vivo*.

1 Introduction

Tissue engineering (TE) is a discipline based on principles of biology, medicine, and engineering aiming at the development of functional constructs comprising cells, biologically active molecules, and biomaterials in the form of porous substrates, scaffolds, or hydrogels to regenerate living tissue where loss or damage has occurred due to injury or disease [1]. The scientific challenge in tissue engineering requires understanding cell behaviour, including cell mass transport requirements, and the physiological environment, e.g. the extracellular matrix (ECM), in conjunction with the development of optimal scaffolds, which are usually porous and biodegradable and act as temporary three-dimensional (3D) templates for cell adhesion, proliferation, migration, and ultimately support the formation of new tissue [2]. The structure and properties of these scaffolds are related to the type of tissue concerned and the required mechanical competence for the actual *in vivo* application [2, 3].

The degradation behaviour of porous scaffolds plays an important role in the process of formation of new tissue because cell viability, proliferation, and host tissue response are affected [4, 5]. Once implanted in the body, a porous scaffold should maintain its mechanical properties and sufficient structural integrity until the cells adapt to deposit their own ECM. After the scaffold has provided this supporting function, it should completely degrade and be absorbed by the body, ideally leaving no material residuals. In this regard, the optimal *in vivo* degradation rate should be similar or slightly lower than the rate of tissue formation. Eventually, the three-dimensional space occupied by the porous scaffold is replaced by newly formed tissue.

In the case of bone tissue engineering, the interplay of scaffold porosity, (time-dependant) mechanical properties, and *in vitro* and *in vivo* degradability is key to the success of the scaffold-based TE approach. In this context, porosity is an important parameter defining the properties and suitability of scaffolds for the intended TE applications [6]. For example, scaffolds with higher porosity (>80%) have been shown to promote more tissue ingrowth and neo-tissue formation [7], while the interconnection between macropores (opening size, density, and pathway) is important for cellular activity and bone tissue formation. In addition, scaffolds with suitably open pore structures favour cell and tissue penetration, blood vessel growth, and new bone formation [8].

Synthetic biodegradable polymers are the most widely considered materials for the development of tissue engineering scaffolds. In particular polyesters, such as poly(lactic-acid) (PLA), poly(glycolic-acid) (PGA), and their copolymer poly (lactid-co-glycolic-acid) (PLGA), have been investigated extensively for the regeneration of numerous tissues [9–17]. In several applications, particularly in bone TE, these polymers are combined with inorganic particles or fibres, as fillers or coatings, to develop bioactive composite scaffolds [18, 19]. The present chapter will examine the development of such bioactive and biodegradable scaffolds for bone tissue engineering, focusing on aspects of their mechanical properties and degradation behaviour *in vitro* and *in vivo*. The chapter is organised in the following manner. Section 2 includes general background on biodegradable polymers with focus on the fabrication and properties of scaffolds made from polyesters. Section 3 presents a complete overview of the composite approach to bone tissue engineering, including the development of biodegradable polymer/bioactive glass scaffolds, while Section 4 discusses the degradation of composite polymer/inorganic phase scaffolds. As a case in point, we discuss recent results on a long-term (600 days) degradation study in simulated body fluid of poly(D, L-lactide) (PDLLA)/bioactive glass scaffolds which are being developed for bone tissue engineering.

2 Biodegradable Synthetic Polymer Scaffolds

2.1 General Aspects

Certain biodegradable polymers offer the advantage of being able to be eliminated from the body after fulfilling their intended purpose. As mentioned earlier, highly relevant biodegradable polymers in tissue engineering applications are aliphatic polyesters, such as the family of poly(–hydroxy esters) including PLA, PGA, their copolymers (PLGA), polycaprolactone (PCL), polyhydroxyalkanoates (PHA) [20], polyethylene oxide (PEO), and blends thereof. These biodegradable polymers are of interest because of their biocompatibility and biodegradability, their safety for clinical use, e.g. as suture materials, and the versatility that they offer for producing highly porous scaffolds with different geometries to meet the needs of specific tissue engineering applications [7–20]. The degradation of aliphatic polyesters occurs mainly by hydrolysis, and it is therefore anticipated that the degradation behaviour will not vary from patient to patient [21]. The degradation rate of scaffolds made of aliphatic polyesters can be modulated over a wide range by tailoring the composition, molecular weight, end groups, geometry, and pore wall thickness of the scaffold [9, 22]. Possible risks such as toxicity, immunogenicity, and favouring of infections are lower for synthetic polymers such as PLA, PGA, and PLGA with constituent monomeric units having a well known and simple structure.

2.2 Fabrication of Biodegradable Polymer Scaffolds

A multitude of methods are available for the fabrication of porous scaffolds from biodegradable polymers, as comprehensively discussed in the literature [23, 24], including solid free form fabrication techniques (e.g. fused deposition modelling, 3D printing, selective laser sintering), templating techniques, porogen removal, and gas blowing techniques. A conventional method that has been widely used is solvent-casting/particulate leaching [25]. However, this technique has limited control over pore shape and inter-pore connectivity. Porous structures can also be introduced via phase separation routes, with or without additional porogens. A common approach to produce highly porous scaffolds is thermally induced phase separation (TIPS) [26, 27]. Whereby, a polymer-solvent solution is frozen to induce solvent crystallisation. Separation of the polymer and solvent occurs followed by removal of the solvent. Monolithic structures and porous spheres with controlled structures can be formed using these routes [28]. The porous structure is controlled by the polymer-to-solvent ratio, solvent choice, freeze temperature, freezing rate, thermal transfer, and freezing direction.

Phase separation can be used in combination with other scaffold fabrication techniques such as porogen leaching and solid freeform fabrication. These combined techniques provide better control on the resultant porous architectures and enable the development of hierarchical structures with pore sizes in the macro-, micro-, and nanometric scales [29, 30]. The hierarchical pore structure can be tailored to mimic the natural ECM more closely and, thus, it can enhance tissue regeneration. A typical pore structure of a PDLLA scaffold fabricated by TIPS is shown in Fig. 6.1a [31]. Tubular pores, evident from the scanning electron microscope (SEM) image (Fig. 6.1a), are due to the preferential solvent crystallisation (or freezing direction). Figure 6.1b [32] shows the pore walls of a PDLLA TIPS scaffold in the as-fabricated condition; the thickness is seen to vary from ~1 to 10 μm at junctions of the pore walls.

The phase separation and particle leaching methods have been combined to generate macroporous and nanofibrous 3D polymer scaffolds [33–35]. PLA matrices exhibiting macropores have been prepared by combining phase separation and porogen leaching techniques using PLLA tetrahydrofuran (THF) solution with sugar or salt particles of desired sizes [33]. In order to prepare scaffolds with predesigned pore structures, the porogen components were arranged in the desired architecture using pre-determined assemblies. Some examples of porogen used are uniaxially oriented and orthogonal sugar fibres, helicoidal sugar fibre assembly, and multi-layer sugar disc and particle combinations [34]. Scaffolds developed by this technique exhibit high (interconnected) porosities (>90% porous). The macropore size is determined by the porogen particle size while the interfibre distance was dictated by the concentration of the polymer solution [34]. The sugar fibre diameter is controlled by the phase-separation temperature, solvent used, and processing variables [35]. Recently, combined TIPS and ice microsphere porogen/templating has been used to line pore walls with bacterial cellulose nanofibrils, imparting nanotopology at the scaffold surface [36].

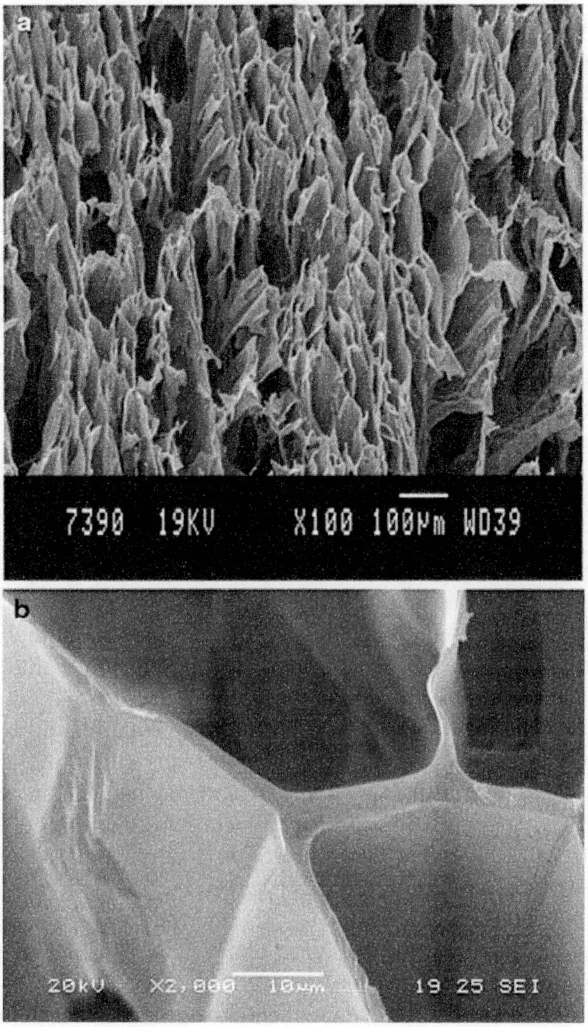

Fig. 6.1 SEM images of PDLLA scaffolds fabricated by TIPS. (**a**) Low-magnification image showing oriented macroporosity [31], (**b**) high-magnification image showing the walls of pores [32] (images reproduced with permission of Elsevier, TBC)

2.3 Biodegradability

Biodegradable polyester degradation occurs by uptake of water followed by hydrolysis of the ester bonds. This section will not present a detailed discussion of the mechanisms of degradation of polyesters, as this topic has been treated comprehensively in the literature (see, for example, [37, 38]). Only aspects of biodegradability related to the application of the materials in tissue engineering are included. Many factors affect the degradation kinetics of these types of polymer scaffolds, such as chemical composition and structure, processing history, molar mass (Mw), polydispersity (Mw/Mn), environmental conditions, crystallinity, device size, morphology (e.g. porosity) and chain orientation, distribution of chemically reactive compounds within the matrix, additives [26, 39–41], presence of original monomers and oligomers, overall hydrophilicity, and environmental conditions (medium, temperature, and pH) [9–18]. Autocatalysis is often associated with scaffolds produced from aliphatic polyesters. Thick samples of these polymers (>100 μm) exhibit heterogeneous degradation, i.e. faster degradation inside than at the exterior of the specimens [26, 41]. Heterogeneous degradation may be related to two phenomena (1) easier diffusion of soluble oligomers from the surface into the external medium than from inside, and (2) neutralisation of carboxylic end-groups located at the surface by the external buffer solution (*in vitro* or *in vivo*). These two mechanisms contribute to reducing the acidity at the surface whereas, in the bulk, degradation rate is enhanced by autocatalysis due to carboxylic end groups. The hydrolysis of amorphous polymers, e.g. PDLLA, is faster than that of partially crystalline polymers due to the lack of crystalline regions and, thus, enhanced water uptake. In general, the amount of absorbed water depends on diffusion coefficients of chain fragments within the polymer matrix, temperature, buffering capacity, pH, ionic strength, additions in the matrix, additions to the medium, and processing history. Different aliphatic polyesters therefore exhibit very different degradation kinetics in aqueous solutions. PGA, for example, is a stronger acid and is more hydrophilic than PLA, which is more hydrophobic due to its methyl groups. The human body contains highly regulated mechanisms for removing monomeric components of lactic and glycolic acids. Based on these favourable properties, PLA and PGA are used in numerous biomedical products, such as degradable sutures and fixation plates, which have approval by the US Food and Drug Administration (FDA). Moreover, PLA and PGA are readily fabricated into porous structures (scaffolds) [9–17], and their degradation rates and physical and mechanical properties are adjustable over a wide range by using various molecular weights and copolymers [42]. As a disadvantage, the degradation of some of these polymers has been associated with inflammatory response, due to the abrupt release of acidic degradation products caused by autocatalysis [43]. Autocatalysis can be counteracted by increasing surface area and reducing pore walls thickness, and by incorporating degradable inorganic fillers, e.g. forming composites, as discussed further below.

It is known that stereochemistry influences the final properties of polyesters. PLA composed of L-lactic repeating units can take >5 years to completely degrade, whereas amorphous PLA (PDLLA) has been reported to degrade faster, in some instances ~1 year [44], and PCL can take several years to degrade *in vivo* [3, 45]. The composition of chains (i.e. contents in L-LA and D-LA and/or GA units) determines the degradation rate of PLGA polymers. It has been found that blends containing the greatest amount of PGA degrade faster [46]. In general, the initial degree of crystallinity of polyesters affects the rate of hydrolytic degradation as crystal segments reduce water permeation in the matrix. Of particular significance for applications in tissue engineering, it has been suggested that debris and crystalline by-products [43], as well as acid degradation products of PLA, PGA, PCL, and their copolymers may be implicated in adverse tissue reactions [3, 47, 48].

In the typical degradation process of polyesters, polymer molecular weight is found to decrease exponentially over the entire degradation process, indicating almost simultaneous degradation on the surface and in the interior of the materials [49]. However, the degradation rate of a porous scaffold is slower than that of a solid block polymer with the same formulation [50], because the acidic degradation products accumulated in the bulk of solid block polymers are difficult to release, causing an autocatalytic effect. However, if the thickness or diameter of the solid block polymer is close to the wall thickness of the porous scaffold, its degradation rate will be similar to that of the porous scaffold [51].

3 Bioactive and Biodegradable Composite Scaffolds

3.1 General Aspects

The combination of degradable polymers and inorganic bioactive particles represents an attractive approach to achieve scaffolds with suitable mechanical and biological properties for bone tissue engineering [18, 19, 52]. The challenge here is forming the aforementioned complex interconnected pore structure with the addition of inorganic phases, which are generally >2 times denser than the polymer, and to disperse the reinforcement adequately. The scaffold surface characteristics (e.g. chemistry and topography) must also be tuned as they determine the suitability of the scaffold for cell attachment and proliferation.

Totally synthetic biodegradable polymer/inorganic bioactive phase composites are particularly interesting as bone tissue engineering scaffold systems due to their flexibility in processing and scope for tailoring structural properties, bioactive behaviour, and biodegradation kinetics. Within this family of composites, those based on polyesters with addition of inorganic bioactive phases such as hydroxyapatite (HA) or bioactive glasses, in the form of particles or fibres, are increasingly being considered [18, 19, 52–54]. In designing these composites, the particulate bioactive phase can be incorporated either as a filler or coating (or both) into the biodegradable polymer matrix. Early developments were based on HA or calcium phosphate particulates in PLA [55], but more recently work has concentrated on bioactive glasses as the added inorganic phase [18].

3.2 The Composites Approach to Tissue Engineering Scaffolds

There are several reasons to combine biodegradable polymers and bioactive inorganic materials for bone tissue engineering, as listed in Table 6.1 and discussed in this section. Being similar to the major inorganic component of natural bone, the addition of HA induces osteoconductivity while the polymer component provides the continuous structure and design flexibility to achieve the high porosity and surface area necessary for anchorage-dependent cells such as bone cells to adhere

Table 6.1 Reasons for combining biodegradable polymers and inorganic bioactive materials for bone tissue engineering scaffolds [18]

A (nanoscale) inorganic bioactive phase ...
Will improve osteoconductivity of a scaffold, enabling the formation of hydroxyapatite (HA) on the surface and a site for bone regrowth
Might be used to control the degradation rate of the polymer and to counteract its acidic degradation
Enhances mechanical properties of the scaffold (composites approach)
Possesses the ability to simulate the surface and/or chemical properties of bone

and differentiate. By standard blending and phase separation techniques, polymer/HA composite scaffolds have been developed with improved mechanical properties and osteoconductivity [56]. These HA-containing scaffolds exhibit improved osteoblastic cell seeding ability and significantly enhanced expression of osteocalcin and other relevant proteins over plain polymer scaffolds. Moreover, bone tissue formation throughout the scaffold has been demonstrated [57]. It was also found that the presence of nanoscale HA particles in a composite scaffold has the ability to promote more effectively apatite deposition in comparison to conventional composite scaffolds, and therefore greater amounts of apatite particles are formed on nanocomposite scaffolds [30]. The addition of bioactive silicate glass particles to a biodegradable polymer has primarily the same goal, i.e. to impart bone bonding ability to the scaffold [18, 45, 52]. It is well known that bioactive glasses develop a carbonated hydroxyapatite surface layer in contact with relevant body fluids, which is pivotal in developing the strong bond to bone [58]. Certain silicate glasses belong to the group of Class A bioactive materials which are characterised by both osteoconduction (i.e. growth of bone at the implant surface) and osteoinduction (i.e. activation and recruitment of osteoprogenitor cells by the material itself, stimulating bone growth) [58–60]. Moreover, in silicate bioactive glasses, such as 45S5 Bioglass®, surface reactions induce the release and exchange of critical concentrations of soluble Si, Ca, P, and Na ions, which lead to favourable intracellular and extracellular responses to promote rapid bone formation [61]. Previous work has indicated also the angiogenic potential of bioactive glasses [62], confirming that these materials are suitable candidates for incorporation into polymer matrices for bone tissue engineering. In this context, the bioactivity induced by bioactive inclusions in biodegradable polymers can be controlled by the volume fraction, size, and shape of the inclusions: increased volume fraction and higher surface area to volume ratio of inclusions, as well as incorporation of nanoscale particles, favour higher bioactivity [18, 63].

The incorporation of basic compounds like bioactive silicate glasses or calcium phosphate ceramics in polyester scaffolds has been also suggested as a means to stabilise the pH of the environment surrounding the polymer and to control its degradation rate [43]. This effect is achieved by allowing rapid exchange of protons in water while the scaffold is degrading. For example, the addition of 45S5 Bioglass® (of composition in wt.%: 45 SiO_2, 24.5 Na_2O, 24.5 CaO, and 6 P_2O_5 [58]) has been shown to increase water absorption compared to neat PDLLA [26, 64] and PLGA [65] foams. Thus, an additional reason for the development of polyester/bioactive glass composites is the possibility of changing and controlling the scaffold degradation kinetics [18, 54, 66, 67].

The use of composites also has advantages in terms of tailoring mechanical properties. While polymers can be processed into complex shapes and structures, they do not usually provide strong bonding to bone and are too weak and flexible. They are therefore unable to meet, in several cases, the mechanical demands associated with applications in bone engineering. On the other hand, bioactive ceramics and glasses are brittle and exhibit poor fracture resistance, being therefore at a disadvantage as scaffold materials when used on their own. However, combination of polymers and inorganic phases leads to composite materials with improved mechanical properties (e.g. stiffness, compression strength), exploiting the "composites approach" [68], i.e. incorporating stiffer and strong inorganic particles or fibres in the softer and flexible polymer matrix. The development of composites offers in this regard an extra degree of flexibility in that composite properties can be tailored to suit the mechanical and physiological demands by controlling not only the pore structure of the polymer matrix, but also the volume fraction, local arrangement, orientation, and shape of the reinforcing inorganic phase [18, 69]. In general, two types of reinforcements can be used for composite scaffolds: fibres and particles. It is known that increased volume fraction and higher surface area to volume ratio of inclusions favour bioactivity [18]. Incorporation of fibres instead of particles is being also investigated [70]. A drawback of fibres is that they are not readily available (e.g. in bioactive silicate or calcium phosphate compositions), and fibre-containing scaffolds are more difficult to manufacture than scaffolds incorporating particles. The mechanical properties of highly porous scaffolds are influenced by the reinforcement shape and size, the distribution of the reinforcement in the matrix, and the reinforcement-matrix interfacial bonding. However, due to extremely high porosity of scaffolds (>90%), the mechanical properties are dominated by porosity, including parameters such as pore volume, size, shape, orientation, and connectivity. Regarding scaffold mechanical properties in comparison to human bone, in general porous polymer/inorganic filler composite scaffolds exhibit insufficient mechanical integrity [18]. According to composites theory [68], efficient load transfer at the filler/matrix interface must be achieved for the reinforcements to strengthen and stiffen the matrix. Thus, the quality of the interfacial bonding between the two phases (filler and matrix) is a key requirement. As discussed elsewhere [18], the mechanical strength of most of today's available porous polymer/bioactive glass composite scaffolds is insufficient for direct bone substitution because they are at least one order of magnitude weaker than natural cancellous bone, and orders of magnitude weaker than cortical bone. Insufficient particle–matrix bonding strength may be a possible reason for this behaviour. Thus, key issues have been identified which must be further investigated to effectively improve the mechanical properties of biodegradable polyesters by adding inorganic particles as fillers. These include (1) improvement of the interfacial bonding; (2) achievement of a proper, homogeneous dispersion of the individual particles in the matrix; (3) confining the reinforcement phase to the polymer matrix such that the stiff phase is not situated within the pores (voids), providing little reinforcement, and (4) techniques to align staggered and aligned reinforcement phases of high aspect ratios in an effort to realise truly hierarchical composites [71]. This last issue can be tackled by suitable particle surface functionalisation and use of nanoparticles, as reviewed recently [63].

To conclude this section, it should be pointed out that numerous studies have been carried out to investigate the cellular response to inorganic particle-containing biodegradable polymer composites *in vitro* and *in vivo*. A previous review has summarised relevant studies focusing on the cellular interactions within polyester/bioactive glass composites [72].

3.3 Fabrication Techniques of Porous Degradable and Bioactive Composites

Many studies have been carried out in the last 10 years in the field of biodegradable and bioactive composite scaffolds for bone tissue engineering. Material combinations, bioactive properties, degradation characteristics, *in vitro* and *in vivo* behaviour, and mechanical properties have been investigated [18, 19, 72–76].

Numerous techniques have been used for the preparation of porous biodegradable and bioactive composite scaffolds, including TIPS, compression moulding and particulate leaching, gas foaming, sintering of composite microspheres, and solid free form fabrication [18, 19, 23, 24, 73]. Table 6.2 gives an overview of typical composites incorporating bioactive glass particles (micro- and nano-sized) fabricated by different methods. It is observed that TIPS has been one of the most popular methods used for developing polyester/BG composite scaffolds [73]. These scaffolds are highly porous, with anisotropic tubular morphology and high pore interconnectivity. Figure 6.2a,b shows SEM micrographs of a TIPS derived PDLLA/10wt.% Bioglass® scaffold in sections parallel and perpendicular to the pore longitudinal direction [31]. The pore structure is identical to that of neat PDLLA foams (Fig. 6.1a). The microporosity of TIPS-produced foams, as well as pore morphology, mechanical properties, bioactivity, and degradation rate, can be controlled by changing the polymer concentration in solution, volume fraction of the reinforcing phase, quenching temperature, and the polymer and solvent used [18, 65]. The technique has been adapted to produce porous composite spheres, which may be viewed as an inverse template to bone, and the pore structure between the spheres forms the continuous space for new bone deposition. Interestingly, these spheres shrink as they degrade, and ultimately their initially well-defined radially orientated pore structure transforms into blister-like pores due to autocatalysis [76].

The pore volume of scaffolds made by TIPS decreases with addition of Bioglass® particles (40 wt.%), with little change observed in the overall pore morphology. TIPS-fabricated PDLLA foams, with or without Bioglass® additions,

Table 6.2 Overview of studies performed on bioactive glass containing composite scaffolds for bone tissue engineering

Bioactive glass	wt.%	Particle size	Matrix	Fabrication technique/ process	References
45S5 m-BG	5, 29, 40	<40 μm	PDLLA	Co-extrusion + compaction; TIPS	[77]
45S5 m-BG	4.8, 28.6	5–20 μm	PDLLA	TIPS	[78]
45S5 m-BG	10	<5 μm	P(3HB)	ST/PL	[79]
45S5 n-BG	10	30 nm	P(3HB)	ST/PL	[79]
S53P4 m-BG	20, 50	90–315 μm	P(CL/DLLA)	ST/PL	[53]
S53P4 m-BG	30	<45 μm	P(CL/DLLA)	ST/PL	[80]
45S5 m-BG	10, 30	<40 μm	PLGA	Microsphere emulsification	[59]
45S5 m-BG	10	4 μm	PDLG	TIPS	[76]
45S5 m-BG	25, 50	50–63 μm	PLA	Freeze extraction technique	[81]
45S5 m-BG	5, 40	>90 μm	PDLLA	Solvent casting	[74]
45S5 m-BG	10, 25, 50	<5 μm	PDLLA	TIPS	[64]
45S5 m-BG	10, 25, 50	<5 μm	PLGA	TIPS	[64]
45S5 m-BG	5, 10, 40	<5 μm	PDLLA	TIPS	[26]
45S5 m-BG	5, 40	<5 μm	PDLLA	TIPS	[82]
45S5 m-BG	10, 25, 50	<5 μm	PLGA	TIPS	[65]
45S5 m-BG	25	<40 μm	PLGA	Solvent casting	[52]
45S5 m-BG	20	<5 μm	P(3HB)	Solvent casting	[83]
45S5 n-BG	10, 20	29 nm	P(3HB)	Solvent casting	[20]
45S5 m-BG	10, 20, 30	<5 μm	P(3HB)	Solvent casting	[84]
45S5 n-BG	10, 20, 30	30–50 nm	P(3HB)	Solvent casting	[84]
45S5 m-BG	5, 30	5 μm	PDLLA	TIPS	[85]
45S5 m-BG	5, 30	5 μm	PDLLA	TIPS	[86]

PDLLA, poly(D,L lactide); P(3HB), poly(3-hydroxybutyrate); P(CL/DLLA), poly(ε-caprolactone/D, L-lactide); PLGA, poly(lactic-co-glycolic acid); PDLG, poly(D, L-lactide-co-glycolide); PLA, poly(L-lactide); S53P4, 53 wt.% SiO_2, 23 wt.% Na_2O, 20 wt.% CaO, 4 wt.% P_2O_5; m-BG, micro-sized bioactive glass; n-BG, nano-sized bioactive glass. ST/PL, sugar template/particulate leaching; TIPS, thermally induced phase separation (adapted from [73])

have been shown to exhibit mechanical anisotropy which is related to the TIPS-induced pore architecture (Fig. 6.1a) [41, 65]. However, the inclusion of stiff inorganic bioactive phases was seen to give only slight improvement in the Young's modulus and compression strength of scaffolds [78]. Maquet et al. [64] have reported highly porous (porosity >90%) PDLLA and PLGA scaffolds, containing 50 wt.% Bioglass®, which exhibit compressive moduli of about 21 MPa and 26 MPa, respectively, i.e. 1.5–2.5 times higher than those for the pure polymer scaffolds.

Microsphere sintering is an alternative technique for producing 3D composites of bioactive glass and degradable polymers [52]. Composite PLGA-Bioglass®

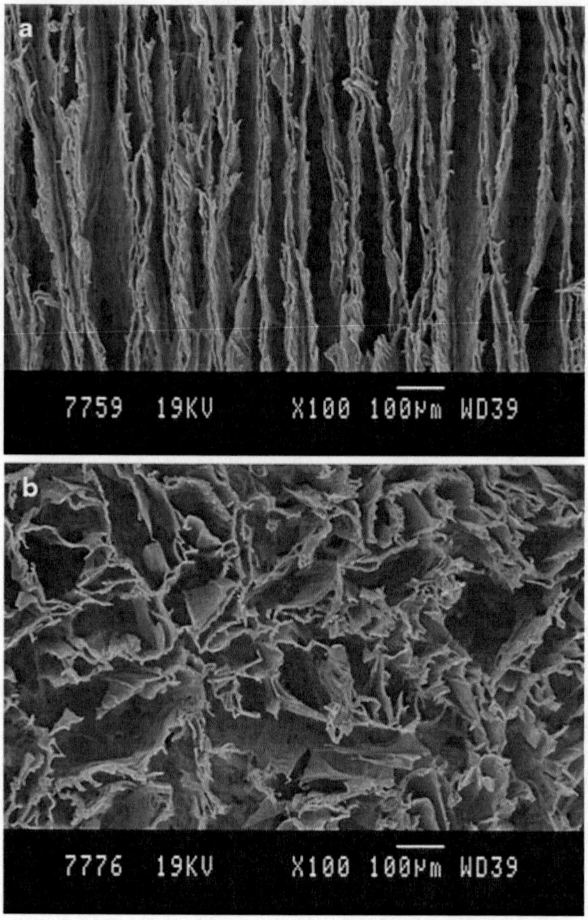

Fig. 6.2 SEM micrographs showing the pore structure of a TIPS-derived PDLLA/10 wt.% Bioglass® scaffold, in sections (**a**) parallel, and (**b**) perpendicular to the longitudinal pore direction [31] (reproduced with permission of Elsevier, TBC)

microspheres obtained through a double emulsion (water-in-oil-in-water) technique were used, and interconnected porous structures were obtained by sintering the microspheres. The average porosity was, however, relatively low (40%), with pore diameters of 90 μm. These mechanical properties were close to those of cancellous bone. The composites were bioactive and were seen to support the adhesion, growth, and mineralisation of human osteoblast-like cells *in vitro*. Porous PLGA microspheres produced via TIPS have also been investigated *in vivo* in adult Wistar rats [87].

CO_2 gas-foaming procedures have been used to create silica/PDLLA composites [77] with relatively low porosities (40–55%). Moreover, composites incorporating calcium phosphate particles [79] or phosphate-based glasses [88] and exhibiting variable and graded porosity have been produced by gas foaming methods. Such composites can be fine-tuned to meet the specific requirements of the host site. Biodegradable polyester matrix composites containing silicate phases other than 45S5 Bioglass® have been also investigated, for example PDLLA with wollastonite [80].

4 Degradation of Composite Polymer/Inorganic Phase Scaffolds

4.1 General Aspects

The degradation behaviour of composite polymer/inorganic phase scaffolds has been investigated in relation to the application of the scaffolds in bone tissue engineering [18]. This section will concentrate on the family of polyester (e.g. PDLLA) composites incorporating bioactive glass as inclusion phase. Degradation in both phosphate-buffered saline (PBS) and simulated body fluid (SBF) is considered. *In vitro* studies in PBS at 37°C have shown that addition of Bioglass® to PDLLA foams increased water absorption and weight loss in comparison to pure polymer foams in the early stages of degradation (~1 week); the massive weight loss is associated with glass particle dissolution, erosion, and particle loss [26, 64]. The molecular weight was found to decrease less in the composite foams, which can be explained by the effect of the dissolution of alkaline ions from the Bioglass®, providing a pH buffering effect, facilitating water ingress and diffusion thus allowing a more pronounced surface erosion mechanism to occur. In early experiments, the degradation behaviour was investigated up to relatively short periods of time, e.g. 16 weeks. It was found that both PDLLA/Bioglass® composites and neat PDLLA foams retained the structural integrities until the end of the experiment (16 weeks) and that the degradation was still in the early stages [28]. There is a lack of in-depth understanding regarding the long-term *in vitro* characterisation of this type of biodegradable composites, specifically regarding the long-term effect of the incorporation of inorganic bioactive phases on degradation and ion release kinetics of highly porous systems. As a starting point to remedy this lack of knowledge and reliable data, in the following section we summarise recently published experimental data [32] about the long-term degradation (600 days) of PDLLA/Bioglass® foams, presenting results relevant for the application of the scaffolds in bone tissue engineering.

4.2 Case In Point: PDLLA/Bioglass® Scaffolds: Long-Term Degradation Study

4.2.1 Relevance of the Study and Experimental Approach

For the intended application of Bioglass®-filled poly(α-hydroxyester) composite scaffolds in bone tissue engineering, it is essential that the scaffold degradation behaviour is such that the pore structure is maintained during cell infiltration and tissue growth, in addition to maintaining sufficient structural integrity *in vivo.* Scaffold degradation is a key determinant of its performance, as the degradation rate will affect cell viability and host tissue response. It has been reported that long-lasting implants based on poly(α-hydroxesters) can lead to delayed inflammatory response at implantation sites, due to fibrous encapsulation and abrupt release of acidic degradation products [12]. In general, the scaffold degradation rate can be tuned by altering the composition and porosity of the scaffold structure. In addition, the hypothesis that the presence of Bioglass® inclusions alters the degradation kinetics of poly(α-hydroxyesters) by preventing or retarding autocatalysis has been investigated, as mentioned earlier and discussed in the literature [32].

The long-term degradation of porous poly(α-hydroxyester) scaffolds and the effect of glycolide content (fabricated by compression moulding and particulate leaching) have been reported [9, 82]. Several studies investigating degradation behaviour of related scaffolds are available, including low *M*w PDLLA foams modified by poly(ethylene oxide)-block-poly(PDLLA) [83] and bioactive glass-filled PLGA of varying matrix compositions (fabricated by high-pressure CO_2 processing) [84]. In this section, results of a recent study [32] of the long-term degradation (to 600 days) and *in vitro* compressive mechanical behaviour of PDLLA/Bioglass® composite scaffolds are summarised. *In vitro* studies were conducted dynamically in SBF, whereby the fluid was exchanged at selected time points to better represent the *in vivo* environment.

In the investigated composites, Bioglass® particles had an average particle size of ~5 μm, surface area of 2.7 $m^2\,g^{-1}$, and were of irregular shape. PDLLA foams of 93–94% porosity, filled with 5 and 30 wt.% Bioglass®, referred to as 5BG and 30BG, respectively, were produced by TIPS, as described in the literature [64]. SBF was prepared following the protocol introduced by Kokubo et al. and investigated previously [20]. Foams were immersed individually in SBF, for various periods up to 600 days. It is vital in any degradation study to infiltrate the entire scaffold with fluid; vacuum infiltration methods have been described to this aim [32]. Samples were evaluated in terms of morphological variation, weight change, thermal properties, molecular weight change, and mechanical properties in the wet and dry states, with respect to degradation time [32].

4.2.2 Long-Term Degradation of PDLLA/Bioglass® Composites: Structural Changes

By using the immersion apparatus schemed in Fig. 6.3 [32], complete soaking of the scaffolds with negligible effect to the pore structure was achieved. This is essential for the correct determination of the degradation behaviour, because the foams must be completely wetted, devoid of gas entrapments persisting in the pores; the entire scaffold is then in contact with the SBF solution for accurate measurements. Typical results of weight loss as a function of immersion time in SBF are shown in Fig. 6.4. The mass of the PDLLA scaffolds remained relatively constant for the duration of the study, losing ~9% at 365 days, which increased to ~50% at 600 days. In contrast, scaffolds with Bioglass® inclusions exhibited a rapid weight loss during the first 24 h of incubation (e.g. 12 ± 3% for BG30

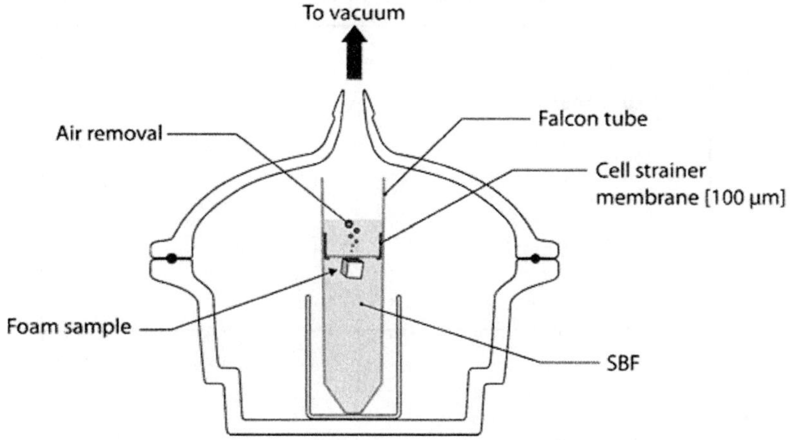

Fig. 6.3 Schematic diagram of the immersion set-up used to fully expose the interior of the relatively hydrophobic foam scaffold to simulated body fluid; for more details, see [32]

foams). At 600 days, the weight loss was ~30%. This is consistent with previous findings: [26, 64] composite foams have a tendency to exhibit rapid then stable weight loss profiles, due to particle loss, particle dissolution, and erosion of the scaffold due to weaknesses in the matrix, possibly caused by particle agglomerations and handling. In these PDLLA/Bioglass® scaffolds, early weight loss occurred, but significant weight loss and erosion were apparent by 600 days, even for neat PDLLA foams.

The dimensional stability of scaffolds was also investigated [32]; it was shown that little change in the height of the scaffolds during the first 70 days occurred. However, scaffolds underwent significant change in dimensions at 365 days, with values of volume reduction of ~70% and height reduction to ~90% [32]. Images of the foams at 600 days are shown in Fig. 6.5: scaffolds appeared whiter once dried (than in their initial, as-fabricated state), and were difficult to handle; they had undergone shrinkage and distortion. The action of water absorption and subsequent plasticisation of the matrix was discussed [32], and the effect was explained considering that the terminal functions, carboxylic acid, on chains of PDLLA facilitate the penetration of water [84]. The influence of water reduces the glass transition temperature (Tg) by up to 12 °C [85], and it was found that the number and weight average molecular weight values of scaffolds decreased exponentially with time throughout the degradation period [32]. The results are in agreement with the findings of Orava et al. [83] and Maquet et al. [26].

SEM observations revealed that the scaffolds' pore structure did not change significantly up to 365 days of immersion in SBF. At 600 days immersion in SBF, the foams were markedly distorted and the porous structure severely altered (Fig. 6.6). The blisters appeared to have originated at the thickest sections of the pore walls, the pore structure collapsed, and it was distorted at the micro-level; this high distortion and blockage of macropores was found not to be affected by

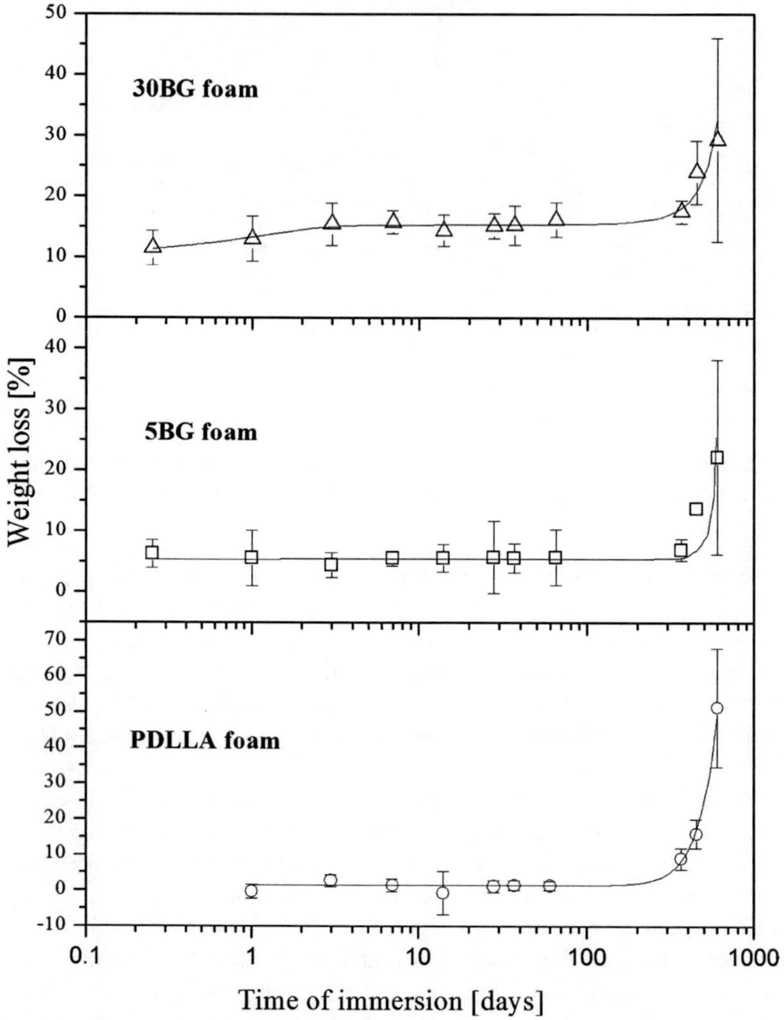

Fig. 6.4 Weight loss of PDLLA, 5BG, and 30BG TIPS foams as a function of immersion time in SBF. The *lines* are used as guide to the eye [32]

Bioglass® content [32]. The relatively thin wall sections (<1–10 μm) of the TIPS-produced foams (e.g. Fig. 6.1b) may retard the onset of autocatalysis, which has been shown to manifest itself in the form of blisters and voids, accompanied by sudden mass loss in these types of materials. Voids appear to nucleate from the thickest polymer sections in these porous scaffolds, which is also evident in the literature for PLGA [5, 9] and for PLGA/bioactive glass scaffolds [65].

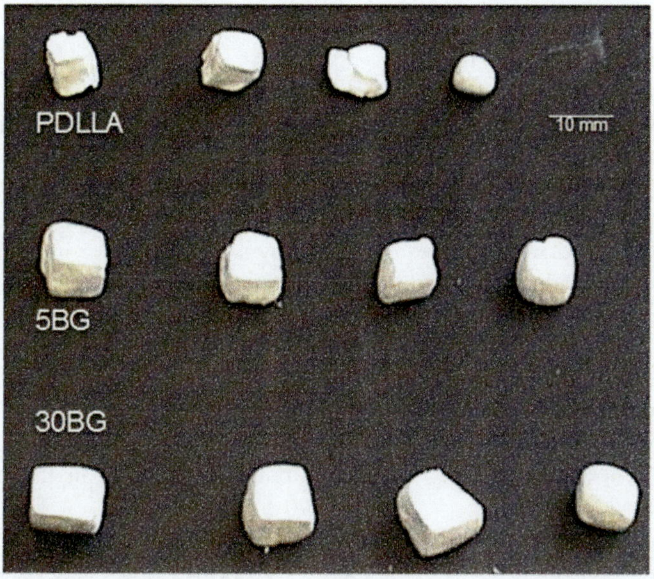

Fig. 6.5 Photograph of PDLLA/Bioglass® scaffolds that have been immersed in SBF for 600 days, post-drying. The scaffolds are in an advanced state of degradation. Distortion of shape and loss of sample mass are more apparent for the neat PDLLA samples (4 scaffolds per sample type are shown) [32] (reproduced with permission of Elsevier, TBC)

Fig. 6.6 SEM micrographs of PDLLA/Bioglass® foam (5 wt.% Bioglass®) after 600 days in SBF in transverse section. The fracture surface indicates embrittlement, the sample fractured on handling, and cutting for SEM imaging (as shown at high magnification, *inset*) [32] (reproduced with permission of Elsevier, TBC)

4.2.3 Long-Term Degradation of PDLLA/Bioglass® Composites: Mechanical Properties

The assessment of the changes in mechanical properties upon degradation in SBF was made by normalising the measured properties for wet and degraded foams to those of the dry foams at time point zero. The results are summarised in Fig. 6.7a, b [32] for normalised compressive modulus and yield strengths, respectively. The yield strength of wet samples decreased by a factor of ~0.5 in comparison to dry samples. This reduction could indicate plasticisation of the matrix and detachment of Bioglass® particles. In the wet state, the mechanical properties were maintained up to 70 days in SBF. The mechanical properties were seen to be most drastically reduced for the 30BG foams, which may be due to the effect of Bioglass® particle loss. Finally, the importance of testing scaffolds in *in vitro* conditions is emphasised, because the effect of water plasticising the polymer and the testing temperature (37 °C) may contribute to reduction in the mechanical properties.

Thus, in the investigated PDLLA/Bioglass® composites, the degradation process seems to proceed in three main stages: [32] (1) the quasi-stable stage, where water absorption and plasticisation occur together with weight loss due to Bioglass® particle loss and dissolution, (2) a second stage of slight increase in the wet mechanical properties and moderate decrease in dimensions, with properties remaining moderately constant until the onset of significant weight loss, and (3) a stage of increased weight loss, distortion of the pore structure, formation of blisters, and embrittlement of the scaffold. PDLLA/Bioglass® scaffolds may be considered for non-load bearing applications in tissue engineering [72–75, 86]. The healing process of bone critical size defects should take 12–16 weeks, and degradation is likely to progress faster *in vivo* than *in vitro* [54]. This is due to enzymatic degradation in conjunction with dynamic loading that the scaffold may undergo *in vivo*, thus causing further erosion and loss of structural integrity. For this application, Bioglass® particles impart bioactivity to the material, but appear less relevant in altering the long-term degradation behaviour of the porous scaffold. In fact, the presence of Bioglass® particles could detrimentally affect the mechanical properties in high-porosity scaffolds if the Bioglass®/PDLLA interfaces are not optimised, as mentioned earlier and discussed in the literature [18].

5 Conclusions

Biodegradable polymer/inorganic phase scaffolds are an important class of materials for bone tissue engineering. While a significant amount of data and knowledge has been generated in the last 10–15 years, and the application of the scaffolds both *in vitro* and *in vivo* has been successfully explored, there are

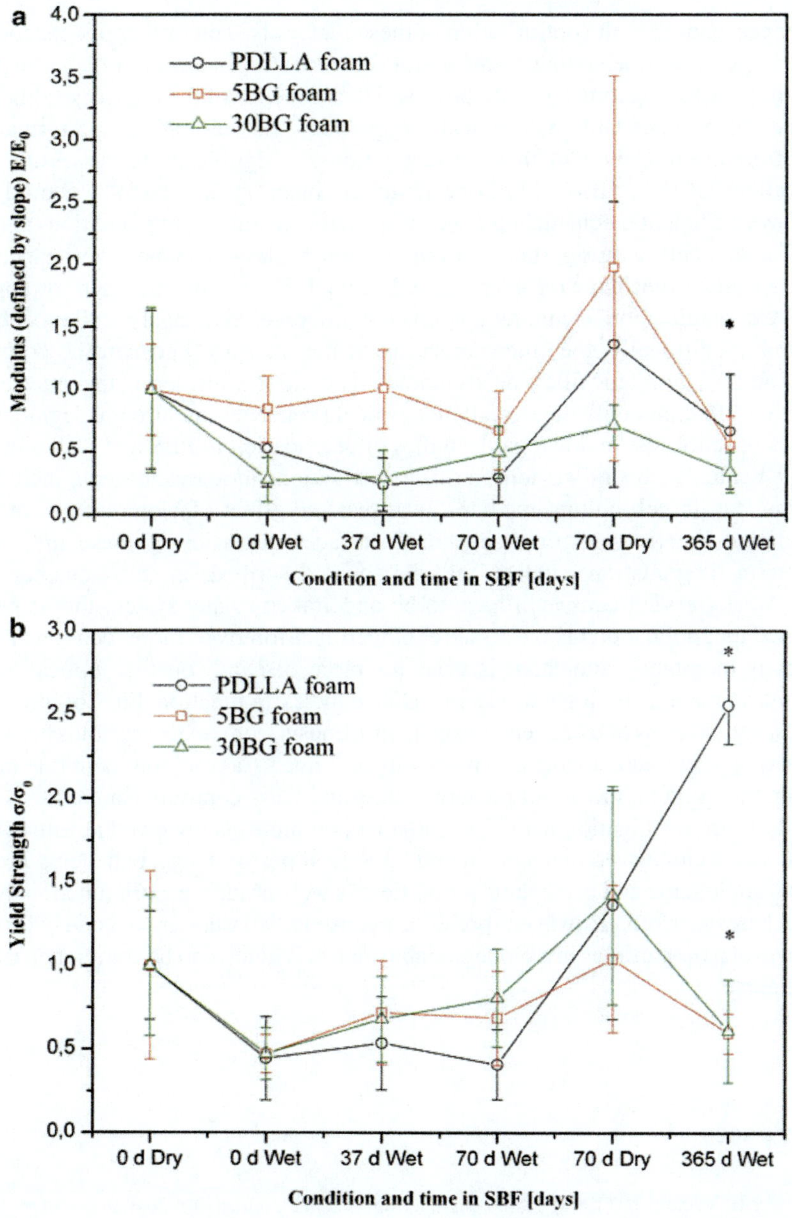

Fig. 6.7 (**a**) Compressive modulus taken from the initial linear portion of the stress vs. strain curve for PDLLA/Bioglass® foams as a function of immersion time in SBF and condition (wet vs. dry), normalised to the modulus of the as-made dry foams; (**b**) compressive yield strength of foams as a function of immersion time in SBF and condition (wet vs. dry), normalised to the yield strength of the as-made dry foams [32]. (The lines connecting the data points are added as aid for the eye)

challenges leading to the optimisation of these materials. The further understanding of the degradation behaviour is mandatory in order to assess the effects of pore structure, scaffolds geometry, fluid flow, and the influence of the bioactive additions on the time-dependant mechanical properties and biological performance. The information gained in those investigations is significant for assessing the applications of the scaffolds in bone tissue engineering, as scaffolds should not only have sufficient mechanical properties to enable handling, implantation (and, in some cases, cell culture), but also they should show adequate resistance to time-dependent mechanical forces exerted *in vivo*. Further development of appropriate characterisation techniques coupled with predictive analytical models is required. Additionally, the characterisation of the interface between the polymer matrix and the inorganic fillers needs increased attention considering that interfacial properties affect not only the overall mechanical properties but also the degradation kinetics of scaffolds. In this regard, studies of degradation in PBS and SBF must be complemented with similar (long-term) investigations in relevant media including proteins, e.g. in cell culture medium, to assess the effect of biomolecules on the degradation kinetics and time-dependant surface properties of these materials. Long-term degradation studies, as the one described in this chapter for PDLLA/Bioglass® foams, will have to be undertaken to any system that is being proposed for application in bone tissue engineering. Moreover, more focus on *in vivo* studies is inevitable, and there is need for more research on the application of scaffolds in realistic biological systems. Due to the extra functionalities of bioactive glasses, e.g. in terms of osteogenesis and angiogenesis, composite scaffolds based on bioactive glass/biodegradable polymers will continue to take a prominent role in the field. In this context, the use of bioactive glass and glass-ceramic nanoparticles and nanofibres, in combination with bioresorbable polymers, as reviewed recently [63], should lead to improved mechanical and biological performance, enhancing proliferation, angiogenic and osteogenic properties, as well as adding extra functionalities to the base scaffold. However, possible cytotoxicity issues associated with the presence of nanoparticles in a biodegradable matrix will have to be comprehensively investigated.

References

1. Langer R, Vacanti JP (1993) Tissue engineering. Science 260(5110):920–926
2. Hutmacher DW (2001) Scaffolds in tissue engineering bone and cartilage. Biomaterials 21 (24):2529–2543
3. Yang S, Leong K, Du Z, Chua C (2001) The design of scaffolds for use in tissue engineering. Part 1. Traditional factors. Tissue Eng 7(6):679–689
4. Babensee JE, Anderson JM, Melntire LV, Mikos AG (1998) Host response to tissue engineered devices. Adv Drug Deliv Rev 33:111–139
5. Holy CE, Dang SM, Davies JE, Shoichet MS (1999) In vitro degradation of a novel poly (lactide-co-glycolide) 75/25 foam. Biomaterials 20:1177–1185

6. Karageorgiou V, Kaplan D (2005) Porosity of 3D biomaterial scaffolds and osteogenesis. Biomaterials 26:5474–5491
7. Roy TD, Simon JL, Ricci JL, Rekow ED, Thompson VP, Parsons JR (2003) Performance of degradable composite bone repair products made via three-dimensional fabrication techniques. J Biomed Mater Res A 66(2):283–291
8. Lu JX, Flautre B, Anselme K, Hardouin P, Gallur A, Descamps M, Thierry B (1999) Role of interconnections in porous bioceramics on bone recolonization in vitro and in vivo. J Mater Sci Mater Med 10(2):111–120
9. Wu L, Ding J (2004) In vitro degradation of three-dimensional porous poly(D, L-lactide-co-glycolide) scaffolds for tissue engineering. Biomaterials 25(27):5821–5830
10. Mikos AG, Thorsen AJ, Czerwonka LA, Bao Y, Langer R, Winslow DN, Vacanti JP (1994) Preparation and characterization of poly(L-lactic acid) foams. Polymer 35(5):1068–1077
11. Borden M, Attawia M, Laurencin CT (2002) The sintered microsphere matrix for bone tissue engineering: in vitro osteoconductivity studies. J Biomed Mater Res 61:421–429
12. Liu X, Ma PX (2004) Polymeric scaffolds for bone tissue engineering. Ann Biomed Eng 32 (3):477–486
13. Agrawal CM, Ray RB (2001) Biodegradable polymeric scaffolds for musculoskeletal tissue engineering. J Biomed Mater Res 55(2):141–150
14. Delabarde C, Plummer CJG, Bourban PE, Månson JAE (2010) Solidification behavior of PLLA/nHA nanocomposites. Compos Sci Technol 70:1813–1819
15. Ara M, Watanabe M, Imai Y (2002) Effect of blending calcium compounds on hydrolytic degradation of poly (DL-lactic acid-co-glycolic acid). Biomaterials 23(12):2479–2483
16. Van Der Meer SAT, De Wijn JR, Wolke JGC (1996) The influence of basic filler materials on the degradation of amorphous D- and L-lactide copolymer. J Mater Sci Mater Med 7(6):359–361
17. Li SM (1999) Hydrolytic degradation characteristics of aliphatic polyesters derived from lactic and glycolic acids. J Biomed Mater Res 48(3):342–353
18. Rezwan K, Chen QZ, Blaker JJ, Boccaccini AR (2006) Biodegradable and bioactive porous polymer/inorganic composite scaffolds for bone tissue engineering. Biomaterials 27:3413–3431
19. Guarino V, Causa F, Ambrosio L (2007) Bioactive scaffolds for bone and ligament tissue. Expert Rev Med Devices 4:405–418
20. Misra SK, Ansari T, Mohn D, Valappil SP, Brunner TJ, Stark WJ, Roy I, Knowles JC, Sibbons PD, Jones EV, Boccaccini AR, Salih V (2010) Effect of nanoparticulate bioactive glass particles on bioactivity and cytocompatibility of poly(3-hydroxybutyrate) composites. J Roy Soc Interface 7:453–465
21. Piskin E (1997) Biomaterials in different forms for tissue engineering: an overview. Mater Sci Forum 250:14–42
22. Griffith LG (2000) Polymeric biomaterials. Acta Mater 48(1):263–277
23. Ma PX, Elisseeff J (2005) Scaffolding in tissue engineering. Taylor and Francis, Boca Raton, FL
24. Boccaccini AR, Blaker JJ (2005) Bioactive composite materials for tissue engineering scaffolds. Expert Rev Med Devices 2:303–317
25. Gross KA, Rodriguez-Lorenzo LM (2004) Biodegradable composite scaffolds with an interconnected spherical network for bone tissue engineering. Biomaterials 25:4955–4962
26. Maquet V, Boccaccini AR, Pravata L, Notingher I, Jerome R (2003) Preparation, characterisation, and in vitro degradation of bioresorbable and bioactive composites based on Bioglass®-filled polylactide foams. J Biomed Mater Res 66(2):335–346
27. Ma PX, Zhang RY (2001) Microtubular architecture of biodegradable polymer scaffolds. J Biomed Mater Res 56(4):469–477
28. Blaker JJ, Knowles JC, Day RM (2008) Novel fabrication techniques to produce microspheres by thermally induced phase separation for tissue engineering and drug delivery. Acta Biomater 4(2):264–272

29. Chen VJ, Smith LA, Ma PX (2006) Bone regeneration on computer-designed nano-fibrous scaffolds. Biomaterials 27(21):3973–3979
30. Wei GB, Ma PX (2006) Macro-porous and nano-fibrous polymer scaffolds and polymer/bone-like apatite composite scaffolds generated by sugar spheres. J Biomed Mater Res A 78(2):306–315
31. Blacher S, Maquet V, Jérôme R, Pirard JP, Boccaccini AR (2005) Study of the connectivity properties of Bioglass®-filled polylactide foam scaffolds by image analysis and impedance spectroscopy. Acta Biomater 1:565–574
32. Blaker JJ, Nazhat SN, Maquet V, Boccaccini AR (2011) Long-term in vitro degradation of PDLLA/Bioglass® bone scaffolds in acellular simulated body fluid. Acta Biomater 7:829–840
33. Ma PX, Zhang RY (1999) Synthetic nano-scale fibrous extracellular matrix. J Biomed Mater Res 46(1):60–72
34. Zhang RY, Ma PX (2000) Synthetic nano-fibrillar extracellular matrices with predesigned macroporous architectures. J Biomed Mater Res 52(2):430–438
35. Wei GB, Ma PX (2008) Nanostructured biomaterials for regeneration. Adv Funct Mater 18 (22):3568–3582
36. Blaker JJ, Lee KY, Mantalaris A, Bismarck A (2010) Ice-microsphere templating to produce highly porous nanocomposite PLA matrix scaffolds with pores selectively lined by bacterial cellulose nano-whiskers. Compos Sci Technol 70(13):1879–1888
37. Vert M, Mauduit J, Li S (1994) Biodegradation of PLA/GA polymers: increasing complexity. Biomaterials 15:1209–1213
38. Middleton JC, Tipton AJ (2000) Synthetic biodegradable polymers as orthopedic devices. Biomaterials 21:2335–2346
39. Thomson RC, Yaszemski MJ, Powers JM, Mikos AG (1998) Hydroxyapatite fiber reinforced poly(α-hydroxy ester) foams for bone regeneration. Biomaterials 19(21):1935–1943
40. Zhang K, Wang Y, Hillmyer MA, Francis LF (2004) Processing and properties of porous poly (L-lactide)/bioactive glass composites. Biomaterials 25(13):2489–2500
41. Roether JA, Boccaccini AR, Hench LL, Maquet V, Gautier S, Jerome R (2002) Development and in vitro characterisation of novel bioresorbable and bioactive composite materials based on polylactide foams and Bioglass® for tissue engineering applications. Biomaterials 23(18):3871–3878
42. Mano JF, Sousa RA, Boesel LF, Neves NM, Reis RL (2004) Bioinert, biodegradable and injectable polymeric matrix composites for hard tissue replacement: state of the art and recent developments. Compos Sci Technol 64:789–817
43. Heidemann W, Jeschkeit S, Ruffieux K et al (2001) Degradation of poly(D, L-lactide) implants of calciumphosphates with or without addition in vivo. Biomaterials 22(17):2371–2381
44. Bergsma JE, Rozema FR, Bos RRM, Boering G, Debruijn WC, Pennings AJ (1995) In-vivo degradation and biocompatibility study of in-vitro pre-degraded as-polymerised polylactide particles. Biomaterials 16(4):267–274
45. Rich J, Jaakkola T, Tirri T, Narhi T, Yli-Urpo A, Seppala J (2002) In vitro evaluation of poly (ε-caprolactone-co-D, L-lactide)/bioactive glass composites. Biomaterials 23(10):2143–2150
46. Dunn AS, Campbell PG, Marra KG (2001) The influence of polymer blend composition on the degradation of polymer/hydroxyapatite biomaterials. J Mater Sci Mater Med 12(8):673–677
47. Bergsma EJ, Rozema FR, Bos RRM, de Bruijn WC (1993) Foreign body reaction to resorbable poly(L-lactide) bone plates and screws used for the fixation of unstable zygomatic fractures. Maxillofac Surg 51:51–66
48. Jagur-Grodzinski J (1999) Biomedical application of functional polymers. React Funct Polym 39(2):99–138
49. Park TG (1995) Degradation of poly(lactic-co-glycolic acid) microspheres: effect of copolymer composition. Biomaterials 16:1123–1130
50. Wu XS, Wang N (2001) Synthesis, characterization, biodegradation and drug delivery application of biodegradable lactic/glycolic acid polymers. Part II: Biodegradation. J Biomater Sci Polym Ed 12:21–34

51. Lu L, Peter SJ, LyMan MD, Lai HL, Leite SM, Tamada JA, Uyama S, Vacanti JP, Langer R, Mikos AG (2000) In vitro and in vivo degradation of porous poly(D, L-lactic-co-glycolic acid) foams. Biomaterials 21:1837–1845
52. Lu HH, El Amin SF, Scott KD, Laurencin CT (2003) Three dimensional bioactive, biodegradable, polymer-bioactive glass composite scaffolds with improved mechanical properties support collagen synthesis and mineralization of human osteoblasts-like cells in vitro. J Biomed Mater Res A 64A:465–474
53. Meretoja VV, Helminen AO, Korventausta JJ, Haapa-aho V, Seppala JV, Narhi TO (2006) Crosslinked poly(e-caprolactone/D, L-lactide)/bioactive glass composite scaffolds for bone tissue engineering. J Biomed Mater Res A 77:261–268
54. Niiranen H, Pyhalto T, Rokkanen P, Kellomaki M, Tormala P (2004) In vitro and in vivo behavior of self-reinforced bioabsorbable polymer and self-reinforced bioabsorbable polymer/bioactive glass composites. J Biomed Mater Res 69A(4):699–708
55. Verheyen CCPM, de Wijn JR, van Blitterswijk CA, de Groot K, Rozing PM (1993) Hydroxyapatite/poly(L-lactide) composites: an animal study on push-out strengths and interface histology. J Biomed Mater Res 27:433–444
56. Zhang R, Ma PX (1999) Poly(alpha-hydroxyl acids)/hydroxyapatite porous composites for bone-tissue engineering I. Preparation and morphology. J Biomed Mater Res 44(4):446–455
57. Ma PX, Zhang R, Xiao G, Franceschi R (2001) Engineering new bone tissue in vitro on highly porous poly(alpha-hydroxyl acids)/hydroxyapatite composite scaffolds. J Biomed Mater Res A 54(2):284–293
58. Hench LL (1991) Bioceramics: from concept to clinic. J Am Ceram Soc 74:1487–1510
59. Yao J, Radin S, Leboy PS, Ducheyne P (2005) The effect of bioactive glass content on synthesis and bioactivity of composite poly (lactic-co-glycolic acid)/bioactive glass substrate for tissue engineering. Biomaterials 26(14):1935–1943
60. Roether JA, Gough JE, Boccaccini AR, Hench LL, Maquet V, Jérôme R (2002) Novel bioresorbable and bioactive composite based on bioactive glass and polylactide foams for bone tissue engineering. J Mater Sci Mater Med 13:1207–1214
61. Xynos ID, Edgar AJ, Buttery LDK, Hench LL, Polak JM (2000) Ionic products of bioactive glass dissolution increase proliferation of human osteoblasts and induce insulin-like growth factor II mRNA expression and protein synthesis. Biochem Biophys Res Commun 276:461–465
62. Gorustovich A, Roether J, Boccaccini AR (2010) Effect of bioactive glasses on angiogenesis: in-vitro and in-vivo evidence. A review. Tissue Eng Part B Rev 16:199–207
63. Boccaccini AR, Erol M, Stark WJ, Mohn D, Hong Z, Mano JF (2010) Polymer/bioactive glass nanocomposites for biomedical applications: a review. Compos Sci Technol 70:1764–1776
64. Maquet V, Boccaccini AR, Pravata L, Notingher I, Jerome R (2004) Porous poly (α-hydroxyacid)/Bioglass® composite scaffolds for bone tissue engineering. I: Preparation and in vitro characterisation. Biomaterials 25(18):4185–4194
65. Boccaccini AR, Maquet V (2003) Bioresorbable and bioactive polymer/Bioglass® composites with tailored pore structure for tissue engineering applications. Compos Sci Technol 63(16):2417–2429
66. Boccaccini AR, Notingher I, Maquet V, Jerome R (2003) Bioresorbable and bioactive composite materials based on polylactide foams filled with and coated by Bioglass® particles for tissue engineering applications. J Mater Sci Mater Med 14(5):443–450
67. Rhee S (2004) Bone-like apatite-forming ability and mechanical properties of poly (ε-caprolactone)/silica hybrid as a function of poly(ε-caprolactone) content. Biomaterials 25(7):1167–1175
68. Matthews FL, Rawlings RD (1994) Composite materials: engineering and science. Woodhead Publishing Limited, CRC Press, Cambridge, UK
69. Boccaccini AR, Roether JA, Hench LL, Maquet V, Jerome R (2002) A composites approach to tissue engineering. Ceram Eng Sci Proc 23(4):805–816

70. Kim H-W, Lee EJ, Jun IK, Kim HE, Knowles JC (2005) Degradation and drug release of phosphate glass/polycaprolactone biological composites for hard-tissue regeneration. J Biomed Mater Res 75B:34–41
71. Blaker JJ, Lee K-Y, Bismarck A (2011) Hierarchical composites made entirely from renewable resources. J Biobased Mater Bioenergy 5(1):1–16
72. Verrier S, Boccaccini AR (2008) Bioactive composite materials for bone tissue engineering scaffolds. In: Polak J, Mantalaris S, Harding SE (eds) Advances in tissue engineering. Imperial College Press, London, UK
73. Gerhardt LC, Boccaccini AR (2010) Review: bioactive glass and glass-ceramic scaffolds for bone tissue engineering. Materials 3(7):3867–3910
74. Day RM, Boccaccini AR, Shurey S, Roether JA, Forbes A, Hench LL, Gabe SM (2004) Assessment of polyglycolic acid mesh and bioactive glass for soft-tissue engineering scaffolds. Biomaterials 25:5857–5866
75. Yang XB, Webb D, Blaker J, Boccaccini AR, Maquet V, Cooper C, Oreffo ROC (2006) Evaluation of human bone marrow stromal cell growth on biodegradable polymer/Bioglass® composites. Biochem Biophys Res Commun 342:1098–1107
76. Keshaw H, Georgiou G, Blaker JJ, Forbes A, Knowles JC, Day RM (2009) Assessment of polymer/bioactive glass-composite microporous spheres for tissue regeneration applications. Tissue Eng Part A 15:1451–1461
77. Korventausta J, Jokinen M, Rosling A, Petola T, Yli-Urpo A (2003) Calcium phosphate formation and ion dissolution rates from silica gel-PDLLA composites. Biomaterials 24 (28):5173–5182
78. Blaker JJ, Maquet V, Jerome R, Boccaccini AR, Nazhat SN (2005) Mechanical properties of highly porous PDLLA/Bioglass® composite foams as scaffolds for bone tissue engineering. Acta Biomater 1:643–652
79. Schiller C, Siedler M, Peters F, Epple M (2001) Functionally graded materials of biodegradable polyesters and bone-like calcium phosphates for bone replacement. Ceram Trans 114:97–108
80. Navarro M, Ginebra MP, Planell JA, Zeppetelli S, Ambrosio L (2004) Development and cell response of a new biodegradable composite scaffold for guided bone regeneration. J Mater Sci Mater Med 15:419–422
81. Hutmacher DW, Schantz JT, Lam CXF, Tan KC, Lim TC (2007) State of the art and future directions of scaffold-based bone engineering from a biomaterials perspective. J Tissue Eng Regen Med 1:245–260
82. Maquet V, Martin D, Scholtes F, Franzen R, Schoenen J, Moonen G, Jerome R (2001) Poly (D, L-lactide) foams modified by poly(ethylene oxide)-block-poly(D, L-lactide) copolymers and a-FGF: in vitro and in vivo evaluation for spinal cord regeneration. Biomaterials 22:1137–1146
83. Orava E, Korventausta J, Rosenberg M, Jokinen M, Rosling A (2007) In vitro degradation of porous poly(D, L-lactide-co-glycolide) (PLGA)/bioactive glass composite foams with a polar structure. Polym Degrad Stab 92:14–23
84. Henry F, Devassine M, Guerin P, Costa LC, Briand X (2005) Biodegradable polymer studied by physical properties measurements. Mater Sci Forum 480–481:165–168
85. Siemann U (1985) The influence of water on the glass transition of poly(D, L-lactic acid). Thermochim Acta 85:513–516
86. Blaker JJ, Gough JE, Maquet V, Notingher I, Boccaccini AR (2003) In vitro evalulation of novel bioactive composites based on Bioglass®-filled polylactide foams for bone tissue engineering scaffolds. J Biomed Mater Res 67A:1401–1411
87. Blaker JJ, Pratten J, Ready D, Knowles JC, Forbes A, Day RM (2008) Assessment of antimicrobial microspheres as a prospective novel treatment targeted towards the repair of perianal fistulae. Aliment Pharmacol Ther 28(5):614–622

88. Georgiou G, Mathieu L, Pioletti DP, Bourban P-E, Månson J-A E, Knowles JC, Nazhat SN (2007) Polylactic acid–phosphate glass composite foams as scaffolds for bone tissue engineering. J Biomed Mater Res 80B(2):322–331
89. Shive MS, Anderson JM (1997) Biodegradation and biocompatibility of PLA and PLGA microspheres. Adv Drug Deliv Rev 28(1):5–24

Biography

Prof. Aldo R. Boccaccini holds the Chair for Biomaterials at the Department of Materials Science and Engineering, University of Erlangen-Nuremberg, Germany, and Visiting Professor of Materials Science and Engineering at Imperial College London, UK. Boccaccini holds a MEng degree from Instituto Balseiro (Argentina), PhD (Dr-Ing.) from RWTH Aachen University (Germany), and Habilitation from Ilmenau University of Technology (Germany). Before joining Imperial College in 2000, he had appointments at the University of Birmingham (UK), the University of California at San Diego (USA) and the Ilmenau University of Technology. The research activities of Boccaccini are in the broad area of glasses, ceramics, and polymer/glass composites for biomedical, functional, and/or structural applications. He is the author or co-author of more than 430 scientific papers and 10 book chapters. Boccaccini is the Editor-in-Chief of the international journal "Materials Letters" and serves in the editorial board of several recognised international journals including

Journal of Materials Science, J. Tissue Engineering and Regenerative Medicine, International Materials Review, J. Biomaterials Applications, Advanced Engineering Materials, Advances in Applied Ceramics, Key Engineering Materials, and J. Mater. Processing Technology. He is the recipient of several awards and prizes, including the Ivor Jenkins Medal 2010 of the Institute of Materials, Minerals and Mining (UK).

Dr. Xanthippi Chatzistavrou is currently working as a post-doctoral research fellow in the group of Professor T. Kasuga at the Department of Materials Science and Engineering, Nagoya Institute of Technology (NIT), Japan. Her research field focuses on the development of new antibacterial bioactive glass composite materials for dental applications. This research project is a continuation of previous work done in the framework of a EU Marie Curie Fellowship and a short-term post-doctoral project under supervision of Professor A.R. Boccaccini at the Department of Materials in Imperial College, London, UK, and the Institute of Biomaterials in the Department of Materials Science and Engineering, University of Erlangen-Nuremberg in Germany, respectively.

Chatzistavrou graduated from the Physics Department of the Aristotle University of Thessaloniki (Greece) in 2001. She then completed an MSc degree in Materials Physics and Technology in the same university. Her Ph.D. research was carried out in the Solid-State Physics Section in the Department of Physics of the same university. The results of her work in the field of composite biomaterials applying glass melting and sol–gel methods have been published in many journals and presented at several international conferences.

Jonny J. Blaker is a Research Associate within the Polymer and Composites Engineering (PaCE) Group, Department of Chemical Engineering, Imperial College London, UK. His current research interests are in the fields of bio(medical)materials/device development, and (hierarchical) renewable nanocomposite materials. Prior to this, he completed a 1-year Medical Research Council (MRC, UK) funded "discipline-hopping" post-doctoral research position within the Burdette Institute of Gastrointestinal Nursing, Kings College London. He holds a Ph.D. in Biomaterials for Bone Tissue Engineering (2007, Imperial College London, UK), an M.Sc. in Composite Materials (2002, Imperial College London, UK) and has a professional background in design engineering.

Showan Nazhat is an Associate Professor at the Department of Mining and Materials Engineering, McGill University. With a BEng degree in Materials Science and Engineering from Queen Mary, University of London, he carried out graduate research studies on bioactive and degradable bone analogous composites at the Interdisciplinary Research Centre in Biomedical Materials. In 2000, he took up a post of Lecturer (Assistant Professor) at the Eastman Dental Institute, University College London, where he developed scaffolds for tissue engineering. Since joining McGill University in 2006, his group's research focus has been on biopolymer tissue models and their interactions with seeded cells in three dimensions.

Chapter 7
Biodegradation of Calcium Phosphate Cement Composites

F.C.J. van de Watering, J.J.J.P. van den Beucken, R.P. Felix Lanao, J.G.C. Wolke, and J.A. Jansen

Abstract Calcium phosphate cements represent a good candidate material to use as bone graft to fill up bone defects, in the field of dental, orthopedic, or reconstructive surgery, because of their biocompatible and osteoconductive properties. A disadvantage of such cements, however, is the poor biodegradation that is required for the replacement by bone tissue. The biodegradation of the cement depends on different factors, including cement properties (e.g., chemical composition, setting reaction, porosity, crystallinity, and particle size of the calcium phosphate compounds) and the patient (e.g., medical condition and implantation site). Small alterations to these factors substantially affect the degradation rate of the cement and thus the formation of new bone. In this chapter, different factors that influence the biodegradation of calcium phosphate cements are discussed.

Abbreviations

ACP	Amorphous calcium phosphate
BCP	Biphasic calcium phosphate
BMP	Bone Morphogenetic Proteins
CA	Carbonated apatite, dahlite
CaP	Calcium phosphate
CPC	Calcium phosphate cement
CDHA	Calcium deficient hydroxyapatite
CS	Calcium sulfate
DCPA	Dicalcium phosphate anhydrous, monetite
DCPD	Dicalcium phosphate dihydrate, brushite

F.C.J. van de Watering • J.J.J.P. van den Beucken • R.P.F. Lanao • J.G.C. Wolke • J.A. Jansen (✉)
Department of Biomaterials (309), Radboud University Nijmegen Medical Centre,
P.O. Box 9101, 6500 HB Nijmegen, The Netherlands
e-mail: J.Jansen@dent.umcn.nl

N. Eliaz (ed.), *Degradation of Implant Materials*, DOI 10.1007/978-1-4614-3942-4_7,

FGF	Fibroblast growth factor
HA	Hydroxyapatite
hBMSCs	Human bone marrow-derived mesenchymal stem cells
MCPM	Monocalcium phosphate monohydrate
OCP	Octacalcium phosphate
pHA	Precipitated hydroxyapatite
PLGA	Poly(D,L-lactic-co glycolic acid)
PMMA	Polymethylmethacrylate
PTMC	Poly(trimethylene carbonate)
pVEGF165	Plasmid encoding vascular endothelial growth factor(165)
Runx2	Runt-related transcription factor 2
SDS	Sodiumdodecyl sulfate
TTCP	Tetracalcium phosphate
VEGF	Vascular endothelial growth factor
TGF	Transforming growth factor
α-TCP	α-Tricalcium phosphate
β-TCP	β-Tricalcium phosphate
μCT	Microcomputed tomography

1 Introduction

Annually, more than 1 million patients are treated worldwide for skeletal complications in the fields of orthopedic surgery, plastic and reconstructive surgery, dental implantology, maxillofacial surgery, and neurosurgery. These bone defects can be caused by congenital skeletal abnormalities, trauma, oncologic surgery, and failures of physiologic osteosynthesis.

A good bone graft has four main characteristics: it is (1) bioactive, which leads to the ability to bond to bone tissue without the intervention of fibrous tissue, (2) osteoconductive, thus able to guide bone tissue growth along the surface and into pores, (3) osteoinductive, i.e., capable of inducing differentiation of pluripotent stem cells into bone-forming osteoprogenitor cells, and (4) it stimulates osteogenesis, i.e., formation of bone by osteoblastic cells within the graft.

The use of autologous bone grafts is the gold standard in treating bone defects. Different donor sites are available depending on the application. For example, free vascularized bone flaps can be harvested from the radial forearm, fibula, iliac crest, and scapula, and for nonvascularized bone autografts donor bone can be obtained from the iliac crest, rib, distal radius, and olecranon [1]. Despite the advantageous clinical efficacy of autologous bone, transplantation with autologous bone grafts has also several drawbacks: low availability of transplantable tissue, donor site morbidity, and rapid depletion of the osteogenic capacity of the graft [2, 3].

Besides autologous bone graft, other treatment modalities can be used, such as allografts (i.e., human cadaveric bone obtained via the bone bank). Allografts are associated with many complications, including risk of viral transmission, bacterial

Table 7.1 Abbreviations of CaP compounds with corresponding chemical formula and Ca/P ratio

Abbreviation	Name	Formula	Ca/P ratio
ACP	Amorphous calcium phosphate	–	$1.25 < x < 1.55$
BCP	Biphasic calcium phosphate	$Ca_3(PO_4)_2 + Ca_{10}(PO_4)_6(OH)_2$	$1.50 < x < 1.67$
CA	Carbonated apatite, dahlite	$Ca_5(PO_5,CO_3)_3$	1.67
CDHA	Calcium deficient hydroxyapatite	$Ca_{10-x}(HPO_4)_x(PO_4)_{6-}(OH)_{2-x}$	$1.50< x < 1.67$
DCPA	Dicalcium phosphate anhydrous, Monetite	$CaHPO_4$	1.00
DCPD	Dicalcium phosphate dihydrate, Brushite	$CaHPO_4{\cdot}2H_2O$	1.00
HA	Hydroxyapatite	$Ca_{10}(PO_4)_6(OH)_2$	1.67
MCPM	Monocalcium phosphate monohydrate	$Ca(H_2PO_4)_2{\cdot}H_2O$	0.50
OCP	Octacalcium phosphate	$Ca_8H_2(PO_4)_6{\cdot}5H_2O$	1.33
pHA	Precipitated hydroxyapatite	$Ca_{10-x}(HPO_4)_x(PO_4)_{6-x}(OH)_{2-x}$	$1.50< x < 1.67$
α-TCP	α-Tricalcium phosphate	α-$Ca_3(PO_4)_2$	1.50
β-TCP	β-Tricalcium phosphate	β-$Ca_3(PO_4)_2$	1.50
TTCP	Tetracalcium phosphate	$CaO{\cdot}Ca_3(PO_4)_2$	2.00

infections, malignancy, toxins, and systemic disorders like rheumatoid arthritis and autoimmune diseases and are therefore not frequently used [4, 5].

An alternative for autologous grafts and allografts are the synthetic biomaterials. In the last decade, different materials have been described, including bioactive glass, polymers, and calcium phosphate (CaP) based ceramics or cements [6].

Synthetic biomaterials can be used as a scaffold or carrier material in the field of tissue engineering. The aim of bone tissue engineering research is to develop a functional bone substitute to regenerate damaged or lost bone tissue. The general approach in tissue engineering is to develop a material composed of one or a combination of (1) a scaffold and/or carrier such as CaP-based ceramics, (2) biological compounds (growth factors and/or bioactive molecules), and (3) cells.

CaP-based materials are of major interest in the bone regenerative therapies because they show a similar biological composition as natural bone mineral. In addition, they are nontoxic, biocompatible, and allow direct bone bonding without intervening soft tissue layers. CaP-based materials can be prepared in the form of granules, blocks, or cements.

CaP ceramics can be classified according to the crystal phase, e.g., hydroxyapatite (HA), β-tricalcium phosphate (β-TCP), biphasic calcium phosphate (BCP), amorphous calcium phosphate (ACP), carbonated apatite (CA), and calcium deficient hydroxyapatite (CDHA). A list of abbreviations of CaP crystals with chemical formula and Ca/P ratio is given in Table 7.1 [7].

CaP ceramics have a compressive strength (300–900 MPa) and tensile strength (40–300 MPa) similar or higher than those of cortical bone. However, due to the lack of elasticity the material is very brittle and therefore not suitable for load-bearing applications. Other disadvantages of CaP ceramics are their low degradation rate [7] and relatively poor handling properties. CaP ceramics are available in a fixed form and, therefore, the material is difficult to shape into the dimensions of the bone defect. These negative properties of CaP ceramics might be improved by using CaP cements (CaPCs).

2 Calcium Phosphate Cements

CPCs consist of a powder phase of calcium and phosphate salts, which after mixing with an aqueous solution will form an injectable paste at body temperature. The advantage of using CaPC as a bone substitute material is that it has a similar chemical composition, the same calcium and phosphate ions ratio as the mineral phase of bone. In addition, the cement paste can be directly injected in bone defects, which leads to less invasive and a faster surgery (Fig. 7.1) as compared to ceramic bone substitutes that have to be shaped to the defect dimensions during operations, which is more labor intensive [8–10].

Self-setting CPC was first described by Brown and Chow (1986) and consisted of only CaP compounds [i.e., tetracalcium phosphate monoxide (TTCP) and dicalcium phosphate dehydrate (DCPD)]. Their study aimed at the development of slurries for remineralizing carious lesions. Studies on the reaction

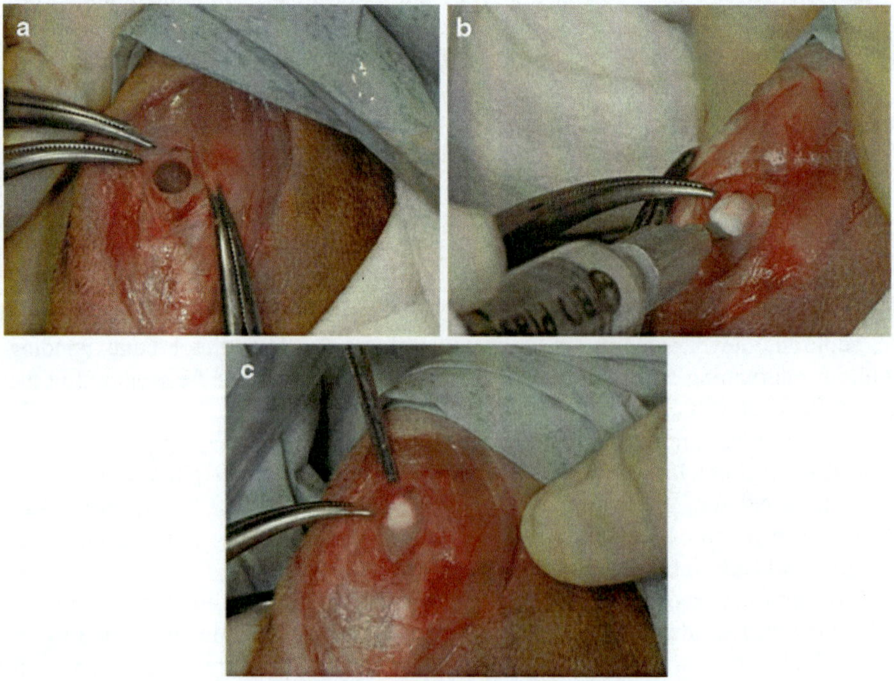

Fig. 7.1 CPC injection in a femoral condyle defect in a rabbit. (**a**) Femoral condyle defect, (**b**) injection of the CPC, (**c**) CPC after *in situ* setting

Table 7.2 Details of currently available commercially available calcium phosphate cements

Cement	Constituents	End product	Setting time	Compression strength (MPa)	Porosity
Biobon (α-BSM)®	ACP, DCPD	Poorly crystalline, HA, Ca/P = 1.45	15–20 min (37°C)	12	50–60% micro
Biopex®	75% α-TCP, 18% TTCP, 5% DCPD 2% HA	HA	8 min	80	Micro
Bone source™	TTCP, DCPD	HA, Ca/P = 1.67	10–15 min (37°C)	36	Micro
Calcibon®	61% α-TCP, 26% DCPA, 10% $CaCO_3$ 3% pHA	Calcium deficient CA Ca/P = 1.53	2–4 min (37°C)	60–70	44% micro
Cementek®	α-TCP, TTCP, Ca $(OH)_2$	HA, Ca/P = 1.64	40 min	20	50% micro
ChronOS inject®	44% β-TCP, 21 % MCPM, 31% β-TCP granules, 5% $MgHPO_4 \cdot 3H_2O$	Brushite (DCPD)	12 min	3	Micro
Mimix™	α-TCP, TTCP, Citric acid	HA	(37°C)	22	Micro
Norian-SRS®	MCPM, α-TCP, $CaCO_3$	Dahlite (CA) Ca/P = 1.67	12 h	28–55	Very low

of TTCP + DCPD to hydroxyapatite (HA) resulted in the finding that some of the aqueous pastes harden in time. This inadvertent discovery resulted in a new type of self-hardening cements consisting only of CaP [8].

In general, it has been demonstrated that different CPC compositions are non-toxic, biocompatible, and osteoconductive [8–10] in different animal models, for example in dogs [11, 12], goats [13, 14], and rabbits [15]. Multiple studies have reported that a temporal increase of newly formed bone can be seen in different CPC [9, 14, 16], indicating that CPC are promising candidates for bone regeneration.

Several CPC formulations have been approved by the Federal Drug Administration for clinical use. Examples of these commercially available CPC are BoneSource™, Cementek®, and Calcibon®. In Table 7.2, a list of commercially available CPC is given [7].

2.1 *Composition of Calcium Phosphate Cements*

Since the study of Brown and Chow, many different formulations have been studied, with variable success. In general, all different types of CPC can be divided into two groups according to their setting product: apatite cements that have

HA, CA, or CDHA as end product, and brushite cements that set in DCPD. In contact with body fluids, brushite cements will eventually transform into apatite. The wide variety of CPC are all a mixture of different CaP compounds (Table 7.1) [7].

There are two different categories of CaP: (1) CaP obtained at low temperature via precipitation from an aqueous solution, and (2) CaP obtained at high temperatures by a thermal reaction. The most common CaP compounds are described in more detail below, starting with the low temperature CaP (DCPD and precipitated hydroxyapatite), followed by the high temperature CPC (β-tricalcium phosphate, α-tri tricalcium phosphate, and HA).

DCPD ($CaHPO_4{\cdot}2H_2O$) is the most easily synthesized CaP compound. DCPD is biocompatible, biodegradable, and osteoconductive, and can be converted into dicalcium phosphate ($pH < 6$), octacalcium phosphate ($pH \approx 6$–7), or precipitated hydroxyapatite (pHA; $pH > 7$). It is observed that DCPD can convert *in vivo* into either pHA [17], or it will be degraded and replaced by bone. When large amounts of DCPD are converted into pHA *in vivo*, a severe inflammatory response can be observed due to the large amounts of acid that are released during this reaction [18].

pHA ($Ca_{10-x}(HPO_4)_x(PO_4)_{6-x}(OH)_{2-x}$) is obtained via precipitation in an aqueous solution above a pH of 7. The formed crystals are poorly crystalline and have dimensions of several nanometers. The specific surface area of pHA is very large (typically around 25–100 m^2/g). It is therefore similar to the apatite form found in natural bone, although bone mineral contains impurities, including carbonate and magnesium. pHA can have different Ca/P molar ratios ranging from 1.50 to 1.67 [19], and even outside this range [20]. All commercially available CPC have pHA as a compound.

β-tricalcium phosphate (β-TCP; β-$Ca_3(PO_4)_2$) is obtained via thermal treatment above 650°C. The production of β-TCP can occur after (a) mixing of equimolar amounts of DCPD and pHA and subsequent calcinations; or (b) calcinations of pHA (Ca/P ratio 1.50). The degradation of β-TCP occurs via cell-mediated processes including osteoclastic activity [21].

α-tricalcium phosphate (α-TCP; α-$Ca_3(PO_4)_2$) is obtained by heating β-TCP above 1,100°C and quenching it to prevent the reverse transformation [18, 22]. α-TCP has the same chemical composition as β-TCP, but differs in crystallographic structure. The difference in crystallinity makes α-TCP much more soluble than β-TCP. It is readily transformed into pHA in an aqueous solution. Due to this property and its biocompatibility and biodegradability, α-TCP is used as the major component of apatitic CPC [21].

Hydroxyapatite (HA; $Ca_5(PO_4)_3OH$) is highly crystalline and the most stable form of CaP in an aqueous environment. HA is the most biocompatible CaP because it resembles the naturally occurring phase of bone mineral. HA can be found in teeth and bones within the human body. Due to its stability, the biodegradation rate of HA is slow: it is estimated to degrade in the range of decades [18, 21].

2.2 Setting Reaction of CPC

As mentioned in Sect. 2.1, CPC consist of a mixture of powder and an aqueous solution that, upon mixing, sets by entanglement of the crystals. For clinical application, a certain initial and final setting time of the CPC is required. These setting times are related to the absence of indentation marks from Gillmore needles (Ø 2.12 mm, 113.4 g for initial setting; Ø 1.06 mm, 453.6 g for final setting) on the cement surface. The clinical meaning of the final setting time relates to the time necessary to set the cement sufficiently so it can be touched without damaging it. For clinical application, it means that after mixing, the cement must be applied before the initial setting time, and that the wound needs to be closed after the final setting time.

Other methods to predict the two setting times are via the Vicat needles method [23], by time after which no disintegration occurs after immersion in water [24], ultrasonically [25], oscillating rheometry [26], or via isothermal differential scanning calorimetry [27]. The first three methods are mostly used because these are cheap and easy to handle, although incorporating some observatory-related subjectivity. These drawbacks are solved by the other methods, which are, however, more complex and expensive to use.

The setting reaction of the CPC controls not only the hardening time and other setting properties, but also determines the nature of the cement products. Therefore, the hardening of the cement influences the physical and biological properties of the hardened cement [8].

Cement setting reactions of the different CaP compounds are quite similar. The driving force of the setting reaction of all CPC is the solubility of the reactants and products. Solubility is normally described as the property of a solid to dissolve in a certain volume of solution. The solubility of CaP can change when the pH changes.

To understand the solubility of different CaP compounds, and hence the setting reaction, a solubility phase diagram can be used to gain information regarding the basic principles concerning the solid-solution equilibrium. In a solubility phase diagram, isotherms visualize the solubility of a salt, expressed as the total calcium and total phosphate concentrations of the saturated solution as a function of pH [8]. In a two-dimensional solubility phase diagram, a salt is less soluble when the isotherm lies below that of another salt. In an acidic surrounding, the solubility of the different CaP compounds does not vary much. In contrast, in a neutral surrounding the solubility of the CaP compounds is more widely spread. In case of tissue damage (accidental or by surgical intervention), the surroundings of the wound become slightly more acidic; thus the solubility of the compounds might be influenced; however, in time the environment will neutralize. As visualized in Fig. 7.2, HA is the most stable (least soluble) CaP compound at neutral pH. Therefore, most neutral pH CPC formulations would be forced to precipitate into HA or another closely related apatitic phase, such as CDHA or carbonated HA.

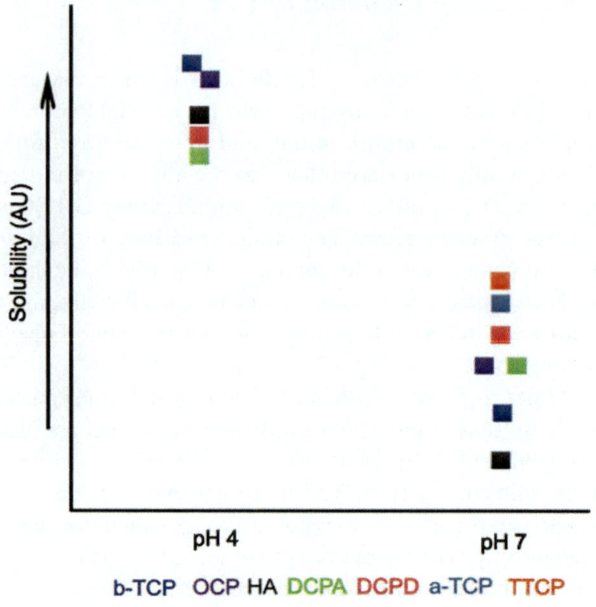

Fig. 7.2 Solubility phase diagram of different CaP compounds in a two-dimensional graph

The setting reactions of CPC can be divided into three types:

1. Cement hardening as a result of reactions among CaP compounds.

   A wide variety of CaP compounds can be used to prepare CPC (see Table 7.1). All reactions of CaP compounds that occur in an aqueous solution are via a dissolution/re-precipitation. For example, when the highly soluble TTCP and DCPA are mixed in an aqueous solution to form a CPC paste, dissolution of TTCP and DCPA leads to a solution that is highly supersaturated with respect to HA. The supersaturated solution will precipitate in HA. In dissolution/re-precipitation reactions, the Ca/P molar ratio is equal for the initial and final CPC. Another example of such a reaction is α-TCP, which transforms precipitated hydroxyapatite after contact with an aqueous solution:

$$3\alpha - Ca_3(PO_4)_2 + H_2O \rightarrow Ca_9(PO_4)_5(HPO_4)OH \tag{7.1}$$

2. Hardening of the cement as a result of a reaction between calcium and a carboxylic acid.

   CaP compound powders mixed with carboxylic acids can rapidly form CaP complexes as well as relatively insoluble and often amorphous Ca-carboxylate compounds. These acids include glycolic, citric, tartaric, malonic, malic, succinic, and maleic acids. The hardening reaction continues until a more stable final product is formed. The end product can be HA, DCPD, or DCPA [8, 28, 29].

3. Cement hardening as a result of a reaction between CaP and an aqueous polymer solution.

Some CaP compounds, especially the more alkaline phases such as TTCP and HA, can react with aqueous solutions of polymers [30]. Examples of such polymer solutions are poly(acrylic)acid [31, 32] and poly(methyl vinyl ether-maleic acid) [33, 34]. The addition of water-soluble polymers to CPC results in a rapid setting reaction. The more basic TTCP will induce the neutralization reaction, resulting in cross-linking of the polymers [31]. This reaction is a result of acid–base reactions between the carboxyl groups of the polymer and the alkaline CaP [8].

2.3 CPC Biodegradability

The replacement of a biomaterial by living bone is preferred because this avoids potential complications later on, such as inflammation, stiffness, pain, and bacterial seeding. To obtain replacement of bone, new bone formation and resorption of the material, e.g., biodegradation of the material, are required. The rate of material degradation and newly formed bone should optimally be in perfect balance to maintain mechanical stability (integrity of the ceramic scaffold) and allow a gradual takeover of mechanical strength by newly formed bone tissue.

Biodegradation of CaP-based material can occur via two processes, namely (1) extracellular liquid dissolution, and (2) disintegration into small particles that can be intracellularly digested or transported to other adjacent tissues. The first process depends on the solubility product of the formed CaP biomaterial, while the second process is mainly dependent on the solubility product of the connecting agents of the powder particles after crystallization [35].

The biodegradation of the material is dependent on different factors, including the final product of the cement setting. For example, apatite is a more stable compound, and therefore the degradation rate is slower as compared to brushite [36]. Due to the slower degradation rate, the material has a relatively good mechanical stability, albeit insufficient for load-bearing applications. On the other hand, a less stable compound like brushite leads to a faster biodegradation. In addition, the CaP chemical composition itself can influence the material degradation. This is exemplified by an approach in which calcium sulfate (CS) was added to CPC, composed of TTCP and DCPA. Although biocompatible, CS is less suitable for bone substitution because of its rapid resorption. Interestingly, a mixture of CS, TTCP, and DCPA leads to a good biocompatibility and enhanced degradability and bone ingrowth as compared to CPC consisting of only TTCP and DCPA [37]. In addition, the chemical composition of the CPC plays also a role in the active resorption of the cement. For example, it has been reported that a cement produced by a mixture of sodium calcium phosphate, $Na_3Ca_6(PO_4)_5$, tetracalcium phosphate and β-tricalcium phosphate powder, and malic acid or citric acid has a different relationship regarding the number of osteoclasts on the cement [28]. Since osteoclastic activity can be a marker for the evaluation of bioresorbability, it was suggested that cement mixed with malic acid is more promising for bone substitution in an active manner than cement mixed with citric acid.

Also, the patient plays an important role in material degradation. Depending on the patient's age, sex, and general metabolic health condition, the speed of degradation can be different. In addition, the implantation site influences scaffold biodegradation and bone formation in the vicinity of the scaffold. For example, ceramic material placed in a femoral or cranial defect will be exposed to different environments, including differences in biomechanical loading, surrounding tissue, and blood/nutrient supply. Besides the environment, the origin of the bone tissue (long bone formation (femur) is endochondral, while cranial bone formation is intramembranous) may play a role in degradation and bone formation due to distinctive signaling properties [38].

Other factors involved in the biodegradability are porosity of the material, crystallinity (crystal size, crystal perfection, and grain size), particle size of the CaP powder, and liquid/powder ratio [35, 39, 40]. These factors influence each other and are influenced by the chemical composition of the material. For example, CaP compounds that transform into relatively highly crystalline hydroxyapatite after hydration will be essentially nonresorbable *in vivo* [35].

3 CPC Porosity

CPC contains an intrinsic high nano/sub-micron-sized porosity. Several parameters, including particle size of the powder phase [39] and liquid/powder ratio [40], can affect the creation of this type of porosity and, hence, material degradation, and consequently allow bone ingrowth.

Nanoporosity is an important parameter to be controlled in CPC materials because it enables fluid flow within the material and can affect protein adsorption. In contrast to nano/sub-micron-sized porosity, macroporosity enables tissue ingrowth and is the key factor for CPC degradation and its substitution by new bone tissue.

3.1 Macroporosity

Macropores have been built into biomedical materials since the early 1970s [41, 42]. Resorbability of a material is an important clinical property; it is preferred that the CPC will degrade in a short period of time parallel to new bone tissue formation. Pores of a suitable diameter and an adequate interconnectivity are essential for bone tissue ingrowth and for enhancing CPC degradation. Debates about the adequate pore size for bone replacement materials are still ongoing. Researchers have suggested that a pore size larger than 100 μm is sufficient for tissue ingrowth [43, 44], whereas other investigations conclude that larger pores (up to 500 μm) have to be preferred [45]. Although highly porous materials will lead to a faster degradation, an adequate balance between degradation and structural integrity should be found. Figure 7.3 gives an overview of typical porous 3D structures.

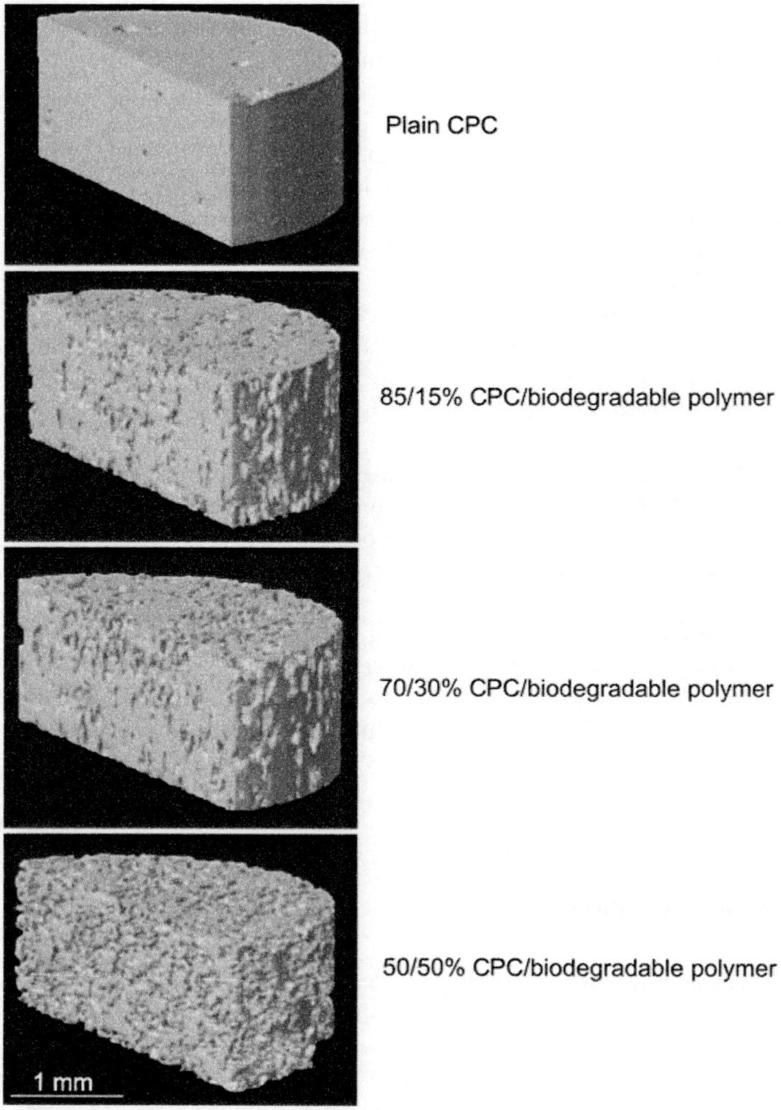

Fig. 7.3 μCT images of composite discs with different CPC/biodegradable polymer ratio after burning out the polymer component. A higher amount of biodegradable polymer leads to an increase in porosity of the scaffold material. Bar represents 1 mm

3.2 Porosity Calculations

Several geometrical methods have been described to quantify porosity. A commonly used approach to calculate porosity is the use of optical methods, such as X-rays or microscopic techniques. With these methods, the pore area can be distinguished from the material area. Furthermore, a direct method for porosity determination consists of determining the bulk volume of the porous sample and subsequently determines the volume of the material with no pores. The pore volume will then be obtained by the subtraction of the material volume from the total volume.

Two procedures based on water volume can also be employed. In the first approach, the sample is immersed in a fluid that wets the pores under vacuum. The pore volume will be calculated by subtracting the volume of water left after the imbibitions from the total volume of the fluid. Second, a water saturation method can be used. The pore volume is here calculated from the weight of the saturated sample minus the weight of the dry sample.

Mercury intrusion porosimetry is based on the premise that a non-wetting liquid (mercury) will only intrude capillaries under pressure. After mercury is forced to go through the material, the total porosity is determined from the total volume intruded.

Last, the gas expansion method can be used. Shortly, a sample of known volume is placed in a close container of known volume and connected to another container with known volume. A valve connected to both containers is opened, and some of the gas passes from the first container to the second. The volume of pores can be calculated by a formula, which takes into account the pressure and the volume values during this process.

The major drawback of most of these techniques is that they are sample destructive. The most suitable method to determine porosity is often chosen depending on the CPC of interest, the pore size, pore distribution, and the characteristics of the porogen agent.

3.3 Interconnectivity

Besides bone ingrowth in the cement pores, interconnectivity is a key factor for new tissue formation within the cement. When the pores form an interconnected network, cellular and vascular penetration will be favored, ensuring new tissue formation in all void areas of the implanted material. Furthermore, variations in interconnectivity will lead to modifications in the architecture and mechanical properties of the CPC. A suitable balance between pore size, pore volume, interconnectivity, and structural integrity is a major issue in the development of CPC-based bone substitution materials.

Mercury intrusion porosimetry, X-ray imaging techniques such as microcomputed tomography (μCT) analysis, or scanning electron microscopy are commonly employed techniques, not only for pore architecture determination as described before, but also as interconnectivity indicators. However, these methods present several drawbacks. For example, mercury intrusion porosimetry cannot detect isolated pores in the scaffold, only the pores of the interconnected network.

Besides, μCT, though it presents the advantage of being a nondestructive technique, cannot consistently discriminate small pore sizes in CPC constructs. In addition, due to the similarity of the chemical composition of CaP cement and natural bone, it is very difficult to distinguish between both materials when using this technique. In order to improve its radiopacity, and therefore improve the X-ray measurements resolution, radiopacifiers have been included in CaP formulations [46]. Furthermore, scanning electron microscope provides high-resolution surface imaging, but its use is limited to two-dimensional measurements, and the results are merely qualitative. Currently, there is a need for more consistent, reproducible, and standardized methods to measure and compare interconnectivity of CPC samples.

3.4 Enhancement of Macroporosity

In general, the degradation rate of plain CPC is limited (Fig. 7.4). For a better bone regeneration, different methods have been used to induce macroporosity in CPC, such as inclusion of water-soluble additives, biodegradable polymeric microspheres, foam-forming agents, and fabrication of scaffolds by additive manufacturing technologies.

3.4.1 Water-Soluble Additives

The inclusion of water-soluble additives in CPC is a common approach to induce macroporosity. Also known as the particle leaching technique, it is often used to create porosity in other bone substitute materials such as preset ceramics or polymeric scaffolds. Saccharides such as sucrose, pectin, chitosan, or mannitol as well as the salts sodium chloride and sodium phosphate are water-soluble compounds to generate macroporosity in CPC.

Sucrose is a disaccharide derived from glucose and fructose. Sucrose crystals of specific sizes have been included in CPC, and macroporosity is generated after dissolution of the crystals during incubation of the samples in aqueous media [47].

A more novel technique is the creation of microspheres from a saccharide. Pectin is a water-soluble polysaccharide that, in the form of microspheres, has been recently introduced in CPC to create round pores after degradation [48].

Chitosan is a linear polysaccharide; it has been processed in the form of fibers which have been embedded within CPC in order to enhance its mechanical properties in an early stage and to generate porosity after the fiber degradation [49]. Mannitol is a biocompatible monosaccharide alcohol often used in biomedical applications. This polyol presents an appropriate solubility to be employed as CPC porogen [50].

Several salts have been used to induce porosity in CPC, such as sodium chloride crystals [51]. Similarly to when saccharides are employed, porosity is generated after the salt is leached out from the CPC in an aqueous environment.

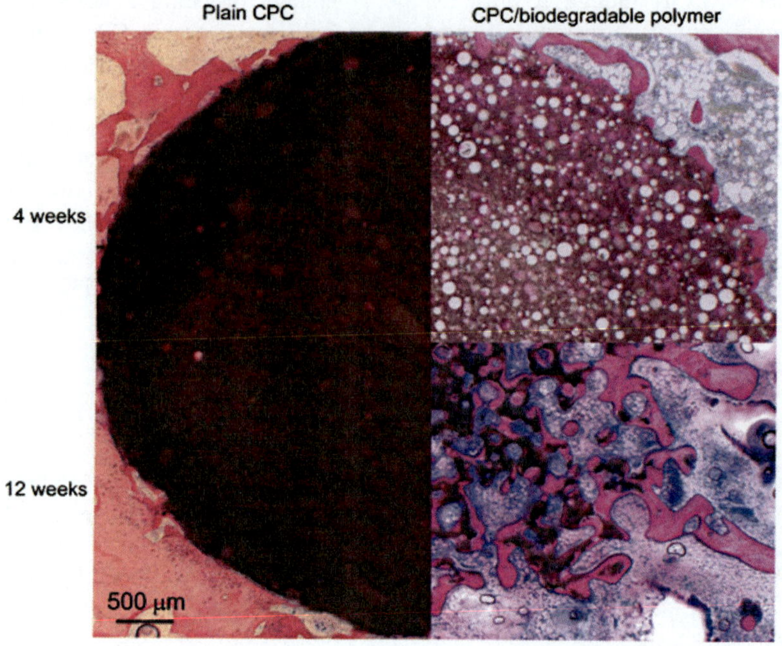

Fig. 7.4 Histological sections of a femoral condyle of a rat injected with plain CPC or CPC with biodegradable polymer, after 4 and 12 weeks of implantation time. In the plain CPC implant, degradation and bone ingrowth is limited and restricted to the edges of the defect. In contrast, CPC with biodegradable polymer shows an increase in implant degradation and bone ingrowth from 4 weeks implantation onwards. Methylene blue and basic fuchsin staining. Bar represents 500 μm

A different approach using a salt as a porogen consists of embedding frozen sodium phosphate particles and CaP powder; while the cement is setting, the sodium phosphate particles will melt generating porosity simultaneously to the hardening process [52].

3.4.2 Biodegradable Polymers

Approximately 30% of bone is composed of organic compounds, of which 90–95% is collagen. Consequently, natural polymers are being widely studied in relation to bone tissue engineering. Collagen presents a slow degradation rate, which makes it unsuitable as porosity inductor for CPC. In contrast, gelatin is a natural enzymatically degradable polymer derived from collagen with a faster degradation rate. For these reasons, gelatin has been used in the form of microspheres to generate macroporosity in CPC after its degradation [53–55].

Besides natural polymers, there is a wide range of synthetic polymers that can be used to induce macroporosity in CPC. The synthetic polymers poly (epsilon-caprolactone) and poly(l-lactic acid) have been used as fibers to enhance resistance of CaP materials in a first stage and porosity after polymer degradation [56].

Poly(DL-lactic-co glycolic acid) (PLGA) is employed for the production of microspheres to be embedded in CPC material and generate porosity after polymer hydrolytic degradation [57–59]. In Fig. 7.5, an overview of PLGA inclusion and degradation in CPC is shown.

In addition, the enzymatically degradable poly(trimethylene carbonate) (PTMC) has been processed also in the form of microspheres and introduced in CPC, generating interconnected macroporosity and minimal inflammatory response [60, 61]. In Fig. 7.5, an overview of PTMC inclusion and degradation in CPC is shown.

3.4.3 Foam-Forming Agents

A different approach to enhance porosity in CPC is to combine it with an agent that, by foaming, will generate gas bubbles that will lead to the generation of pores. Hydrogen peroxide bubbling [62], or the creation of CO_2 bubbles from sodium bicarbonate during the cement setting [63, 64], are examples of foam-forming methods to generate porosity.

Acetic and citric acids have been investigated as porogens by the gas-foaming technique. The final cement presents good injectable properties and higher porosity when combined with citric acid [65]. Furthermore, citric acid has also been used in combination with sodium bicarbonate as effervescent agents to generate porosity [66].

A similar technique is to mix the cement paste with a foam obtained previously from a foaming agent. A foam from the natural protein albumin has been used in this approach, generating macroporous scaffolds with promising *in vivo* results [67–69]. Following this procedure, no potentially toxic gas is liberated after implantation of the cement, since the foam is generated before mixing it with the cement paste.

3.4.4 Additive Manufacturing Techniques

Rapid prototyping is an advanced manufacturing technique that allows the automatic construction of physical objects using additive manufacturing technology. Using this concept, CPC scaffolds with highly controlled internal pore architectures have been obtained [70, 71]. Stereolithography is another additive manufacturing process which has been applied for CaP scaffold fabrication with a controlled internal channel architecture [72]. However, a major drawback of these techniques is that the scaffolds have to be fabricated prior to implantation in the body, meaning that this approach is not suitable for injectable and/or minimally invasive procedures.

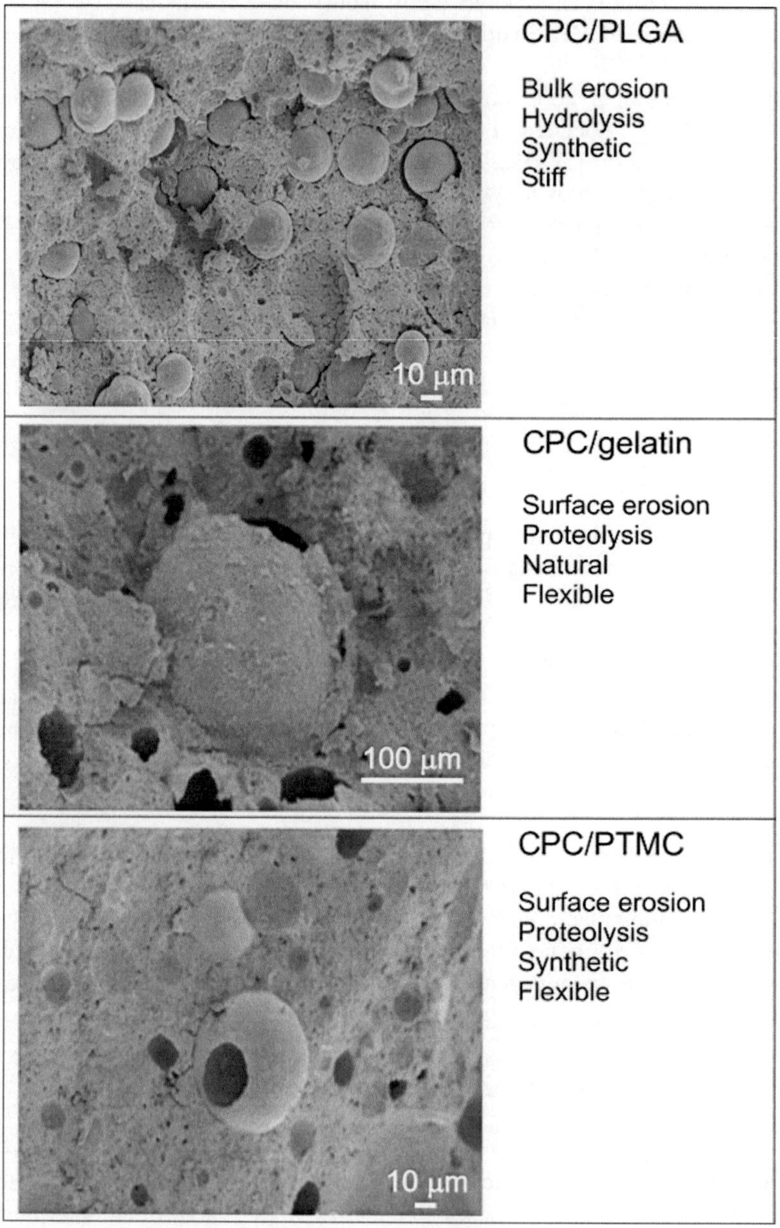

Fig. 7.5 CPC/PLGA, CPC/gelatin, and CPC/PTMC samples. The different characteristics of the polymers lead to differences in material degradation of the CPC

3.4.5 Surfactants

Sodium dodecyl sulfate (SDS) [73] has been applied within CPC because of its air-entrapping properties. Similar to other surfactants, SDS presents a low surface tension. Air bubbles are incorporated during the mixing of the powder and the liquid component of the cement. These bubbles become covered by a sheath of SDS molecules, creating macroporosity after the CPC solidification. Alternatively, cetyltrimethyl ammonium is a surfactant that has been applied as CPC porosity inducer [74].

3.4.6 Combinations

Finally, it has to be mentioned that, in several studies, more combinations of the previously mentioned methods have been investigated to generate porosity in CPC. For example, CO_2 foaming has been used to induce porosity in combination with PLGA-microparticles embedded in CPC for the creation of secondary porosity at a later time point [75]. Furthermore, mannitol crystals have been used to generate early macroporosity in combination with slower resorbable chitosan fibers that will improve the mechanical properties of the cement at initial time points, but subsequently will degrade, enhancing the already generated macroporosity after mannitol dissolution [76, 77].

4 CPC as a (Drug) Delivery System

Besides the role as a bone substitute, CPC can be used as a carrier for local and controlled drug release. This makes the material very attractive because it enables (a) directed bone healing, and (b) treatment of the potential disease underlying the bone defect. Examples of skeletal diseases to treat are bone tumors, osteoporosis, or osteomyelitis [78].

Several aspects need to be considered regarding the incorporation of drugs to CPC. First of all, it needs to be verified that the setting reaction and hardening mechanisms and physicochemical properties do not alter after the addition of drugs. Second, the drug release kinetics *in vitro* needs to be characterized, e.g., by *in vitro* release experiments with radioiodinated drugs of interest [79]. Subsequently, a release profile is obtained by measuring in time the remaining activity in a gamma counter. This needs to be followed by an assessment of the *in vivo* efficacy of the drug delivery carrier [79], after which the clinical use must be evaluated [78]. The efficacy needs to be analyzed because the activity of the drugs and/or bioactive molecule might be lost due to the chemical reactions during cement setting, and therefore the drug and/or bioactive molecule will have no clinical effect [80, 81]. The incorporation of the drugs or other bioactive molecules can be in the liquid phase [82] and/or powder phase [83] of the CPC.

4.1 The Release Profile of a Drug

The release of drugs from the carrier depends on (a) the drug solubility, (b) the interaction (e.g., type of bonding) between the drugs and the material, and (c) the degradation characteristics of the carrier.

In general, CPC are nondegradable, or at least the rate of the material degradation is much lower than the rate of drug liberation. Thus, the initial diffusion coefficient of the CPC depends on the porosity and microstructural parameter of the material, as can be described in the adapted Higuchi's law [78, 84].

$$M_t = AM_0\left[\frac{D's}{\tau}Cs(2C_0 - \varepsilon C_s)t\right]^{1/2}. \tag{7.2}$$

In the adapted Higuchi's law, M_t is the amount of drug released in time (t); M_0 is the total amount of drugs incorporated; A is the surface area of the carrier; D is the diffusion coefficient of the drug in the carrier; C_s is the solubility of the drugs; and C_0 is the initial concentration of the drug in the matrix [78]. In later stages (after 60% of the drug is released), the release pattern is proportional to time, followed by stabilization of the release depending on the increased concentrations in the environment.

4.2 Loading Growth Factors and/or Bioactive Molecules

As mentioned before in Sect. 3.4, CPC degradation is slow and can be increased by the incorporation of macroporosity to CPC via polymer microparticle addition [53, 59, 63, 85, 86]. If the porosity of the material increases in time, then the release profile of the drugs will not only be dependent on the diffusion through the material. Increasing the porosity is therefore not only beneficial for the material degradation, but also for the release profile of incorporated drugs.

Another advantage of the inclusion of polymers is that these polymers can serve as a carrier system to obtain a more sustained release of drugs and/or bioactive molecules. For example, the sustained release of rhBMP-2 is tightly controlled and self-limiting via the combination of different signal transducing and inhibiting proteins [87]. It is shown that a sustained release of rhBMP-2 from PLGA-microparticles resulted in more and faster bone formation in rabbit calvarial defects as compared to immediate release of rhBMP-2 [88]. In addition, other CPC carrier systems based on different polymeric microparticles such as gelatin [89] or the natural occurring pectin have demonstrated to be beneficial for the retention of drugs [48].

The drugs and/or bioactive molecules can be either absorbed on the surface of the microparticle after microparticle fabrication or entrapped in the microparticles during the microparticle fabrication. Different types of molecules can be loaded or entrapped in the microparticles, including antibiotics, anti-inflammatory drugs [90, 91], and growth factors [92–95]. In addition, these delivery systems can be used to deliver non-proteinaceous compounds (e.g., genes [96, 97]) and even cells [98]. Drugs and/or bioactive molecules incorporated in CPC (either directly or via the incorporation of polymeric microparticles) that are known to induce bone formation are described in more detail below.

4.3 Growth Factors Delivery

Bone healing is a complex physiological process that is initiated and controlled by growth factors, such as members of the transforming growth factor (TGF) super-family, which includes TGF-β that plays a significant role in the repair of skin and bone [92], as well as Bone Morphogenetic Proteins (e.g., BMP-2), which play an important regulatory role in many steps of bone morphogenesis [93]. BMPs seem to play an important role in increasing the material degradation and bone forming capacity of CPC. For example, it has been reported that addition of BMP-2 to the PLGA-microparticles in CPC composites placed in a rat critical sized cranial defect model significantly increases the degradation and the bone forming capacity of the material as compared with plain CaP/PLGA cement composites, from 4-weeks implantation period onwards (Fig. 7.6) [94]. Another study reported that injectable rhBMP-2/CPC stimulated bone formation and blood vessel ingrowth after 2 months in vertebroplasty, as compared to rhBMP-2 /polymethylmethacrylate (PMMA) bone substitute [99]. In addition, an effect of rhBMP-2 has been observed in an equine model, where surgically osteotomies and ostectomies of the accessory metatarsal bones were induced. The rhBMP-2/CPC increased early bone formation as compared to nontreated bone defects [100].

Another example of a growth factor used to induce material degradation and bone formation is TGF-β. It have been demonstrated that addition of TGF-β to CPC within the liquid component is released in a bioactive form and still capable of stimulating the differentiation of preosteoblastic cells *in vitro* [101] and results *in vivo* in an increased bone formation when implanted in 5-mm rat cranial defects [95]. However, depending on the loading of the TGF-β and the chemical composition, different results may be obtained. For example, addition of TGF-β to gelatin microparticles in CPC composites significantly accelerated implant degradation, and only a tendency toward increased bone ingrowth 12 weeks postoperatively as compared to plain CaP/gelatin cement composites was observed [102].

Bone is a highly vascularized tissue, and therefore the formation of new blood vessels, angiogenesis, is essential for bone regeneration [103]. Thus, besides an important role for BMPs, angiogenesis stimulating factors seem to play an essential role in bone formation.

Vascular endothelial growth factor (VEGF) is a potent angiogenic growth factor, which is known to stimulate blood formation from several studies [104, 105]. The half-life time of VEGF is less than 1 h following *in vivo* injection [106], and

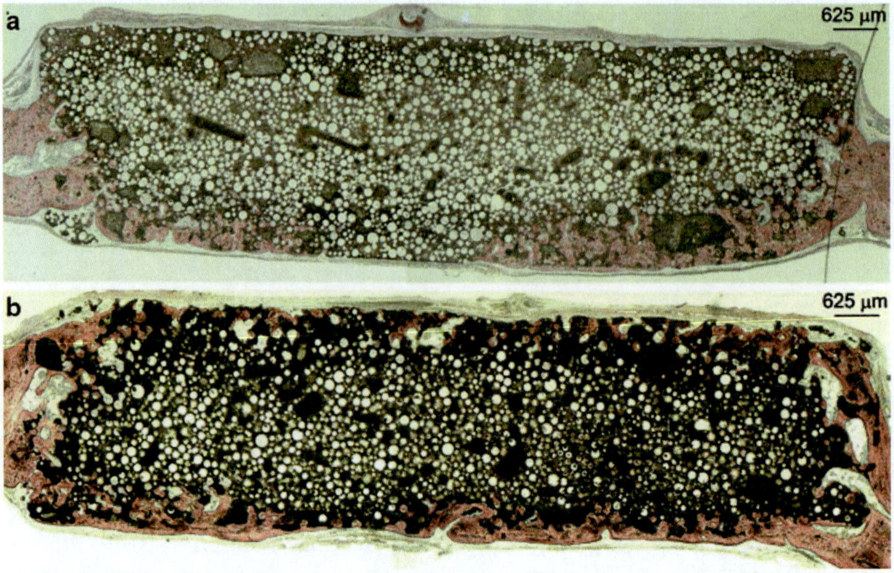

Fig. 7.6 Histological sections of cranial defects, consisting of either (1) plain CPC, or (2) CPC loaded with BMP-2, after 12 weeks. CPC/PLGA shows limited ingrowth of bone after 12 weeks of implantation. In the CPC/PLGA loaded with BMP-2, more bone formation is observed. Methylene blue and basic fuchsin staining. Bar represents 625 μm

therefore a carrier system is required to have a controlled release of the growth factor. The method of delivery seems to influence the efficacy of VEGF. For example, when VEGF is directly injected *in vivo*, poorly perfused fragile capillaries without connection to the preexisting circulation as well as agiomas are observed [107]. In contrast, sustained release of VEGF results in the formation of regularly organized vasculature [108]. The CPC/polymer microsphere carrier system is a promising candidate for sustained release VEGF and has shown to promote vascularization within these bone substitute materials [109]. In addition, local sustained VEGF release from carrier biomaterials improved the healing of noncritical sized bone defects [110]; however, this is contradictory with another study which reported no difference in bone formation when VEGF was released locally from a carrier system [111]. The different VEGF release kinetics may be a result of differences in carrier system, VEGF concentration, and animal model used.

To mimic bone healing in a more natural fashion, a combination of different growth factors, such as angiogenic and osteogenic growth factors, may be beneficial. It is reported that dual delivery of VEGF and BMP-2 loaded on different scaffold materials can increase both capillary density and bone formation ectopically [112], whereas it leads to more bone formation orthotopically as compared to single delivery of BMP-2 [111]. The dual delivery of BMP-2 and fibroblast growth factor (FGF) by a CPC carrier has also been reported to more substantially induce new bone formation [113, 114].

4.4 Other Products and/or Molecules

From a biological point of view as well as considering cost effectiveness, research is focusing on either the reduction of included growth factors or the replacement of the growth factors by less expensive, alternative molecules. Previous studies on alternative molecules with angiogenic and osteogenic activity revealed some agents, which effectively increased bone formation, such as statins [115], isoflavone derivates [116], copper ions [117], and TAK-778 [118]. Another alternative might be Icariin, a flavonoid glycoside isolated from the herb *Epimedium pubescen* [119]. Icariin has been reported to significantly increase bone formation after 4 weeks in a mouse calvarial defect model.

Instead of delivering bioactive molecules, it is possible to deliver cells via CaP scaffolds [120]. In particular, the delivery of stem cells is promising because stem cells are undifferentiated cells with the capacity to differentiate into one or more cell types and have the ability of self-renewal [121]. In bone tissue engineering, human bone marrow-derived mesenchymal stem cells (hBMSCs) are of interest because they have the ability to differentiate into osteoblasts, chondrocytes, and endothelial cells [121]. The relative easy access to isolate hBMSCs, their capacity for expansion, and the minimal immunogenic and tumorigenic hazards make these cells an excellent cell source for dental craniofacial and orthopedic repair.

Culturing preostoblast cells on CPC composites was shown to be non-cytotoxic and supported cell proliferation and growth [98]. However, due to pH changes and ionic activity in the CPC setting reaction, this may result in adverse effects when hBMSCs are seeded directly inside the CPC paste [122]. To overcome the negative effects of the cement setting, it has been reported that cells can be encapsulated in alginate hydrogel beads [120]. When encapsulated, the hBMSCs remained viable, osteodifferentiated, and synthesized bone minerals. Hence, hBMSCs-encapsulating CPC show promising results for bone tissue engineering applications [98].

In addition, the delivery of genetic material (e.g., DNA, RNA) can induce bone formation [96]. It has been reported that adenoviral vectors can be incorporated in synthetic CPC to provide locally effective and sustained release of genetic material [123]. The delivery of a plasmid encoding vascular endothelial growth factor(165) (pVEGF165) via collagen/CaP carrier resulted *in vivo* in an increase in bone volume [97]. Besides the delivery of plasmids encoding growth factors, the delivery of numerous signaling molecules involved in bone development and regeneration, such as the transcription factors Runt-related transcription factor 2 (Runx2) and Osterix, shows promise for the regeneration of bone *in vivo* [124].

5 Clinical Use of CPC Composites

Depending on its application, different types of CPC can be prepared by variations in CaP compounds powders, the liquid phase, and additional compounds such as biodegradable polymers. These different types of CPC have promising characteristics (biocompatibility, injectability, moldabillity, ability to load bioactive molecules, and osteoconductive and osteoinductive properties) for various clinical applications in the field of dentistry and orthopedics. An overview of various (pre)clinical applications of CPC is described below.

5.1 Dental Applications

The forces applied on a biomaterial during dental application are complex and include bending, torsion, tension, and compression. The poor mechanical properties of CaP are a challenge to overcome; CPC usually has diametral tensile strength lower than 20 MPa. However, different clinical applications of CPC are studied (e.g., the use of CPC for alveolar ridge augmentation [125], or for dental pulp capping [126], CPC as filler around dental implants [127], and the use of CPC for the treatment of periodontal defects [128]) in different animal models. The outcome of the different studies is promising, although further studies to determine the clinical value for dental applications are necessary [9].

5.2 Orthopedic Applications

To date, the best documented use of CPC in the field of orthopedics is its use in augmentation of the tibial plateau fractures. Often after fracturing the tibial plateau, depression of the articular fragments and instability occur, which makes surgery necessary. A metaphyseal void outlined by crushed cancellous bone is often observed after elevating the depressed articular fragments. Over the years, new angle-stable plate systems and rafting techniques decreased the need to fill the void. However, in certain cases, the void still needs to be filled to support the articular fragments. Several clinical trials show fewer complications and better stability of the articular fragment with the use of CPC [129, 130]. In recent years, several (pre) clinical studies of the use of CPC at other anatomic locations have been reported; for example, distal radius, calcaneus, proximal humerus, hip fractions, and tibial plateau [129]. In addition, CPC are used in vertebroplasty [129]. In osteoporotic patients, vertebral compression fractures are common and often painful. Early findings suggest a role for CPC because of their high compression strength.

5.3 Other Applications

Loading antibiotics and/or anti-inflammatory drugs to CPC makes it a good candidate material to treat dental, periodontal, and bone infections. For instance, dideoxy-kanamycin B [90] or vancomycin [91] impregnated CPC seems promising to treat (chronic) osteomyelitis. Another application for CPC may be the treatment of the bone defect created after the removal of a musculoskeletal tumor [131]. In addition, CPC can be loaded with chemotherapeutic agents to prevent the risk of recurrence [132, 133].

5.4 Future Perspectives

The limitation to use CPC clinically is its limited degradation and its relatively low mechanical strength regarding shear, bending, and tensile strength. To overcome the limitation in material degradation, different compounds (water soluble additives, biodegradable polymeric microspheres, foam-forming agents) can be added to increase the porosity of the material. In addition, bioactive molecules can be incorporated into the CPC for an increased bone regeneration.

Nevertheless, the low mechanical strength is a severe drawback for the application in load-bearing applications. However, by the inclusion of compounds to improve the mechanical strength, such as fibers, the CPC may ultimately be used for load-bearing applications.

When polymer/cement degradation is fully controllable, the combination of added compounds, biodegradable polymer microparticles, and therapeutic and/or bioactive agents to the CPC may result in the most attractive composition for the regeneration of bone.

In addition, the injectability and moldability of the CPC are promising for less invasive and faster surgery as compared to preset ceramic bone substitutes. At present, the preparation of CPC is not standardized and should be optimized to create an off-the-shelf product that is easily applicable by clinicians. A mixing system consisting of two (or more) compartments for the liquid and powder phases of CPC would be of interest to develop. With the automatic mixing cartridge, a clinician can automatically mix the different compounds, after which the obtained ceramic paste can be directly injected into the bone defect.

In conclusion, the treatment of bone defects with CPC in dental, orthopedic, and other applications is promising. Therefore, it is likely that CPC will rapidly become the treatment modality of choice in the armamentarium for the treatment of bone defects.

References

1. Hartman EHM, Spauwen PHM, Jansen JA (2002) Donor-site complications in vascularized bone flap surgery. J Invest Surg 15(4):185–197
2. Bodde EW, de Visser E, Duysens JEJ et al (2003) Donor-site morbidity after free vascularized autogenous fibular transfer: subjective and quantitative analyses. Plast Reconstr Surg 111(7):2237–2242

3. Pape HC, Evans A, Kobbe P (2010) Autologous bone graft: properties and techniques. J Orthop Trauma 24(Suppl 1):S36–S40
4. Graham SM, Leonidou A, Aslam-Pervez N et al (2010) Biological therapy of bone defects: the immunology of bone allo-transplantation. Expert Opin Biol Ther 10(6):885–901
5. Vandevord P, Nasser S, Wooley P (2005) Immunological responses to bone soluble proteins in recipients of bone allografts. J Orthop Res 23(5):1059–1064
6. Ruhé P, Wolke J, Spauwen P et al (2005) Calcium phosphate ceramics for bone tissue engineering. In: Bronzino J (ed) Biomedical engineering handbook, section tissue engineering, 3rd edn. CRC Press, Connecticut
7. Habraken WJEM, Wolke JGC, Jansen JA (2007) Ceramic composites as matrices and scaffolds for drug delivery in tissue engineering. Adv Drug Deliv Rev 59(4–5):234–248
8. Chow L (2009) Next generation calcium phosphate-based biomaterials. Dent Mater J 28:1–10
9. Ambard A, Mueninghoff L (2006) Calcium phosphate cement: review of mechanical and biological properties. J Prosthodont 15(5):321–328
10. LeGeros R (2002) Properties of osteoconductive biomaterials: calcium phosphates. Clin Orthop Relat Res 395:81–98
11. Yuan H, Li Y, de Bruin J et al (2000) Tissue responses of calcium phosphate cement: a study in dogs. Biomaterials 21(12):1283–1290
12. Arisan V, Ozdemir T, Anil A et al (2008) Injectable calcium phosphate cement as a bone-graft material around peri-implant dehiscence defects: a dog study. Int J Oral Maxillofac Implants 23(6):1053–1062
13. Ooms E, Wolke J, van der Waerden J et al (2002) Trabecular bone response to injectable calcium phosphate (ca-p) cement. J Biomed Mater Res 61(1):9–18
14. Ooms EM, Wolke JGC, van de Heuvel MT et al (2003) Histological evaluation of the bone response to calcium phosphate cement implanted in cortical bone. Biomaterials 24(6):989–1000
15. Cavalcanti SCSXB, Pereira CL, Mazzonetto R et al (2008) Histological and histomorphometric analyses of calcium phosphate cement in rabbit calvaria. J Craniomaxillofac Surg 36(6):354–359
16. Youji M, Kunio I, Masaaki T et al (1997) Tissue response to fast-setting calcium phosphate cement in bone. J Biomater Mater Res 37(4):457–464
17. Constantz B, Barr B, Ison I et al (1998) Histological, chemical, and crystallographic analysis of four calcium phosphate cements in different rabbit osseous sites. J Biomed Mater Res 43(4):451–461
18. LeGeros (1991) Calcium phosphate in oral biology and medicine. Karger, Basel
19. Brown PW, Martin RI (1999) An analysis of hydroxyapatite surface layer formation. J Phys Chem B 103(10):1671–1675
20. Constantz BR, Ison IC, Fulmer MT et al (1995) Skeletal repair by in situ formation of the mineral phase of bone. Science 267(5205):1796–1799
21. Bohner M (2000) Calcium orthophosphates in medicine: from ceramics to calcium phosphate cements. Injury 31(Supplement 4):D37–D47
22. Apelt D, Theiss F, El-Warrak AO et al (2004) In vivo behavior of three different injectable hydraulic calcium phosphate cements. Biomaterials 25(7–8):1439–1451
23. Bohner M, Merkle HP, Landuyt PV et al (2000) Effect of several additives and their admixtures on the physico-chemical properties of a calcium phosphate cement. J Mater Sci Mater Med 11(2):111–116
24. Driessens FCM, Boltong MG, Maeyer EAPD et al (2000) Effect of temperature and immersion on the setting of some calcium phosphate cements. J Mater Sci Mater Med 11(7):453–457
25. Viano AM, Auwarter JA, Rho JY et al (2001) Ultrasonic characterization of the curing process of hydroxyapatite-modified bone cement. J Biomed Mater Res 56(4):593–599
26. Nicholson JW, Abiden F (1998) Studies on the setting of polyelectrolyte cements: part vi the effect of halide salts on the mechanical properties and water balance of zinc polycarboxylate and glass–ionomer dental cements. J Mater Sci Mater Med 9(5):269–272

27. Hofmann MP, Nazhat SN, Gbureck U et al (2006) Real-time monitoring of the setting reaction of brushite-forming cement using isothermal differential scanning calorimetry. J Biomed Mater Res B 79B(2):360–364
28. Doi Y, Shimizu Y, Moriwaki Y et al (2001) Development of a new calcium phosphate cement that contains sodium calcium phosphate. Biomaterials 22(8):847–854
29. Yokoyama A, Yamamoto S, Kawasaki T et al (2002) Development of calcium phosphate cement using chitosan and citric acid for bone substitute materials. Biomaterials 23(4):1091–1101
30. Miyazaki K, Horibe T, Antonucci JM et al (1993) Polymeric calcium phosphate cements: analysis of reaction products and properties. Dent Mater 9(1):41–45
31. Majekodunmi A, Deb S (2007) Poly(acrylic acid) modified calcium phosphate cements: the effect of the composition of the cement powder and of the molecular weight and concentration of the polymeric acid. J Mater Sci Mater Med 18(9):1883–1888
32. Takechi M, Miyamoto Y, Ishikawa K et al (1998) Initial histological evaluation of anti-washout type fast-setting calcium phosphate cement following subcutaneous implantation. Biomaterials 19(22):2057–2063
33. Matsuya Y, Antonucci JM, Matsuya S et al (1996) Polymeric calcium phosphate cements derived from poly(methyl vinyl ether-maleic acid). Dent Mater 12(1):2–7
34. Matsuya Y, Matsuya S, Antonucci JM et al (1999) Effect of powder grinding on hydroxyapatite formation in a polymeric calcium phosphate cement prepared from tetracalcium phosphate and poly(methyl vinyl ether-maleic acid). Biomaterials 20(7):691–697
35. Lu J, Descamps M, Dejou J et al (2002) The biodegradation mechanism of calcium phosphate biomaterials in bone. J Biomater Mater Res B 63(4):408–412
36. Charrière E, Terrazzoni S, Pittet C et al (2001) Mechanical characterization of brushite and hydroxyapatite cements. Biomaterials 22(21):2937–2945
37. Hu G, Xiao L, Fu H et al (2010) Study on injectable and degradable cement of calcium sulphate and calcium phosphate for bone repair. J Mater Sci Mater Med 21(2):627–634
38. Chung U, Kawaguchi H, Takato T et al (2004) Distinct osteogenic mechanisms of bones of distinct origins. J Orthop Sci 9(4):410–414
39. Ginebra MP, Driessens FCM, Planell JA (2004) Effect of the particle size on the micro and nanostructural features of a calcium phosphate cement: a kinetic analysis. Biomaterials 25(17):3453–3462
40. Espanol M, Perez RA, Montufar EB et al (2009) Intrinsic porosity of calcium phosphate cements and its significance for drug delivery and tissue engineering applications. Acta Biomater 5(7):2752–2762
41. Hulbert S, Morrison S, Klawitter J (1972) Tissue reaction to three ceramics of porous and non-porous structures. J Biomed Mater Res 6:347–374
42. Eggli P, Muller W, Schenk R (1988) Porous hydroxyapatite and tricalcium phosphate cylinders with two different pore size ranges implanted in cancellous bone of rabbits. A comparative histomorphometric and histologic study of bony ingrowth and implant substitution. Clin Orthop Relat Res (232):127–138
43. De Groot K (1988) Effect of porosity and physicochemical properties on the stability, resorption, and strenght of calcium phosphate ceramics. Ann N Y Acad Sci 523:227–233
44. el-Ghannam A, Ducheyne P, Shapiro 1 (1995) Bioactive material template for in vitro synthesis of bone. J Biomed Mater Res 29:359–370
45. Chang BS, Lee CK, Hong KS et al (2000) Osteoconduction at porous hydroxyapatite with various pore configurations. Biomaterials 21(12):1291–1298
46. Wang X, Ye J, Wang Y (2007) Influence of a novel radiopacifier on the properties of an injectable calcium phosphate cement. Acta Biomater 3(5):757–763
47. Takagi S, Chow LC (2001) Formation of macropores in calcium phosphate cement implants. J Mater Sci Mater Med 12(2):135–139

48. Girod Fullana S, Ternet H, Freche M et al (2010) Controlled release properties and final macroporosity of a pectin microspheres-calcium phosphate composite bone cement. Acta Biomater 6(6):2294–2300
49. Lian Q, Li D, He J et al (2008) Mechanical properties and in vivo performance of calcium phosphate cement-chitosan fibre composite. Proc Inst Mech Eng H 222:347–353
50. Tajima S, Kishi Y, Oda M et al (2006) Fabrication of biporous low-crystalline apatite based on mannitol dissolution from apatite cement. Dent Mater J 25(3):616–620
51. Hu G, Xiao L, Fu H et al (2009) Degradable and bioactive scaffold of calcium phosphate and calcium sulphate from self-setting cement for bone regeneration. J Porous Mater 17 (5):605–613
52. Barralet JE, Grover L, Gaunt T et al (2002) Preparation of macroporous calcium phosphate cement tissue engineering scaffold. Biomaterials 23(15):3063–3072
53. Habraken W, de Jonge L, Wolke J et al (2008) Introduction of gelatin microspheres into an injectable calcium phosphate cement. J Biomed Mater Res A 87A(3):643–655
54. Habraken WJEM, Wolke JGC, Mikos AG et al (2009) Porcine gelatin microsphere/calcium phosphate cement composites: an in vitro degradation study. J Biomed Mater Res B 91B (2):555–561
55. Link DP, van den Dolder J, van den Beucken JJJP et al (2009) Evaluation of an orthotopically implanted calcium phosphate cement containing gelatin microparticles. J Biomed Mater Res A 90A(2):372–379
56. Zuo Y, Yang F, Wolke JGC et al (2010) Incorporation of biodegradable electrospun fibers into calcium phosphate cement for bone regeneration. Acta Biomater 6(4):1238–1247
57. Bodde E, Cammaert C, Wolke J et al (2007) Investigation as to the osteoinductivity of macroporous calcium phosphate cement in goats. J Biomed Mater Res B 83B(1):161–168
58. Habraken W, Wolke J, Mikos A et al (2006) Injectable plga microsphere/calcium phosphate cements: physical properties and degradation characteristics. J Biomater Sci Polym Ed 17(9):1057–1074
59. Link D, van den JD, Beucken Jvd et al (2008) Evaluation of the biocompatibility of calcium phosphate cement/plga microparticle composites. J Biomater Mater Res A 87A(3):760–769
60. Habraken WJEM, Zhang Z, Wolke JGC et al (2008) Introduction of enzymatically degradable poly(trimethylene carbonate) microspheres into an injectable calcium phosphate cement. Biomaterials 29(16):2464–2476
61. Habraken WJEM, Liao HB, Zhang Z et al (2010) In vivo degradation of calcium phosphate cement incorporated into biodegradable microspheres. Acta Biomater 6(6):2200–2211
62. Almirall A, Larrecq G, Delgado JA et al (2004) Fabrication of low temperature macroporous hydroxyapatite scaffolds by foaming and hydrolysis of an α-tcp paste. Biomaterials 25 (17):3671–3680
63. del Real R, Wolke J, Vallet-Regí M et al (2002) A new method to produce macropores in calcium phosphate cements. Biomaterials 23(17):3673–3680
64. del Real R, Ooms E, Wolke J et al (2003) In vivo bone response to porous calcium phosphate cement. J Biomed Mater Res A 65A(1):30–36
65. Hesaraki S, Sharifi D (2007) Investigation of an effervescent additive as porogenic agent for bone cement macroporosity. Biomed Mater Eng 17(1):29–38
66. Saeed H, Ali Z, Fatollah M (2008) The influence of the acidic component of the gas-foaming porogen used in preparing an injectable porous calcium phosphate cement on its properties: acetic acid versus citric acid. J Biomater Mater Res B 86B(1):208–216
67. Miño-Fariña N, Muñoz-Guzón F, López-Peña M et al (2009) Quantitative analysis of the resorption and osteoconduction of a macroporous calcium phosphate bone cement for the repair of a critical size defect in the femoral condyle. Vet J 179(2):264–272
68. del Valle S, Miño N, Muñoz F et al (2007) In vivo evaluation of an injectable macroporous calcium phosphate cement. J Mater Sci Mater Med 18(2):353–361
69. Ginebra M-P, Delgado J-A, Harr I et al (2007) Factors affecting the structure and properties of an injectable self-setting calcium phosphate foam. J Biomed Mater Res A 80A(2):351–361

70. Dagang G, Kewei X, Yong H (2009) The in situ synthesis of biphasic calcium phosphate scaffolds with controllable compositions, structures, and adjustable properties. J Biomater Mater Res A 88A(1):43–52
71. Xu S, Li D, Wang C et al (2007) Cell proliferation in cpc scaffold with a central channel. Biomed Mater Eng 17(1):1–8
72. Li X, Li D, Lu B et al (2007) Fabrication and evaluation of calcium phosphate cement scaffold with controlled internal channel architecture and complex shape. Proc Inst Mech Eng H 221:951–958
73. Sarda S, Nilsson M, Balcells M et al (2003) Influence of surfactant molecules as air-entraining agent for bone cement macroporosity. J Biomed Mater Res A 65A(2):215–221
74. Wang X, Ye J, Li X et al (2008) Production of in-situ macropores in an injectable calcium phosphate cement by introduction of cetyltrimethyl ammonium bromide. J Mater Sci Mater Med 19(10):3221–3225
75. Ruhé P, Hedberg-Dirk E, Padron NT et al (2006) Porous poly(dl-lactic-co-glycolic acid)/calcium phosphate cement composite for reconstruction of bone defects. Tissue Eng 12(4):789–800
76. Xu HHK, Weir MD, Simon CG (2008) Injectable and strong nano-apatite scaffolds for cell/ growth factor delivery and bone regeneration. Dent Mater 24(9):1212–1222
77. Xu H, Carey L, Simon C (2007) Premixed macroporous calcium phosphate cement scaffold. J Mater Sci Mater Med 18(7):1345–1353
78. Ginebra MP, Traykova T, Planell JA (2006) Calcium phosphate cements as bone drug delivery systems: a review. J Control Release 113(2):102–110
79. Ruhé P, Boerman O, Russel F et al (2006) In vivo release of rhbmp-2 loaded porous calcium phosphate cement pretreated with albumin. J Mater Sci Mater Med 17(10):919–927
80. Tessmar JK, Göpferich AM (2007) Matrices and scaffolds for protein delivery in tissue engineering. Adv Drug Deliv Rev 59(4–5):274–291
81. Kretlow JD, Klouda L, Mikos AG (2007) Injectable matrices and scaffolds for drug delivery in tissue engineering. Adv Drug Deliv Rev 59(4–5):263–273
82. Alkhraisat MH, Rueda C, Cabrejos-Azama J et al (2010) Loading and release of doxycycline hyclate from strontium-substituted calcium phosphate cement. Acta Biomater 6(4):1522–1528
83. Masaaki T, Youji M, Kunio I et al (1998) Effects of added antibiotics on the basic properties of anti-washout-type fast-setting calcium phosphate cement. J Biomater Mater Res 39(2):308–316
84. Siepmann J, Peppas NA (2001) Modeling of drug release from delivery systems based on hydroxypropyl methylcellulose (hpmc). Adv Drug Deliv Rev 48(2–3):139–157
85. Xu H, Quinn J, Takagi S et al (2001) Strong and macroporous calcium phosphate cement: effects of porosity and fiber reinforcement on mechanical properties. J Biomed Mater Res 57(3):457–466
86. Ruhé P, Hedberg E, Padron NT et al (2005) Biocompatibility and degradation of poly (dl-lactic-co-glycolic acid)/calcium phosphate cement composites. J Biomed Mater Res A 74A(4):533–544
87. Chen D, Zhao M, Mundy GR (2004) Bone morphogenetic proteins. Growth Factors 22(4):233–241
88. Woo BH, Fink BF, Page R et al (2001) Enhancement of bone growth by sustained delivery of recombinant human bone morphogenetic protein-2 in a polymeric matrix. Pharm Res 18(12):1747–1753
89. Habraken WJEM, Boerman OC, Wolke JGC et al (2009) In vitro growth factor release from injectable calcium phosphate cements containing gelatin microspheres. J Biomater Mater Res A 91A(2):614–622
90. Kisanuki O, Yajima H, Umeda T et al (2007) Experimental study of calcium phosphate cement impregnated with dideoxy-kanamycin b. J Orthop Sci 12(3):281–288
91. Urabe K, Naruse K, Hattori H et al (2009) In vitro comparison of elution characteristics of vancomycin from calcium phosphate cement and polymethylmethacrylate. J Orthop Sci 14(6):784–793

92. Centrella M, Horowitz MC, Wozney JM et al (1994) Transforming growth factor-β gene family members and bone. Endocr Rev 15(1):27–39
93. Cao X, Chen D (2005) The bmp signaling and in vivo bone formation. Gene 357(1):1–8
94. Bodde E, Boerman O, Russel F et al (2008) The kinetic and biological activity of different loaded rhbmp-2 calcium phosphate cement implants in rats. J Biomed Mater Res A 87A (3):780–791
95. Blom E, Klein-Nulend J, Yin L et al (2001) Transforming growth factor-β1 incorporated in calcium phosphate cement stimulates osteotransductivity in rat calvarial bone defects. Clin Oral Implants Res 12(6):609–616
96. Phillips JE, Gersbach CA, García AJ (2007) Virus-based gene therapy strategies for bone regeneration. Biomaterials 28(2):211–229
97. Keeney M, van den Beucken JJJP, van der Kraan PM et al (2010) The ability of a collagen/calcium phosphate scaffold to act as its own vector for gene delivery and to promote bone formation via transfection with vegf165. Biomaterials 31(10):2893–2902
98. Michael DW, Hockin HKX, Carl GS Jr (2006) Strong calcium phosphate cement-chitosan-mesh construct containing cell-encapsulating hydrogel beads for bone tissue engineering. J Biomater Mater Res A 77A(3):487–496
99. Bai BMD, Yin Z, Xu Q et al (2009) Histological changes of an injectable rhbmp-2/calcium phosphate cement in vertebroplasty of rhesus monkey. Spine 34(18):1887–1892
100. Perrier M, Lu Y, Nemke B et al (2008) Acceleration of second and fourth metatarsal fracture healing with recombinant human bone morphogenetic protein-2/calcium phosphate cement in horses. Vet Surg 37(7):648–655
101. Blom EJ, Klein-Nulend J, Klein CPAT et al (2000) Transforming growth factor-beta1 incorporated during setting in calcium phosphate cement stimulates bone cell differentiation in vitro. J Biomed Mater Res 50(1):67–74
102. Link DP, van den Dolder J, van den Beucken JJ et al (2008) Bone response and mechanical strength of rabbit femoral defects filled with injectable cap cements containing tgf-[beta]1 loaded gelatin microparticles. Biomaterials 29(6):675–682
103. Gerber HP, Ferrara N (2000) Angiogenesis and bone growth. Trends Cardiovasc Med 10 (5):223–228
104. Soker S, Machado M, Atala A (2000) Systems for therapeutic angiogenesis in tissue engineering. World J Urol 18(1):10–18
105. Sheridan MH, Shea LD, Peters MC et al (2000) Bioabsorbable polymer scaffolds for tissue engineering capable of sustained growth factor delivery. J Control Release 64(1–3):91–102
106. Lee KY, Peters MC, Mooney DJ (2003) Comparison of vascular endothelial growth factor and basic fibroblast growth factor on angiogenesis in SCID mice. J Control Release 87 (1–3):49–56
107. Drake C, Little C (1995) Exogenous vascular endothelial growth factor induces malformed and hyperfused vessels during embryonic neovascularization. Proc Natl Acad Sci USA 92 (17):7657–7661
108. Zisch AH, Lutolf MP, Hubbell JA (2006) Biopolymeric delivery matrices for angiogenic growth factors. Cardiovasc Pathol 12(6):295–310
109. Patel Z, Ueda H, Yamamoto M et al (2008) In vitro and in vivo release of vascular endothelial growth factor from gelatin microparticles and biodegradable composite scaffolds. Pharm Res 25(10):2370–2378
110. Darnell K, Zhuo W, Kim H et al (2006) Vegf scaffolds enhance angiogenesis and bone regeneration in irradiated osseous defects. J Bone Miner Res 21(5):735–744
111. Patel ZS, Young S, Tabata Y et al (2008) Dual delivery of an angiogenic and an osteogenic growth factor for bone regeneration in a critical size defect model. Bone 43(5):931–940
112. Kakudo N, Kusumoto K, Wang YB et al (2006) Immunolocalization of vascular endothelial growth factor on intramuscular ectopic osteoinduction by bone morphogenetic protein-2. Life Sci 79(19):1847–1855

113. Wang L, Huang Y, Pan K et al (2010) Osteogenic responses to different concentrations/ratios of bmp-2 and bfgf in bone formation. Ann Biomed Eng 38(1):77–87
114. Alam S, Ueki K, Marukawa K et al (2007) Expression of bone morphogenetic protein 2 and fibroblast growth factor 2 during bone regeneration using different implant materials as an onlay bone graft in rabbit mandibles. Oral Surg Oral Med Oral Pathol Oral Radiol Endod 103(1):16–26
115. Mundy G, Garrett R, Harris S et al (1999) Stimulation of bone formation in vitro and in rodents by statins. Science 286(5446):1946–1949
116. Notoya K, Yoshida K, Tsukuda R et al (1994) Effect of ipriflavone on expression of markers characteristic of the osteoblast phenotype in rat bone marrow stromal cell culture. J Bone Miner Res 9(3):395–400
117. Barralet J, Gbureck U, Habibovic P et al (2009) Angiogenesis in calcium phosphate scaffolds by inorganic copper ion release. Tissue Eng Part A 15(7):1601–1609
118. Notoya K, Nagai H, Oda T et al (1999) Enhancement of osteogenesis in vitro and in vivo by a novel osteoblast differentiation promoting compound, tak-778. Pharmacol Exp Ther 290 (3):1054–1064
119. Zhao J, Ohba S, Komiyama Y et al (2010) Icariin: a potential osteoinductive compound for bone tissue engineering. Tissue Eng Part A 16(1):233–243
120. Weir MD, Xu HHK (2010) Human bone marrow stem cell-encapsulating calcium phosphate scaffolds for bone repair. Acta Biomater 6(10):4118–4126
121. Shanti RM, Li W-J, Nesti LJ et al (2007) Adult mesenchymal stem cells: biological properties, characteristics, and applications in maxillofacial surgery. J Oral Maxillofac Surg 65(8):1640–1647
122. Link DP, van den Dolder J, Wolke JGC et al (2007) The cytocompatibility and early osteogenic characteristics of an injectable calcium phosphate cement. Tissue Eng 13(3):493–500
123. Kirschner RE, Karmacharya J, Ong G et al (2001) Synthetic hybrid grafts for craniofacial reconstruction: sustained gene delivery using a calcium phosphate bone mineral substitute. Ann Plast Surg 46(5):538–545
124. Takeuchi H, Nagayama M, Imaizumi Y et al (2009) Immunohistochemical analysis of osteoconductivity of beta-tricalciumphosphate and carbonate apatite applied in femoral and parietal bone defects in rats. Dent Mater J 28(5):595–601
125. Sugawara A, Fujikawa K, Kusama K et al (2002) Histopathologic reaction of a calcium phosphate cement for alveolar ridge augmentation. J Biomater Mater Res 61(1):47–52
126. Qingyi S, Jiao S, Jie W et al (2010) An in vitro investigation of the mechanical–chemical and biological properties of calcium phosphate/calcium silicate/bismutite cement for dental pulp capping. J Biomater Mater Res B 94B(1):141–148
127. Comuzzi L, Ooms E, Jansen JA (2002) Injectable calcium phosphate cement as a filler for bone defects around oral implants: an experimental study in goats. Clin Oral Implants Res 13(3):304–311
128. Hayashi C, Kinoshita A, Oda S et al (2006) Injectable calcium phosphate bone cement provides favorable space and a scaffold for periodontal regeneration in dogs. J Periodontol 77(6):940–946
129. Larsson S (2010) Calcium phosphates: what is the evidence? J Orthop Trauma 24(Supplement 1):S41–S45
130. Russell TA, Leighton RK (2008) Comparison of autogenous bone graft and endothermic calcium phosphate cement for defect augmentation in tibial plateau fractures. A multicenter, prospective, randomized study. J Bone Joint Surg Am 90(10):2057–2061
131. Matsumine A, Kusuzaki K, Matsubara T et al (2006) Calcium phosphate cement in musculoskeletal tumor surgery. J Surg Oncol 93(3):212–220
132. Verron E, Khairoun I, Guicheux J et al (2010) Calcium phosphate biomaterials as bone drug delivery systems: a review. Drug Discov Today 15(13–14):547–552
133. Tani T, Okada K, Takahashi S et al (2006) Doxorubicin-loaded calcium phosphate cement in the management of bone and soft tissue tumors. In Vivo 20(1):55–60

Biography

Floor CJ van de Watering obtained her Bachelor degree in Applied Science in 2004 at the Avans Hogeschool in Breda, The Netherlands. In 2004, she continued education at the Radboud University Nijmegen (Master Medical Biology), the Netherlands, and graduated in 2008. During her study she conducted research internships at the department of Cellular Animal Physiology, Radboud University Nijmegen, and with the Erasmus program at the department of Cellular Biology, Physiology and Immunology, University of Cordoba, Spain. In May 2008, she started as a PhD-student dealing with the development and biological evaluation of novel degradable calcium phosphate cement as bone substitute material in the department of Biomaterials at the Radboud University Nijmegen, The Netherlands.

Prof. Jeroen JJP van den Beucken studied Animal Husbandry & Health (1998; Den Bosch, the Netherlands) and Medical Biology (2002; Nijmegen, the Netherlands). In 2002, he started as a PhD-student at the Dept. Periodontology & Biomaterials (Radboud University Nijmegen Medical Center, Nijmegen, the Netherlands) on a research project involving the development and biological evaluation of DNA-based coatings for biomedical implants. In 2007, he was appointed as scientific researcher at the Dept. Biomaterials (Radboud University Nijmegen Medical Center, Nijmegen, the Netherlands). Since 2008, he is appointed as Assistant Professor at the same department.

Rosa P. Félix Lanao received her Master degrees in Medical Biology (2004) and in Biochemistry (2005), both from the University of Navarre, Pamplona, Spain. After an internship in the department of Pharmaceutics at NV Organon (now MSD) Oss, The Netherlands, in September 2005 she started working as a researcher in the same department. In April 2008 she began her Ph.D. research dealing with degradable calcium phosphate as bone substitute material in the department of Biomaterials at the Radboud University Nijmegen, The Netherlands.

Prof. Joop GC Wolke worked from 1976 as a research assistant at the department of Biomaterials of the Free University Medical Center in Amsterdam. In 1989 he moved to Leiden University Medical Center. He studied from 1990 part-time chemistry at the University of Amsterdam and graduated in 1995. From Augustus 1992 he worked at the Radboud University Nijmegen Medical Center Nijmegen, Department of Biomaterials and Periodontology. His research project dealt with sputtering of implant material. After obtaining a PhD degree, he continued to work in the department as assistant professor. He has contributed more than 100 publications.

Prof. John A. Jansen studied Dentistry at Radboud University Nijmegen and graduated in 1977. In the same year he started a part-time dental practice in Maassluis, the Netherlands. He completed his Ph.D. dealing with the adhesion of epithelial cells to dental implant materials in 1984 at the Radboud University Nijmegen. After working as assistant professor in Amsterdam and Leiden he returned to Nijmegen in 1991 to become associate professor of Biomaterials and Implantology before being appointed Full Professor of the department Periodontology and Biomaterials in April 1996. He has contributed over 400 publications, is the owner of seven patents and editorial board member/editor of eight international scientific journals. Currently, 25 Ph.D. students work under his supervision.

Chapter 8
Enzyme-Promoted Degradation of Polymeric Matrices for Controlled Drug Delivery: Analytical Model and Numerical Simulations

Tomer Gold, Rosa Azhari, and Noah Lotan

Abstract Controlled drug delivery is one of the main avenues along which the multibillion dollar pharmaceutical industry is concentrating its efforts today. This emphasis is dictated by attempts to improve therapy efficiency and patient compliance, as well as by most stringent economic constraints.

This chapter presents the design of an implantable, degradable, drug delivery device, the function of which is controlled by the concentration and activity of a given enzyme present at the site of implantation. Also, using an appropriately developed analytical model, the performance of this device is assessed in terms of its geometrical characteristics and functional parameters. The engineering procedures thus developed provide the efficient tools required for both the rational design and performance analysis of such devices.

1 Introduction

1.1 Background

Nowadays, the world of therapeutics undergoes profound transformations. During the coming 5 years, a number of brand drugs—of combined market value estimated at about \$50B—will lose their patent coverage. In addition, developing new drugs—particularly in the biopharmaceuticals class—becomes increasingly costly

This chapter is taken in part from the M.Sc. thesis of Tomer Gold, submitted to the Senate of the Technion—Israel Institute of Technology, in 2001.

T. Gold • N. Lotan (✉)
Faculty of Biomedical Engineering, Technion—Israel Institute of Technology, Haifa, Israel
e-mail: Noah@bm.technion.ac.il

R. Azhari
Department of Biotechnology Engineering, ORT Braude College, Karmiel, Israel

N. Eliaz (ed.), *Degradation of Implant Materials*, DOI 10.1007/978-1-4614-3942-4_8,

and time consuming, the pertinent estimates being some $1B and 10 years, respectively, per drug.

Under these circumstances, intensive efforts of the pharma industry are being invested on alternative avenues, the most promising of which are controlled delivery of already approved drugs, development of biosimilars, and drugs repositioning.

Drug delivery systems (DDSs) are therapeutic modalities aimed at achieving efficient regimens for dosing active pharmaceutical ingredients. Various families of biodegradable polymers have been widely suggested for use in DDSs. In the category of polymers undergoing chemical hydrolysis *in vivo*, homopolyesters derived from glycolic acid, lactic acid, hydroxybutyric acid, hydroxyvaleric acid, ε-hydroxycaproic acid, and their copolymers are most commonly applied [1–9] and are already used in commercial products (e.g., Risperdal Consta [10]). Polyanhydrides have also been approved by the US Food and Drug Administration (FDA), leading to the use of biodegradable wafers for the release of 1,3-Bis(2-chloroethyl)-1-nitrosourea for treatment of brain tumors [11–13]. Polyorthoesters [14] and polyphosphazenes [15] are additional synthetic degradable polymers suggested for use in drug delivery devices. In certain polymers, hydrolysis occurs via surface erosion [13], in which various layers of the polymeric matrix are gradually exposed to the physiological environment, thus resulting in a sequential degradation of the matrix. Alternatively, bulk erosion [16] occurs when hydration of the polymer is faster than the hydrolysis rate.

Enzyme-based DDSs are also intensively considered. In stimuli-responsive devices, the enzymes react with physiological molecules to induce changes in the system (pH, temperature) that lead to a change in polymer properties (e.g., swelling) which, in turn, causes the release of drug [17–21]. For example, glucose oxidase has been used in numerous glucose-sensitive DDSs to induce a pH change in the presence of glucose, triggering swelling in a pH-sensitive polymer that leads to the release of insulin [20, 21]. Some other devices are based on enzyme-promoted degradation of a polymeric matrix or of a link between a polymeric carrier and a pendant drug.

The present report addresses drug delivery devices based on enzyme-promoted degradation of polymeric matrices [22]. Enzyme-degradable matrices for drug delivery include natural polymers such as polysaccharides [23–28], proteins and peptides [29–33], pseudo polyamino acids [34], and synthetic polymers in which enzyme-labile moieties (e.g., peptides) were introduced during polymer synthesis [35–43]. Drug molecules can be conjugated to the polymeric backbone as a pendant chain, incorporated within the polymeric backbone or polymerized into drug chains. Drug release is achieved by enzyme-promoted hydrolysis of the pendant bond or of the polymeric backbone.

Enzyme-promoted degradation of polymeric matrices for controlled drug delivery involves a sequence of events. First, a solution penetrates the polymeric matrix, a wetting front is produced, which may induce matrix swelling. Subsequently, the enzyme penetrates the wetted matrix and initiates cleavage of labile bonds in the polymer, thus leading to the release of drug. Diffusion of the drug moieties outside the system completes the drug delivery process.

The rational design of such systems requires appropriate analytical models that rely on sound physical, chemical, and biochemical principles. Due to their predictive capabilities, such models are also highly desirable for assessing the viability of the concepts involved and the ability of the system considered to perform as required.

Mathematical models have been developed for describing drug delivery from various systems. Empirical as well as analytical models for diffusion, swelling, and erosion-controlled systems have been suggested and reviewed extensively in several publications [44–52]. Diffusion and reaction models of erosion-controlled systems usually concentrate on ones in which bulk erosion or surface erosion proceeds by hydrolysis. No model, yet, has been suggested for enzyme-promoted release from drug delivery devices. On the other hand, analytical models describing the progress of enzymatic degradation of polymers have also been developed [43, 53–58].

In our studies, reported herein, an analytical approach is developed, integrating both the diffusion and the enzymatic degradation phenomena occurring in the drug delivery device. Relying on this approach, we established a computational procedure that allows one to describe the performance of such devices.

In particular, this report addresses the design of a DDS containing a device which is to be fully implanted and which will release the contained drug in a controlled manner, following an enzymatic degradation process. Furthermore, for the system at hand (a) an analytical model is developed that accounts for its performance; (b) the sensitivity of this model is assessed with regard to variations in the values of the geometrical characteristics of the device and of its operational parameters; and (c) the ranges of realistic values acceptable for the operational parameters involved are delineated.

It should be mentioned that polymeric matrices containing enzyme-labile sequences are finding new applications as tissue-engineering scaffolds and also in the design of environmentally friendly plastics [59, 60]. Therefore, the use of the developed models can be extended into other fields.

1.2 Terminology

In order to avoid misunderstandings, the terminology employed herein considers the DDS envisaged as encompassing both the device itself and the surrounding fluids.

2 The Drug Delivery System

The system considered is presented here in terms of the device itself, its mode of operation, and the mass transfer characteristics affecting its performance.

2.1 The Device

The device is a box with a cross-section area "A" and length "a" (see Fig. 8.1). Except for the end facet, with cross-section area A, all other facets of the device are

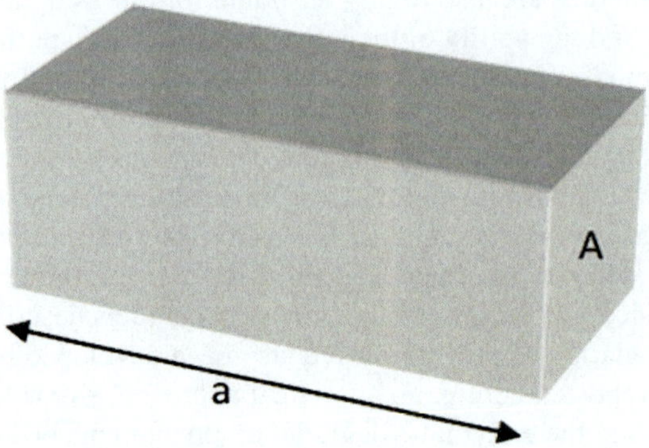

Fig. 8.1 Geometric characteristics of the drug delivery device. "*A*"—The surface area of the facet through which mass transfer occurs; "*a*"—Length of the device

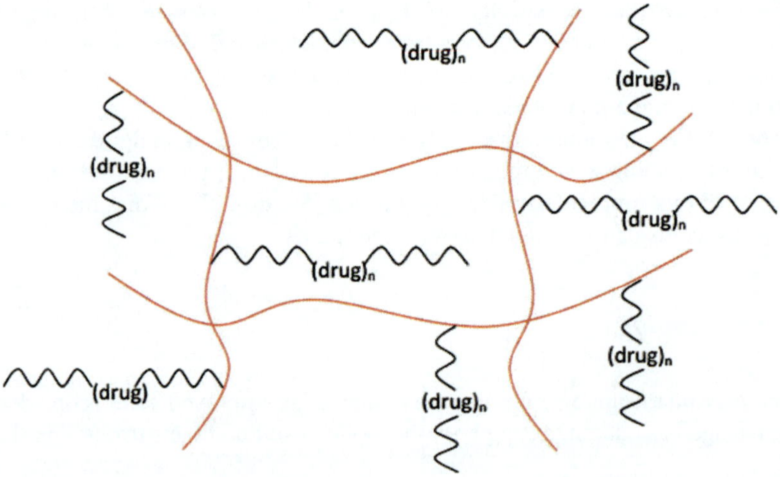

Fig. 8.2 Schematic representation of the content of the drug delivery device. Chains of polymeric drug [~~~(drug)$_n$ ~~~~] are tethered to the inert polymeric network [—] via covalent bonds

impermeable and do not allow the transfer of surrounding fluids from the environment into the device, or of materials from the device outward into the environment. The device contains an inert, nondegradable polymeric network. To this network, linear chains of polymeric drug are covalently tethered at one of their ends (see Fig. 8.2). Before the device is implanted, all its components are in their solid and dry state.

2.2 The Surrounding Fluids

The device (Sect. 8.2.1) is implanted in the body of a patient, at a location where there are fluids around (the "surrounding fluids"). The relevant characteristic of these fluids is that they penetrate the device and, in so doing, they trigger the drug release process.

2.3 The Drug-Releasing Enzyme

This is the critical component of the surrounding fluids. It is its hydrolytic action on the polymeric drug chains that causes the release of free drug moieties within the device and, subsequently, out of the device.

2.4 Mass Transfer Characteristics of the System

Mass transfer into and out of the device takes place only through the end facet of the device of cross-section area A. The inert polymeric network allows the mass transfer within the device to take place freely. The environment outside the device serves as the source of the drug releasing enzyme. The enzyme-containing surrounding fluids penetrate the device and wet its content. The environment outside the device also serves as an "infinite sink" for the released drug.

2.5 System Operation Mode

The system operation is envisaged in terms of three main processes:

(a) Wetting of the device content: As soon as the device is implanted, surrounding fluids start penetrating and wet the device content, also bringing with them the drug-releasing enzyme (see Fig. 8.3).
(b) Enzymatic activity: Once in the device, the enzyme starts hydrolyzing the polymeric drug, as indicated in eq. (8.1). As a result of this process, free drug is produced within the device.
(c) Outward diffusion of the free drug from the device and into the surrounding fluids.

$$R-[\mathrm{Drug}]_n+E\leftrightarrow R-[\mathrm{Drug}]_nE\xrightarrow{k_3}R-[\mathrm{Drug}]_{n-1}+\mathrm{Drug}+E,\qquad(8.1)$$

where R is the inert, nondegradable polymeric network to which polymeric drug chains are tethered, $[\mathrm{Drug}]_n$ is a polymeric drug chain containing n drug moieties, where $1\leq n\leq N$, N is the number of drug moieties originally composing a polymeric drug chain (i.e., before the device is implanted), E is the drug-releasing enzyme, and k_3 is the reaction rate constant for the elementary stage shown.

2.6 Basic Assumptions

For modeling the performance of the system, the following assumptions are made (see Figs. 8.3 and 8.4):

- The polymeric drug chains are linear and do not have cross-linking bonds in between them.
- All of the polymeric drug chains initially contain the same number of residues.
- Within the device, mass transfer occurs along one dimension only, namely along the longitudinal axis of the device.
- The surrounding fluids penetrate the device in a plug-flow mode. Hence, a step-type wetting front is created, and its position along the device is indicated as $r(t)$.
- The enzyme penetration rate depends on its diffusion coefficient and temperature, as well as on the packing of the inert polymeric network and of the polymeric drug chains. Throughout the entire drug release process, there is essentially no significant change in these parameters. Accordingly, the enzyme concentration is assumed to be constant and uniform in the section of the device where it is present.
- The total enzyme concentration, C_{ET}, is constant, and this assumption is valid both inside and outside the device.
- The enzyme hydrolyses the chains of polymeric drug by the exo-single attack mechanism [54, 55]. Accordingly, upon encountering the free end of a polymeric drug chain, the enzyme releases only one drug moiety and, subsequently, leaves that site.
- As a direct consequence of the assumed exo-single attack mechanism of action of the enzyme, the effective concentration of the substrate, C_{eff}, is the concentration of chain ends, which is independent of the polymeric drug chain lengths. C_{eff} is constant and maintains its maximal (i.e., initial) value during most of the enzymatic degradation process. C_{eff} declines when polymeric drug chains undergo complete hydrolysis. In Figs. 8.3 and 8.4, $l(t)$ denotes the point where C_{eff} begins to decline.
- As all polymeric drug chains undergo complete hydrolytic cleavage by the enzyme, an erosion front is established at position $s(t)$ along the device, as shown in Figs. 8.3 and 8.4.
- For the enzymatic process considered, K_{M} and k_3 are the Michaelis–Menten constant and reaction rate constant, respectively. These parameters are assumed not to vary for the entire period of system operation.
- Only free drug moieties diffuse within the device and out of it. Dimers, trimers, and larger oligomers cannot do so, as they are being covalently attached to the inert polymeric network.
- It is also assumed that the diffusion coefficient of the free drug within the device is constant, as long as the concentration and size of the polymeric drug chains do

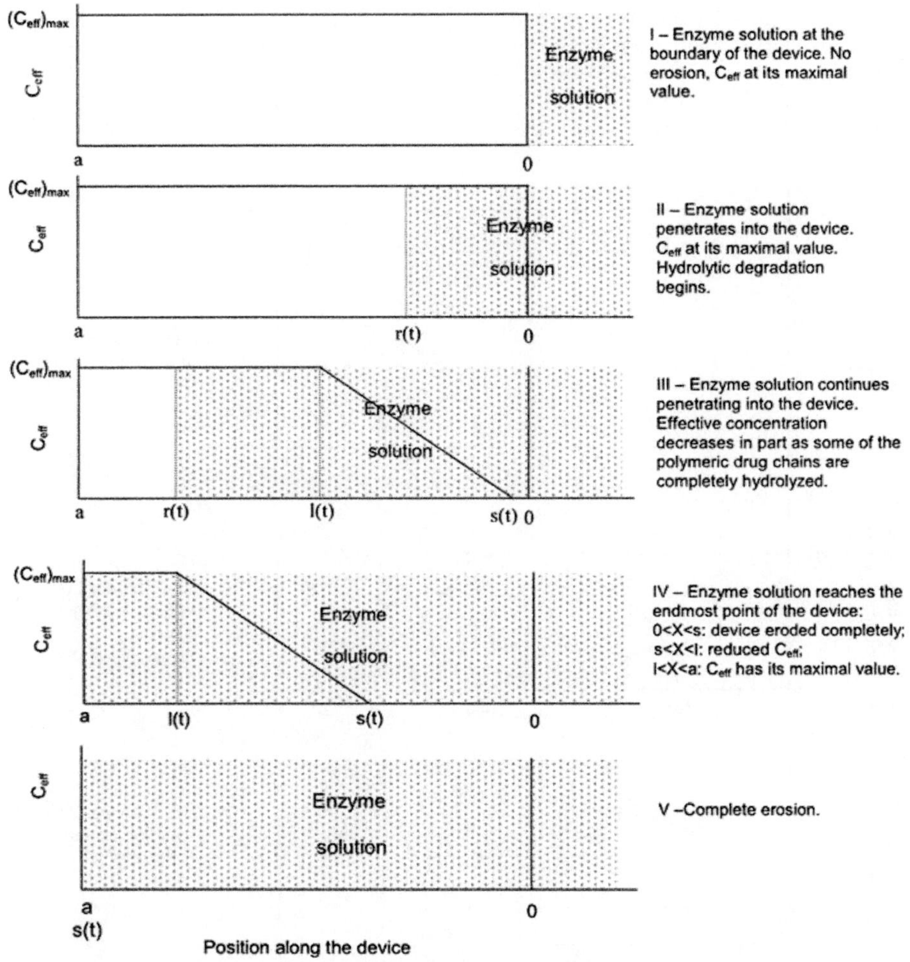

Fig. 8.3 Schematic representation of the progress of the wetting process, enzyme penetration into the device, and the effective concentration of bonds that are actually available to the enzyme. C_{eff} *Panel* I depicts the state of the device at $t = 0$, while *panels* II, III, IV, and V are presentations of subsequent points in time. The concentration profile depicted here is a schematic representation only and is not the outcome of actual calculations. *Shaded area* represents the wetted parts of the device. The *symbols* in this figure indicate: *a*—the endmost point of the device, i.e., the last to be wetted; 0—the point of entrance to the device; *r*(*t*)—position of the wetting boundary; *l*(*t*)—position where the effective substrate concentration starts to decline; *s*(*t*)—position of the erosion front

not decrease below a predetermined threshold. When that threshold is reached, it is considered that all the drug moieties still left in the pertinent part of the device—both in free form and as still part of the tethered polymer chains—are released from there at once.

- No diffusion processes take place in the dry space of the device.

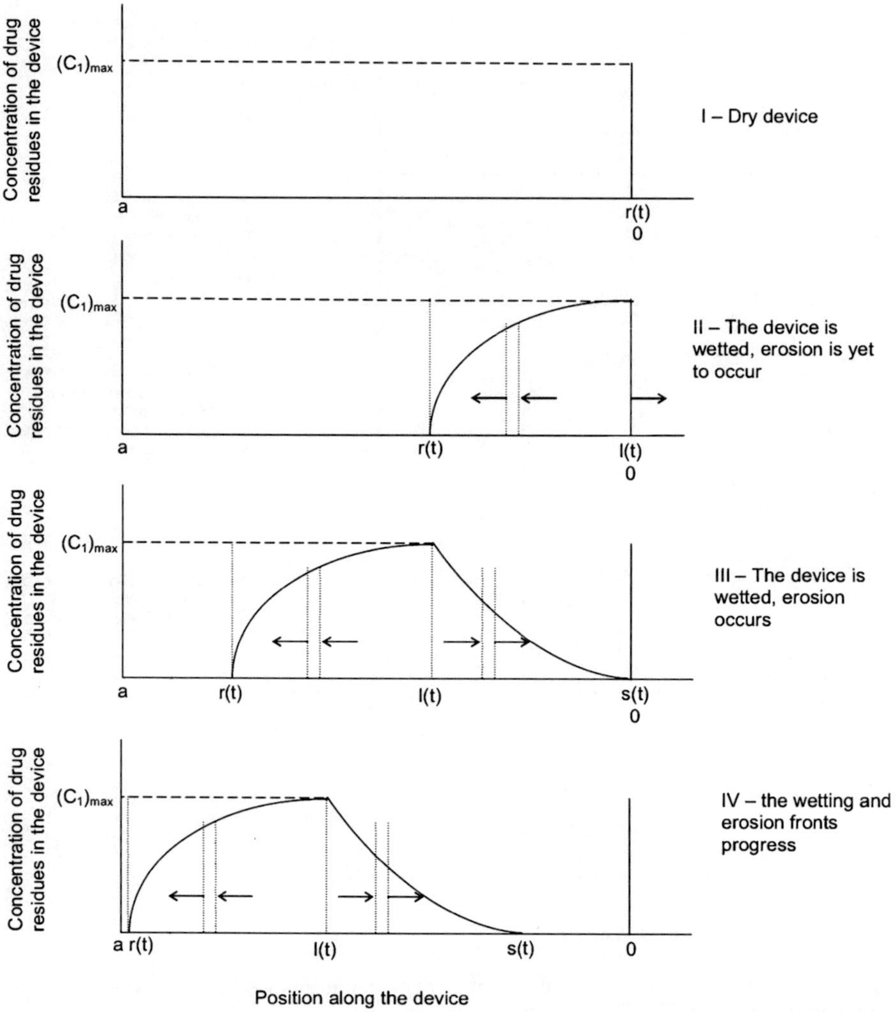

Fig. 8.4 Schematic representation of the concentration of free drug moieties in the device. *Panel* I depicts the state of the device at $t = 0$, while *panels* II, III, and IV are presentations of subsequent points in time. The *arrows* (→ and ←) indicate the directions of diffusion of free drug moieties into and out of the respective slabs. The other symbols are: *a*—the endmost point of the device, i.e., the last to be wetted; 0—the point of entrance to the device; *r(t)*—position of the wetting boundary; *l(t)*—position where the concentration of free drug moieties is at its maximum; *s(t)*—position of the erosion front

2.7 Analysis of the System Performance

In order to analyze the performance of the system, the content of the device was divided along the "a" dimension into infinitesimal slices (i.e., slabs), of thickness Δx. Each of these slabs is considered to be homogeneous, meaning that, within a given slab, there is no concentration gradient for any of the components contained.

The analysis performed accounts for the following stages:

1. *Wetting of the device by the enzyme-containing fluid*: Once the device is implanted, surrounding fluids containing the enzyme start crossing the permeable facet of the device and the dry content of the latter is being wetted. The wet space behind the wetting front at position $r(t)$ allows the enzyme to diffuse into it and to act on the polymeric drug chains.
2. *Enzymic degradation of the polymeric drug*: This process begins as soon as the polymeric drug is wetted by the enzyme-containing fluids. Consequently, in each slab of the device the enzymatic hydrolysis starts at a different point in time, according to the advancement of the wetting front.

One can now consider the concentration profile, C_{eff}, of bonds that are effectively available to the enzyme, at a particular point in time. As depicted in Figs. 8.3 and 8.4, the wetted parts of the device can be divided into two domains—a domain in which C_{eff} is constant, and one in which the value of C_{eff} declines. $l(t)$ denotes the point where C_{eff} begins to decline. We note that, in the space between positions $r(t)$ and $l(t)$, the value of C_{eff} is constant, and at its original (i.e., maximal) value $(C_{eff})_{max}$, even though in this domain the enzyme is present and active. This is due to the fact that the enzyme acts by the exo-single attack mechanism, i.e., by hydrolyzing a drug monomer from the polymeric chain ends only, and the concentration of these ends does not change throughout the degradation process. C_{eff} declines when polymeric drug chains undergo complete hydrolysis, reducing the number of chain ends in the system.

3. Changes in the *concentration of free drug in the device*: Once released by the enzymatic cleavage, the free drug moieties diffuse between the slabs, as dictated by the concentration gradient formed. The position of the erosion front, $s(t)$, defines the location where the entire polymeric drug has been hydrolyzed and, hence, the concentration of the latter is zero. Therefore, at both the wetting front, at position $r(t)$, and at the erosion front, at position $s(t)$, the concentration of free drug is zero. On the other hand, a complex picture emerges for the concentration of free drug, C_1, within the device (see Fig. 8.4). This is due to two processes occurring concomitantly, namely the hydrolytic cleavage at the end of the polymeric drug chains and the diffusion of the liberated drug moieties out of the device. Consequently, at any given point in time, the concentration C_1 will exhibit a maximal value, $(C_1)_{max}$. The location of this point will be identical to the position in which C_{eff} and reaction duration are maximal, i.e., point $l(t)$.

   According to the local gradient of free drug concentration, the molecules present between the wetting front, at position $r(t)$, and the position of highest free drug concentration, at position $l(t)$, will diffuse towards the wetting front, *into* the device, where its concentration is lower.

   Also, according to the local gradient of free drug concentration, the molecules present between the erosion front, $s(t)$, and the position of highest free drug concentration, at $l(t)$, will diffuse towards the degradation front, i.e., in the opposite direction, *out of* the device. Thus, throughout the operation of the system, diffusion of free drug occurs in both directions, into and out of the device.

Figures 8.3 and 8.4 contain schematic representations of the sequence of events following the implantation of the device. These figures summarize the relationship between penetration of the surrounding fluids and the contained enzyme on the one hand, and the formation and release of free drug moieties, on the other hand. It can be seen that, in the device, there are two distinctive domains, namely:

(a) Between the positions $r(t)$ and $l(t)$:

- The effective concentration of polymeric drug chain ends available to enzyme attack, C_{eff}, is constant, and at its maximal (i.e., initial) value
- Since C_{eff} is constant, the rate of production of free drug moieties is also constant
- The direction of diffusion of free drug moieties is *into* the slab

(b) Between the positions $l(t)$ and $s(t)$:

- The effective concentration of polymeric drug chain ends available to enzyme attack, C_{eff}, decreases and becomes zero at position $s(t)$
- Due to the decrease in C_{eff}, the rate of production of free drug moieties also decreases
- The direction of diffusion of free drug moieties is *out of* the slab

In spite of these differences, the two domains have a common characteristic, namely in both of them the total enzyme concentration, C_{ET}, is the same and constant.

3 Analytical Model

Based on the assumptions in Sect. 8.2, an analytical model was devised for describing the concomitant enzymatic hydrolysis of the polymeric drug and the diffusion of the active drug monomers produced by the enzymatic reaction. As mentioned earlier, the device in the model is divided into slices of thickness Δx and a mass balance on the monomeric residues (drug) in each slice (see also Fig. 8.5), thus obtaining

$$\frac{\mathrm{d}m_1}{\mathrm{d}t} = \frac{\mathrm{d}m_{\text{in}}}{\mathrm{d}t} + \frac{\mathrm{d}m_{\text{p}}}{\mathrm{d}t} - \frac{\mathrm{d}m_{\text{out}}}{\mathrm{d}t}, \tag{8.2}$$

where m_1 is the amount of free drug moieties in the slab, m_{in} is the amount of free drug moieties diffusing into in the slab, m_{out} is the amount of free drug moieties diffusing out of the slab, and m_{p} is the amount of free drug moieties produced within the slab. The pertinent boundary conditions are:

At position $x = r(t)$ and for $t > 0$: $m_1 = 0$
At position $x = s(t)$ and for $t > 0$: $m_1 = 0$

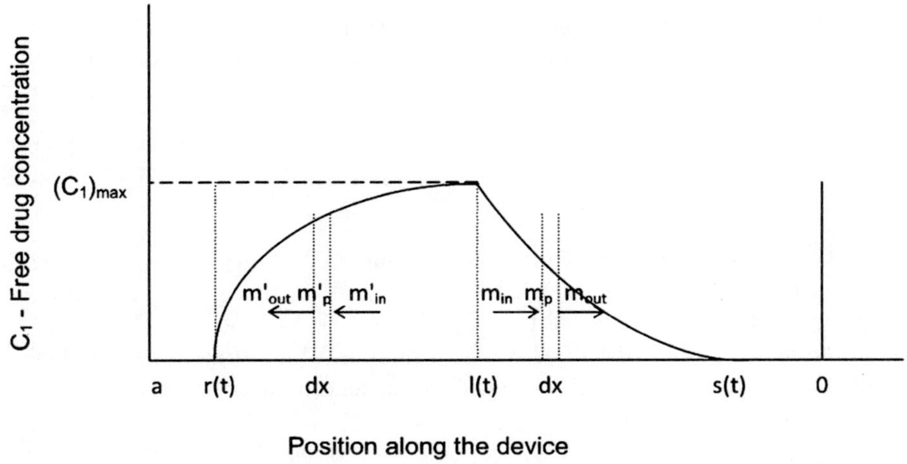

Fig. 8.5 Mass balance on free drug moieties in the device. The concentration profile depicted here is a schematic representation only and is not the outcome of actual calculations. The symbols in this figure indicate: *a*—the endmost point of the device, i.e., the last to be wetted; 0—the point of entrance to the device; Δx—the thickness of a slab in the device; m_{in}, m'_{in}—amount of free drug diffusing into slabs of thickness Δx. The associated *arrows* indicate the directions of mass transfer; m_{out}, m'_{out}—amount of free drug diffusing out of slabs of thickness Δx. The associated *arrows* indicate the directions of mass transfer; m_{p}, m'_{p}—amount of free drug produced in slabs of thickness; Δx; $r(t)$—position of the wetting boundary; $l(t)$—position where the concentration of free drug moieties is at its maximum; $s(t)$—position of the erosion front

Dividing the mass by the volume of one slab yields eq. (8.3), which describes the change in concentration of drug in the slab:

$$\frac{\mathrm{d}C_1}{\mathrm{d}t} = \left(\frac{\mathrm{d}C_1}{\mathrm{d}t}\right)_{\text{in}} + \left(\frac{\mathrm{d}C_1}{\mathrm{d}t}\right)_{\text{P}} - \left(\frac{\mathrm{d}C_1}{\mathrm{d}t}\right)_{\text{out}}. \tag{8.3}$$

The pertinent boundary conditions are:

At position $x = r(t)$ and for $t > 0$: $C_1 = 0$
At position $x = s(t)$ and for $t > 0$: $C_1 = 0$

The diffusion of drug moieties through the slice boundaries can be described by Fick's second law, using eq. (8.4):

$$\frac{\partial C_1}{\partial t} = D\frac{\partial^2 C_1}{\partial x^2}. \tag{8.4}$$

Production of monomeric drug units is performed in an exo-type enzymatic hydrolysis of the polymeric drug and is assumed to follow the Michaelis–Menten mechanism, as described in eq. (8.1). It is also assumed that the kinetic constants of the hydrolysis are not dependent on the number of moieties in the polymeric drug and remain constant throughout the reaction. Therefore, as we have previously shown [53, 54], the rate of production of monomeric residues is described by eq. (8.5).

$$\left(\frac{\mathrm{d}C_1}{\mathrm{d}t}\right)_{\mathrm{p}} = \frac{k_3}{K_{\mathrm{M}}} C_{\mathrm{eff}} C_{\mathrm{E}}, \tag{8.5}$$

where C_1 is the concentration of monomeric residues, K_{M} is the Michaelis–Menten constant, k_3 is a rate constant defined in eq. (8.1), C_{E} is the concentration of free enzyme, and dt is the time interval.

As discussed earlier, the effective concentration of substrate (C_{eff}, available bonds for enzymatic hydrolysis) is equal to the concentration of polymeric drug chains in the slice (eq. 8.6).

$$C_{\mathrm{eff}} = \sum_{j=1}^{N} C_j, \tag{8.6}$$

where C_j is the concentration of a polymeric drug chain containing j residues, and N is the initial number of residues in each polymeric drug chain.

Hence, the rate of production of monomeric residues can be described by eq. (8.7):

$$\left(\frac{\mathrm{d}C_1}{\mathrm{d}t}\right)_{\mathrm{p}} = \frac{k_3}{K_{\mathrm{M}}} C_{\mathrm{E}} \sum_{j=1}^{N} C_j. \tag{8.7}$$

The free enzyme concentration is dependent upon the available number of polymeric chain substrates and can be calculated from a mass balance on the enzyme in the slice (eq. 8.8):

$$C_{\mathrm{ET}} = C_{\mathrm{E}} + C_{\mathrm{ES}}, \tag{8.8}$$

where C_{ET} is the concentration of all enzyme forms in the slice, and C_{ES} is the concentration of enzyme–polymeric drug complexes.

Assuming a steady-state concentration for the enzyme–substrate complexes and summing up all the possible enzyme–polymeric drug complexes, the concentration of free enzyme is described by eq. (8.9) [53, 54]:

$$C_{\mathrm{E}} = \frac{1}{K_{\mathrm{M}}} \frac{C_{\mathrm{ET}}}{\left(1 + \sum_{j=1}^{N} C_j\right)}, \tag{8.9}$$

where C_{ET} is the total enzyme concentration in the slice.

Substituting eqs. (8.4), (8.7), and (8.9) into eq. (8.3), the change in concentration of the drug residue, due to monomer drug production and diffusion, can be described by eq. (8.10):

$$\frac{\mathrm{d}C_1}{\mathrm{d}t} = \left(D\frac{\partial^2 C_1}{\partial x^2}\right)_{\mathrm{in}} + \frac{k_3 C_{\mathrm{ET}} \sum_{j=1}^{N} C_j}{K_{\mathrm{M}} + \sum_{j=1}^{N} C_j} - \left(D\frac{\partial^2 C_1}{\partial x^2}\right)_{\mathrm{out}}. \tag{8.10}$$

It should be noted that the product $k_3 \times C_{\mathrm{ET}}$ is actually v_{max}, the maximal velocity of the enzymatic reaction.

The concentration of polymer drug chains of any length, j, changes as these chains are degraded by enzymatic attack or produced when a longer chain (containing $j + 1$ residues) undergoes enzymatic degradation. Hence, the concentration C_j can be described by eq. (8.11) which takes into consideration the inert gel effect and is based on a previously published model [53, 54]:

$$\frac{\mathrm{d}C_j}{\mathrm{d}t} = -\frac{k_3 C_{\mathrm{ET}} C_j}{K_{\mathrm{M}} + \sum_{i=1}^{N} C_i} + \frac{k_3 C_{\mathrm{ET}} C_{j+1}}{K_{\mathrm{M}} + \sum_{i=1}^{N} C_i} \quad N \geq j > 1. \tag{8.11}$$

The final analytical model comprises of a set of N differential equations, $N - 1$ equations describing the concentration of drug oligomers in the slice (eq. 8.11), and one mass balance equation (eq. 8.10), combining drug production and diffusion. This set of equations is solved numerically, as described in Sect. 8.4, to yield the time course of drug release from the system.

4 Numerical Solution

The numerical solution of the analytical model outlined earlier was carried out using two main numerical methods: Euler numerical integration [61] and the Smith method [62], which was further developed by Crank [63] to cover diffusion phenomena. Using the superposition assumption, the equation was solved, step by step, using a MATLAB program. In the first step, each of the phenomena (diffusion and biopolymer hydrolytic degradation) was carried out separately, and then a combined solution was offered. The basic set of parameters used in solving the equations is listed in Table 8.1. Unless otherwise stated, the values in Table 8.1 were the ones used for the simulation process.

In the numerical simulation, the enzymatic hydrolysis of the polymeric drug chains in a slab is considered to continue as long as their concentration exceeds a given threshold, ε. When the polymeric drug chains' concentration falls below the set value for ε, all the remaining drug moieties, both free and within the polymeric drug chains, are considered as released from the system at once.

Table 8.1 Values of parameters used in the numerical simulations

Value	Parameter
20	Initial number of drug moieties in a polymeric drug chain, N (dimensionless)
25	Initial concentration of polymeric drug chains in the device, C_0 (mM)
5	Maximal reaction rate of the process delineated in eq. (8.10), v_{max} (mM/min)
60	Michaelis–Menten constant for the process delineated in eq. (8.5), K_M (mM)
0.12	Total enzyme concentration in the system, C_{ET} (Units/mL)
5	Integration step, Δt (min)
0.025	Thickness of a slab in the device, Δx (cm)
5×10^{-5}	Penetration rate of surrounding fluids into the device, r (cm/min)
2.22×10^{-5}	Diffusion coefficient of the free drug moiety, D (cm^2/min)
1	Device length, a (cm)
0.5	Cross-section area of the device, A (cm^2)
1×10^{-10}	Limiting concentration of polymeric chains, below which the drug moieties still contained are released from the device at once, ε (mM)

5 Results of Numerical Simulation

5.1 Kinetics of Drug Release

The kinetics of drug release from the system was calculated by numerical solution of eq. (8.6), using the parameters given in Table 8.1. As shown in Fig. 8.6, the release profile is comprised of two linear release rates. After 4 days of constant release, an increase in the release rate is observed. The enhanced rate is maintained until the total amount of residues in the system (250 μmol) is released.

To check the stability of the numerical solution, the sensitivity of the procedure to the size of the integration intervals (Δx and Δt) was investigated. As given in Table 8.2, the solution is stable in the range of 1–6 min integration steps. Further reduction of the interval results in no convergence of the numerical solution.

The procedure is more sensitive to the size of the slab (Δx). Reducing the size of the interval means that the number of slabs in the device is increased. As given in Table 8.3, when this integration interval was increased, in the range of 0.025–0.25 cm, the calculated amount of drug released due to diffusion was reduced, indicating a reduced accuracy of the simulation. The remainder of the drug is considered to be released in a "burst" that occurs when the concentration of polymer drug chains reaches a value which is lower than ε. The "burst" release is minimized when the integration interval is reduced. Division of the device to 40 slabs was used for further simulations.

The effect of enzyme characteristics was studied using the numerical simulations. As shown in Fig. 8.7, as the maximal reaction rate $v_{\max}$ decreases, the time for complete release from the device increases. The release profile contains a long period of an almost linear drug release rate, but there is a point (easily detected for the lowest $v_{\max}$), where a change in the releasing rate is detected. This changing point can be correlated to the progress of the erosion front, as shown in Fig. 8.8, and is probably

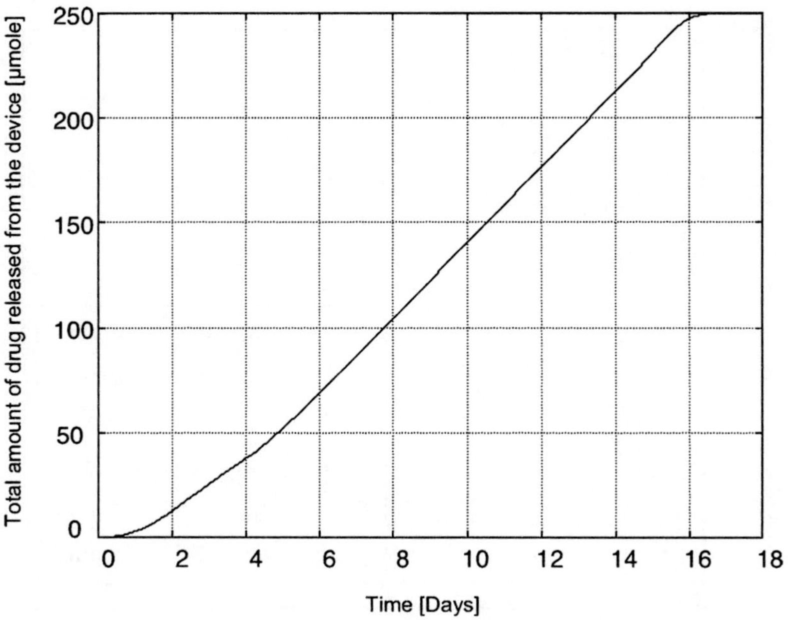

Fig. 8.6 Kinetics of drug release from the device. Parameters used for numerical simulation are listed in Table 8.1

Table 8.2 Time for complete drug release from the device for various time integration intervals (Δt)

Time step, Δt (min)	Number of iterations	Complete drug release from the device (days)
1	25,774	17.9
2	12,880	17.9
4	6,433	17.9
6	4,285	17.9
8	NC[a]	NC[a]

[a]*NC* No convergence of the numerical solution was achieved

Table 8.3 Drug released by diffusion for various Δx values. $\Delta t = 5$ min

Slice size, Δx (cm)	Number of slices in the device	Drug released by diffusion (%)
0.025	40	96
0.05	20	96
0.1	10	92
0.25	4	91

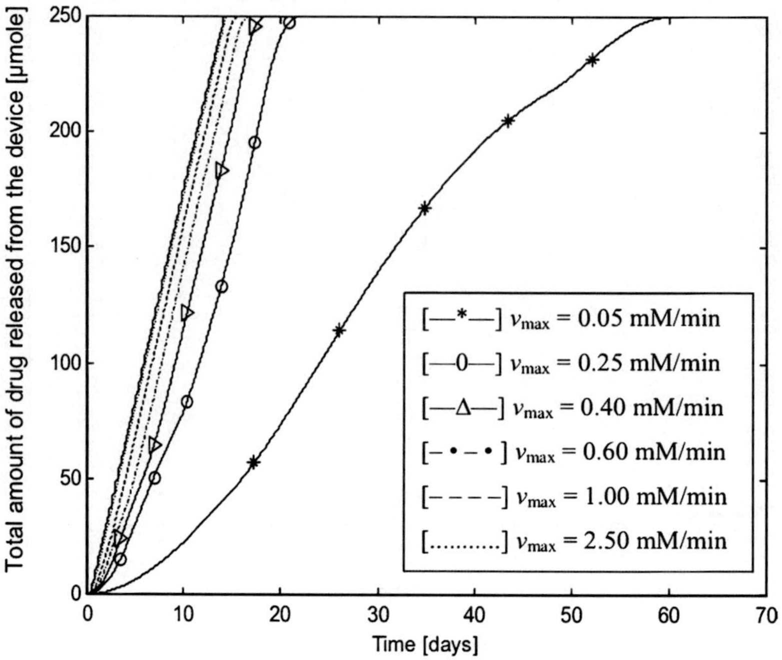

Fig. 8.7 Effect of enzyme characteristics on the time course of drug release from the device. Other parameters used for numerical simulation are listed in Table 8.1

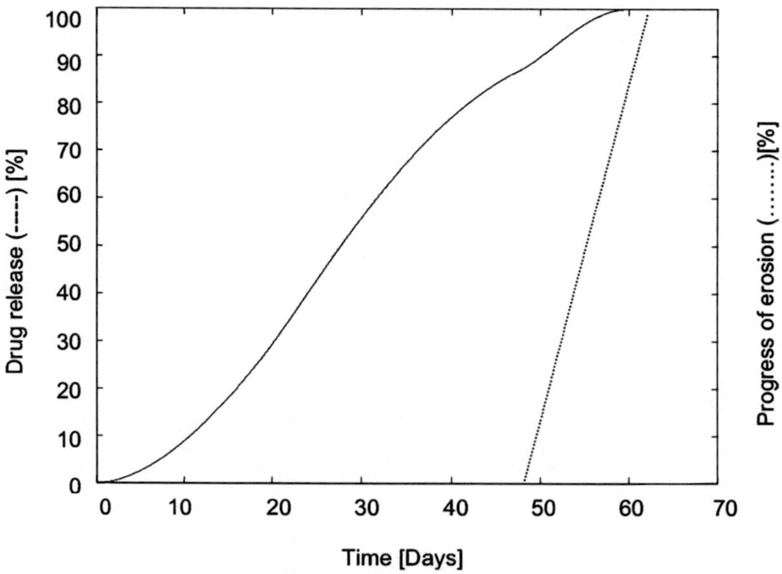

Fig. 8.8 Cumulative fraction of drug released from the device and progress of the erosion front (i.e., position $s(t)$). All the calculations were performed using $v_{max} = 0.05$ mM/min. Other parameters used for numerical simulation are listed in Table 8.1

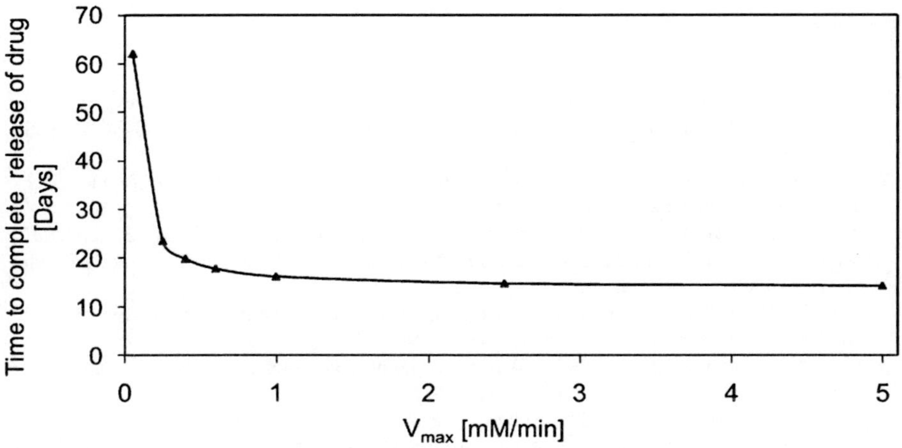

Fig. 8.9 Effect of v_{max} on the duration of complete release of the drug from the device. Parameters used for numerical simulation are listed in Table 8.1

due to the "burst" release which occurs when the polymer concentration is lower than ε. The maximal velocity of the reaction, v_{max}, can change due to changes in either enzyme concentrations (C_{ET}) or enzyme kinetic parameters (k_3).

The time to full release of the drug from the device (Fig. 8.9) is reaching an asymptotic value as v_{max} is increased. This is probably due to the fact that above a certain rate of production of drug moieties, the diffusion rate will be more significant. Thus, an increase in the degradation rate will shorten the time for complete release only until the point where diffusion becomes the rate limiting step.

6 Conclusions

An implantable system, capable of releasing a contained drug in a controlled manner, was designed. The system is based on a polymeric drug immobilized within an inert polymeric framework. The controlling element that affects the drug release rate is the status of the surrounding physiological environment, as expressed by the level of activity of a specific enzyme, capable of hydrolyzing free drug moieties from the polymeric chain.

In order to assess the performance of the drug releasing system, an analytical model was developed. The model takes into account enzyme-promoted degradation of the polymeric drug as well as mass transfer of free drug in the device. Using the analytical model and extensive numerical simulations, it could be shown that the drug release rate is sensitive to the enzyme characteristics, mostly the enzyme concentration and kinetic parameters. Thus, using enzyme-promoted degradation, drug release devices which are sensitive to changes in the physiological environment can be designed.

References

1. Reed AM, Gilding DK (1981) Biodegradable polymers for use in surgery—Poly (glycolic)-poly(lactic acid) Homo and co-polymers.2. In vitro degradation. Polymer 22:494–498
2. Shive MS, Anderson JM (1997) Biodegradation and biocompatibility of PLA and PLGA microspheres. Adv Drug Deliv Rev 28:5–24
3. Heller J, Hoffman AS (2004) Drug delivery systems. In: Ratner BD, Hoffman AS, Schoen FJ, Lemons JE (eds) Biomaterials science, an introduction to materials in medicine, 2nd edn. Elsevier Academic, Amsterdam
4. Wischke C, Schwendeman SP (2008) Principles of encapsulating hydrophobic drugs in PLA/PLGA microparticles. Int J Pharm 364(2):298–327
5. Kumari A, Yadav SK, Yadav SC (2010) Biodegradable polymeric nanoparticles based drug delivery systems. Colloids Surf B Biointerfaces 75:1–18
6. Giteau A, Venier-Julienne MC, Aubert-Pouëssel A, Benoit JP (2008) How to achieve sustained and complete protein release from PLGA-based microparticles? Int J Pharm 350(1–2):14–26
7. Mohamed F, Van der Walle C (2008) Engineering biodegradable polyester particles with specific drug targeting and drug release properties. J Pharm Sci 97(1):71–87
8. Benny O, Menon LG, Gilert A, Goren E, Kim SK, Stewman C, Black PM, Carroll RS, Machluf M (2009) Local delivery of polylactic-co-glycolic acid microspheres containing Imatinib mesylate inhibits intracranial xenograft glioma growth. Clin Cancer Res 15:1222–1231
9. Bobo WV, Shelton RC (2010) Risperidone long-acting injectable (Risperdal Consta®) for maintenance treatment in patients with bipolar disorder. Expert Rev Neurother 10(11):1637–1658
10. Brem H, Kader A, Epstein JI, Tamargo RJ, Domb A, Langer R, Leong KW (1989) Biocompatibility of a biodegradable, controlled-release polymer in the rabbit brain. Sel Cancer Ther 5(2):55–65
11. Attenello FJ, Mukherjee D, Datoo G, McGirt MJ, Bohan E, Weingart JD, Olivi A, Quinones-Hinojosa A, Brem H (2008) Use of Gliadel (BCNU) wafer in the surgical treatment of malignant glioma: a 10-year institutional experience. Ann Surg Oncol 15(10):2887–2893
12. Göpferich A, Tessmar J (2002) Polyanhydride degradation and erosion. Adv Drug Deliv Rev 54(7):911–931
13. Heller J, Barr J, Ng SY, Abdellauoi KS, Gurny R (2002) Poly(ortho esters): synthesis, characterization, properties and uses. Adv Drug Deliv Rev 54(7):1015–1039
14. Eng NF, Garlapati S, Gerdts V, Potter A, Babiuk LA, Mutwiri GK (2010) The potential of polyphosphazenes for delivery of vaccine antigens and immunotherapeutic agents. Curr Drug Deliv 7(1):13–20
15. Lakshmi S, Katti DS, Laurencin CT (2003) Biodegradable polyphosphazenes for drug delivery applications. Adv Drug Deliv Rev 55(4):467–482
16. Körber M (2010) PLGA erosion: solubility- or diffusion-controlled? Pharm Res 27 (11):2414–2420
17. Katz JS, Zhong S, Ricart BG, Pochan DJ, Hammer DA, Burdick JA (2010) Modular synthesis of biodegradable diblock copolymers for designing functional polymersomes. J Am Chem Soc 132(11):3654–3655
18. Bawa P, Pillay V, Choonara YE, du Toit LC (2009) Stimuli-responsive polymers and their applications in drug delivery. Biomed Mater 4(2):22001
19. Miyata T, Uragami T, Nakamae K (2002) Biomolecule-sensitive hydrogels. Adv Drug Deliv Rev 54(1):79–98
20. Qiu Y, Park K (2001) Environment-sensitive hydrogels for drug delivery. Adv Drug Deliv Rev 53(3):321–339
21. Traitel T, Cohen Y, Kost J (2000) Characterization of glucose-sensitive insulin release systems in simulated in vivo conditions. Biomaterials 21(16):1679–1687

22. Tomer Gold (2001) Drug controlled release device based on enzymic degradation of a polymer matrix and diffusion: analytical models. M.Sc. Thesis, Technion—Israel Institute of Technology, Haifa
23. Singh S (2010) Nanomedicine-nanoscale drugs and delivery systems. J Nanosci Nanotechnol 10(12):7906–7918
24. Gaffney J, Matou-Nasri S, Grau-Olivares M, Slevin M (2009) Therapeutic applications of hyaluronan. Mol Biosyst 6(3):437–443
25. Singh M, Briones M, O'Hagan DT (2001) A novel bioadhesive intranasal delivery system for inactivated influenza vaccines. J Control Release 70(3):267–276
26. Challa R, Ahuja A, Ali J, Khar RK (2005) Cyclodextrins in drug delivery: an updated review. AAPS Pharm Sci Tech 6(2):E329–E357
27. Heise T, Brugger A, Cook C, Eckers U, Hutchcraft A, Nosek L, Rave K, Troeger J, Valaitis P, White S, Heinemann L (2009) PROMAXX inhaled insulin: safe and efficacious administration with a commercially available dry powder inhaler. Diabetes Obes Metab 11(5):455–459
28. Lim ST, Martin GP, Berry DJ, Brown MB (2000) Preparation and evaluation of the in vitro drug release properties and mucoadhesion of novel microspheres of hyaluronic acid and chitosan. J Control Release 66(2–3):281–292
29. Maham A, Tang Z, Wu H, Wang J, Lin Y (2009) Protein-based nanomedicine platforms for drug delivery. Small 5(15):1706–1721
30. Sehgal PK, Srinivasan A (2009) Collagen-coated microparticles in drug delivery. Expert Opin Drug Deliv 6(7):687–695
31. Basu SK, Kavitha K, Rupeshkumar M (2009) Evaluation of Ketorolac romethamine Microspheres by Chitosan/Gelatin B Complex Coacervation. Sci Pharm 78(1):79–92
32. Golumbek PT, Azhari R, Jaffee EM, Levitsky HI, Lazenby A, Leong K, Pardoll DM (1993) Controlled release, biodegradable cytokine depots: a new approach in cancer vaccine design. Cancer Res 53(24):5841–5844
33. Hanes J, Sills A, Zhao Z, Suh KW, Tyler B, DiMeco F, Brat DJ, Choti MA, Leong KW, Pardoll DM, Brem H (2001) Controlled local delivery of interleukin-2 by biodegradable polymers protects animals from experimental brain tumors and liver tumors. Pharm Res 18 (7):899–906
34. Bourke SL, Kohn J (2003) Polymers derived from the amino acid L-tyrosine: polycarbonates, polyarylates and copolymers with poly(ethylene glycol). Adv Drug Deliv Rev 55(4):447–466
35. Duncan R, Ringsdorf H, Satchi-Fainaro R (2006) Polymer therapeutics–polymers as drugs, drug and protein conjugates and gene delivery systems: past, present and future opportunities. J Drug Target 14(6):337–341
36. Yang J, Jacobsen MT, Pan H, Kopecek J (2010) Synthesis and characterization of enzymatically degradable PEG-based peptide-containing hydrogels. Macromol Biosci 10(4):445–454
37. Ohya Y, Takamido S, Nagahama K, Ouchi T, Katoono R, Yui N (2009) Polyrotaxane composed of poly-L-lactide and alpha-cyclodextrin exhibiting protease-triggered hydrolysis. Biomacromolecules 10(8):2261–2267
38. Barbato F, La Rotonda MI, Maglio G, Palumbo R, Quaglia F (2001) Biodegradable microspheres of novel segmented poly(ether-ester-amide)s based on poly(epsilon-caprolactone) for the delivery of bioactive compounds. Biomaterials 22(11):1371–1378
39. Giammona G, Pitarresi G, Cavallaro G, Buscemi S, Saiano F (1999) New biodegradable hydrogels based on a photocrosslinkable modified polyaspartamide: synthesis and characterization. Biochim Biophys Acta 1428(1):29–38
40. Yoshida M, Asano M, Kumakura M, Katakai R, Mashimo T, Yuasa H, Imai K, Yamanaka H (1990) Sequential polydepsipeptides as biodegradable carriers for drug delivery systems. J Biomed Mater Res 24(9):1173–1184
41. Moon HJ, Choi BG, Park MH, Joo MK, Jeong B (2011) Enzymatically degradable thermogelling poly(alanine-co-leucine)-poloxamer-poly(alanine-co-leucine). Biomacromolecules 12(4):1234–1242

42. Tokatlian T, Shrum CT, Kadoya WM, Segura T (2010) Protease degradable tethers for controlled and cell-mediated release of nanoparticles in 2- and 3-dimensions. Biomaterials 31(31):8072–8080
43. Yaacobi Y, Sideman S, Lotan N (1985) A mechanistic model for the enzymic degradation of synthetic biopolymers. Life Support Syst 3(4):313–326
44. Gopferich A, Langer R (1995) Modeling monomer release from bioerodible polymers. J control Release 33:55–69
45. Lemaire V, Belair J, Hildgen P (2003) Structural modeling of drug release from biodegradable porous matrices based on combined diffusion/erosion process. Int J Pharm 258:95–107
46. Arifin DY, Lee LY, Wang CH (2006) Mathematical modeling and simulation of drug release from microspheres: implications to drug delivery systems. Adv Drug Delivery Rev 58:1274–1325
47. Lao LL, Venkatraman SS, Peppas NA (2008) Modeling of drug release from biodegradable polymer blends. Eur J Pharm Biopharm 70:796–803
48. Yu R, Chen H, Chen T, Zhou X (2008) Modeling and simulation of drug release from multi-layer biodegradable microstructure in three dimensions. Simul Model Prac Theory 16:15–25
49. Siepmann J, Siepmann F (2008) Mathematical modeling of drug delivery. Int J Pharm 364:328–343
50. Barba AA, d'Amore M, Chirico S, Lamberti G, Titomanlio G (2009) A general code to predict the drug release kinetics from different shaped matrices. Eur J Pharm Sci 36:359–368
51. Soares J, Zunino P (2010) A mixture model for water uptake, degradation, erosion and drug release from polydisperse polymeric networks. Biomaterials 31:3032–3042
52. Brandl F, Kastner F, Gschwind RM, Blunk T, Tebmar J, Gopferich A (2010) Hydrogel-based drug delivery systems: comparison of drug diffusivity and release kinetics. J Control Release 142:221–228
53. Azhari R, Sideman S, Lotan N (1991) A generalized model for enzymic depolymerization processes: Part I—Reaction pathways and kinetics. Polym Degrad Stab 33(1):35–52
54. Azhari R, Lotan N (1991) Enzymic hydrolysis of biopolymers via single-scission attack pathways: a unified kinetic model. J Mater Sci Mater Med 2:9–18
55. Azhari R, Lotan N (1991) Enzymic depolymerization processes: reaction pathways as a basis for a new classification and nomenclature. J Mater Sci Lett 10:243–245
56. Tzafriri AR, Bercovier M, Parnas H (2002) Reaction diffusion model of the enzymatic erosion of insoluble fibrillar matrices. Biophys J 83(2):776–793
57. Tayal A, Khan SA (2000) Degradation of a water-soluble polymer: molecular weight changes and chain scission characteristics. Macromolecules 33(26):9488–9493
58. Watanabe M, Kawai F (2006) Mathematical modelling and computational analysis of enzymatic degradation of xenobiotic polymers. Appl Math Model 30(12):1497–1514
59. Klemchuka PP (1990) Degradable plastics: a critical review. Polym Degrad Stab 27 (2):183–202
60. Maeda H, Ymagata Y, Abe K, Hasegawa F, Machida M, Ishioka R, Gomi K, Nakajima T (2005) Purification and characterization of a biodegradable plastic-degrading enzyme from Aspergillus oryzae. Appl Microbiol Biotechnol 67:778–788
61. Maron MJ (1982) Numerical analysis, a practical approach, 2nd edn. MacMillan publishing, New York
62. Crank J (1975) The mathematics of diffusion, 2nd edn. Clarendon Press, Oxford
63. Smith GD (1986) Numerical solution of partial differential equations: finite difference method, 3rd edn. Oxford University Press, Oxford

Biography

Mr. Tomer Gold is a Chemical Engineer by basic training and was granted a M.Sc. degree in Biomedical Engineering from the Technion—Israel Institute of Technology in Haifa, Israel. He subsequently worked in the Global Generic Research and Development branch of Teva Pharmaceuticals Ltd., as the Pharmaceutical Development Manager of the European Union team. His activities focused on the development of generic solid dosage forms, from the initiation stage until the successful launch of the product. As such, he developed controlled release products as well as enteric and immediate release products. Today, he is the Vice president for Research and Development of the Dexcel Pharma Co., in charge of a group of 120 scientists and engineers, and responsible for the development, registration and launching of generic and innovative products in the entire world.

Rosa Azhari is an Associate Professor at the Professor Ephraim Katzir Department of Biotechnology Engineering at ORT Braude College, Karmiel, Israel. She also serves as chairperson of the sub-committee for Engineering and Technological Institutions, at the Council for Higher Education in Israel, and was the President of the Israeli Chapter of the Controlled Release Society. She holds a B.Sc. degree in Chemical Engineering from the Technion—Israel Institute of Technology in Haifa, Israel, a M.Sc. degree in Chemistry from the Weizmann Institute of Science, Rehovot, Israel, and a D.Sc. degree in Biomedical Engineering from the Technion. She has been a Post-doctoral research fellow in the Biomedical Engineering Department at Johns Hopkins University in Baltimore, MD, USA and a visiting scientist at the Harvard-MIT Health, Science and Technology program in Cambridge, MA, USA. Her research interests include: biomaterials, biodegradable polymers, modeling of enzymic degradation of biopolymers, tissue engineering, mechanotransduction phenomena in cells, microencapsulation, and vaccine delivery.

Professor Noah Lotan is a Chemical Engineer by basic training, he received a Ph.D. degree (with Distinction) from the Weizmann Institute of Science in Rehovot, Israel, and was appointed Professor of Biomedical Engineering at the Technion—Israel Institute of Technology, in Haifa, Israel. He was a visiting scientist at universities in the USA, France, Germany, Italy, Switzerland, UK, and Mexico, as well as at the industrial companies—Israel Aircraft Ind. (Israel) and Ciba-Geigy Pharmaceutical Co. (Switzerland). His academic activities focus on biomaterials and, particularly, on nanotechnology, physiologically controlled drug delivery systems, corrosion of polymeric biomaterials, biomolecular electronics, biocomputers, tissue engineering, as well as chemical and electrochemical biosensors. Together with one of his former Ph.D. students, he recently published the book "Information processing by biochemical systems: Neural network-type configurations."

Chapter 9
Degradation of Bioceramics

L. Gremillard, S. Meille, J. Chevalier, J. Zhao, V. Fridrici, Ph. Kapsa, J. Geringer, and J. Uribe

Abstract After roughly 100 years of controlled clinical use, the *in vivo* and *in vitro* degradation mechanisms of ceramic materials are still largely unknown. In bioinert ceramics such as alumina and zirconia used in orthopedics, crack propagation mechanisms are well known, but their interactions with other degradation mechanisms (low-temperature degradation, shocks, wear, dissolution, etc.) and the *in vivo* environment remain to be firmly established. In bioactive ceramics like calcium phosphates and bioactive glasses, dissolution–precipitation processes play a major role on both degradation of the implant and biological efficiency. Even without the ambition to be exhaustive, it is the purpose of this chapter to present the degradation mechanisms of ceramic implants, both inert and bioactive, and the interactions between them and with their environment.

L. Gremillard • S. Meille • J. Chevalier (✉)
Université de Lyon, INSA-Lyon, MATEIS, UMR CNRS 5510, 20 avenue Albert Einstein, 69621 Villeurbanne cedex, France
e-mail: jerome.chevalier@insa-lyon.fr; jeromechevalier@me.com

J. Zhao • V. Fridrici • P. Kapsa
Laboratoire de Tribologie et Dynamique des Systèmes, UMR 5513 CNRS ECL ENISE, École Centrale de Lyon, 36, Avenue Guy de Collongue, 69134 Ecully cedex, France

J. Geringer • J. Uribe
Center for Health Engineering, Biomechanics and Biomaterials Department, UMR CNRS 5146, IFR 143, Ecole Nationale Supérieure des Mines de Saint-Etienne, 158 cours Fauriel, 42023 Saint-Etienne, France

N. Eliaz (ed.), *Degradation of Implant Materials*, DOI 10.1007/978-1-4614-3942-4_9,

1 Introduction

Ceramics have been used as biomaterials for millennia: "artificial nacre teeth" have been found in 4,000 years old Mayan skulls [1] (nacre is a natural composite consisting of 95–98 wt% of calcium carbonate—aragonite, the "ceramic" phase). The controlled, clinical implantation of bioceramics started late eighteenth century in dentals with porcelain crowns and late nineteenth in orthopedics [bone filling with gypsum (calcium sulfate dihydrate)] [2]. More and more "technical" ceramics became available for medical purpose in the twentieth century [3]: in particular, tricalcium phosphate (TCP) was first proposed in 1920 as a bioresorbable substance to fill bone gaps, and (not before 1965) the first strong and tough alumina (Al_2O_3) material dedicated to hip joints was patented [4]. Synthetic calcium phosphate (CaP) ceramics (together with calcium and/or phosphorus containing ceramics and glasses) and zirconia (ZrO_2) were then proposed as alternatives to TCP and alumina, respectively. However, after roughly 100 years of controlled clinical use, the *in vivo* and *in vitro* degradation mechanisms of ceramic materials are still largely unknown. Crack propagation mechanisms are well known, but their interactions with other degradation mechanisms (low-temperature degradation, shocks, wear, dissolution–reprecipitation, etc.) and the *in vivo* environment remain to be firmly established. Even without the ambition to be exhaustive, it is the purpose of this chapter to present the degradation mechanisms of ceramic implants, both inert and bioactive, and the interactions between them and with the environment.

2 Degradation of Structural Ceramics in Orthopedics

2.1 A Short Introduction on Zirconia Ceramics

2.1.1 Different Phases of Zirconia

The use of zirconia ceramics for orthopedic implants derives from the exceptional mechanical properties of this family of ceramics: high strength and high toughness. In particular, yttria-stabilized zirconia ceramics present the highest strength (tensile strength up to 1,200 MPa) and toughness (up to 6 MPa $m^{1/2}$) of all single-phase oxide ceramics. These properties are a consequence of the phase transformation toughening mechanism that is particular to zirconia ceramics, and that we will expose below. Indeed, the evolution from one metastable, tetragonal polymorph to the stable monoclinic one explains both the phase transformation toughening mechanism responsible for the high mechanical properties of zirconia and its sensitivity to low-temperature degradation (aging).

From high temperature to low temperature, pure zirconia exists under three stable phases: cubic (between the melting point, 2,680°C and 2,370°C), tetragonal

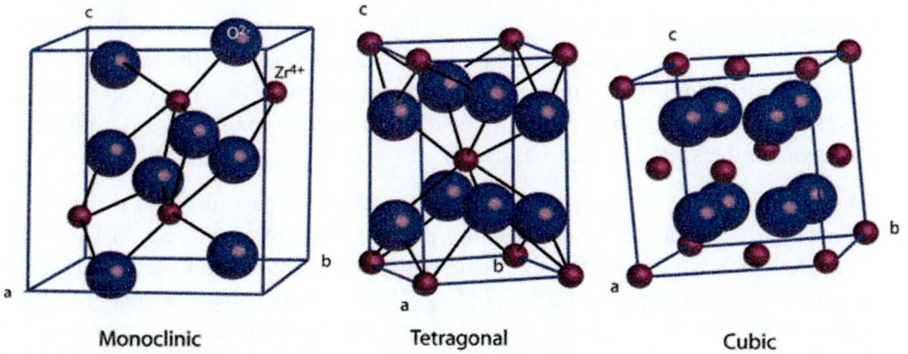

Fig. 9.1 Crystallographic structures of zirconia. *From left to right*: monoclinic, tetragonal and cubic

(down to 1,170°C), and monoclinic (the room-temperature stable phase). The monoclinic phase can be described in the P21/c space group, with the Zr atoms in sevenfold coordination with the O-sublattice (eightfold in the CaF_2 structure). The tetragonal phase presents a distorted calcium fluorite structure (P42/nmc). The cubic phase is a calcium fluorite type structure (face centered cubic, Fm3m), with the O atoms displaced from the (1/4, 1/4, 1/4) position toward a higher z (around 0.28), which may be due to the tendency of the Zr atoms to form a sevenfold coordination [5]. The three unit cells are illustrated in Fig. 9.1. The processing of zirconia ceramics by sintering involves temperatures up to 1,500°C, temperatures at which the tetragonal phase is stable. Upon cooling to room temperature, pure zirconia transforms to its monoclinic stable phase. The ~5% volume increases accompanying the tetragonal to monoclinic (t–m) transformation inevitably lead to extensive cracking and ruin of the pure zirconia pieces. Thus, pure zirconia can retain neither tetragonal nor cubic phase at room temperature and is never used alone for structural applications. Doping zirconia with a few percent of other oxides (the most used ones being Y_2O_3 and CeO_2) allows the retention of an intact body after sintering, at room temperature.

2.1.2 Stabilization of Tetragonal Zirconia

Zirconia ceramics can be stabilized under two different forms:

- Partially stabilized zirconia (PSZ) are obtained with low solubility stabilizers (such as MgO and CaO) and are precipitated systems with tetragonal or monoclinic precipitates inside a cubic matrix. Magnesia partially stabilized zirconia (Mg-PSZ) was the first zirconia introduced in orthopedics, but was abandoned due to modest strength and fracture toughness.
- If the solubility of the stabilizer (Y_2O_3, CeO_2) is high, a solid solution forms at high temperature and is retained upon cooling. Zirconia is then called TZP, for

tetragonal zirconia polycrystal. Today, only 3Y-TZP (stabilized with 3 mol% Y_2O_3) is used in clinical application. 3Y-TZP is composed of metastable tetragonal grains with a few cubic grains, in accordance with the Y_2O_3–ZrO_2 phase diagram at sintering temperature (the exact cubic–tetragonal phase partitioning depends on the sintering thermal cycle).

The mechanism of stabilization of the tetragonal phase is still under discussion. Some authors point out an easier ordering of the anions when the cation lattice is disordered, which would allow the zirconia lattice to achieve a more stable state and thus remain cubic or tetragonal after cooling. Others suggest that the change in the electronic charge balance allows the cubic or tetragonal structure to survive [5]. The roles of the oxygen vacancies and of the internal stresses due to the differences in ionic radii are also evoked. The most extensive work published on this topic [6, 7] explains the stabilization by a decrease of the "oxygen overcrowding" around the zirconium cations.

The stability of the different zirconia phases may be shown in a metastable phase diagram [8] that plots the transformation temperature T_0 t–m (for tetragonal-to-monoclinic transformation) versus the stabilizer content. These diagrams are called metastable, because they are valid for compositions that are out of equilibrium (for example, they give the T_0 t–m temperature of a tetragonal zirconia containing 3 mol% Y_2O_3, whereas the stable phase distribution with 3 mol% Y_2O_3 is a mix of yttrium-rich cubic phase and yttrium-poor tetragonal phase).

2.2 *Slow Crack Growth and Delayed Fracture*

2.2.1 Corrosion-Assisted Crack Propagation

Slow crack growth (SCG, also called Subcritical Crack Growth) is a phenomenon by which a crack can propagate in a material even below the critical stress intensity factor (crack propagation below K_{IC}). It involves speeds between 10^{-12} and 10^{-2} m s^{-1}. It was first observed in glass by Grenet [9] at the beginning of the twentieth century. Orowan [10] later evidenced the key role of water in the propagation: having observed that the strength of glass is three times lower in air than in vacuum, he proposed that corrosive molecules (such as water) accumulate (adsorb) at the crack tip and provoke a large decrease of the surface energy of the material (thus, easier crack propagation). He proposed also that the diffusion rate of water to the crack tip can limit the crack velocity. Later works showed that, for given material and environment, there is a unique relationship between the crack velocity and the stress intensity factor at the crack tip. The sensitivity of a material to SCG is thus represented in a graph showing the crack velocity versus either the stress intensity factor (V–K_I diagram) or the energy release rate (V–G_I diagram).

A model of corrosion-assisted cracking (derived from Charles' model [11]) was proposed by Wiederhorn [12] and Lawn [13] to describe the crack propagation in glass. It readily applies to zirconia and alumina ceramics, whose ionic component of the chemical bonds is close to the one of glass. This model supposes two serial phenomena: first the transport of the corrosive species (water) to the crack tip (reaction site) and then a chemical reaction between water and ceramic (leading to a rupture of the chemical bonds in the ceramic). This reaction can be summarized as follows (eq. 9.1):

$$\text{HOH} + -\text{MOM'}- \rightarrow -\text{MOH} + \text{HOM'}-, \quad (9.1)$$

where M is a metallic or nonmetallic ion of the ceramic: Si for glass, Zr, Y, or Al for yttria-stabilized zirconia or alumina. Note that the specific presence of water is not mandatory: any other polar molecule (more precisely containing a pair of free electrons versus a proton donor site) can react in the same way (the case of ammonia and hydrazine).

In terms of Griffith criterion [14], crack propagation will occur when the applied energy release rate is higher than the surface energy of the ceramic: $G \geq 2\gamma$. However, the surface energy mentioned here is the solid–gas interfacial energy, thus depending on the environment.

The competition between transport and reaction explains the presence of three propagation stages (Fig. 9.2). Stage I corresponds to crack velocities controlled by the reaction speed between water and the ceramic bonds. In stage II, the reaction speed is higher than the transport speed of the corrosive molecules to the reaction site: transport is the limiting phenomenon. In stage III, the applied K_I is high enough to provoke fracture without water corrosion assistance: the crack propagation occurs in vacuum conditions. Stages I and II are serial processes: first transport, then reaction, and the slower process controls the propagation. Stages I and II on one side, and stage III on the other side, occur in parallel (at high K_I, rupture of the bonds can occur in both vacuum and with the assistance of water); thus the fastest phenomenon controls the crack propagation.

One of the most critical points of this analysis is the existence of a crack propagation threshold, K_{I0}, below which no crack propagation can occur. The threshold obviously depends on the material and on the environment (chemical composition, temperature, etc.). If no toughening mechanism is present, K_{I0} is related to the surface energy γ by the relation: $\gamma = \frac{K_{I0}^2}{E}$, where E is the Young's modulus of the material. Although it has been much debated, the existence of the threshold makes no doubt in zirconia [15] and has even been proven theoretically in silica by ab initio calculations [16]. Through the surface energy, it depends strongly on the environment: for example, in zirconia γ is around 25 J m^{-2} in water, but 31 J m^{-2} in air [17]. Figure 9.3 shows that not only the threshold, but also the whole V–K_I curve is sensitive to the environment, which can be also deduced from the equations that give the crack propagation speed in the different stages [18]:

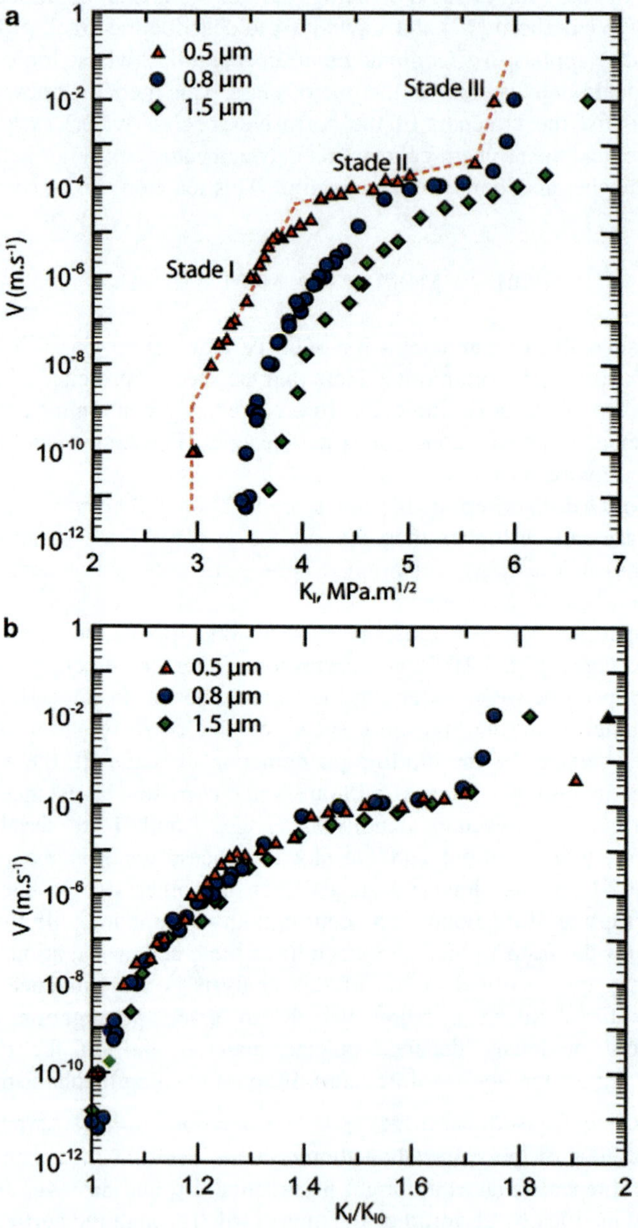

Fig. 9.2 Slow crack growth in 3Y-TZP zirconia with different grain sizes

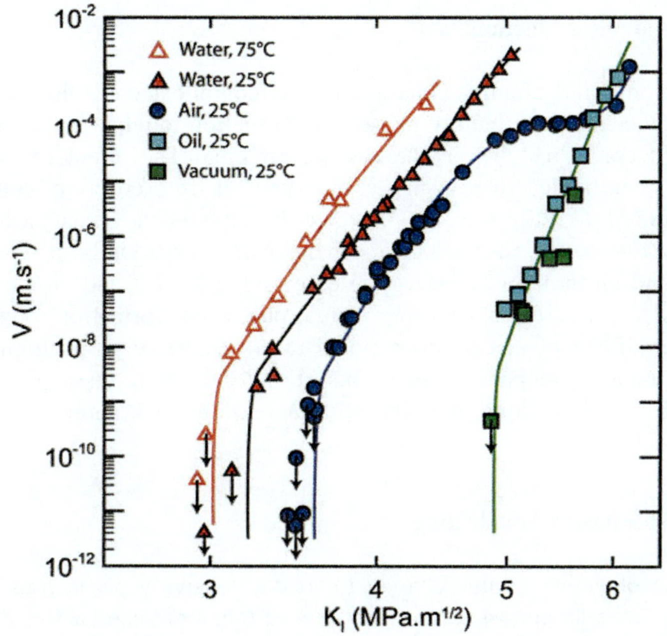

Fig. 9.3 The effect of environment on slow crack growth of zirconia (from Chevalier et al. [8])

$$V_{\mathrm{I}} = 2v_0 a_0 \exp\left(-\frac{E_{0\mathrm{I}}}{kT}\right) \cdot \sinh\left(\frac{\alpha_I (G^* - 2\gamma)}{kT}\right), \tag{9.2a}$$

$$V_{\mathrm{II}} = \frac{64G^*}{3\pi E a_0 \ln\left(\frac{l}{a_0}\right)} \left[\frac{a_0^3 p_E}{\sqrt{2\pi M k T}}\right], \tag{9.2b}$$

$$V_{\mathrm{III}} = 2v_0 a_0 \exp\left(-\frac{E_{0\mathrm{III}}}{kT}\right) \cdot \sinh\left(\frac{\alpha_{\mathrm{III}} (G^* - 2\gamma_S)}{kT}\right), \tag{9.2c}$$

where $E_{0\mathrm{I}}$ and $E_{0\mathrm{III}}$ are activation energies for the fracture of ceramic bonds (respectively in water and in vacuum), v_0 is the fundamental reaction frequency ($v_0 = kT/h$, with h and k being Planck's and Boltzmann's constants, respectively), G^* the mechanical energy that is due to the stresses at the crack tip, a_0 a characteristic bond separation, α_{I} and α_{III} the activation areas, γ and γ_{S} the fracture surface energy (in the presence of water molecules and in vacuum, respectively), E is the Young's modulus, p_E the water pressure, M the mass of water molecules, and l the crack-opening displacement. Note that the crack velocity in stage II is almost constant versus K_{I}, and mainly depends on the stiffness of the studied material.

2.2.2 Toughening Mechanisms

In both zirconia and alumina, this analysis is complicated by the existence of toughening mechanisms: mainly phase transformation toughening in the case of zirconia and crack bridging in the case of alumina. The effect of toughening mechanisms can be described using a "modified" Griffith criterion: crack propagation occurs if $G \geq 2\gamma + R_{\Phi}(c,G)$, where $R_{\Phi}(c,G)$ is the resistance to crack propagation due to the toughening mechanism that depends both on the crack length (c) and on the applied energy release rate. Alumina and zirconia present two very different cases; indeed in zirconia, phase transformation toughening is almost independent of c, but proportional to G. Conversely, in alumina, crack bridging is almost independent of G, but strongly depends on c, giving rise to R-curve effects (toughening increases with increasing crack length).

Phase Transformation Toughening

It is not the objective of this chapter to treat extensively phase transformation toughening, since an almost exhaustive view of this topic can be found in Green book [5].

This toughening mechanism allows zirconia to be one of the toughest oxide ceramics. It occurs when, around a propagating crack, the stresses are high enough to destabilize the tetragonal phase. The tetragonal-to-monoclinic phase transformation then occurs. McMeeking and Evans [19] developed a model in the 1980s, in which this transformation superposes a shielding stress intensity factor K_{Ish}, which means that the real stress intensity factor at the crack tip (K_{Itip}) is smaller than the one applied by external loads:

$$K_{\mathrm{Itip}} = K_{\mathrm{I}} - K_{\mathrm{Ish}}. \tag{9.3}$$

Theoretical models [19] and experimental results [20] show that increasing the applied K_{I} increases the size of the transformed zone, thus shielding K_{I}, which leads to Eqs. (9.4a) and (9.4b):

$$K_{\mathrm{Ish}} = C_{\mathrm{sh}} K_{\mathrm{I}} \tag{9.4a}$$

with:

$$C_{\mathrm{sh}} = \frac{E}{\sigma_{\mathrm{m}}^{\mathrm{c}}} \frac{(1+\nu)}{(1-\nu)} V_{\mathrm{f}}\, e^{\mathrm{T}} \frac{0.214\sqrt{3}}{12\pi}, \tag{9.4b}$$

where E is the Young's modulus, ν the Poisson ratio, V_{f} the volume fraction of transformable particles, e^{T} the dilatation strain associated to the transformation of one particle, and $\sigma_{\mathrm{m}}^{\mathrm{c}}$ the critical local stress that triggers the t–m transformation.

Hence, C_{sh} depends directly on σ_m^c, which itself depends directly on the temperature (and mainly on the difference between room temperature and temperature T_0 (t–m) of the metastable transformation line).

Phase transformation toughening and its dependence on materials parameters are illustrated in Fig. 9.2 that summarizes the V–K_I curves of zirconia ceramics with different thermal treatments, thus with different stabilities (i.e., different T_0 t–m temperatures) [15]. These curves show an increase with grain size of the K_I necessary for a given crack propagation speed. Increase of the grain size was obtained only through thermal treatment and thus is associated with phase partitioning and depletion of Y out of the tetragonal phase. Thus, higher grain size means less stable tetragonal grains, smaller σ_m^c, and more phase transformation toughening. Plotting the crack velocity versus the K_I/K_{I0} ratio brings all the curves back to a single, "master" curve, intrinsic to zirconia, illustrating Eq. (9.4a).

Toughening by Crack Bridging

This mechanism is more specific to coarse-grained ceramics [21–23]. In orthopedics, it can be found mainly in alumina ceramics. Coarse-grained alumina shows an R-curve behavior which is characterized by an increase in crack resistance with crack extension. It was demonstrated experimentally [24, 25] that this effect is caused by crack surface interactions in the crack wake. R-curve behavior arises because additional energy is required besides that needed at the crack tip to propagate the crack by simple stress corrosion mechanism. The additional energy is required to overcome the tractions of interlocking grains and pullout or the restraining faces of unbroken ligaments in the wake of the crack. Different studies have shown the influence of an R-curve behavior on SCG in ceramics by a shift towards higher stress intensity factor values with crack extension [26, 27]. Fett et al. [28] have even shown that a crack velocity decrease could be observed at the beginning of the crack propagation of small macrocracks in alumina.

R-curve behavior gives rise to very specific artifacts on the V–K_I curves: the measured curves depend very strongly on the way they are measured. This is explained as follows: at the beginning of the crack propagation test, the crack is short and the toughening by crack bridging is small. At the end of the crack propagation, the crack is necessary longer and the toughening is more important. This means that the V–K_I curves are shifted toward higher K_I at the end of the test. When tests involve decreasing crack propagation velocities (for example, double torsion tests in the relaxation mode), the foot of the curve is shifted to higher K_I and the curve is much steeper than it would be without toughening, with in particular higher values of crack propagation threshold K_{I0}. On the contrary, when tests involve increasing crack propagation velocities, the top of the V–K_I curve is shifted to higher K_I, and the curve is dilated. In both cases, intrinsic V–$K_{I\text{–Tip}}$ curves must be obtained by subtracting the R-curve effect from the experimental curve [29] ($K_{I\text{–Tip}}$ designating the real stress intensity factor at the crack tip).

2.3 *Low-Temperature Degradation of Zirconia-Containing Implants*

Low-temperature degradation is a mode of degradation specific to zirconia ceramics and composites, especially metastable, tetragonal zirconia.

2.3.1 Tetragonal-to-Monoclinic Phase Transformation and Low-Temperature Degradation

In tetragonal zirconia polycrystals, since tetragonal grains are metastable, bringing them enough energy can provoke their transformation to the stable state, i.e., the monoclinic phase. The t–m phase transformation in 3Y-TZP is martensitic in nature. As a consequence, crystallographic correspondences exist between the parent (tetragonal) and the product (monoclinic) phase, and can be described by habit planes and directions (shape strain) [30]. The t–m transformation is accompanied by shear strain (amplitude of ~0.16) and volume expansion (around 0.05).

The energy necessary for t–m transformation can be brought into the system either as mechanical energy (for example, at the tip of a propagating crack, as shown in the Sect. 2.2), or as chemical energy (for example, by water molecules adsorbed at the surface). Hydrothermal aging of zirconia (also called Low-Temperature Degradation, LTD) is directly related to the latter case: the presence of water in the environment of zirconia pieces triggers the t–m transformation of some grains on the surface.

It has to be mentioned that LTD is more important for zirconia stabilized by trivalent ions such as yttrium: indeed, trivalent cations in the network of tetravalent zirconium ions necessitate the presence of oxygen vacancies for charge compensation. During LTD, water-derived species diffuse through the network of oxygen vacancies [31] and occupy the vacancies. This results in ionic rearrangements—thus internal stresses—around the vacancies that can trigger the t–m transformation. The first grains to transform are the less stable ones (for example, the ones with lower yttrium content) and whose crystallographic orientation allows the whole volume change to be relaxed through a surface uplift: this is the case of the grains with their *c*-axis perpendicular to the surface [8, 32]. The transformation of such grains does not give rise to residual stresses, but to roughening of the surface. However, with a slight disorientation, given the stiffness and lack of plasticity of zirconia ceramics, the mechanical stresses arising from such a large volume increase are not entirely relaxed through a surface uplift: most of the time surface uplift is accompanied by microcracking and residual stresses in the neighboring zones. Neighboring zones then also transform under the combined actions of stresses and water. The surface is thus progressively transformed into monoclinic phase, following nucleation and growth kinetics. Then, the microcracks allow the access of water in deeper and deeper layer of the material, allowing the transformed layer to progress into the bulk.

Different parameters control the kinetics of low-temperature degradation. The most important ones are probably residual stresses and yttria content (or more generally stabilizing oxide content). Both parameters are directly related to the stability of the tetragonal phase: the more yttria, the higher stability, whereas residual stresses (especially tensile residual stresses) decrease the stability. Parameters that influence the above-mentioned ones have an indirect influence on aging. The presence of the cubic phase is detrimental to aging (cubic grains pump yttria from tetragonal ones, leaving them less stable). A large grain size is also detrimental (it often results from high sintering temperature that enables fast diffusion of yttria and an in-homogeneous repartition of stabilizer). On the other hand, compressive residual stresses (such as the ones left on zirconia after strong grinding) prevent aging, as well as the presence of glass phase, or the presence of high amount of stabilizer. One can almost completely suppress LTD by stabilizing zirconia by tetravalent cations [33] (such as Ce^{4+}) or by a mix of trivalent and pentavalent cations [34] (such as a combination of Y^{3+} and Ta^{5+}).

2.3.2 Impact of Aging on the Durability of Implants

LTD was first described by Kobayashi in 1980 [35], in a paper that showed that the t–m transformation could take place on the surface of zirconia pieces at 250°C, and results in a decrease of the mechanical properties. Since the first reports dealt with temperatures significantly higher than body temperature, LTD was first considered as negligible at 37°C. However, it was soon to be discovered that aging is a thermally activated phenomenon: accelerated tests [36] correlated with *in vivo* observations [37] made apparent that LTD cannot be neglected during the lifetime of an orthopedic implant. Indeed, this phenomenon may be involved in at least two sequences of the lifetime of an implant: during sterilization and during *in vivo* residence.

The first alerts of the possible effect of aging on zirconia prostheses came from the Food and Drugs Administration (FDA). In 1997, the FDA issued a warning cautioning the surgeons against the use of steam sterilization on zirconia implants [38]: this procedure was proven to roughen the surface and provoke increased wear of polyethylene cups. More important, recent clinical studies showed an increased wear rate of polyethylene cups wearing against zirconia head as compared to alumina heads [39]. This was later correlated to the expected kinetics of LTD at 37°C [40]. This increased wear rate due to LTD resulted in increased risks of aseptic loosening of the hip prostheses, thus shorter lifetime.

In both previous cases, surface roughening is the most visible result of t–m phase transformation. However, aging may give rise to other problems, mainly related to fracture of the ceramic. Indeed, LTD results in microcracking of the aged surface and the microcracked area may act as potentially critical defects. This was illustrated, for example, in a paper by Lamghari et al. [41] who showed that long aging times (100 h at 131°C) lead to a slight decrease of the strength (930–900 MPa) of 3Y-TZP, while the Weibull modulus increased (due to the fact that the critical defect was no longer intrinsic to the ceramic pieces, but related to the thickness of the microcracked surface layer). This decrease of the mechanical strength because of aging was dramatically put in evidence in 2001, when several hundreds of zirconia Prozyr® hip heads fractured *in vivo* around 2 years after implantation [42–44]. It was then shown that a slight decrease in the density of the heads (although the density was way above the legal threshold) was sufficient to make the residual porosity percolative, and to allow access of water inside the material, where aging took place [40].

However, this mechanism (decrease of the mechanical properties due to LTD) must not be considered as a general one. In some cases, the t–m transformation occurring during aging can put the surface in compression, thus increasing the apparent strength [45]. This might be valid for short aging times, while the transformed area is not yet microcracked. As soon as microcracking appears, one can expect a decrease of the strength and of the lifetime. Thus, short aging of zirconia ceramics must not be considered as an easy way to increase their strength.

The impact of aging on implants' lifetime has been evaluated by some simple calculations based on two hypotheses: the surface monoclinic layer thickness is proportional to the aging time, and this microcracked layer is assimilated to a defect (the thickness of the layer being the size of the defect). Under these assumptions, with growth rates of the monoclinic layer taken from the literature [36], one may calculate that the lifetime of a zirconia piece falls to a few years or a few tens of years [40], even under mechanical solicitations well below the crack propagation threshold (and for which no crack propagation should ever occur). In fact, in this case the defects grow because of aging, until they reach a size sufficient for crack propagation to occur. Once this stage is reached, the failure of the zirconia piece occurs by SCG in a matter of days or weeks.

2.3.3 Aging Assessment and Relevance of Accelerated Tests

Different techniques to assess LTD are currently in use. Some focus on the "definition" of aging: since LTD is a phase transformation, techniques able to detect phase changes are employed [such as X-Ray Diffraction (XRD) or Raman Scattering (RS)]. Others focus on the morphological changes of the material: quantification of the surface roughening or thickness of the microcracked layer [Scanning Electron Microscopy (SEM), Optical interferometry, Atomic Force Microscopy (AFM)]. Finally, some techniques evaluate the evolution of the mechanical properties, with the aim of quantifying the degradation due to aging. Note that because LTD may introduce compressive surface stresses, aging is not always accompanied by a loss of strength, at least during the earliest stages. In our opinion, XRD and RS are the most convenient techniques to determine aging kinetics. Since these tests are nondestructive, kinetics can be measured on only one sample. In these cases, aging is characterized by the proportion of monoclinic phase, given by the expressions:

$$V_m = \frac{1.311 \cdot X_m}{1 + 0.311 \cdot X_m} \tag{9.5a}$$

with

$$X_m = \frac{I^m_{111} + I^m_{\bar{1}11}}{I^m_{111} + I^m_{\bar{1}11} + I^t_{101}} \tag{9.5b}$$

for XRD, where I^f_{hkl} is the intensity of the peak of the (*hkl*) plane of the phase *f* [46, 47], or by

$$R_m = \frac{I^m_{181} + I^m_{192}}{I^m_{181} + I^m_{192} + 0.97\left(I^t_{148} + I^t_{264}\right)} \tag{9.5c}$$

for RS, where I^f_{xyz} is the intensity of the band at the subscript attributed to the phase *f* [48].

As mentioned before, aging is a thermally activated phenomenon (at least, between room temperature and 200°C for 3Y-TZP). Therefore, knowing the aging kinetics at different temperatures, one can extrapolate the kinetics at room or body temperature. Indeed, aging kinetics measured by XRD can be expressed by Mehl–Avrami–Johnson laws:

$$f_m = 1 - \exp(-(b \cdot t)^n). \tag{9.6a}$$

In this expression, *t* is the time, *f* is the monoclinic fraction (the degree of aging is assimilated to *f*), f_m is the maximum possible monoclinic fraction (equals to the initial tetragonal fraction), *n* is an exponent (between 1 and 4, but constant versus temperature for a given material), and *b* is a thermally activated coefficient:

$$b = b_0 \cdot \exp\left[-\frac{Q}{RT}\right], \tag{9.6b}$$

where b_0 is a constant, *T* is the absolute temperature, *R* the ideal gas constant, and *Q* is the activation energy. Plotting ln(*b*) versus 1/*T* gives a straight line of slope *Q*/*R*, allowing the calculation of *b* for either room or body temperature.

This method allows accelerated tests to determine the sensitivity of a given material to LTD. The most used accelerated test is aging in water vapor in autoclave. For 3Y-TZP, activation energies around 100 kJ mol^{-1} are generally measured, resulting in 1 h of autoclaving at 134°C (standard sterilization temperature), corresponding to 2–3 years at 37°C. These tests can give good results: for example, it has been shown that accelerated tests on one given batch of zirconia heads for hip prostheses were very predictive of the *in vivo* sensitivity to LTD of zirconia heads from the same batch [40]. However, they are not predictive from one zirconia-based material to the other: since aging kinetics depend so much on extrinsic material properties (residual stresses, porosity, grain size, etc.), values of b_0, n, and even Q must be determined for every new material. For instance, data obtained on 3Y-TZP from one prostheses manufacturer are not safely usable to predict the behavior of a 3Y-TZP processed by another manufacturer (even if they use the same powder), much less to predict the aging behavior of zirconia-toughened alumina.

Accelerated tests by autoclave are also limited by the fact that the tested material is free from any mechanical stress. Since it has been shown that tensile stresses may accelerate aging [49], the accuracy of the kinetics predicted from stress-free materials might not be so good for materials in use situation, where the zirconia component is subjected to wear, shocks, and mechanical loading. However, to our knowledge, this has never been quantified.

2.3.4 LTD in Alumina–Zirconia Composites

Perhaps the most promising future of zirconia as a biomaterial is through its use as reinforcement in zirconia toughened alumina (ZTA), where zirconia grains are distributed in an alumina matrix. These composites can offer an exceptional balance between SCG resistance and aging resistance [50, 51]. Two kinds of composites can be found, depending on the stabilization mechanism of tetragonal zirconia grains. Indeed, yttria can be used for stabilization (for example, in the Biolox® Delta material). But, since the residual stresses due to the stiff alumina matrix may stabilize the tetragonal phase, yttria is not always necessary, and yttria-free materials exist.

When zirconia grains are stabilized by yttria, the control of the size and homogeneity of the repartition of the zirconia grains is of primary importance [52, 53]. Indeed, it has been shown that above a critical amount (~16 vol.%), geometrical percolation of zirconia grains leads to a continuous path between zirconia particles, making the composite sensitive to aging [51]. Below this critical amount, only the largest zirconia grains (or the zirconia clusters in a nonhomogeneously dispersed material) undergo aging, the smallest ones being over-stabilized by the stiff alumina matrix. LTD can thus be completely suppressed in a ZTA containing fine-grained, well-dispersed zirconia particles below the geometrical percolation threshold. The commercially available Biolox® Delta materials contains Y-stabilized zirconia just above the percolation threshold, and is thus slightly sensitive to aging without it being detrimental to their durability in the time frame of the hip prostheses application [54–56].

On the other hand, LTD can be completely suppressed in composites with no yttria (Fig. 9.4): in this case, since zirconia grains do not contain oxygen vacancies, water cannot access to the grain, and LTD cannot proceed [57]. Monoclinic phase can still be present in the material due to residual stresses (depending on amount

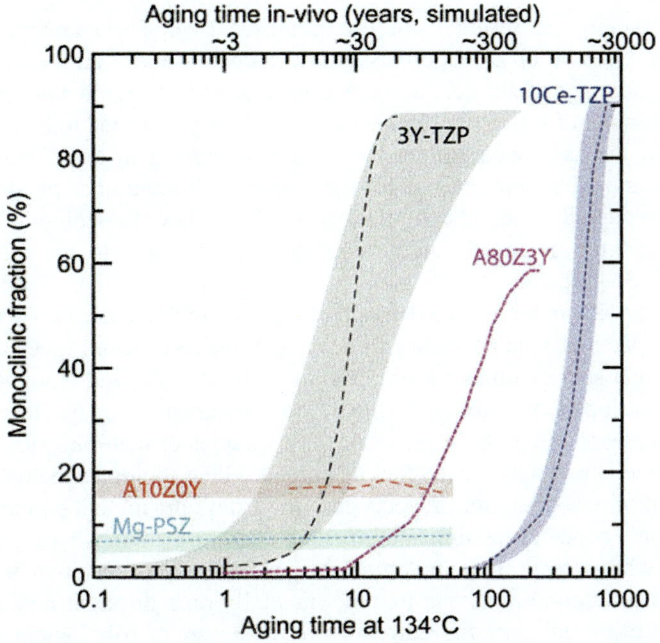

Fig. 9.4 Low-temperature degradation of different zirconia ceramics and composites (from Chevalier et al. [8])

and spatial distribution of zirconia grains), but its quantity will be constant with time, thus it cannot affect the durability.

There is another kind of zirconia–alumina composites: alumina toughened zirconia (ATZ) in which a major phase of zirconia contains some alumina grains (typically, a few percent up to 20%). In these composites, "chemical" stabilization of tetragonal zirconia is mandatory. It is currently achieved either by yttria doping or by ceria doping. Stabilizing with ceria in such composites leads to LTD-free materials [58], and a careful control of the microstructure can result in very high mechanical properties (strength up to 1,400 MPa [58], toughness and crack propagation threshold around 9 and 4.5 MPa $m^{1/2}$, respectively [59]).

2.4 Wear Degradation of Ceramics in Orthopedics

2.4.1 Introduction

Total hip replacement (THR) is one of the most successful surgeries in orthopedic surgery. For the last five decades, in THR, the dysfunctional or damaged hip joint is replaced by an artificial hip joint prosthesis. It can alleviate pain, reconstruct function of hip joint, and improve patients' daily living ability, thus bringing benefits to patients suffering from osteoarthritis, rheumatoid arthritis, and hip fractures as a result of osteoporosis or mechanical shock. However, although implant technology and surgical technique have advanced significantly, to date an increasing number of hip replacements fail prematurely and need to be revised. It is reported that the aseptic loosening of implant, occurring in over 70% of cases, is one of the most frequent reasons for the long-term limitation of prostheses [60]. Wear is considered to be the major cause of the late loosening of hip joint prostheses and the osteolysis of the surrounding bone due to the formation of wear debris [61–63].

Nowadays, with an increasing demand of hip joint replacements for younger and more active patients, it is necessary to develop prostheses with higher wear resistance for longer service time [64, 65]. Metals, polyethylene, and bioinert ceramics have been developed to be used in THR. Before they are used *in vivo*, it is necessary to determine the tribological behavior of these couples of materials, for heads and inserts. It is well known that many factors influence the tribological behaviors of hip joint implant *in vivo*, including aspects deriving from implant and patients, such as materials, load, contact pressure, and kinetics. Therefore, there is an urgent need to understand what affects the tribological response of a prostheses material under given tribological conditions and to experimentally rank different new developed materials. We present here the current state of the art of tribological laboratory evaluation of hip joint prostheses materials. The use of tribological test devices and the influence of test parameters on the tribological behaviors of articulating surfaces between cup and femoral head are presented and discussed.

2.4.2 The Current Situations of Hip Joint Implant

Wear usually occurs in the articulating interface of acetabular cup and femoral head due to the relative motion during activities of daily life. The sliding between ball head and cup generate wear debris, which will cause inflammatory tissue reactions and subsequently lead to the osteolysis around the implant. Such osteolysis is directly related to the loosening of the fixation that is the most frequent short-term failure mode of the prosthetic joints. Many factors are involved in the performances and lifetime of hip prosthesis when implanted *in vivo*. Next, we will summarize the most widely studied contributing factors for hip joint replacement.

Bearing Materials

Bearing material is regarded as a very critical factor to affect the tribological behavior of hip joint implant in wear process. Since Charnley's hip implant was successfully used *in vivo* in 1967, several types of materials have been widely employed as hip prosthetic materials, such as polymers (e.g., ultra-high-molecular-weight polyethylene, UHMWPE), metals (cobalt chromium alloy and stainless steels), and ceramics (alumina, zirconia, and composites of these two ceramics)

[66, 67]. At present, the most commonly used bearing couples include metal-on-metal (M/M), metal-on-UHMWPE (M/P), ceramic-on-UHMWPE (C/P), and ceramic-on-ceramic (C/C).

M/M implant was the first bearing couple to be implanted *in vivo* for hip joint replacement. However, the first generation of M/M implants had many shortcomings in their design and quality and thus were phased out as a result of higher loosening rates by 1975. However, the second generation of M/M, offering better design and materials properties, was available by the end of 1980s. Dumbleton and Manley have summarized the clinical experience of the second-generation M/M implants [68]. The superiority of second-generation M/M implants in the clinical studies with regard to the implant survivorship was not established. Indeed, there were large differences between the results obtained by laboratory wear studies and retrieval analyses [69, 70]. Moreover, the potential harms of M/M implant, including metal ion release, biological response, hypersensitivity, mutagenicity, and carcinogenicity, are not yet completely known [71, 72]. Therefore, M/M implants appear not to be a satisfying candidate in hip joint replacement for long-time implantation.

M/P implants are the most commonly used artificial joints in total hip arthroplasty, and the most clinically successful long-term bearings, with a survivorship of over 75% at 20 years follow-up for cemented bearings [73]. They replaced the first-generation M/M bearing couple because of their lower frictional torques. The success of total hip joint arthroplasty in the past 30 years is mainly due to the introduction of UHMWPE as a bearing surface, which possesses notably high abrasion resistance, low friction, high impact strength, excellent toughness, low density, ease of fabrication, biocompatibility, and biostability [74]. However, in recent years, there has been a great number of literatures reports on the osteolysis and loosening of the prosthesis due to the wear particles from UHMWPE [75]. It can be found that there is a direct relationship between the presence of macrophages and giant cells in the tissues surrounding the implant, and UHMWPE wear debris [76]. Some retrieval studies have verified a correlation between the number of macrophages and the volume of UHMWPE debris. The size range of UHMWPE debris is about 0.1–1 μm, which lies in the size range able to trigger biological activity [77]. The key problem for M/P bearing is to reduce the volume of UHMWPE wear debris, which is the base to slow down the osteolysis and loosening, and thus expand the lifetime of hip prosthesis.

C/C bearings have been introduced in hip joint replacement in order to prevent the osteolysis from UHMWPE wear debris and expanding the lifetime of hip prosthesis due to the high wear resistance and bioinertness. There has been increased interest in the development and use of low wearing alternative bearing surfaces such as alumina ceramic-on-ceramic hip prostheses. The first use of Al_2O_3 C/C bearing in THR was reported in 1970 [78]. In *in vitro* wear screening test, the wear rate of alumina/alumina ceramic bearing is reported to be 4,000 times lower than that of M/P bearing [79]. In *in vitro* hip joint wear simulator test, the wear rate of alumina/alumina hip prostheses is in the order of 0.01–0.1 mm^3/million cycles

[80, 81]. However, these *in vitro* wear rates are far lower than those obtained *in vivo*, typically 1–5 mm^3 per year [82]. Even *in vivo*, the wear rate of C/C is much lower than that of M/M or M/P bearings. Therefore, alumina/alumina bearing is of great potential as hip prosthetic materials when considering wear rate and biological properties. Implanted from the 1980s to around 2002, zirconia ceramics (ZrO_2) were generally believed to possess better wear resistance than Al_2O_3. However, LTD of zirconia might lead to a reduction of fracture toughness and an increase of roughness in the presence of water [83]. C/C prostheses may lead to stiffness of walking and specific wear damage [84]. A major disadvantage of ceramics is their susceptibility to brittle fracture, which is perhaps the most dangerous failure mode found in THA. The top priorities for C/C bearing are to prevent the fracture and further improve the reliability.

C/P bearing is a hard-on-soft bearing adopted to further reduce the amount of polyethylene wear debris. However, many complications of C/P bearing were found, including osteolysis, loosening, dislocation, and component failure. Catastrophic acetabular component failure involves the fracture of the ceramic femoral head and wear-through of the polyethylene liner [85]. Some reports that studied the wear rates have demonstrated that C/P components exhibit superior wear characteristics to metal–polyethylene [86]. However, it is difficult to compare individual studies because of differences in patient activity, surgical technique, component sizes, component positioning, and component materials. Most *in vivo* studies have demonstrated lower linear wear rates (0.03 mm per year after wearing in, compared to 0.07 mm per year with standard metal-on-polyethylene), offering the prospect of increased polyethylene longevity and a decreased potential for osteolytic response.

Mechanics

The contact force has an important influence on the tribological behaviors of hip prosthesis and is also the most widely investigated factor in THA. During the most frequent activities of daily life, there are several kinds of gait patterns, such as walking, going upstairs and downstairs, which must result in various hip contact forces to hip joint head. Thus, it is significant to know the real contact pressure of hip joint implant *in vivo* for design and materials selection of hip prosthesis. In hip mechanics, the average contact force is 238% BW (percent of body weight) for walking, 251% BW for upstairs climbing, and 260% BW for going downstairs. Therefore, an appropriate contact force can be applied in *in vitro* wear tests according to the real range of contact force. Similarly, peak contact pressure is diverse in different kinds of gait patterns: it goes from 3.26 MPa for walking, to 5.71 MPa for going upstairs (and 3.77 MPa for going downstairs). These data provide useful evidences for *in vitro* wear tests and design of hip prosthesis. So, in pin-on-disk or flat-on-flat wear tests, the contact pressure is set in the range of 1–10 MPa.

Kinetics

THR is a very complex mechanical–biological system when working *in vivo*. It can be simplified as a ball-in-socket joint with three degrees of freedom (DOF): flexion–extension (FE), abduction–adduction (AA), and internal–external rotation (IER). It is reported that the amplitudes are $\pm 23°$ for FE, $\pm 5°$ for AA, and $\pm 7.5°$ for IER in given wave forms during normal gait [87].

In order to ensure the safety of hip prosthesis *in vivo*, the movement is recommended to be minimized as far as possible. In *in vitro* wear test, the velocity of sliding is usually reported to be less than 1.5 m s^{-1} in pin-on-disk motion. In wear simulators, the angular range of movements of the femoral head is carried out by use of DOF in kinematics motion. Various DOF have been reproduced in different hip simulators, including single-axis, two-axis, and three-axis movement [88].

Synovial Lubricant

It is reported that most artificial joints at best are lubricated by mixed lubrication. Thus, it is known that wear takes place throughout the life of the prosthesis. The synovial fluid (SF) of natural joints normally acts as a biological lubricant as well as a biochemical pool through which nutrients and regulatory cytokines traverse. SF contains many molecules that provide low-friction and low-wear properties to articulating cartilage surfaces. It has been found that four components of SF (albumin, hyaluronic acid, lubricin and other polyelectrolytes, and phospholipids multilayers) are critical to the tribological performance of hip joint replacement. They play important roles to protect the hip interface, increase the viscosity, or modify the boundary lubricant regime [89–91]. Mishina and Kojima adopted a pin-on-disk wear test to determine the effect of human serum albumin on the friction and wear of different bearing surfaces [92]. The results showed that protein adsorption leads to differences in the hydrophilicity of the prostheses materials, changes of lubricant regime, and finally influence its friction and wear behaviors. However, it is still an unsolved problem due to the complication of *in vivo* conditions. In addition, the effects of SF also depend on the load condition and hip prosthesis.

2.4.3 Classification of Wear

As mentioned before, wear of the articulated surface is the main factor to limit the lifetime of hip prosthesis due to the aseptic loosening by wear particles. Defining the wear mechanism of hip prosthesis is very important and helpful in preventing its recurrence. The wear mechanisms of the contact between the ball and the cup in hip implants *in vivo* mainly include adhesion, abrasion, and fatigue. Adhesion involves bonding of the surfaces when they are pressed together under load. Sufficient relative motion results in material being pulled away from one or more surfaces, usually from the weaker material. Abrasion is a mechanical process wherein asperities on the harder surface cut and plow through the softer surface, resulting in removal of material. When local stresses exceed the fatigue strength of a material, it would fail after a certain number of loading cycles, releasing material from the surface. One or more of these classical mechanisms of wear may be operating on the prosthesis, and prosthesis may function in several wear modes over its *in vivo* service life.

2.4.4 Tribological Evaluation Tests

The aim of the tribological study of hip prosthetic materials is the development of bearing couples that minimize wear and friction in order to improve the long-term performance of hip joint prostheses.

Wear Screening Tests

In order to get a quick evaluation of the hip prosthesis materials, simple wear tests are developed to obtain the basic data like friction coefficient and wear rate *in vitro*. It is a kind of method to quickly examine the material properties by using simplified test devices. The wear rate, friction coefficient, and wear mechanism of hip prosthetic materials can be obtained by such wear screening tests. After that, the availability of prosthetic materials according to these data can be speculated. The commonly used wear test devices are pin-on-disk, pin-on-flat, and ring-on-disk. Usually, the pin sample can be replaced by ball sample. Reciprocating pin-on-disk wear test is the only test recommended by the American Society for Testing Materials (ASTM) for evaluating polymeric bearing materials for THA [93]. However, the wear screening tests have inevitable shortcomings due to the simplified wear mode. It is different from the real wear conditions of hip joint prosthesis *in vivo*. There are many factors involved in this difference, including contact forces, contact pressure, motion, velocity, and so on. However, the advantages of wear screening tests are the short experimental period and easy to use aspects. Therefore, they are still widely used to obtain the basic materials properties *in vitro*.

Wear Simulators

Wear simulators are developed to simulate the real condition *in vivo*, with complex load profile, velocity, kinetics, and positions. They are closer to the real environment of hip prosthesis in human body than wear screening tests. Therefore, they can provide very useful and valuable data about the wear behaviors of the hip prosthesis in dry or lubricated conditions, under static or dynamic conditions. However, the most important shortcoming of wear simulator is the long period and expensive cost. That is why, recently, more and more studies are related to numerical simulation. Liu et al. [94] have developed a computational wear simulation of metal-on-metal hip resurfacing replacements. The computational prediction of the volumetric wear and the wear geometry shows close agreement with that of the simulator tests. It is believed that finite element modeling (FEM) will become a practical method to predict the wear behaviors of hip joint prosthesis *in vivo* with the development of the simulation techniques.

2.4.5 Conclusions

As the best successful arthroplasty surgery, THA have brought many benefits for patients to function on the damaged hip joint. With developments of materials and techniques, the materials and design of hip prosthesis have gained great process. The lifetime of the current hip prosthesis is usually 15–20 years, before the need for a painful, dangerous, and expensive revision surgery. Therefore, longer lifetime (>20 years) is needed. Many hip prosthetic materials are developed for THA. One critical step is to evaluate these materials or hip prosthesis *in vitro* to predict the wear behavior when implanted *in vivo*. Various devices have been adopted to investigate the wear behavior *in vitro* by simulating the condition *in vivo*. However, up to now, no ideal device or parameters are completely satisfying due to the complexities in the hip joint.

2.5 Degradation by Shock Degradation in Ceramic–Ceramic Couples for Artificial Hip Joints

Ceramic–ceramic couples have been implanted as hip joints since the 1970s. The first implantation was carried out by Boutin et al. [95–98]. As was mentioned in the previous section, these materials are characterized by very low wear rates in common tribological investigations, i.e., hip simulator. These experiments are as close as possible to the ISO 14242 [99]. Moreover, due to the progress in manufacturing alumina, the measured wear rates of C/C have been decreasing since the first implanted systems. Since the 1990s, the fractures of C/C bearings are scarce, 0.004% for femoral heads [100]. The complications for this typical bearing are: unanticipated bearing fractures and bearing noise (squeaking and clicking) [64]. Moreover, loosening of alumina sockets can occur due to elastic mismatch between ceramic materials and bone or sealing cement, or due to metallic debris generated by degradation of the femoral stem [101]. This kind of degradation is due to the impingement (contact) between the femoral neck of the stem and the cup and/or metal back. It is due to extreme positions reached by the prosthesis during gait, for example positions needed to exit from a car. This phenomenon could accelerate some specific degradation of the alloys composing the femoral stem [101].

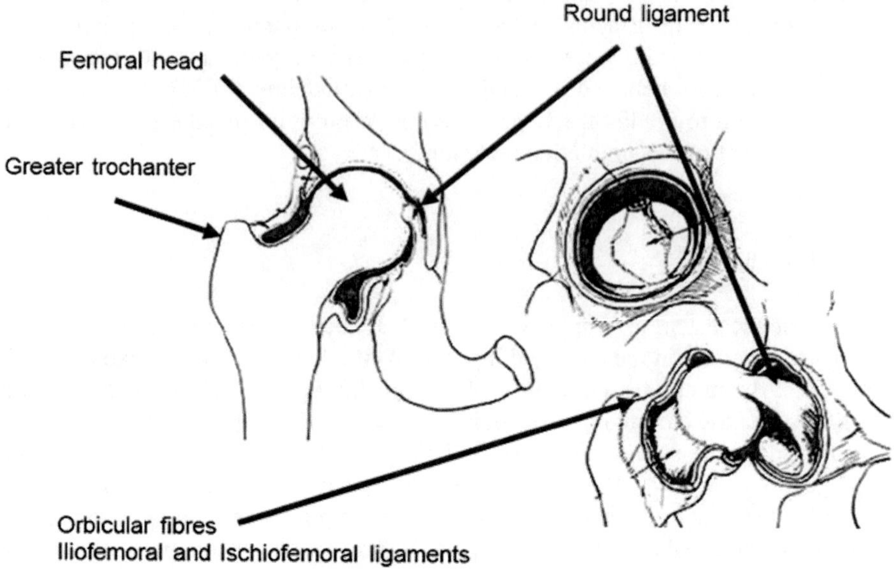

Fig. 9.5 Anatomical scheme of a hip joint

Nevertheless, the main mechanism of shock-related degradation originates from the microseparation between head and cup [102]. Indeed, during arthroplasty, the round ligament (that joins the femoral head and the acetabulum) is necessarily sectioned, and the orbicular fibers (around the joint) and articular capsule [103] are damaged, Fig. 9.5 [104, 105]. Thus, during everyday movements with a hip prosthesis, the contact between cup and head is not always maintained. This microseparation involves shocks during gait, at heel-ground contact. This typical phenomenon will degrade the head–cup assembly in ceramics (Fig. 9.6). Indeed, the ceramic materials are brittle, and if the elastic limit is reached, the material could be broken. Moreover, local degradation can be involved by microseparation.

2.5.1 Analysis of Explants: Specifications for Lab Tests

Ceramic implants were implanted as hip joint since the 1970s. The study of retrieved ceramic explants has shown particular degradation patterns. Nevelos et al. investigated alumina–alumina artificial hip joint: 11 explants were analyzed and exhibited wear stripes with roughness of approximately 200 nm, to compare to the 20 nm measured on pristine alumina surface [82]. These values and the wear patterns were confirmed by other works [106, 107]. Thus, the point is to reproduce *in vitro* this typical stripe wear. Interesting investigations revealed discrepancies between *in vivo* (explants) and *in vitro* tests. Indeed, tests from hip walking simulator (conditions close to ISO standard 14242) show no significant degradation of ceramic hip implants [108]. As mentioned before, since the round ligament and

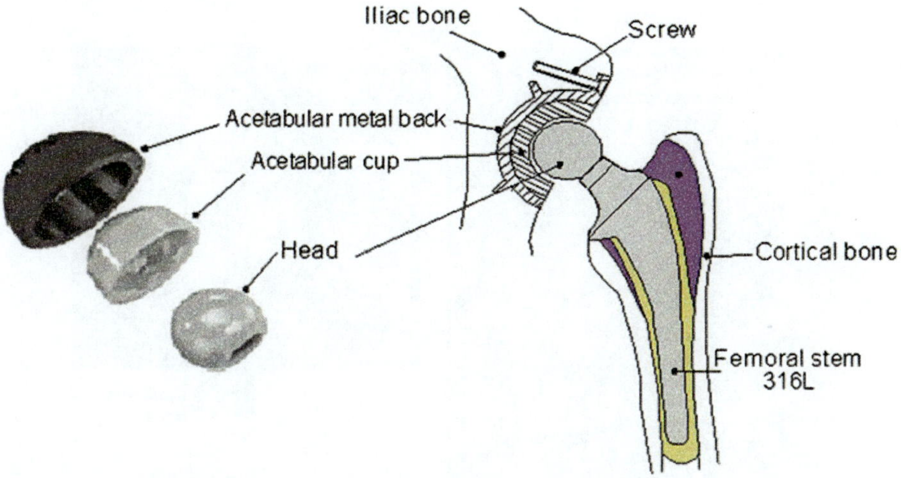

Fig. 9.6 Hip prosthesis. The head–cup assembly is made of ceramics

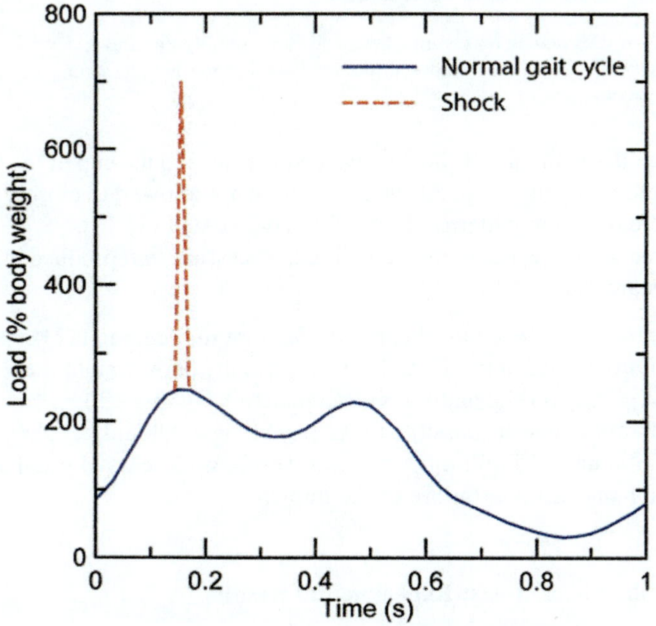

Fig. 9.7 Contact load during usual gait and the specific signal related to shock

orbicularis zone are damaged during surgery [103], a microseparation might occur during gait. It is worth noting that, at the heel-ground contact, the normal load applied in the joint can go up to 7 times the usual body weight (usually when walking down stairs [109, 110]), at the beginning of the gait cycle [111] (Fig. 9.7).

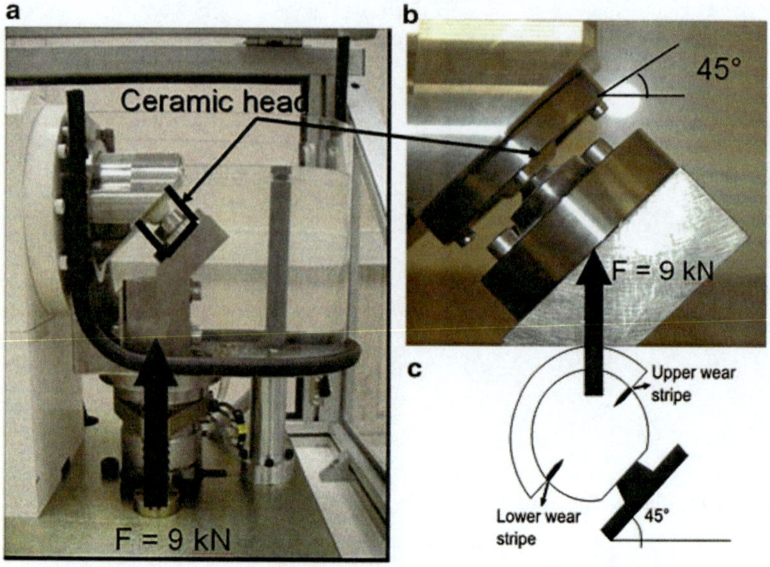

Fig. 9.8 Shock testing device for ceramic head and cup assembly. (**a**) Shock device. (**b**) Head and mechanical assembly, *bold line* represents the container of the bovine serum. (**c**) Wear stripes related to the shocks tests

All these conditions involve a shock between the head and the cup of the prosthesis. Thus, implementing microseparation in hip simulator allowed to closely reproduce *in vitro* the degradation patterns observed *in vivo* [102, 112].

To summarize, the experimental conditions necessary to reproduce the degradation patterns are:

- Microseparation between head and cup (3.3 mm maximum [113]).
- Applied normal load, a minima of seven times the body weight, i.e., 7 kN for a patient weighing 100 kg and 9 kN for a patient weighing 130 kg.
- Test medium as close as possible to the physiological liquid, i.e., bovine serum.
- Test duration up to 15 million cycles (representing 15 years of usual service life of hip implant), or up to failure of the implants.

2.5.2 Specific Shock Tests: Experimental Results

Based on the previous section, a specific device (Quiri™) was manufactured for testing the ceramic materials to shocks. Figure 9.8a, b show the mechanical assembly, with head and the cup inclined at 45°. Figure 9.8c shows the upper and lower wear stripes obtained after test.

In this study, 32 mm alumina heads were exposed to shocks in air or in calf serum (at $T = 36 \pm 2$°C according to ISO standard 14242-1, changed weekly to avoid protein degradation); a standard value of 1.3 mm for the microseparation was

Table 9.1 Test conditions and results for the failure of ceramic head–cup assemblies, 32 mm in diameter

Materials	Alumina head and cup	
Tests conditions	Dry	Wet (calf serum)
Force/frequency	9 kN/2 Hz	
Microseparation	1.3 mm	
Number of cycles	245,000 ± 43,000	>800,000

The number of cycles is related to the failure of the cup

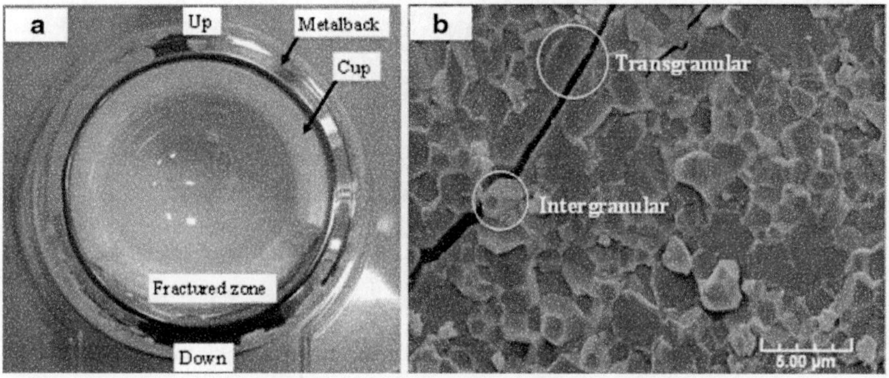

Fig. 9.9 (**a**) Damaged cup failed in dry conditions. (**b**) SEM image of the failed cup

applied (distance between head and cup along a radius), at either 1.5 or 2 Hz frequency, for 800,000 cycles or until implant failure; loading of 9 kN, and duration of shock of about 30 ms, red peak in Fig. 9.7. The results of the shock tests are presented in Table 9.1.

From Table 9.1, it is clear that the presence of calf serum attenuates the shock severity: no failure of ceramic cup in calf serum was noted (although the presence of high amount of water in the serum might have made the ceramic more sensitive to SCG). *In vivo*, the joint is not always lubricated: since the synovial sac disappears during surgery, the lubrication of the prosthesis could be alternative. Consequently, studying both behaviors of tests conditions (with or without lubrication) is close to the probable actual lubrication.

Figure 9.9a shows a failed cup (dry shock conditions). The failure occurs at the lower rim of the cup. Three cups were broken at the same location in dry conditions. Figure 9.9b shows a SEM image of fracture after failure. It is worth noting that both intragranular and intergranular fractures occur. This fact is related to severe embrittlement because the intragranular fracture occurs with high level of fracture energy. This kind of fracture does not occur during usual friction tests on the hip simulator.

One should pay attention to the wear stripe generated during these tests. This typical wear is specific to shock-related degradations. It does not occur on a usual hip simulator. Figure 9.10 presents the 3D roughness of alumina head.

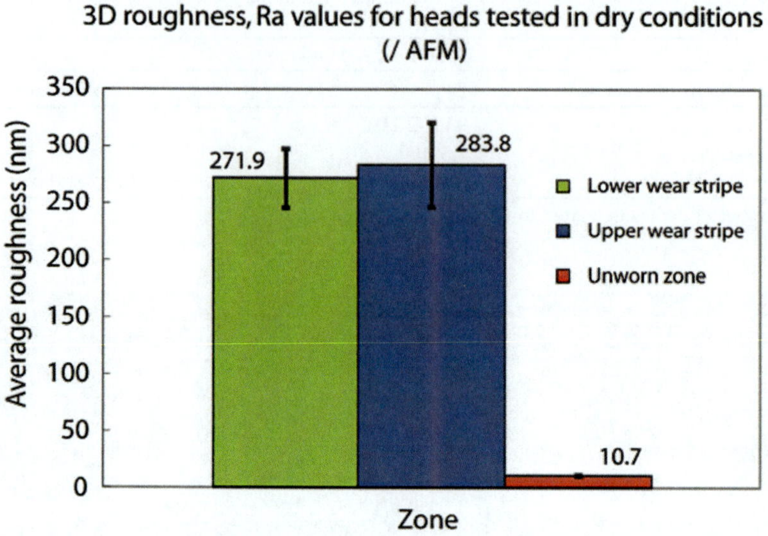

Fig. 9.10 Average roughness, R_a, in different zones of the head after shock test

The roughness was obtained by AFM (Digital Instruments, multimode AFM, Si_3N_4 tips, stiffness of 0.12 N m^{-1}). The analyzed area by AFM was 40 × 40 μm^2. Five measurements were carried out in both unworn and worn surfaces. The raw data were filtered by a filter order 1 in order to remove the tilting of the samples. These results show that wear stripes are related to significant degradation in comparison with the pristine surface of alumina. These measurements are in accordance with the previous ones proposed by Nevelos et al. [82].

As a conclusion, the shock testing device presented in this study allows reproducing the *in vivo* degradation, wear stripes, and 3D roughness. The wear measurements will be described in the next section.

2.5.3 Wear Rate Due to Shock Degradation: Methodology Assessment

Measurements of wear volume were also conducted on samples after shock tests, so as to assess different methods of wear volume measurement. The weight loss of the ceramic head was investigated, as well as the wear volume obtained from 3D profilometry, and the results were compared to results from the literature. This kind of measurements is a key point for validating the shock tests. Table 9.2 is divided into two parts [99, 108, 114, 115]: the first is related to the experimental results from the shock device described in this paper, the second is the comparison with bibliographical results.

The results of both methods (weight loss and the 3D profilometry) are similar, with an error of approximately 25%. The measured wear volumes may differ slightly, depending on the measurement methodology, but the orders of magnitude

Table 9.2 Wear volume generated in shock tests and the comparison with results from bibliography

	Wear volume mm^3/million cycles		
	Weight loss	Profilometer	% Error
Test 1	0.29	No measurement	–
Test 2	0.19	0.15	21.2
Test 3	0.31	0.2	34.6
Stewart et al. [113]	0.2–1.84, with microseparation		
Tipper et al. [98]	1.24–1.74, with microseparation		
Essner et al. [114]	0.02–0.08, without microseparation		
Walter et al. [108]	0.7 per year from explants (1 year could correspond to 1 million of cycles)		

The error is defined as $\%\,\text{Error} = \frac{\text{wear}_{\text{weight loss}} - \text{wear}_{\text{profilometer}}}{\text{wear}_{\text{weightloss}}} \times 100$. The profilometer used: Veeco NT 9100

are the same. The comparison with results from the literature shows that shock tests are also relevant to characterize wear behavior: the wear volumes are in the same range as those extracted from works where microseparation was imposed. It is worth noting that the wear volume reported by Stewart et al. and by Tipper et al. is related to a device combining microseparation, not specifically controlled, and friction (same kinematics as a hip simulator). Thus, one may suggest that combining shock and friction should promote wear of prosthetic elements.

On the other hand, results from hip simulator (Essner et al. [108]) do not fit well with the specific wear due to shock. This shows that hip simulator without microseparation is not the right mechanical device for testing ceramic materials for hip joint. Microseparation is a key point for testing ceramic material, with results as close as possible to the ones obtained *in vivo*. One might even suggest that the typical device for testing the ceramic hip implant should be a shock device with applied microseparation and friction. To check the relevance of the combined test, different couples of materials (C/P, C/C, M/P, M/M) should be investigated. Finally, where zirconia-based ceramics are concerned, aging (LTD) combined with shock–friction should be added in the test protocol. Especially, ZTA is more and more investigated for hip implants, and the effect of aging should be characterized in more depth for studying the service life of this material.

2.5.4 Conclusion

Shock is a relevant degradation mechanism of ceramic assemblies dedicated to hip implant. This artificial joint is promising because of lowest wear rate in comparison with other systems such as M/P, etc. However, damages on ceramic implants during everyday life might not be negligible. Wear mechanisms are not exclusively due to friction: we show that *in vitro* degradation produced by shock (experimental device with microseparation) is close to degradation patterns observed on retrieved explants, contrary to basic friction tests. Roughness and wear rate from *in vitro* tests are in accordance with explants. The service lifetime of ceramic implants is principally related to the shock degradation.

New trends in ceramic materials are under investigation. Developments in ceramic bearings are leading to new material formulations such as ZTA. This new hybrid material couple is now proposed by different companies (Depuy, Zimmer, etc.).

Finally, one may suggest that some investigations have to be carried out on the biocompatibility of alumina or composite grains removed from the ceramic during the shocks. It is worth noting that no deleterious or biological reactions have been highlighted for alumina and zirconia grains [116]. Moreover, *in vivo* tests with cells show no significant signs of inflammation. This shows that ceramic wear debris might not be as deleterious as polyethylene wear debris.

3 Implants of Bioactive and Resorbable Ceramics

3.1 General Description of Bioactive and Resorbable Ceramics

3.1.1 Context

The previous sections of this chapter focused on ceramic implants made of so-called bioinert ceramics, with typical examples of zirconia, alumina, and zirconia–alumina composites. These materials are often associated with the "first generation" of bioceramics, developed since the 1960s. They have been chosen for their biocompatibility and also to limit the interaction with the body. They are still widely used for applications demanding high mechanical properties (toughness, hardness, wear, and fatigue resistance), such as hip and knee prosthesis. Interaction with living tissue always leads to the formation of a non-adherent soft tissue interface around the implant, even if it can be very thin, as in the case of ceramic hip heads. Such an interface might be partly responsible for the long-term possible degradation.

Some clinical applications demanding a strong interface between the implant and the body need other biomaterials, as for bone defect treatment and for the fixation of prosthesis (for example, femoral stems and dental implants with the application of coatings). In the case of bone substitution, conventional approaches involve bone grafting, with patient autograft, human donor (allograft), or animal bone (xenograft). The limited availability of natural bone, the increase of clinical risk associated with a second operation site, and the possible disease transmission limit the applicability of bone grafting. Bioactive ceramics have been largely developed to elicit a favorable response of the living tissues, and especially to form a mechanically strong bond with calcified tissues, such as bone and teeth. Such ceramics can also resorb in the body with time, leading to the formation of natural bone. This ability has the interest of improving the long-term implant survivability as compared to bioinert ceramics. Many literature reviews have been published on such bioceramics [117–121], commercially available since the 1980s and often referred to as the "second generation" of bioceramics. These materials must be biocompatible, i.e., non-cytotoxic, and also avoid the release of particle during degradation.

Since the end of the twentieth century, a "third generation" of bioceramics is being developed more specifically to induce an enhanced bioactivity, an ability to deliver biological agents and to favor rapid tissue regeneration. The materials associated with this need are usually based on the bioactive materials of the second generation in the form of highly porous scaffolds. Such structures possess interconnected macro-pores (easy access for cells) and microporosity that increases the specific surface, accelerates the degradation processes, but also provides the cells with attachment sites. Scaffolds can also be seeded with cells, hormones, or growth factors, or may be a vector for drug delivery. The objective is to shift from tissue replacement to tissue regeneration [122] by combining engineering materials, living cells, and biochemical factors. Such an approach is often called Tissue Engineering [123].

We will focus in this part of the chapter on implants made of bioactive and resorbable ceramics. Some commercially available bone substitute materials are described in [117]. Composite scaffolds and CaP cements, as well as dental restoration parts, are described comprehensively in other chapters in this book and will therefore not be discussed in detail herein.

3.1.2 Bioactive and Resorbable Ceramics

History

Attempts to fill bone defect started with bone transplantation in the seventeenth and eighteenth centuries, followed by the use of man-made ceramics: calcium sulfate in the late nineteenth century, and Tri Calcium Phosphate (TCP) in 1920. The clinical use of so-called bioinert ceramics started in the 1960s [124]. After these initial attempts, the first success to develop a bioactive material able to form a strong bond with bone is the discovery of Bioglass® by Hench in 1969 [125], based on the hypothesis that an implant forming a hydroxyapatite (HAp) layer *in vivo* may not be rejected by the body. The formation of this surface layer has been confirmed *in vivo* and *in vitro* for nearly all bioactive ceramics. Hydroxyapatite, a hydrated calcium phosphate, is known for a long time as being the mineral component of natural bone. Bone mineral is in fact a heterogeneous material in constant evolution in the human body through a remodeling process. It can be defined as a poorly crystallized carbonated hydroxyapatite (CHA). Its chemical composition also varies, but can be represented as similar to hydroxyapatite: $Ca_{10}(PO_4)_6(OH)_2$, with possible calcium substitution (e.g., potassium and sodium) and variable amounts of HPO_4 and CO_3 in substitution of the phosphate.

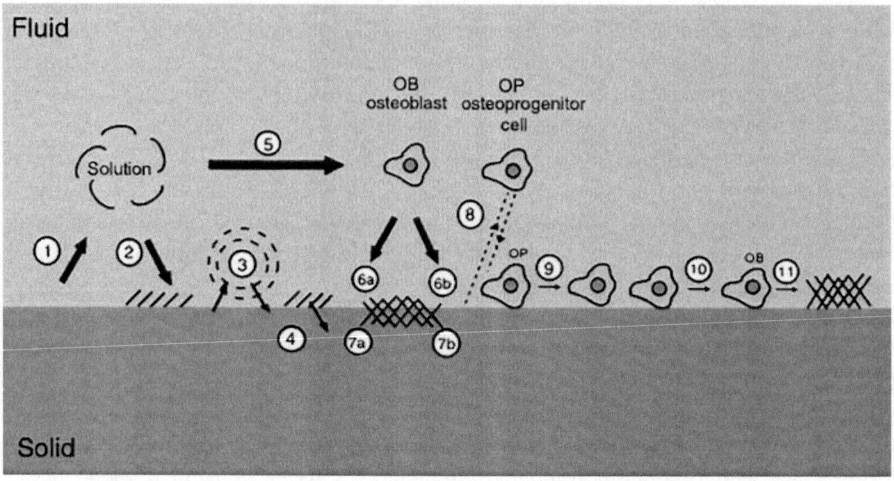

Fig. 9.11 Schematic diagram representing the events which take place at the interface between bioactive ceramics and the surrounding biological environment. (1) Dissolution of the ceramic, (2) precipitation from solution, (3) ion exchange and structural arrangement of the interface, (4) interdiffusion from the surface layer into the bioceramic, (5) solution-mediated effect on cellular activity, (6) deposition of either the mineral phase (a) or the organic phase (b) without integration into the bioceramic, (7) deposition with integration into the bioceramic, (8) chemotaxis to the bioceramic surface, (9) cell attachment and proliferation, (10) cell differentiation, (11) extracellular matrix formation. Reprinted with permission from [126]

Bioactivity: Bond with Bone

A bioactive material is defined as a material that elicits a specific biological response at its interface, which results in the formation of a bond between the tissue and that material [119]. For bioceramics, bioactivity is often characterized in terms of bone bonding ability through the formation of an apatite layer at their surface. The mechanisms underlying the apatite layer formation are described by different models, Hench's [119] and Ducheyne's [126] being the most cited. They have common steps, with initial chemical reaction at the surface of the bioceramic leading to the precipitation of an apatitic layer, followed by cellular interactions with this layer (see Fig. 9.11). Proteins adsorption is likely to occur rapidly on the ceramics surface and may have an influence of the first stages. Some differences are noted between bioceramics, as will be discussed below.

To ensure such bone bonding ability, several physicochemical parameters are noted as important: the presence of functional groups at the surface such as Si–OH or Zr–OH, the presence of soluble silica, the presence of negative surface charge and of nanostructured or rough surface, etc [122]. As an example, the release of soluble silica stimulates stem cells differentiation, cells proliferation and division, production of growth factors, and extracellular matrix proteins. This enhances bioactivity and has been proven effective in the case of bioglasses.

All bioactive ceramics are considered as osteoconductive (defined as the apparent growth of bone tissue "along" an implant's surface). Some of them are also reputed osteoinductive, meaning that the material stimulates bone formation even in non-osseous sites. This phenomenon has not yet been fully understood, but the presence of nanostructured rough surface is also postulated to have an important role in cell differentiation and in new bone formation [127]. The ability of bioceramics to concentrate bone growth factors circulating in the biological fluids is also supposed to favor osteoinductivity.

Characterization of Bioactivity

In vivo bioactivity can be characterized by a single parameter, the I_B factor, directly linked to the time $t_{0.5bb}$ required to achieve more than 50% of the implant surface bonded to bone after *in vivo* implantation: $I_B = \frac{100}{t_{0.5bb}}$ with $t_{0.5bb}$ expressed in days. Experimental values for bioceramics range from 0 for bioinert ceramics (alumina), 2.3 for HA, 3.2 for apatite wollastonite glass ceramic (A–W), and above 8 for some bioactive glasses. This factor is mainly used for bioglasses (Fig. 11 in [119]), showing different bioactivity domains depending on their SiO_2 content.

In vitro study of the bioactivity of ceramics is usually carried out in a solution, so-called simulated body fluid (SBF), with ion concentrations nearly equal to those of the human blood plasma. This method has been initially developed by Kokubo and co-workers in the beginning of the 1990s [128]. Nearly all the current bioceramics clinically used for bone grafts show the formation of an apatite-like layer on their surface when immersed in SBF. Correlation of kinetics of bone formation *in vivo* and apatite layer forming ability *in vitro* in SBF showed satisfactory tendencies for Bioglass®, glass ceramics and HAp [129], and for TCP, apatite–wollastonite (A–W) glass ceramics, bioactive glasses (BG), HAp, and mixture of HAp and TCP [130]. Typical immersion times for apatite formation *in vitro* in SBF are 1 day for BG, 5–7 days for A–W, 14 days for β-TCP, and 28 days for HAp. In some cases, *in vitro* test in SBF and *in vivo* implantation lead to contrary results, for example in the case of highly resorbable ceramics, such as β-TCP and calcite that do not show apatite formation *in vitro* but can bond to bone *in vivo*.

This *in vitro* method to characterize bioactivity is now generalized (ISO standard 23317), but is currently still in debate [131]. The SBF solution is supersaturated with respect to apatite; it is therefore difficult to prepare and is very sensitive to the presence of impurities or surface defects that can act as nuclei for apatite crystallization. It also does not contain any proteins found in human plasma, which may act as crystal growth inhibitors [132, 133] and that tend to favor the formation of amorphous calcium phosphate (ACP), and not crystalline HAp [134]. Parameters such as the volume ratio of ceramic-to-SBF, the specific surface area of the tested material, the static or dynamic (shaking or circulation) conditions during test, etc. have a large influence on the results, especially in the case of powders or for bioceramics with high solubility. However, an *in vitro* method to simulate the *in vivo* bonding ability is of great interest.

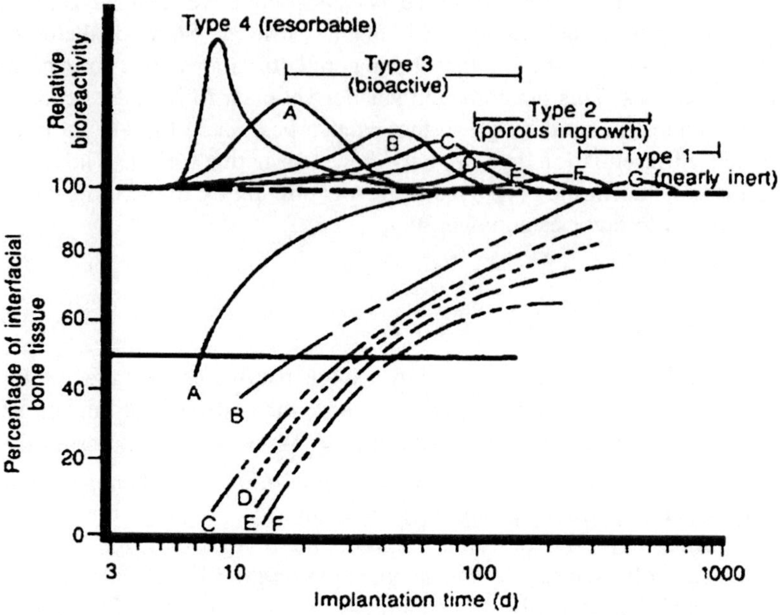

Fig. 9.12 Bioactivity spectrum for various bioceramic implants. (A) Relative rate of bioactivity, and (B) time dependence of formation of bone bonding at an implant surface (A: 45S5 Bioglass®, B: KGS Ceravital, C: 55S4.3 Bioglass®, D: A–W glass–ceramic, E: HAp, F: KGX Ceravital, G: Al_2O_3–Si_3N_4). Reprinted with permission from [119]

Resorbable Ceramics

A resorbable biomaterial is designed to degrade gradually over a period of time and to be replaced by the natural host tissue. A highly resorbable material will tend to aid bone formation, thus recent developments tend to focus on such materials (see Fig. 9.12).

The mechanisms of resorption of bioceramics *in vivo* are related to both physico-chemical processes (with dissolution and precipitation stages) and cellular processes, as noted for bioactivity mechanisms. The cell-mediated resorption refers to the action of osteoclasts involved in the natural resorption of bone, and of inflammatory multinucleated giant cells formed by the fusion of macrophages, which degrades the material by phagocytosis. HA and TCP are more quickly resorbed in the bone than under the skin due to the specific action of osteoclasts. The properties of the bioceramics, such as the rate of solubility, can affect the cell action. Materials with high solubility (calcium sulfate, some CaPs) release a high amount of Ca ions that may cause the detachment of osteoclasts from the material surface. In that case, resorption of implants is mainly controlled by chemical dissolution.

Understanding of the chemical resorption of bioceramics *in vivo* requires consideration of three different aspects [135]: the solubility of the material in the human body, the kinetics of the dissolution, since dissolution is not necessary observed even if is possible thermodynamically, and the precipitation of another compound. The composition (Ca/P ratio for CaPs) and the structure of ceramics logically influence their solubility and dissolution rate. From the work of Driessen [136], it is known that the phase that controls the solubility in living bone is octacalcium phosphate (OCP), $Ca_8H_2(PO_4)_6{\cdot}5H_2O$. This means that phases that are more soluble than OCP tend to dissolve in physiological conditions, whereas less soluble phases can be dissolved only through cell-meditated mechanisms. Cell action is supposed to induce a local decrease of the pH, which favors the dissolution of calcium phosphates and carbonates. Ceramics more soluble than OCP in physiological conditions are sometimes referred to as "biosoluble," whereas ceramics less soluble are referred to as "bioresorbable." Protein adsorption can affect the dissolution and precipitation steps and can act as a precipitation inhibitor, like some trace elements in the ceramic, such as soluble magnesium. It is also known that geometrical factors can affect the solubility of materials, e.g., crystal size and surface area. The difference in resorption time between sintered hydroxyapatite and calcium phosphate cements is mainly attributed to the difference in crystal size, around the micrometer for HAp and at the nanometer range for CaP cements. The role played by porosity amount and morphology in resorption is logical at first order and explains in part the development of scaffolds to ensure rapid interaction with the body and possible resorption.

Resorbable and Bioactive Ceramics

Bioactive implants need to have a controlled dissolution rate: lower in the case of a coating on metal prosthesis, to ensure a long-term bond between bone and the metal, and higher for bone substitution. The resorption process has indeed a strong dependence on structural factors (porosity, specific surface, etc.), and even low solubility materials such as synthetic HAp can show resorption *in vivo* when in the form of tiny particles or scaffolds. The concepts of resorbable and bioactive ceramics have then converged in the last years [137]. Both properties are needed for good integration in the human body and for long-term stability of implants, as noted in the case of middle ear bone implants (see Fig. 9.13).

The principal classes of bioactive and resorbable ceramics developed and clinically used in the last two decades are: calcium sulfate, sintered HAp, β-TCP, HAp/TCP ceramics, BG, and A–W glass ceramics. In the following section, a short description of the bioceramics currently used in clinical applications is presented.

3.2 Different Families of Resorbable and Bioactive Ceramics

3.2.1 Calcium Sulfate [138]

This material is one of the first used for synthetic bone grafts (since the end of the nineteenth century). It is made of gypsum, a hydrated calcium sulfate $CaSO_4{\cdot}2H_2O$, largely used in the construction industry. It can be used as a self-setting cement

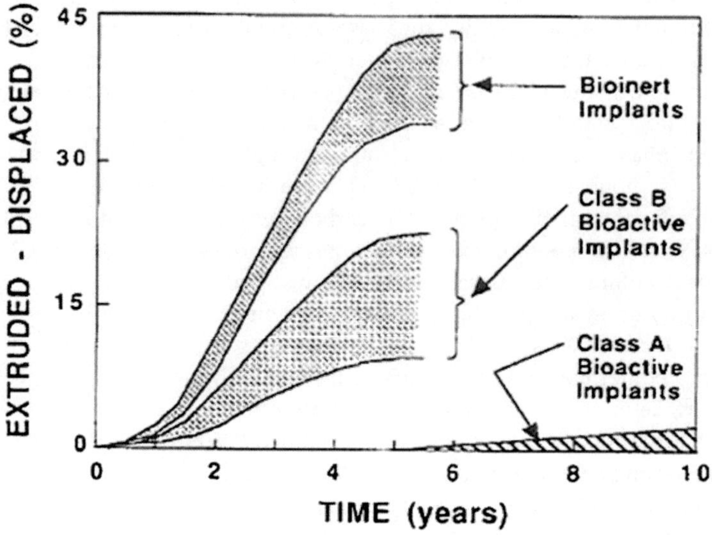

Fig. 9.13 Survivability comparison of bioinert implants. Class B bioactive implants (synthetic HAp) and class A bioactive glass implants (45S5) used to replace middle-ear bones. Class A and Class B refer to osteoproductive and osteoconductive bioceramics. Osteoproduction occurs when the material shows enhanced osteoblast activity and corresponds to $I_B > 8$ (mainly, bioactive glasses). Reprinted with permission from [119]

(mixing calcium sulfate hemihydrate with water), or in the solid form. The mechanical properties of calcium sulfate (CaS) are controlled by the amount of porosity, usually ranging from 20 to 60 vol.%, with a typical compressive strength varying from 1 to 20 MPa. In biomedical applications, the highest densities and strengths are often required, thus the hemihydrate used to produce CaS is in the α-form. CaS also shows a high sensitivity to the presence of liquid phase, with a 50% reduction of the mechanical strength.

Bioactivity

The bioactivity has been seldom characterized *in vitro*; the few studies in SBF have demonstrated the ability of CaS to precipitate CaP [139]. The important and rapid release of calcium ions is supposed to help osteoblast differentiation.

Resorption

It is the most rapidly dissolving bioceramic, with a classical *in vivo* resorption rate of several weeks. Indeed, different resorption rates can be noted, from 3 to 8 weeks, depending on the porosity level of the material and the implantation site.

The resorption mechanism *in vivo* seems to be more related to chemical dissolution than to cell-mediated resorption with giant cells or osteoclasts.

Clinical Applications

The use of CaS has been constant during the twentieth century, and developments are still being made, mainly for filling bone defect in non-load-bearing areas (mandible, maxilla, craniofacial, in the form of self-setting cement or pellets), and as a barrier material [138]. Peltier's work in 1959 concluded that CaS is a biocompatible material that can be used to prevent fibrous tissue formation in bone defects. It is unambiguously admitted that no inflammatory reactions are associated with the use of calcium sulfate. It is currently used for periodontal defects and extraction sockets.

Some authors state from clinical applications that CaS may have pseudo-antiseptic properties, i.e., less infection than other materials. This is possibly related to the presence of sulfate ions or to a local pH drop around the implant due to the rapid dissolution of CaS.

Limits

The relative poor mechanical properties and fast resorption *in vivo* of CaS limit its use to some applications: large defect are not always filled with bone after resorption, as compared to the high clinical success for small defects [140]. Treatment of periodontal defects is usually successful, but bone graft of lumbar spinal fusion has failed with CaS [141] due to a too high resorption rate.

Recent Developments

New implants based on CaS are developed to improve its mechanical resistance and lower its resorption rate. This is done with biphasic composites with a secondary phase that dissolves slowly, such as HAp, BG, or CaP, in order to form a 3D support for bone growth after CaS resorption.

3.2.2 Calcium Phosphate Ceramics

CaPs represent a family of materials with a wide range of chemical compositions (see Table 9.3 [142]). The main advantage in the use of these materials is their similarity in composition to the mineral phase of bone and teeth, which ensures good compatibility *in vivo*. The first clinical application of TCP was in 1920, and of HAp in 1951, but the development of CaPs for clinical application really started in the 1980s.

Table 9.3 Existing calcium orthophosphates and their major properties

Ca/P molar ratio	Compound	Formula	Solubility at 25 °C, −log (Ks)	Solubility at 25 °C, g L^{-1}	pH stability range in aqueous solutions at 25 °C
0.5	Monocalcium phosphate monohydrate (MCPM)	$Ca(H_2PO_4)_2 \cdot H_2O$	1.14	~18	0.0–2.0
0.5	Monocalcium phosphate anhydrous (MCPA)	$Ca(H_2PO_4)_2$	1.14	~17	c
1	Dicalcium phosphate dihydrate (DCPD), mineral brushite	$CaHPO_4 \cdot 2H_2O$	6.59	~0.088	2.0–6.0
1	Dicalcium phosphate anhydrous (DCPA), mineral monetite	$CaHPO_4$	6.90	~0.048	c
1.33	Octacalcium phosphate (OCP)	$Ca_8(HPO_4)_2(PO_4)_4 \cdot 5H_2O$	96.6	~0.0081	5.5–7.0
1.5	α-Tricalcium phosphate (α-TCP)	α-$Ca_3(PO_4)_2$	25.5	~0.0025	a
1.5	β-Tricalcium phosphate (β-TCP)	β-$Ca_3(PO_4)_2$	28.9	~0.0005	a
1.2–2.2	Amorphous calcium phosphate (ACP)	$Ca_XH_Y(PO_4)_z \cdot nH_2O$ $N = 3$–4.5: 15–20% H_2O	b	b	~5–12
1.5–1.67	Calcium-deficient hydroxyapatite (CDHA)[e]	$Ca_{10-x}(HPO_4)_x(PO_4)_{6-x} \cdot (OH)_{2-x}$[f] $(0 < x < 1)$	~85.1	~0.0094	6.5–9.5
1.67	Hydroxyapatite (HA or HAp)	$Ca_{10}(PO_4)_6(OH)_2$	116.8	~0.0003	9.5–12
1.67	Fluorapatite (FA or FAp)	$Ca_{10}(PO_4)_6F_2$	120.0	~0.0002	7–12
2.0	Tetracalcium phosphate (TTCP or TetCP), mineral hilgenstockite	$Ca_4(PO_4)_2O$	38–44	~0.0007	a

Reprinted from Dorozhkin [142]

[a]These compounds cannot be precipitated from aqueous solutions

[b]Cannot be measured precisely

[c]Stable at temperature above 1,000 °C

[d]Always metastable

[e]Occasionally, CDHA is named as precipitated HAp

[f]In the case $x = 1$ (the boundary condition with Ca/P = 1.5), the chemical formula of CDHA looks as follows: $Ca_9(HPO_4)(PO_4)_5(OH)$

CaPs can be found as ceramics or as injectable cements [143] made of a mixture of CaP powders that harden when mixed with water. Calcium phosphate cements (CaPC) would not be described in detail here, as they are the subject of another chapter in this book.

It has to be noted that CaP ceramics with a Ca/P atomic ratio less than 1 are not suitable for implantation due to their high solubility and acidity. Tetracalcium phosphate (TTCP) with the highest Ca/P ratio, on the other hand, shows a too high basicity for implantation. Thus, clinically used CaP bioceramics are mainly HAp, β-TCP, and a mixture of these two components, named biphasic calcium phosphate (BCaP). OCP, also implanted, is considered as a CaP cement and will be discussed in the chapter of this book related to cements.

These materials have shown a very good biocompatibility: no cell toxicity, no fibrous tissue around the implants, and no inflammation even after long implantation times. This systemic biocompatibility of CaP is related to their chemical composition, only calcium and phosphorous ions participating in normal metabolic process. Some cases of inflammation after implantation *in vivo* have been reported, but are possibly related to other causes rather than the material itself (e.g., edge of implants, migration of granules due to a lack of mechanical stability, etc.). The main families of CaP ceramics are presented below.

Hydroxyapatite [144]

Hydroxyapatite (HAp) used as a biomaterial can have a synthetic or a natural origin. Its chemical composition at stoichiometry is: $Ca_{10}(PO_4)_6(OH)_2$. Possible substitution in the structure of either calcium or phosphate will affect its microstructure, its chemical and thermal stability. Natural HAp, obtained from animal bones, has more minor constituents, and its composition is usually described as carbonated apatite (CHAp). Synthetic HAp can be elaborated using different methods: sintering, solid-state reactions, sol–gel, plasma spraying, precipitation from solution, electrochemical deposition, etc.

Bioactivity

The mechanism of apatite formation on synthetic HAp has been documented [145], even if it is less frequently described than for bioactive glasses. It is a four-stage process, with accumulation of Ca^{2+} ions close to the negatively charge HAp surface, then the deposition of a calcium-rich amorphous CaP layer. As the calcium ions accumulate, the surface becomes positively charged and reacts with phosphate ions of the solution, forming a calcium-deficient CaP layer. This amorphous layer can then be transformed into a more stable bone-like apatite that will help cellular proliferation and differentiation to form native bone tissue. The time to crystallize an apatite layer at the surface of HAp is, however, relatively important, typically after 28 days of immersion in SBF.

Cellular response of HAp has also been reported with osteoblasts, osteoclasts, and odontoblasts. The acidic condition produced by osteoclasts may help local dissolution of HAp, thus promoting the precipitation of apatite layer. It is also shown that carbonate-substituted HAp has a better stimulation influence on cells than the synthetic HAp. An enhanced cell proliferation and differentiation is also noted on porous HAp as compared to dense HAp, a common feature observed on different bioceramics.

Resorbablility

The ability of HAp to resorb *in vivo* is usually lower than other bioceramics. Stoichiometric crystalline HAp has a very low solubility, and no resorption after several years of implantation *in vivo* is commonly noted [146]. However, poorly crystalline HAp with a high specific surface area can show high levels of calcium and phosphate release. *In vivo* resorption of HAp scaffolds and thin powders has also been noted [142]. Cell-mediated dissolution mechanisms exist, with a small amount of phagocytized HAp observed after *in vivo* implantation.

Clinical Applications of HAp

- Dense HAp implants have been developed for spinal fusion, cranio-maxillofacial reconstruction, and fracture treatments in the 1980s and 1990s.
- HAp granules are used as fillers in bone defect. A study on 6 mm holes drilled in rabbit condyles [129] showed the importance to have small granules, i.e., small spaces between granules, to ensure bonding with newly formed bone.
- Coatings: HAp is largely used as a coating on titanium implants to combine the high strength of metal and the bioactivity of HAp. The analysis of the composition of HAp coatings from plasma sprayed techniques reveals the presence of a composite structure with HAp, ACP, and traces of TCP [147].

Limits

HAp implants have been largely developed since the 1970s. Many of them are currently limited by a relatively low bioactivity. Recent developments aim at increasing the bioactivity of HAp, e.g., as Si-substituted HA, based on the recognized role of silicon ions in stimulating cellular activity and enhancing apatite formation (see Sect. 3.2.3).

Dense HAp implants have also limited mechanical properties for load-bearing applications: the fracture toughness of HAp is approximately 1 MPa $m^{1/2}$. To improve both mechanical properties and bioactivity, dense A–W glass–ceramic implants have been developed and will be presented in a following part.

The nearly non-resorbable behavior of dense HAp leads to a limited survivability of implants in time. The use of HAp dense cone to replace roots of extracted teeth to preserve the edentulous alveolar ridge of denture wearers results in less than

50% survivability after 5 years. One can also mention the potential inflammation risk due to the release of non-resorbable HAp particles after a long implantation time *in vivo*.

TCP

TCP is one of the oldest synthetic biomaterials; it is still a common material for implants with a high resorption rate, as compared to HAp. β-TCP is most widely used for ceramic implants; other TCP forms are present as components in CaP cements and as coatings on metal prostheses [148].

Bioactivity

The ability of β-TCP to form an apatite layer on its surface is still debated, with in some cases a poor ability and in others no formation of apatite *in vitro* and *in vivo* [130]. The reason for its relatively low bioactive behavior is the poor nucleation ability of apatite on β-TCP. As for CaS, the *in vitro* testing conditions are supposed to have an influence on the ability to precipitate apatite on this highly resorbable ceramic.

Cellular cultures have been largely carried out on β-TCP. It is considered as non-cytotoxic, but few cases of osteoblast death have been noted, maybe related to the presence of small particles after erosion. Studies with osteoclasts and bone marrow cells show good interaction with β-TCP, favoring cellular activity. In some cases, an absence of cellular activity was observed, possibly related to a release of ions and an alkaline surface reaction.

The tendency to show contradictory results in bioactivity and cellular culture tests can be partly explained by the limited characterization of samples, since differences in composition, crystallization, grain size, or surface properties are possible and can largely affect the results.

Resorption

The resorption rate for β-TCP is much higher than for HAp, and it is related to its high solubility in physiological fluids [149]. Cell-mediated resorption has also been put forward for β-TCP, through two different mechanisms: degradation by osteoclasts, especially for dense materials, and phagocytosis by giant cells, favored in the case of porous materials with a higher solubility, more prone to the release of particles.

Clinical Applications

β-TCP is used when high resorbability is needed for non-load-bearing applications, mainly as bone substitutes for the treatment of defect after tooth extraction, sinus reconstruction, etc. The implants are usually transformed into bone, starting from the surface to the bulk.

The *in vitro* resorption time is usually several months. After 9–12 months of implantation in human mandible, and for sinus floor augmentation, 50–75% of the implanted β-TCP is resorbed, the highest resorption rate being noted for the most porous implants [150]. After 24 months in round defects (5 mm diameter, 4 mm height) in the mandible of dogs, β-TCP particles are completely resorbed [151].

Recent Developments

Recent developments involve mainly high-purity β-TCP with a homogeneous solubility to prevent premature separation of microparticles that may induce inflammatory response [134]. Ion-substituted TCP, mainly with Mg, Sr, and Zn, are also developed to tailor the resorption rate of TCP *in vivo*.

HA-TCP Composites [152]

These composites, also named biphasic calcium phosphates (BCaP), are made of a mixture of HAp and β-TCP, mainly to control both the resorption rate *in vivo* and the bioactivity by controlling the HAp-to-TCP ratio. These composites are obtained from a thermal treatment above 700°C of calcium deficient apatite, i.e., with a Ca/P ratio lower than 1.67, the stoichiometric ratio for HAp.

Resorption

The resorption rate *in vivo* largely depends on the particle size and to a lesser extent to the HAp/TCP ratio: BCP granules of 40–80 μm diameter are totally resorbed after 3 months of implantation in dog mandible whereas 200–300 μm particles are still present in the same conditions of implantation. Particles less than 10 μm are even faster resorbed but may trigger signs of inflammation.

Applications

The applications are mainly focused on bone substitution in dentistry (prevention of bone resorption, sinus lift augmentation) and orthopedics (inserts for cage fusion in cervical spine arthrodesis).

Recent Developments

Attempts are made to associate BCaP granules to a polymer or to a resorbable cement to form an injectable bone substitute.

3.2.3 Bioactive Glasses

The idea of developing a synthetic material that will quickly bond to bone originated from the numerous amputations in the Vietnam War. These were due to the lack of available material, non-rejected by the human body at that time, as mainly metal and plastic implants were available. The development of a specific glass composition that leads to the formation of a HAp layer in physiological conditions was done by Hench in 1969, using calcium and phosphorous oxides. The 45S5 composition (45 mol% SiO_2, 24.5 mol% Na_2O and CaO, 6 mol% P_2O_5) is known as being the most bioactive material, in terms of speed of an apatite layer formation in SBF.

Bioactivity

The mechanism of bioactivity for Bioglass® has been largely described in the literature. Hench has proposed five chemical reaction stages at the surface of bioactive glass immersed in SBF:

1. Exchange of alkaline ions (Na^+, Ca^{2+}) with the H^+ and H_3O^+ from the solution, causing hydrolysis of silica that creates silanol, with an increase of the pH by hydroxyl release.
2. Attack of the silica glass network, with loss of soluble silica and continued formation of silanol.
3. Condensation of silanol groups into a silica-rich layer on the Bioglass® surface.
4. Migration of a Ca^{2+} and PO_4^{3-} through the silica surface to form a $CaO–P_2O$ rich surface on top of the SiO_2 layer.
5. Crystallization of this film as it incorporates OH^- and CO_3^{2-} ions from the solution to form a CHAp layer.

These initial stages are followed by six stages involving cellular interaction, respectively: adsorption of biological moieties, action of macrophages, stem cells attachment, stem cells differentiation, matrix generation, and crystallization.

The high bioactivity of Bioglass® is associated with the release of calcium and silicon ions, elements shown as favoring the activation of the cell cycle of osteogenic precursor cells, and controlling cell differentiation [153]. *In vitro* cell culture tests of Bioglass® with osteoblasts from patients aged 50–70 have confirmed the activation of several gene families that stimulate proliferation and differentiation of mature osteoblast phenotype.

High bioactivity of Bioglass® has also been confirmed *in vivo* as compared to A–W glass ceramic and sintered HAp [129], the rate of bone regeneration being linked to the rate of dissolution of soluble Si and Ca ions in the corresponding biomaterials (see Fig. 9.14). For very short times of implantation, within the first week, a quicker bone formation on Bioglass® particles has also been noted as compared to A–W and HAp.

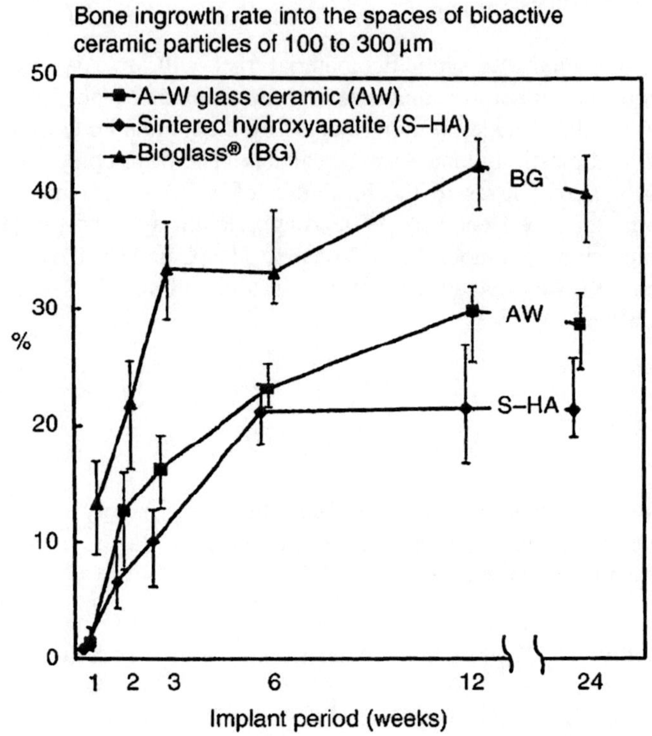

Fig. 9.14 Quantitative comparison of percentage of bone growth into the bone defect, from 1 to 24 weeks, due to 45S Bioglass®, A–W glass–ceramic and sintered HAp particles of 100–300 μm in diameter. Reprinted from [129]

Resorption

In vivo implantations of Bioglass® have shown its capacity of resorption. In a clinical case, the resorption of 300 μm diameter granules implanted for sinus floor elevation is noted, with a relative left volume of BG particles from the biopsies varying from 29% after 4 months to 15% at 6 months, and 8% at 15 months [154], with an associated graduated bone ingrowth. The BG particles appear to resorb after 1–2 years of implantation, with no signs of osteoclastic activity, such as resorption cavities. The authors therefore suggest that chemical dissolution is the main mechanism for BG resorption.

Clinical Applications

The first clinical use of Bioglass® dates from the 1980s, with two applications: the replacement of bone from the middle ear (see Fig. 9.14) and a device to provide a more stable ridge for denture construction after tooth extraction. These applications showed a higher clinical success than other solutions. For ear, nose, and throat surgery, the ability of Bioglass® to bond with soft tissue is an important advantage, as well as its ability to be drilled or machined.

Since the middle 1990s, Bioglass® particles are available in the market for the treatment of periodontal disease in infrabony defect and of frontal sinusitis. Long-term (9 years) study of the use of BG particles for sinus obliteration shows a clear bone formation and no inflammatory response. Several studies suggest a possible antibacterial effect of the BG, related to the increase of pH close to the BG implants, as OH^- ions are released in aqueous solutions. BG plates are also used for the treatment of orbital and cranial defects, with promising results.

Recent Developments

Bioactive glasses can currently be elaborated using a sol–gel approach at ambient temperature. These glasses show a higher reaction rate *in vitro* due to higher nanoporosity content and greater concentration of silanols at their surface. Bioactive glasses containing more than 80% of SiO_2 are then possible, as compared to 55% for the traditional glass elaboration method.

Another development in progress is the modification of the bioactivity of glasses through thermal treatments. Such treatments may cause partial or total crystallization, reducing the bioactivity and the release of alkaline ions. This is a possible way to tailor the bioactivity of bioglass-derived materials [155, 156].

3.2.4 Apatite–Wollastonite (A–W) Glass–Ceramic [157]

Glass ceramics are made of a mixture of a glassy phase and crystalline precipitates. Different bioactive glass ceramics exist; the most widely used clinically is A–W. It is made of precipitates of apatite and wollastonite ($CaO{\cdot}SiO_2$) in a MgO–CaO–SiO_2–P_2O_5 glass and has been developed in Japan after the observation of a clinical case. In 1981, HAp granules were used for a giant tumor treatment in the right ilium of a young patient. This material lead to a complete gap filling with good bone bonding, but it took one entire year to ensure a complete filling. This experience led to the development of a biomaterial with higher bioactivity than HAp and better mechanical properties to be used in case of applied loads in the body.

The material is usually fully dense (but can be made porous), machinable, with crystalline phases in the form of nanometer apatite and fibrous wollastonite that reinforce the glass by a composite effect. Its mechanical properties are well above dense HAp and Bioglass®, especially for bending strength and fracture toughness, close to 2 MPa $m^{1/2}$.

Bioactivity

Dense and porous A–W glass–ceramic exhibits a good bond with bone after implantation, with a quicker recovery compared to HAp.

In SBF, calcium and silicate ions are dissolved from the glassy phase and wollastonite, whereas phosphate is supposed to come from the SBF itself, allowing the formation of an apatite layer at the surface. It is expected that osteogenic cells can proliferate and differentiate on such a layer, then producing collagen and apatite. Hydrated silica at the surface of the glass possibly helps nucleation of apatite crystals.

A difference compared to Bioglass® is the absence of a silica gel layer on which the apatite layer forms. The good bond with bone of A–W glass–ceramic is attributed to the absence of this supposedly weak layer.

Resorption

Dense A–W shows nearly no resorption after several years. However, it has been shown that porous A–W partially resorbs after 12 weeks in a rabbit's tibia. This confirms the structural influence (shape, porosity) on the resorbability of bioceramics.

Clinical Applications

A–W implants are mostly load-bearing implants in combination with structural plates and screws for vertebral prostheses (nearly 3,000 cases treated in Japan since 1983), inter-vertebral spacers for lumbar spine and cervical spine (more than 1,000 cases for each application), and implants to treat giant bone tumors in load-bearing parts of long bones. Coating of metallic prostheses has also been developed using A–W glass–ceramic. No serious complications, such as loosening, inflammation, or fracture of the prosthesis, have been noted.

3.3 Conclusion

Implants made of bioactive and resorbable ceramics, often called "second generation" bioceramics, bring a clear improvement in survivability as compared to "bioinert" ceramics for non-load-bearing applications. No major problem in the clinical use of bioactive and resorbable ceramics implants have been noted, and they are both biocompatible and osteoconductive.

A large range of bioceramics are now available in the market for implants design, with different bone-bonding ability and resorption rate. The mechanisms of resorption and bioactivity are governed by chemical (dissolution, precipitation), physical (porosity, specific surface), and cell-mediated processes. They can be tailored by controlling implant geometry (granules, dense blocks, etc.) and material nature, porosity, grain size, level of crystallinity, etc.

The resorption kinetics *in vivo* of these ceramics can be globally ranked as follow: CaS > β-TCP > BG > A–W > HAp. The bone-bonding capacity evaluated *in vivo* and *in vitro* gives the following general ranking: BG > A–W > HAp.

Depending on the application, different implants are possible:

- For applications demanding significant service load, such as vertebras prosthesis in combination with structural plate sand screws, nearly non-resorbable ceramics can be used (A–W glass–ceramic, for example). A strong bond with bone is required, with good inherent mechanical properties of implants. However, the mechanical properties of dense bioactive ceramics are usually 5–7 times lower than "bioinert" ceramics used for hip or knee prostheses.
- For nonbearing applications, such as cancellous bone substitute, implants with a resorption rate compatible with kinetics of tissue regeneration are preferred. CaS, β-TCP, HAp/β-TCP, and BG can be used.

In the case of Bioglass®, the positive influence on cellular activity and gene activation properties has been demonstrated. Bioglass® implants are also found to have a better tolerance with time *in vivo*.

The long-term survivability of implants is a key issue, especially when taking into account the aging world population. Bioactive surface alone does not bring directly long-term survivability of implants. Possible mismatch in elastic properties, presence of wear debris, and limited tissue remodeling can limit the durability of implants. Bioactive and resorbable implants still lack self-repair, maintenance of blood supply, and auto-adaptation [122].

New developments to increase the survivability of implants focus on the control of cellular interaction. This can be done by developing highly porous scaffolds, seeded before implantation with progenitor cells or growth factors (tissue engineering). New bioceramics can also be designed to stimulate the local tissue repair (tissue regeneration). Ion substitution in bioglasses is one of the trends currently being explored [158].

4 Concluding Remarks

This chapter highlights an important idea: knowledge on *in vivo* degradation mechanisms of ceramics is very difficult to extract from knowledge of degradation mechanisms *in vitro*. This is due to the strong interactions between the different degradation mechanisms that occur concurrently *in vivo*. Two examples follow. For bioinert, structural ceramics such as alumina, zirconia and their composites, auto-accelerating mechanisms can be seen: shocks increase wear, that triggers low-temperature degradation, that increases wear, and so on, until the defects created this way are large enough for SCG to proceed. For bioactive, degradable ceramics, the desired degradation (or dissolution) is difficult to control due to very important interactions with cells (e.g., osteoclasts).

Separately, each of these degradation phenomena can be reproduced and accelerated in laboratory more or less easily. However, combined degradation mechanisms are much more difficult to reproduce, and a strong effort of the scientific community should be devoted to this kind of study.

Acknowledgements The authors acknowledge the Agence Nationale de la Recherche, ANR, for financial support through the "Opt-Hip" project (grant # ANR-MAPR07-0014). The authors are grateful to N. Curt for technical help.

References

1. Bobbio A (1970) The first endosseous alloplastic implant in the history of man. Bull Hist Dent 20(1):1–6
2. Dreesman H (1892) Uber Knochenplombierung. Beitr Klin Chir 9:804–810
3. Rieger W (2001) Ceramics in orthopedics—30 years of evolution and experience. In: Rieker C, Oberholzer S, Wyss U (eds) World tribology forum in arthroplasty. Hans Huber, Bern
4. Rock M (1933) German Patent, DRP n°583589
5. Green DJ, Hanninck RHJ, Swain MV (1989) Transformation toughening of ceramics. CRC, Boca Raton, FL
6. Li P, Chen IW, Penner-Hahn JE (1994) Effect of dopants on zirconia stabilization—an X-ray absorption study: I, trivalent dopants. J Am Ceram Soc 77(1):118–128
7. Li P, Chen IW, Penner-Hahn JE (1994) Effect of dopants on zirconia stabilization—an X-ray absorption study: II, Tetravalent dopants. J Am Ceram Soc 77(5):1281–1288
8. Chevalier J, Gremillard L, Virkar AV, Clarke DR (2009) The tetragonal-monoclinic transformation in zirconia: lessons learned and future trends. J Am Ceram Soc 92:1901–1920
9. Grenet L (1889) Recherches sur la résistance mécanique des verres. Bull Soc Encour Ind Nat 4:838–848
10. Orowan E (1944) The fatigue of glass under stress. Nature 154:341–349
11. Charles RJ, Hillig WB (1962) The kinetics of glass failure by stress corrosion, in symposium sur la résistance mécanique du verre et les moyens de l'améliorer. Union Sciences Continentales du Verre, Charleroi, pp 502–511
12. Wiederhorn SM (1969) Fracture of ceramics. In: Mechanical and thermal properties of ceramics (NBS special publication), vol 303. NBS, Washington, DC, pp 217–241
13. Lawn B (1993) Fracture of brittle solids, 2nd edn. Cambridge University Press, Cambridge, 378 p
14. Griffith RW (1920) Phenomena of rupture and flow in solids. Philos Trans R Soc Lond A 221:163–198
15. Chevalier J (1996) Etude de la propagation des fissures dans une zircone 3Y-TZP pour applications biomédicales, thèse, INSA de Lyon, 161 p
16. Zhu T, Li J, Lin X, Yip S (2005) Stress-dependent molecular pathways of silica-water reaction. J Mech Phys Solids 53:1597–1623
17. Chevalier J (2001) Caractérisation mécanique et optimization des biocéramiques, Habilitation à Diriger des Recherches (HDR), INSA-Lyon and Université Lyon 1
18. Chevalier J, Olagnon C, Fantozzi G, Cales B (1997) Subcritical crack growth and thresholds in a 3Y-TZP ceramic under static and cyclic loading conditions. Ceram Int 23(3):263–266
19. McMeeking RM, Evans AG (1982) Mechanics of transformation-toughening in brittle materials. J Am Ceram Soc 65(5):242–246

20. Chevalier J, Olagnon C, Fantozzi G (1999) Subcritical crack propagation in 3Y-TZP ceramics: static and cyclic fatigue. J Am Ceram Soc 82(11):3129–3138
21. Knehans R, Steinbrech RW (1984) Effect of grain size on the crack resistance curves of Al_2O_3 bend specimens. In: Vincenzini P (ed) Science of ceramics, vol 12. Research Institute for Ceramics Technology, Faenza, pp 613–619
22. Chantikul P, Bennison S, Lawn BR (1990) Role of the grain size in the strength and R-curve properties of Alumina. J Am Ceram Soc 73(8):2419–2427
23. Vekinis G, Ashby MF, Beaumont PWR (1990) R-curve behaviour of Al_2O_3 ceramics. Acta Metall Mater 38(6):1151–1162
24. Swanson P, Fairbanks C, Lawn BR, Mai Y, Hockey B (1987) Crack – interface bridging as a fracture resistance mechanism in ceramics: experimental study on alumina. J Am Ceram Soc 70(4):279–289
25. Pezzotti G, Sbaizero O, Sergo V, Muraki N, Maruyama K, Nishida T (1998) In situ measurements of frictional bridging stresses in alumina using fluorescence spectroscopy. J Am Ceram Soc 81(1):187–192
26. Osaka A, Hirosaki A, Yoshimura M (1990) Subcritical crack growth in sintered silicon nitride exhibiting a rising R-curve. J Am Ceram Soc 73(7):2095–2096
27. Deuhlerc F, Knehans K, Steinbrech R (1985) Fortschrittsberichte der Deutschen Keramischen Gesellschaft 1:51
28. Fett T, Munz D (1992) Subcritical crack growth of macro-cracks in alumina with R-curve behaviour. J Am Ceram Soc 75(4):958–963
29. Ebrahimi ME, Chevalier J, Fantozzi G (2000) Slow crack growth behavior of alumina ceramics. J Mater Res 15(1):142–147
30. Deville S, Guenin G, Chevalier J (2004) Martensitic transformation in zirconia. Part I. Nanometer scale prediction and measurement of transformation induced relief. Acta Mater 52:5697–5707
31. Duong T, Limarga AM, Clarke DR (2009) Diffusion of water species in yttria-stabilized zirconia. J Am Ceram Soc 92(11):2731–2737
32. Deville S, Guenin G, Chevalier J (2004) Martensitic transformation in zirconia. Part II. Martensite growth. Acta Mater 52:5709–5721
33. El Attaoui H (2003) Influence du renforcement sur le comportement en fatigue statique et cyclique des céramiques monolithiques de type alumine et zircone, PhD Thesis, INSA-Lyon, France. http://docinsa.insa-lyon.fr/these/pont.php?id=el_attaoui
34. Shen Y, Clarke DR (2010) Resistance to low-temperature degradation of equimolar YO 1.5-TaO2.5 stabilized tetragonal ZrO2 ceramics in air. J Am Ceram Soc 93(7):2024–2027
35. Kobayashi K, Kuwajima H, Masaki T (1980) Phase change and mechanical properties of ZrO2-Y2O3 solid electrolyte after ageing. Solid State Ionics 3(4):489–493
36. Chevalier J, Cales B, Drouin JM (1999) Low-temperature aging of Y-TZP ceramics. J Am Ceram Soc 82:2150–2154
37. Chevalier J (2006) What future for zirconia as a biomaterial? Biomaterials 27:535–543
38. US Food and Drug Administration (1997) Steam re-sterilization causes deterioration of zirconia ceramic heads of total hip prostheses. http://www.fda.gov/MedicalDevices/Safety/AlertsandNotices/PublicHealthNotifications/ucm062472.htm
39. Hernigou P, Bahrami T (2003) Zirconia and alumina ceramics in comparison with stainless-steel heads. J Bone Joint Surg Ser B 85(4):504–509
40. Chevalier J, Gremillard L, Deville S (2007) Low temperature degradation of zirconia and its implication on biomedical implants. Annu Rev Mater Res 37:1–32
41. Maro FG, Mestra A, Lamghari M, Anglada M (2010) Weibull statistical study of the flexural strength of hydrothermally degraded 3Y-TZP. In: Proceedings of the twelfth inter-regional conference on ceramics, Mons, 7–9 Sept 2010
42. Agence Française de Sécurité Sanitaire et des produits de Santé (2001) Décision portant sur la suspension d'utilisation de certaines têtes de prothèse de hanche en céramique de zircone. http://agmed.sante.gouv.fr/htm/alertes/filalert/dm010811.htm

43. Maccauro G, Piconi C, Burger W, Pilloni L, De Santis E, Muratori F, Learmonth ID (2004) Fracture of a Y-TZP ceramic femoral head: analysis of a fault. J Bone Joint Surg Ser B 86:1192–1196
44. Masonis JL, Bourne RB, Ries MD, McGalden RW, Salehi A, Kelman DC (2004) Zirconia femoral head fractures: a clinical and retrieval analysis. J Arthroplasty 19:898–905
45. Chevalier J, Loh J, Gremillard L, Meille S, Adolfson E (2011) Low temperature degradation in zirconia with a porous surface. Acta Biomater 7(7):2986–2993. doi:10.1016/j.actbio.2011.03.006
46. Garvie RC, Nicholson PS (1972) Phase analysis in zirconia systems. J Am Ceram Soc 55:303–305
47. Toraya H, Yoshimura M, Somiya S (1984) Calibration curve for quantitative analysis of the monoclinic-tetragonal ZrO2 system by X-ray diffraction. J Am Ceram Soc 67(6):C119–C121
48. Clarke DR, Adar F (1982) Measurement of the crystallographically transformed zone produced by fracture in ceramics containing tetragonal zirconia. J Am Ceram Soc 65:284–288
49. Li J, Zheng Q, Hashida T (2001) Degradation of yttria-stabilized zirconia at 370K under a low applied stress. Mater Sci Eng A 297:26–30
50. De Aza AH, Chevalier J, Fantozzi G, Schehl M, Torrecillas R (2002) Crack growth resistance of alumina, zirconia and zirconia toughened alumina ceramics for joint prostheses. Biomaterials 23:937–945
51. Pecharroman C, Bartolome JF, Requena J, Moya JS, Deville S, Chevalier J, Fantozzi G, Torrecillas R (2003) Percolative mechanism of aging in zirconia-containing ceramics for medical applications. Adv Mater 15:507–511
52. Schehl M, Diaz LA, Torrecillas R (2002) Alumina nanocomposites from powder-alkoxide mixtures. Acta Mater 50:1125–1139
53. Deville S, Chevalier J, Fantozzi G, Bartolome JF, Requena J, Moya JS, Torrecillas R, Diaz LA (2003) Low-temperature ageing of zirconia-toughened alumina ceramics and its implication in biomedical implants. J Eur Ceram Soc 23:2975–2982
54. Pezzotti G, Saito T, Padeletti G, Cossari P, Yamamoto K (2010) Nano-scale topography of bearing surface in advanced alumina/zirconia hip joint before and after severe exposure in water vapor environment. J Orthop Res 28(6):762–766
55. Chevalier J, Grandjean S, Kuntz M, Pezzotti G (2009) On the kinetics and impact of tetragonal to monoclinic transformation in an alumina/zirconia composite for arthroplasty applications. Biomaterials 30(29):5279–5282
56. Nakanishi T, Sasaki M, Ikeda J, Miyaji F, Kondo M (2007) Mechanical and phase stability of zirconia toughened alumina. Key Eng Mater 330–332(II):1267–1270
57. Gutknecht D, Chevalier J, Garnier V, Fantozzi G (2007) Key role of processing to avoid low temperature ageing in alumina zirconia composites for orthopaedic application. J Eur Ceram Soc 27(2–3):1547–1552
58. Ban S, Sato H, Suehiro Y, Nakanishi H, Nawa M (2008) Biaxial flexure strength and low temperature degradation of Ce-TZP/Al 2O3 nanocomposite and Y-TZP as dental restoratives. J Biomed Mater Res B Appl Biomater 87(2):492–498
59. Benzaid R, Chevalier J, Saâdaoui M, Fantozzi G, Nawa M, Diaz LA, Torrecillas R (2008) Fracture toughness, strength and slow crack growth in a ceria stabilized zirconia-alumina nanocomposite for medical applications. Biomaterials 29(27):3636–3641
60. Wimmer MA, Artelt D, Schneider E, et al (2001) Friction and wear properties of metal/metal hip joints: application of a novel testing and analysis method. Mat. wiss. U. Werkstofftech 32:891–896
61. Marshall A, Ries MD, Paprosky W (2008) How prevalent are implant wear and osteolysis and how has the scope of osteolysis changed since 2000? J Am Acad Orthop Surg 16(suppl 1): S1–S6
62. Hall RM, Unsworth A (1997) Friction in hip prostheses. Biomaterials 18(15):1017–1026
63. Schmalzried TP, Callaghan JJ (1999) Wear in total hip and knee replacements. J Bone Joints Surg Am 81(1):115–136

64. Manley MT, Sutton K (2008) Bearings of the future for total hip arthroplasty. J Arthroplasty 23(7 Suppl):47–50
65. Blunt L, Bills P, Jiang X, Hardaker C, Chakrabarty G (2009) The role of tribology and metrology in the latest development of bio-materials. Wear 266:424–431
66. Santavirta S, Konttinen YT, Lappalainen R, Anttila A, Goodman SB, Lind M (1998) Materials in total joint replacement. Curr Orthop 12:51–57
67. Piconi C, Maccauro G (1999) Zirconia as a ceramic biomaterial. Biomaterials 20:1–25
68. Dumbleton JH, Manley MT (2005) Metal-on-metal total hip replacement: what does the literature say? J Arthroplasty 20:174–188
69. Clarke IC, Good V, Williams P et al (2000) Ultra-low wear rates for rigid-on-rigid bearings in total hip replacements. Proc Inst Mech Eng 214:331
70. Goldsmith AAJ, Dowson D, Isaac GH et al (2000) A comparative joint simulator study of the wear of metal-on-metal and alternative material combinations in hip replacements. Proc Inst Mech Eng 214:39
71. Masse A, Bosetti M, Buratti C et al (2003) Ion release and chromosomal damage from total hip prostheses with metal-on-metal articulation. J Biomed Mater Res 67:750
72. Tharani R, Dorey FJ, Schmalzreid TP (2001) The risk of cancer following total hip or knee arthroplasty. J Bone Joint Surg Am 83A:774
73. Malchau H, Herberts P, Eisler T, Garellick G, Soderman P (2002) The Swedish Total Hip Replacement Register. Friction of total hip replacements with different bearings and loading conditions. J Bone Joint Surg Am 84:2–20
74. Lewis G (1997) Polyethylene wear in total hip and knee arthroplasties. J Biomed Mater Res 38(1):55–75
75. Ingham E, Fisher J (2005) The role of macrophages in osteolysis of total joint replacement. Biomaterials 26(11):1271–1286
76. Schmalzried TP, Jasty M, Harris WH (1992) Periprosthetic bone loss in total hip arthroplasty: polyethylene wear debris and the concept of the effective joint space. J Bone Joint Surg Am 74A:849–863
77. Revell PA (1997) Biological reaction to debris in relation to joint prostheses. Proc Inst Mech Eng J Eng Med 211:187–197
78. Boutin P, Christel P, Dorlot JM et al (1988) The use of dense alumina-alumina ceramic combination in total hip replacement. J Biomed Mater Res 22:1203
79. Hamadouche M, Boutin JD, Bolander ME, Sedel L (2002) Alumina-on-alumina total hip arthroplasty. J Bone Joint Surg 84:69–77
80. Taylor SK, Serekian P, Manley M (1998) Wear performance of a contemporary alumina–alumina bearing couple under hip joint simulation. In: Proceedings of the 44th orthopaedic research society, vol 51, New Orleans, LA
81. Oonishi H, Nishida M, Kawanabe K, Yamamoto K, Downs B, Sorensen K, Good V, Braham A, Clarke IC (1999) In vitro wear of Al_2O_3/Al_2O_3 implant combination with over 10 million cycles duration. In: Proceedings of the 45th orthopaedic research society, Anaheim, CA, p 50
82. Nevelos JE, Ingham E, Doyle C, Fisher J, Nevelos AB (1999) Analysis of retrieved alumina ceramic components from Mittelmeier total hip prostheses. Biomaterials 20:1833–1840
83. Teoh SH (2000) Fatigue of biomaterials: a review. Int J Fatigue 22:825–837
84. Elsner JJ, Mezape Y, Hakshur K, Shemesh M, Linder-Ganz E, Shterling A, Eliaz N (2010) Wear rate evaluation of a novel polycarbonate-urethane cushion form bearing for artificial hip joints. Acta Biomater 6:4698–4707
85. Ries MD (2003) Complications in primary total hip arthroplasty: avoidance and management: wear. Instr Course Lect 52:257
86. Skinner HB (1999) Ceramic bearing surfaces. Clin Orthop 369:83
87. Johnston RC, Smidt GL (1969) Measurement of hip-joint motion during walking—evaluation of an electrogoniometric method. J Bone Joint Surg 51-A:1083–1094
88. Ortega-Saenz JA, Hernandez-Rodriguez MAL, Perez-Unzueta A, Mercado-Solis R (2007) Development of a hip wear simulation rig including micro-separation. Wear 263:1527–1532

89. Wimmer MA, Sprecher C, Hauert R, Tager G, Fischer A (2003) Tribochemical reaction on metal-on-metal hip joint bearings. A comparison between in-vitro and in-vivo results. Wear 255:1007–1014
90. Jay GD, Harris DA, Cha C-J (2001) Boundary lubrication by lubricin is mediated by O-linked b(1–3)Gal-GalNAc oligosaccharides. Glycoconj J 18:807–815
91. Schwarz IM, Hills BA (1998) Surface-active phospholipid as the lubricating component of lubricin. Br J Rheumatol 37:21–26
92. Mishina H, Kojima M (2008) Changes in human serum albumin on arthroplasty frictional surfaces. Wear 265:655–663
93. ASTM (1996) Standard practice for reciprocating pin-on-flat evaluation of friction and wear properties of polymeric materials for use in total joint prostheses. ASTM F732, sect 13, p 171
94. Liu F, Leslie I, Williams S, Fisher J, Jin Z (2008) Development of computational wear simulation of metal-on-metal hip resurfacing replacements. J Biomech 41:686–694
95. Boutin P (1972) Arthroplastie totale de hanche par prothèse en alumine frittée. Revue de Chirurgie Orthopédique et Réparatrice de l'Appareil moteur 58:229–246
96. Boutin P, Blanquaert D (1981) Le frottement alumine-alumine en chirurgie de la hanche. 1205 arthroplasties totales: Avril 1970 - juin 1980. Revue de Chirurgie Orthopédique et Réparatrice de l'Appareil moteur 67:279–287
97. Boutin P, Christel P, Dorlot JM, Meunier A, de Roquancourt A, Blanquaert D, Herman S, Sedel L, Witvoet J (1988) The use of dense alumina-alumina ceramic combination in total hip replacement. J Biomed Mater Res 22:1203–1232
98. Bizot P, Nizard R, Sedel L (2010) Le couple alumine-alumine dans les prothèses totales de hanche. de la théorie à la pratique, Online edition. http://www.maitrise-orthop.com/viewPage.do?id=619
99. Tipper JL, Firkins PJ, Besong AA, Barbour PSM, Nevelos J, Stone MH, Ingham E, Fisher J (2001) Characterisation of wear debris from UHMWPE on zirconia ceramic, metal-on-metal and alumina ceramic-on-ceramic hip prostheses generated in a physiological anatomical hip joint simulator. Wear 250:120–128
100. Schmalzried TP (2004) How I choose a bearing surface for my patients. J Arthroplasty 19:50–53
101. Brown TD, Callaghan JJ (2008) (ii) Impingement in total hip replacement: mechanisms and consequences. Curr Orthop 22:376–391
102. Nevelos J, Ingham E, Doyle C, Streicher R, Nevelos A, Walter W, Fisher J (2000) Microseparation of the centers of alumina-alumina artificial hip joints during simulator testing produces clinically relevant wear rates and patterns. J Arthroplasty 15:793–795
103. Callaghan JJ, Rosenberg AG, Rubash HE (2007) The adult hip. Lippincott Williams & Wilkins, Philadelphia, PA, p 53
104. Kapandji AL (2009) Anatomie fonctionnelle-membre inférieur. Ed. Maloine, Paris
105. Thompson JC, Netter FH (2008) Netter-précis d'anatomie clinique d'orthopédie-French edition coordinated by Duparc J. Ed. Masson, Issy-les-Moulineaux
106. Magnissalis EA, Xenakis TA, Zacharis C (2001) Wear of retrieved ceramic THA components–four matched pairs retrieved after 5–13 years in service. J Biomed Mater Res 58:593–598
107. Shishido T, Yamamoto K, Tanaka S, Masaoka T, Clarke IC, Williams P (2006) A study for a retrieved implant of ceramic on ceramic total hip arthroplasty. J Arthroplasty 21:294–298
108. Essner A, Sutton K, Wang A (2005) Hip simulator wear comparison of metal-on-metal, ceramic-on-ceramic and crosslinked UHMWPE bearings. Wear 259:992–995
109. Gruen T, Mc Neice GM, Amstutz HC (1979) Modes of failures of cemented stem type femoral components: a radiographic analysis of loosening. Clin Orthop Relat Res 141:17–27
110. Paul JP (1979) Advances in hip and knee joint technology. In: Schaldach M, Hohmann D (eds) Loading on normal hip and knee joints and joint replacements. Springer, Berlin, pp 53–70

111. Jobard B (2003) Contribution à l'étude des impacts transmis à l'articulation de la hanche humaine, PhD thesis. Lyon University I
112. Manaka M, Clarke IC, Yamamoto K, Shishido T, Gustafson A, Imakiire A (2004) Stripe wear rates in alumina THR—comparison of microseparation simulator study with retrieved implants. J Biomed Mater Res B Appl Biomater 69B:149–157
113. Dennis DA, Komistek RD, Northcut EJ, Ochoa JA, Ritchie A (2001) In vivo determination of hip joint separation and the forces generated due to impact loading conditions. J Biomech 34:623–629
114. Stewart TD, Tipper JL, Insley G, Streicher RM, Ingham E, Fisher J (2003) Severe wear and fracture of zirconia heads against alumina inserts in hip simulator studies with microseparation. J Arthroplasty 18:726–734
115. Walter WL, Insley GM, Walter WK, Tuke MA (2004) Edge loading in third generation alumina ceramic-on-ceramic bearings: stripe wear. J Arthroplasty 19:402–413
116. Roualdes O, Duclos M-E, Gutknecht D, Frappart L, Chevalier J, Hartmann DJ (2010) In vitro and in vivo evaluation of an alumina–zirconia composite for arthroplasty applications. Biomaterials 31:2043–2054
117. Bongio M, Van Den Beucken JJP, Leeuwenburgh SCG, Janson JA (2010) From 'biocompatible' to 'instructive'. J Mater Chem 20:8747–8759
118. Chevalier J, Gremillard L (2009) Ceramics for medical applications: a picture for the next 20 years. J Am Ceram Soc 29:1245–1255
119. Hench LL (1998) Bioceramics. J Am Ceram Soc 81:1705–1728
120. Plannell JA (ed) (2009) Bone repair biomaterials. CRC, Boca Raton, FL
121. Kokubo T (ed) (2008) Bioceramics and their clinical applications. CRC, Boca Raton, FL
122. Jones JR, Hench LL (2001) Biomedical materials for the next millennium: perspective on the future. Mater Sci Technol 17:891–900
123. Langer R, Vacanti JP (1993) Tissue engineering. Science 260:920–926
124. Yaszemski MJ, Payne RG, Hayes WC, Langer R, Mikos AC (1996) Evolution of bone transplantation: molecular, cellular and tissue strategies to engineer human bone. Biomaterials 17:175–185
125. Hench LL (2006) The story of Bioglass®. J Mater Sci 17:967–978
126. Ducheyne P, Qiu Q (1999) Bioactive ceramics: the effect of surface reactivity on bone formation and bone cell function. Biomaterials 20:2287–2303
127. Habibovic P, de Groot K (2007) Osteoinductive biomaterials – properties and relevance in bone repair. J Tissue Eng Regen Med 1:25–32
128. Kokubo T, Takadama H (2006) How useful is SBF in predicting in vivo bone bioactivity? Biomaterials 27:2907–2915
129. Oonishi H, Hench LL, Wilson J, Sugihara F, Tsuji E, Matsuura M (2000) Quantitative comparison of bone growth behaviour in granules of Bioglass®, A-W glass ceramic and hydroxyapatite. J Biomed Mater Res 51:37–46
130. Xin R, Leng Y, Chen J, Zhang Q (2005) A comparative study of calcium phosphate formation on bioceramics in vitro and in vivo. Biomaterials 26:6477–6486
131. Bohner M, Lemaître J (2009) Can bioactivity be tested in vitro with SBF solution? Biomaterials 30:2175–2179
132. Orly I, Gregoire M, Menanteau J, Heughebaert M, Kerebel B (1989) Chemical changes in hydroxyapatite biomaterial under *in vivo* and *in vitro* biological conditions. Calcif Tissue Int 45:20–26
133. Eliaz N, Eliyahu M (2007) Electrochemical processes of nucleation and growth of hydroxyapatite on titanium supported by real-time electrochemical atomic force microscopy. J Biomed Mater Res A 80:621–634
134. Knabe C, Ducheyne P (2008) Cellular response to bioactive ceramics, chap 6, Bioceramics and their clinical applications. CRC, Kokubo
135. Bohner M (2008) Bioresorbable ceramics, chap 5. In: Fraser Buchanan (ed) Degradation rate of bioresorbable materials. CRC, Boca Raton

136. Driessens FCM, Verbeeck RM (eds) (1988) Relation between physico-chemical solubility and biodegradability of calcium phosphates. Elsevier, Amsterdam
137. Hench LL (2005) Challenges for bioceramics in the 21st century. Am Ceram Soc Bull 84:18–22
138. Thomas MV, Puleo DA (2008) Review calcium sulfate: properties and clinical applications. J Biomed Mater Res B 88(2):597–610
139. Chan H, Mijares D, Ricci JL (2004) In vitro dissolution of calcium sulfate: evidence of 1662 bioactivity. In: Transactions – 7th World Biomaterials Congress (Sidney, Australia, may 17–21, 2004): 627; ISBN: 1877040193;978-187704019-1
140. Stubbs D, Deakin M, Chapman-Sheath P, Bruce W, Debes J, Gillies RM, Walsh WR (2004) In vivo evaluation of resorbable bone graft substitutes in a rabbit tibial defect model. Biomaterials 25:5037–5044
141. Glazer PA, Specner UM, Alkalay RN, Schwardt JS (2001) In vivo evaluation of calcium sulfate as a bone graft substitute for lumbar spinal fusion. Spine J 1:395–401
142. Dorozhkin SV (2010) Bioceramics of calcium orthophosphate. Biomaterials 31:1465–1485
143. Bohner M (2000) Calcium orthophosphate in medicine: from ceramics to calcium phosphate cements. Injury 31(4):D37–D47
144. LeGeros RZ, LeGeros JP (2008) Hydroxyapatite, chap 16. In: Kokubo T (ed) Bioceramics and their clinical applications. CRC, Boca Raton
145. Kim HM, Himeno T, Kokubo T, Nakamura T (2005) Process and kinetics of bonelike apatite formation on sintered hydroxyapatite in a simulated body fluid. Biomaterials 12:155–163
146. Beekmans HCS, Meijer GJ, Barkhuysen R, Blijdorp PA, Merkx MA, Jansen J (2008) The hydroxylapatite-bone-interface: 10 years after implant installation. Int J Oral Maxillofac Surg 37:768–772
147. Eliaz N (2008) Electrocrystallization of calcium phosphates. Isr J Chem 48:159–168
148. Rey C, Combes C, Drouet C, Somrani S (2008) Tricalcium phosphate ceramics, chap 15. In: Kokubo T (ed) Bioceramics and their clinical applications. CRC, Boca Raton
149. Klein CPAT, Driessen AA, de Groot K, van den Hoff A (1983) Biodegradation behaviour of various calcium phosphate materials in bone tissue. J Biomed Mater Res 17:769–784
150. Knabe C, Koch C, Rack A, Stiller M (2008) Effect of β-tricalcium phosphate particle with varying porosity on osteogeneis after sinus floor augmentation in humans. Biomaterials 29:2249–2258
151. Artzi Z, Weinreb M, Givol N, Rohrer MD, Nemcovsky CE, Prasad HS, Tal H (2004) Biomaterial resorption rate and healing site morphology of IBB and β-TCP in the canine: a 24-months longitudinal histologic study and morphometric analysis. Int J Oral Maxillofac Implants 19:357–368
152. Daculsi G, Laboux O, Malard O, Weiss P (2003) Current state of the art of biphasic calcium phosphate ceramics. J Mater Sci Mater Med 14:195–200
153. Hench LL (2009) Genetic design of bioactive glass. J Eur Ceram Soc 29:1257–1265
154. Tadjoedin ES, de Lange GL, Lyaruu DM, Kuiper L, Burger EH (2002) High concentration of bioactive glass material vs. autogenous bone for sinus floor elevation. Clin Oral Implants Res 13:428–436
155. Boccaccini AR, Chen Q, Lefebvre L, Gremillard L, Chevalier J (2007) Sintering, crystallisation and biodegradation behaviour of Bioglass (R)-derived glass-ceramics. Faraday Discuss 136:27–44
156. Magallanes-Perdomo M, Meille S, Chenal J.-M, Pacard E, Chevalier J, Bioactivity modulation of Bioglass® powder by thermal treatment, J Euro Ceram Soc, http://dx.doi.org/10.1016/j.jeurceramsoc.2012.03.018
157. Yamamuro T (2008) Clinical application of bioactive glass-ceramics, chap 26. In: Kokubo T (ed) Bioceramics and their clinical applications. CRC, Boca Raton
158. Hoppe A, Güldal NS, Boccaccini AR (2011) A review of the biological response to ionic dissolution products from bioactive glasses and glass-ceramics. Biomaterials 32:2757–2774

Biography

Laurent Gremillard is a Research Scientist at the National Centre for Scientific Research (CNRS). He received his engineering degree and his Ph.D. from the National Institute of Applied Science (INSA) in Lyon (France) in 1998 and 2002, respectively. He then was a postdoctoral fellow at the Lawrence Berkeley National Laboratory for 2 years, where his research was focused on the behavior of lead-free solders on ceramic surfaces. He joined CNRS in October 2004, and is currently working in the ceramic group of the Materials, Engineering and Science laboratory (joint laboratory between INSA-Lyon and CNRS). He is interested in every aspect of bioceramics, from processing to mechanical, microstructural and biological characterization. He currently works on zirconia ceramics and composites and on organic–inorganic porous composites for bone substitution. He has authored or co-authored around 35 scientific papers in international peer-reviewed materials science journals.

Dr. Sylvain Meille graduated from the National Institute of Applied Sciences (INSA) in Lyon, France in 1997, with an engineering diploma and a MS degree in Materials Science. His PhD, completed in 2001, focused on the characterization of the mechanical behavior of a porous material, gypsum. In 2000 he spent 6 months as a guest scientist in the Building and Fire Research Laboratory at NIST, Gaithersburg, USA. After defending his thesis, he held for more than 6 years a position of research engineer in Lafarge Research Center, France, on the development of new construction materials. He currently holds an Assistant Professor position at MATEIS laboratory, INSA Lyon. He is working on the relation between microstructure and mechanical properties in porous ceramics and organic--inorganic composites for biomedical applications.

Born in 1970, Jérôme Chevalier is currently a Full Professor at the National Institute of Applied Sciences (University of Lyon), in France. After receiving his PhD in 1996 (Mechanical properties of biomedical grade zirconia), he first became Ceramic Engineer in Saint Gobain Group, responsible for the R&D activity on biomedical applications of ceramics. In 1997 he joined the National Institute of Applied Sciences, in Villeurbanne, as Assistant Professor, and then became a Full Professor in 2004. He is in charge of the Ceramic Group of the Materials Department (14 permanent researchers, 20 PhD's). His research activity is now mainly focused on bioceramics, from their processing to their mechanical and biological properties. His research interests include new generations of zirconia-based composites in orthopedics and dentistry, nanostructured ceramics and composites, hybrid organic–inorganic composites, and bioactive glasses and ceramics for tissue engineering. He has published more than 80 original peer reviewed papers, and 6 review papers in journals and encyclopedia. He is junior member of the Institut Universitaire de France since 2009, and since 2011 Editor of the Journal of the European Ceramic Society.

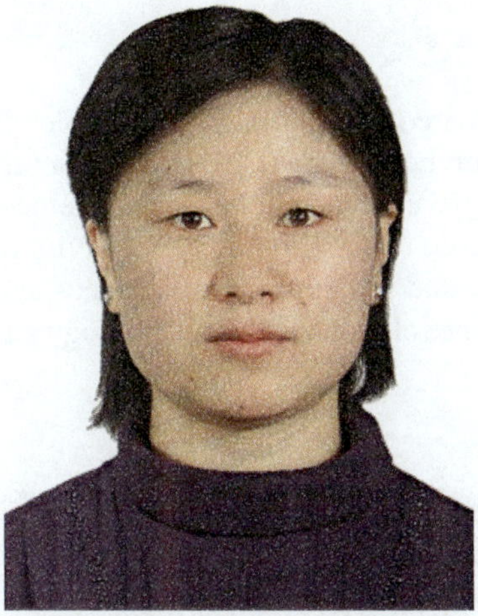

Dr. Jing Zhao obtained her Doctor degree in Mechanical Design and Theory with honors from Southwest Jiaotong University in China in 2009. She was hired as a Post-doc in Ecole Centrale de Lyon in a joint programme sponsored by ANR in 2009. Now she is a lecturer in School of Mechanical Engineering at Southwest Jiaotong University. Her research is focused on biotribology and porous biomaterials.

Dr. Vincent Fridrici graduated in Tribology from Ecole Centrale de Lyon (Master Degree in 1999, Ph.D. with honors in 2002). He is currently Associate Professor in "Laboratoire de Tribologie et Dynamique des Systèmes" at Ecole Centrale de Lyon. His research is focused on the tribological behavior of materials and coatings, with two main aspects related to durability of surfaces and coatings and wear and friction of biomaterials. The relations between tribology and materials and mechanics are put forward.

Prof. Philippe Kapsa is Research Director of CNRS working as a group leader in "Laboratoire de Tribologie et Dynamique des Systèmes." He graduated from Ecole

Centrale de Lyon in 1976 and received a Ph.D. from Lyon University in 1982. He is active in the field of Tribology–Materials–Mechanics, and his main research activities concern friction and wear behavior of bulk materials and thin coatings in dry or lubricated conditions, mechanical properties measurements of thin coatings, experimentation and modeling of surface damage, high-temperature solid lubricants, and biotribology. The fundamental aspects of tribological problems are studied considering also the applications in relation with industry. His main interests concern the study and prediction of tribological behavior from the material and mechanical point of view.

Dr. Jean Geringer obtained a Master Degree with honors from Ecole Nationale Supérieure de Chimie de Toulouse in 1997. After working in high school as professor, he earned the "agrégation" of physics and chemistry, the highest-rated competitive academic examination in the French system, in 2001. He obtained a second Master Degree in materials science, and received his Ph.D. in 2005. He spent more than one year at Penn State University-USA during his PhD. After working at INSA Lyon as a teaching assistant; he currently holds the position of assistant professor at the Ecole Nationale Supérieure des Mines de Saint-Etienne-France. He focuses his attention on research with ceramic materials as well as in the field of tribocorrosion. He has published more than 30 articles and book chapters.

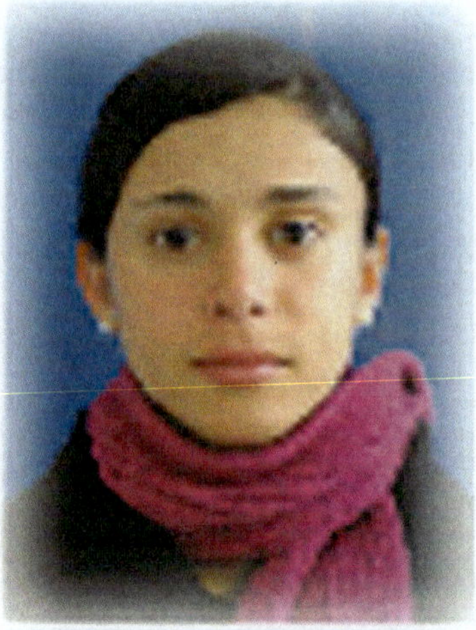

Dr. Julianna Uribe completed her Ph.D in January 2012 in shock degradations of ceramic materials. She obtained her Master degree in Biomedical Engineering in 2007, in Colombia. She received a Master Degree in materials science in France in 2008. She wants to dedicate her career to bio-engineering topics, i.e. using her researcher skills for improving diagnosis and curing in implants fields. In this optics, she is now pursuing a master degree on imaging diagnosis.

Chapter 10
Fundamentals of Tribology and the Use of Ferrography and Bio-Ferrography for Monitoring the Degradation of Natural and Artificial Joints

Noam Eliaz and Keren Hakshur

Abstract Here, we summarize the principles of tribology and demonstrate the importance of wear particle analysis in condition monitoring. The principles of Ferrography, a technique that was originally developed for condition monitoring of engineering systems, are explained. A new modification of this technique, known as Bio-Ferrography, is reviewed and its application in magnetic isolation of target cells or tissues is demonstrated. Routes for magnetic labeling of biological matter and synthetic polymers are also presented. The hip and knee natural joints are discussed, along with osteoarthritis—their most common noninflammatory disorder. The use of Ferrography and Bio-Ferrography for isolating bone and cartilage particles from the synovial fluids in osteoarthritic joints is reviewed. Artificial joints and the use of Ferrography and Bio-Ferrography for monitoring their wear, either during the development stage or during service, are also reviewed. The concept of soft bearing materials is presented.

1 Fundamentals of Tribology

Tribology is the science and technology concerned with the friction, lubrication, and wear of contacting surfaces that move relative to each other [1–3]. The three elements of tribology are shortly summarized in Sects. 1.1–1.3.

N. Eliaz (✉) • K. Hakshur
Biomaterials and Corrosion Laboratory, School of Mechanical Engineering and the Materials and Nanotechnologies Program, Tel-Aviv University, Ramat Aviv, Tel-Aviv 69978, Israel
e-mail: neliaz@eng.tau.ac.il

N. Eliaz (ed.), *Degradation of Implant Materials*, DOI 10.1007/978-1-4614-3942-4_10,

1.1 Friction

Friction is the resisting force tangential to the interface between two bodies when, under the action of an external force, one body moves or tends to move relative to the surface of the other. One may distinguish between different types of friction, e.g., static and dynamic (kinetic). The *coefficient of friction*, μ, is a dimensionless ratio of the friction force, F, between two bodies to the normal force, N, pressing these bodies together. When two surfaces are in contact with each other, the load is carried by many high points, or *asperities*, on the surfaces. During sliding, the total tangential force required to shear these asperity junctions is usually high, causing unacceptable friction, wear, and surface damage. To reduce the frictional force, thus allowing easier sliding, a lubricant is deliberately introduced to separate the asperities either totally or partially [1–3].

1.2 Lubrication

A *lubricant* is added at the sliding interface in order to reduce friction and wear, and to carry away heat and debris generated during the sliding process. The two basic geometries for lubricated surfaces are conformal and counterformal (Fig. 10.1). In *conformal bearings*, surfaces are usually separated by a thick oil or gas film generated hydrodynamically by the surface velocities, or hydrostatically by an externally pressurized lubricant. These surfaces usually operate in the regime of thick-film hydrodynamic or hydrostatic lubrication. Lubrication performance can be predicted readily from analyses based on the *Reynolds equation*. In *counterformal bearings*, on the other hand, lubrication takes place in an extremely small concentrated area known as the Hertzian conjunction, the film thickness is very thin (of the same order of the surface roughness), and lubricant pressure is very high. Lubricant performance is affected strongly by the elastic deformation of the bearing surface. The lubricant film thickness and pressure and their distribution in the conjunction can be determined by elastohydrodynamic theories. Roughness effects are particularly important in counterformal contacts because the majority of these contacts operate in mixed lubrication [1–3].

Let us define briefly several types of lubrication. Lubrication that involves the use of a solid substance in the form of powder or thin film to reduce friction and wear is called *solid lubrication* (or *dry-film lubrication*). Examples of solid lubricants include graphite and molybdenum disulfide (MoS_2). In contrast, the use of liquid or gas lubricants is known as *fluid-film lubrication* [1–3].

There are three major lubrication regimes. *Full-film lubrication* (or *thick-film lubrication*) is the ideal tribological condition. In this regime, the mating surfaces are completely separated by a relatively thick film of lubricant, thus the effect of

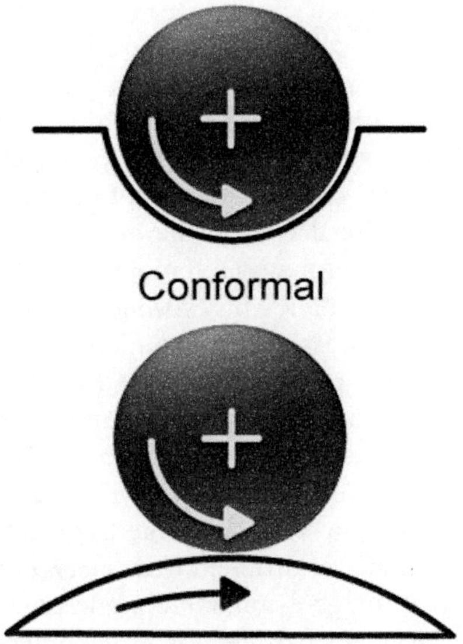

Fig. 10.1 Geometry of conformal and counterformal contacts. Reproduced after [1]

asperities is not noticeable. In *thin-film lubrication* (or *mixed lubrication*), on the other hand, there is incomplete separation of the mating surfaces by the lubricant film. This regime occurs when the load is sufficiently high, the temperature is sufficiently high, or the speed is sufficiently low to significantly reduce lubricant viscosity. Consequently, the tallest asperities of the mating surfaces will protrude through the film and occasionally come in contact. Under thin-film conditions, the coefficient of friction is often 10–100 times greater than under thick-film conditions, and wear is no longer negligible. In the most severe lubrication regime, the entire load is carried by surface asperities rather than by the lubricant, which is only molecularly thin. Thus, the properties of the bulk fluid lubricant become insignificant. This condition is known as *boundary lubrication* [1–3].

There are three major lubrication modes for full-film lubrication. In *hydrostatic lubrication*, the lubricant is supplied under sufficient external pressure; thus, it is maintained where it would otherwise be squeezed out. In *hydrodynamic lubrication*, the shape and relative motion of the sliding surfaces causes the formation of a fluid film that has sufficient pressure to separate the surfaces. Such bearings might wear when started or stopped because of breakdown of the lubricant film. In *elastohydrodynamic lubrication* (EHL), the load is high enough for the surfaces to elastically deform during the hydrodynamic action. As the load-bearing area increases, the viscous resistance of the lubricant becomes capable of supporting the load at the prevailing pressure, temperature, and rate of shear.

1.3 Wear

Wear is a process in which a solid surface is damaged due to relative motion between that surface and a contacting body or medium; this process generally involves progressive loss of material. *Wear rate* is the rate of material removal or dimensional change due to wear per unit of exposure parameter—for example, quantity of material removed (mass, volume, thickness) in unit distance of sliding or unit time. *Wear resistance* is the resistance of a body to removal of material by wear processes. It is commonly expressed as the reciprocal of wear rate [1–3].

Wear can be categorized according to its level of severeness. *Normal wear* means loss of material within the design limits expected for the specific intended application. *Mild wear* is a form of wear characterized by the removal of material in very small fragments. *Severe wear* is a form of wear characterized by removal of material in relatively large fragments [1–3].

There are different *wear mechanisms*, which are briefly defined below. *Abrasive wear* (or, abrasion or *cutting wear*) refers to the removal of material from a surface when hard particles or protuberances are forced against and moving along this surface. Such hard particles may be loose or may be part of another surface in contact with the surface being abraded. When the abrasive slides along the surface, the process is called *two-body wear*. In contrast, when the abrasive is caught between one surface and another, the process is called *three-body wear*. Two-body wear typically involves a significantly higher mass loss compared to three-body wear for a given load and path length of wear. *Scoring* is a severe form of wear characterized by the formation of extensive grooves and scratches in the direction of sliding [1–3].

Erosion (or, *erosive wear*) denotes destruction of materials by the abrasive action of moving fluids, usually accelerated by the presence of solid particles carried with the fluid. Erosion in which the relative motion of particles is nearly parallel to the solid surface is called *abrasive erosion*. Erosion in which the relative motion of the solid particles is nearly normal to the solid surface is called impingement erosion, or *impact erosion*. *Erosion–corrosion* refers to a conjoint action involving corrosion and erosion in the presence of a corrosive medium [1–3].

In a broader sense, *corrosive wear* denotes a wear process in which chemical or electrochemical reaction with the environment predominates. The combined effects of wear and corrosion can result in total material losses that are much greater than the additive effects of each process taken alone, which indicates synergism between the two processes. Sometimes, the term *electrocorrosive wear* is used to denote wear of a solid surface which is accelerated by the presence of a corrosion-inducing electrical potential across the contact interface. *Oxidative wear* is a type of wear resulting from the sliding action between two metallic components that generates oxide films on the metal surfaces. These oxide films prevent the formation of a metallic bond between the sliding surfaces, resulting in fine wear debris and mild wear [1–3].

Adhesive wear denotes removal or displacement of material from a surface by the solid-phase welding together and subsequent shearing of minute areas of two surfaces that slide across each other under pressure. *Scuffing* is a form of adhesive wear that produces superficial scratches or a high polish on the rubbing surfaces. It is observed most often on inadequately lubricated parts. *Seizure* denotes the stopping of relative motion as result of interfacial friction. *Galling* is a severe form of scuffing associated with gross damage to the surfaces or failure [1–3].

Thermal wear denotes removal of material due to softening, melting, or evaporation during sliding or rolling. *Cavitation* is defined as the repeated nucleation, growth, and violent collapse of cavities or bubbles in a liquid. In general, it originates from a decrease in the static pressure in the liquid [1–3].

Fatigue wear refers to removal of particles detached by fatigue arising from cyclic stress variations [1–3]. *Surface fatigue wear*, also known as *contact fatigue*, is common in bearings, gears, and rolling parts under loads. This failure mechanism is associated with Hertzian contact (shear) stresses that are maximal at the subsurface. Therefore, cracks are initiated at the subsurface and propagate toward the outer surface. The final stage is removal of a particle from the material [4]. The term *rolling-contact wear* is often used to indicate that the surface fatigue wear results from rolling contact.

Fretting is a small-amplitude oscillatory motion, usually tangential, between contacting surfaces. The movement is usually the result of external vibration, but in many cases it is the consequence of one of the members of the contact being subjected to a cyclic stress (that is, fatigue), which gives rise to another and usually more damaging aspect of fretting, namely the early initiation of fatigue cracks. This is termed *fretting fatigue* or *contact fatigue*. If particles of wear debris are generated, then the term *fretting wear* is sometimes used [1–3]. Fretting increases the tensile stress as well as the shear stress at the contact interface and generates flaws, which lead to premature crack nucleation. Furthermore, fretting fatigue results in a drastic reduction (by a factor of ≥ 2) in the fatigue endurance limit and orders of magnitude decrease in fatigue lifetime from that seen under pure axial cyclic loading alone. The generation of multiple cracks is a feature of fretting fatigue failures [5]. *Fretting corrosion* is a form of fretting in which chemical reaction predominates. The chemical reaction products may be, or may not be, abrasive [1–3].

1.3.1 Wear Particles

The act of rupture due to wear is localized in a small volume of material, which is removed in the form of *wear particles* (or, *wear debris*). Different types of wear result in generation of wear particles with different shapes, sizes, surface morphologies, and colors [1–3, 6]. A wear particle atlas was constructed [6], providing information for the identification of various wear particle types, the description of wear modes that generate these particles, and as a guide to the prediction of machine operating condition based on the identified modes. Since then, several new atlases have been published, some of which in electronic format [1–3].

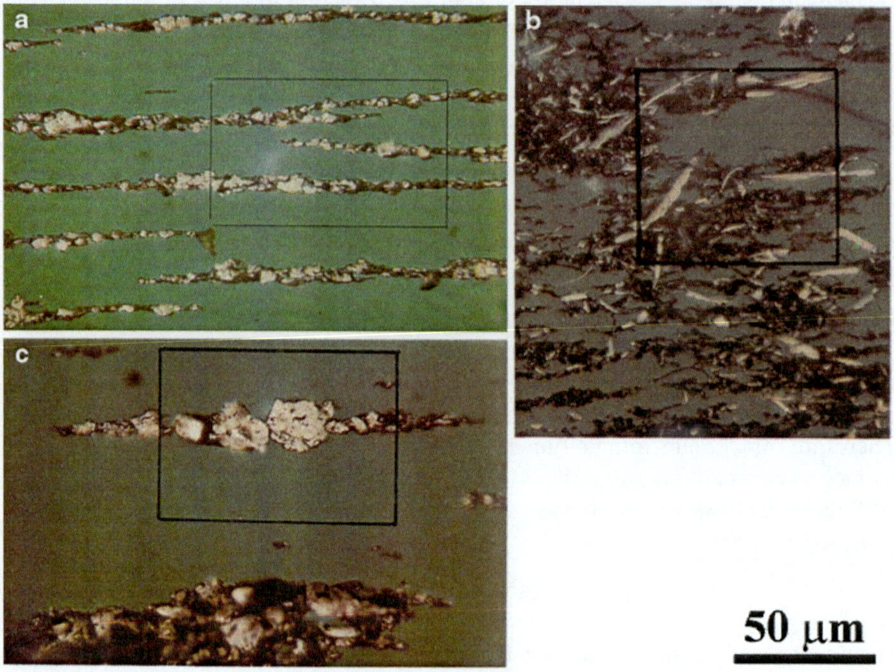

Fig. 10.2 Bichromatic microscope images of (**a**) rubbing, (**b**) cutting, and (**c**) fatigue wear particles that were isolated from a lubricant on microscope slides (ferrograms) by means of Analytical Ferrography. Reproduced after [6]

Few types of wear particles are described herein. *Rubbing wear particles* (Fig. 10.2a) are typical of abrasive wear. They consist of flat platelets, typically 5 μm or smaller, although they may range up to 15 μm before their wear effect is considered to be severe. Their thickness is 1 μm or less, and there is little or no visible texturing of the surface. *Sliding wear particles* are identified by parallel striations on their surfaces. They are generally larger than 15 μm, with a length-to-thickness ratio between 5 and 30. Severe sliding wear particles sometimes show evidence of temper colors, which may change the appearance of the particle after heat treatment. *Cutting wear particles* (Fig. 10.2b) may resemble wire, drill turnings, whittling chips, or gouged-out curls. *Fatigue wear debris* (Fig. 10.2c) in the form of micro-spall, laminar, chunky, and spherical particles are removed from the metal surface as a pit or crack opens up. Small spalls are distinguished from normal rubbing wear by slightly greater thickness and surface texture. Fatigue chunks are generally larger than 5 μm, with a length-to-thickness ratio of less than 5. They usually exhibit some surface texture. Spalls are similar in appearance to chunks, but are thinner. *Spheres* are caused by wear, fatigue, or contamination. Their formation due to wear is generally associated with rolling elements. Spheres formed by wear mechanisms are generally less than 5 μm in diameter, with very smooth surfaces. Such spheres are often a precursor of fatigue spalling. If the diameter exceeds 5 μm, or if the surfaces appear rough or oxidized, the spheres were probably caused by cavitation or contamination. *Inorganic crystalline minerals* that are commonly associated with dirt and construction materials will depolarize light that has passed through a polarizer. This phenomenon is called *birefringence*. Birefringent materials usually show some degree of crystallinity. *Birefringent organic crystalline materials* include certain plastics, wood, Teflon, and cotton. These materials will generally char or lose birefringence upon heating to 345°C [1–3].

2 The Bathtub Curve and Condition Monitoring

Lubricated machines are usually characterized by three regimes in their wear rate versus time ("Bathtub") curve (Fig. 10.3): wearing-in (running-in), normal (stationary), and severe (abnormal) wear. Wearing-in occurs during the start-up stages of a new machine and is therefore also known as "infant mortality." During this regime, increased normal rubbing wear occurs, forming a finished surface and a match between opposing components. The quantity of large particles quickly increases in this regime, but then settles to an equilibrium concentration during normal running conditions. Machines wearing in an abnormal (i.e., wearing-out) mode will produce unusually large amounts of particles and a particle distribution with proportionally more large particles. The different regimes of wear, from mild to severe, are characterized by different size particles, the most severe being associated with particles larger than 1 mm [7, 8].

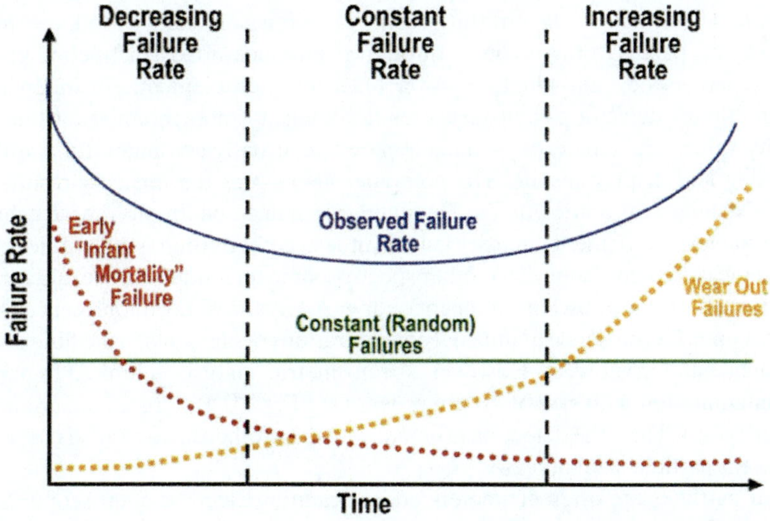

Fig. 10.3 The "Bathtub" curve, illustrating three wear regimes typically observed in lubricated engineering systems (1) wearing-in ("infant mortality"), (2) stationary (normal) wear, and (3) wearing-out (severe wear)

Condition monitoring is the process of monitoring a parameter of condition in machinery, such that a significant change is indicative of a developing failure. It is a major element of *predictive maintenance*, which refers to predictive, periodic, and planned maintenance actions taken to maintain a component within design operating conditions and extend its life. The use of condition monitoring allows maintenance to be scheduled, or other actions to be taken, in order to avoid the consequences of failure, before it occurs. Condition monitoring is typically much more cost effective than allowing the machinery to fail [4, 7].

2.1 Wear Particle Analysis

Wear particle analysis is one procedure used for condition monitoring which allows prediction of the imminent behavior of the machine. It is a powerful procedure for nonintrusive examination of the lubricant (e.g., oil or grease). A major advantage of this procedure is that disassembly of the system is not required a priori.

Spectrometric oil analysis program (SOAP) is used for predictive maintenance in many closed-loop lubricating systems. Atomic emission spectroscopy (AES) is used to determine the concentrations of different elements in used lubricants. A high-temperature source raises atomized elements to an excited energy state, and characteristic emission lines are produced when one of the outer shell electrons is expelled from the excited atoms. The empty electron orbit is filled by a nearby electron, which drops into it and emits its excess energy as a photon at a specific wavelength in or near the visible light range. The wavelength of the photon (emission line) thus identifies the chemical element of the emitting atom. The emission energy produced by the excited atoms passes from the source through an entrance slit to a diffraction grating, which separates the individual emission lines for each element. An independent photomultiplier detector circuit measures the intensity of each emission line. The intensity values are transferred to a readout device, usually computer, for additional processing and display results. The computer also stores the intensity results of a series of standard calibration fluids. The standards define a calibration curve, which is used to convert the element intensity values into parts per million (ppm) values over a specific measurement range. Two other spectroscopic techniques in use, although to less extent, are atomic absorption spectroscopy (AAS) and X-ray fluorescence (XRF) spectroscopy. The presence of unusual concentrations of elements in the fluid sample can indicate abnormal wear. However, spectrometric analysis is limited to particulate contamination with size of 10 μm or less (see Fig. 10.4); larger contaminations are overlooked. Thus, the detection efficiency of spectrometric oil analysis decreases as the wear particle size increases [4, 7].

Wear particles are often characterized by scanning electron microscopy (SEM) coupled with energy dispersive X-ray spectroscopy (EDS). In SEM, a high-energy electron beam is focused and rastered on a solid specimen. The interaction of the electron beam with the material yields low-energy electrons associated with inelastic scattering processes (secondary electrons, SE), electrons with energy similar to

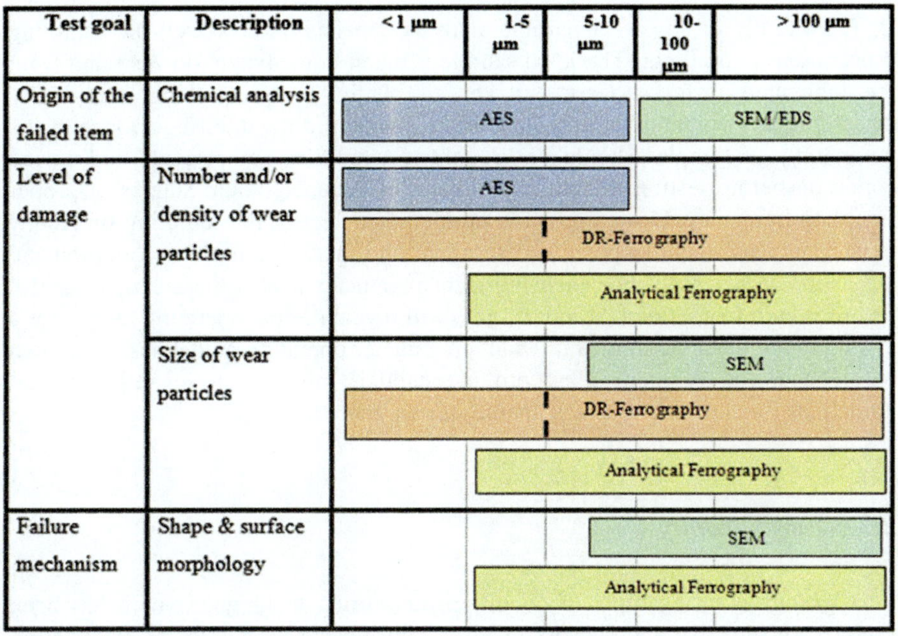

Fig. 10.4 Comparison between the identification capabilities of various techniques for wear particle analysis

that of the incident beam that are associated with elastic scattering processes (backscattered electrons, BSE), electromagnetic radiation with large wavelength (cathodoluminescence), XRF, etc. Advantages of imaging by SEM include high depth of field, high resolution, possibly high magnification, and three-dimensional (3D) images. While a SE detector provides information on the topography of the surface, a BSE detector is sensitive to the chemical composition. The energy of each X-ray photon emitted from the sampling volume is characteristic of the element which produced it. The EDS microanalysis system collects the X-rays, sorts and plots them by energy, and automatically identifies and labels the elements responsible for the peaks in this energy distribution. The EDS data are typically compared with either known or computer-generated standards to produce a full quantitative analysis showing the sample composition. EDS analysis is usually possible for all elements with an atomic number greater than 5 (Boron); most elements are detected at concentrations on the order of 0.1%.

Characterization of wear particles by SEM/EDS requires that the particles are large enough (say, above 10 μm in depth, of course depending on the chemical composition of the material and on operating conditions such as acceleration voltage) in order to extract a spectrum that represents the chemistry of the particle per se (see Fig. 10.4). The great advantage of this method is the ability to acquire the chemical composition of each particle separately. This information can aid in identifying the origin of the particles, i.e., from which part in the machine the particle was removed.

The validity of any wear particle analysis depends on how well the sampling procedure is carried out. The ideal sample is taken immediately downstream from the lubricated surfaces. Care is taken to obtain a representative sample by discarding any volume that may have been stagnant in the drain line. Once in the laboratory, all samples should be brought to a uniform temperature and stirred condition before testing. Machine maintenance records should suggest a proper sampling frequency. Once a possible problem is detected, the sampling frequency must be increased, until a positive determination is made on machine condition and the action to be taken. For each lubricant parameter that is measured, a control record is built that, after a period of time, will reveal normal operating ranges for a given type of machine/lubricant. Unless parameter operating guidelines are known beforehand, the program must provide enough information to set statistical guidelines for acceptable versus abnormal parameter limits [7].

2.2 *Ferrography*

Ferrography is an example of condition monitoring technique, which has been found very sensitive and successful in monitoring the wear state of engineering systems, including in aeronautical and aerospace applications. This is a nondestructive method of capturing ferromagnetic and paramagnetic particles on a glass slide based upon the interaction between an external magnetic field and the magnetic moments of the particles suspended in a flow stream [9–11]. The method was developed by Westcott et al. in the early 1970s to investigate the occurrence of ferrous wear particles in lubricated dynamic components [12–16], and is used either as the primary analytical method or in conjunction with spectrometric analysis [7, 17]. In ferrographic examination, four types of Ferrographs may be used, as described in the following subsections.

2.2.1 Direct-Reading (DR) Ferrograph

The *Direct-Reading Ferrograph* (Fig. 10.5) uses optical density to quantitatively measure the concentration of wear particles in a lubricating oil or hydraulic fluid [16, 18, 19]. A powerful magnetic-gradient field causes particle deposition into a glass precipitator tube, in descending order of size. Light attenuation at two locations along the path—at the entry deposit (C_L, particles larger than 5 μm) and at a point several millimeters farther down the tube (C_S, particles smaller than 5 μm)—is used to quantify the relative concentration C_L/C_S particles. An increase in the C_L value indicates that the system has entered into an abnormal wear mode. Values of wear particle concentration (WPC) and the percent of large particles (PLP) are thereby derived, establishing machine wear baselines and trends in wear condition [7]. When an abnormal wear mode is detected, Analytical Ferrography is used for more in-depth analysis [4].

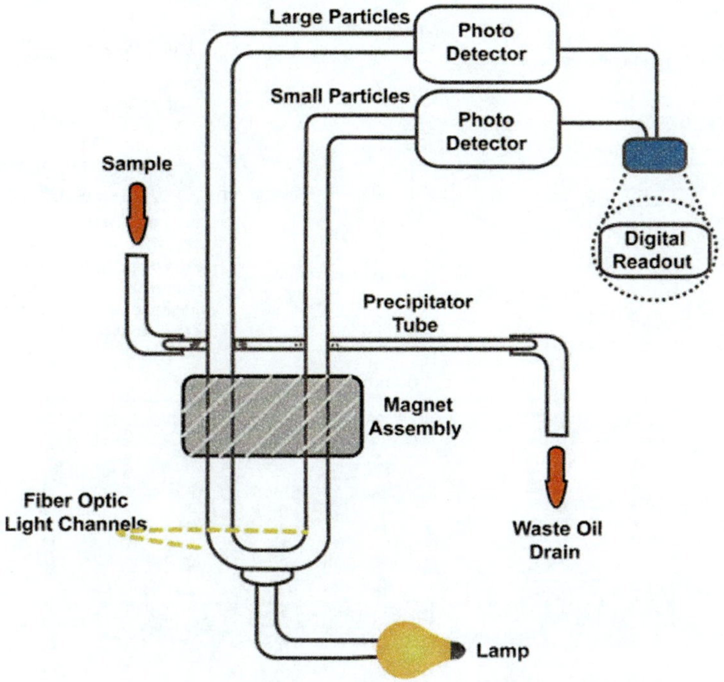

Fig. 10.5 Scheme of a Direct-Reading Ferrograph

2.2.2 Analytical Ferrograph

In *Analytical Ferrography* (Fig. 10.6a–c), a ferrogram (i.e., microscope slide with deposit of captured particles) is prepared by pumping a fluid sample (e.g., oil or diluted grease) that contains wear particles through Teflon tubing. The pumping is onto a specially prepared glass substrate, which is mounted above two permanent magnets (Fig. 10.6a) and has a non-wetting barrier painted on one surface to centrally channel the liquid (Fig. 10.6c). The magnets are separated with their magnetic poles counterposed, so that a strong magnetic field gradient is created in the vertical direction above an aluminum strip. Three types of forces act on the suspended particles: magnetic, hydrodynamic (drag), and gravitational. Depending on the size of the particles, they may behave either as saturated magnetically (small particles) or as soft magnets (large particles). In addition to the magnetic force tending to accelerate the particles, drag forces retard their motion. The magnet of the ferrograph was designed to develop an extremely high gradient of magnetic field near the magnet poles. The ferrogram is mounted at a slight angle to the horizontal, with the entry end elevated so that the fluid flows downward towards a waste bottle (Fig. 10.6a). Because the distance from the magnet to the substrate is slightly greater at the entry side than at the exit side, the magnetic field strength is weaker at the entry side, causing only the largest (magnetically affected)

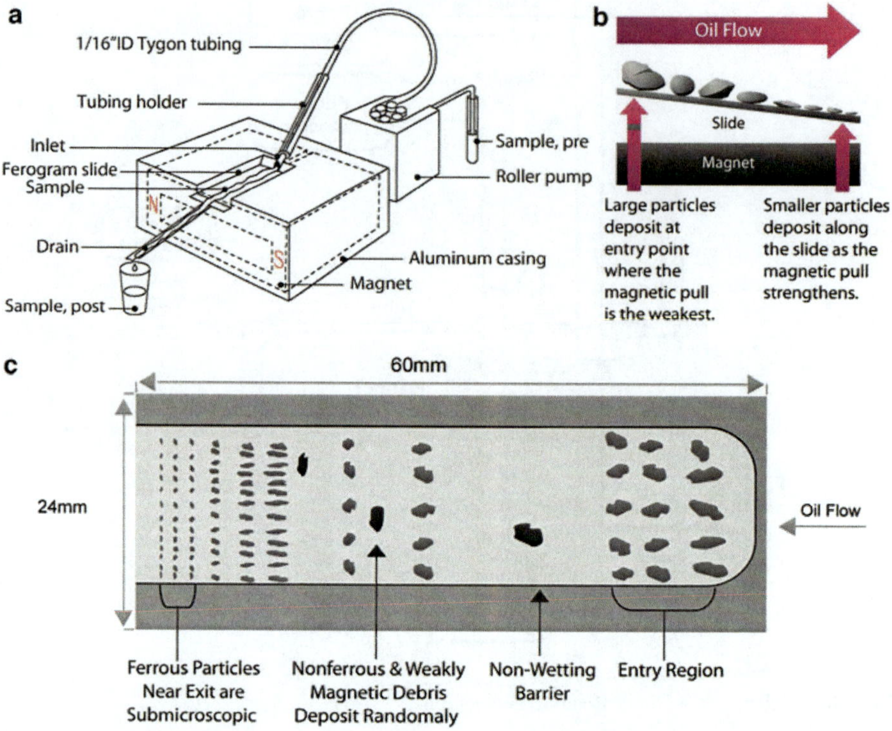

Fig. 10.6 The principles of Analytical Ferrography. (**a**) The magnet assembly, slide, and flow scheme. Reproduced after [20]. (**b**) Tilting the slide results in deposition of wear particles with gradual size. (**c**) The deposition scheme of ferrous and nonferrous particles on the ferrogram

particles to deposit. Farther down the ferrogram the progressively stronger magnetic field deposits progressively smaller particles (Fig. 10.6b). For the same particle shape, motion downwards through the fluid as a function of size is governed by the ratio of the particle diameter cubed to the particle diameter squared. Ferrous particles are deposited on the ferrogram in strings perpendicular to the direction of fluid flow. In contrast, nonferrous particles and contaminants travel downfield in a random distribution pattern, not being oriented by the magnetic field. They often appear between the strings of ferrous particles (Fig. 10.6c). After all the fluid in a given sample has been run across the ferrogram, a fixer solution is run to remove residual fluid [4, 7, 12].

After the fixer dries, the ferrogram is ready for observation under the microscope. This is often a special microscope, equipped with both filtered-red reflected and filtered-green transmitted light sources that may be used simultaneously (Fig. 10.7). This lighting scheme, called bichromatic illumination, will show a metal as red and a nonmetal as green. Chemical analysis of wear particles can be carried out by various analytical techniques. Heat treatment of the particles on the ferrogram is another quick, inexpensive method of identification. Heating the particles, for example, at

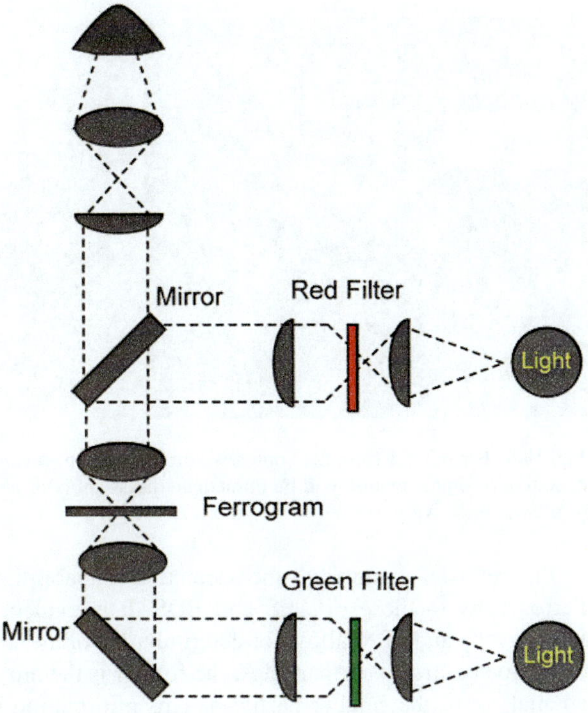

Fig. 10.7 The bichromatic microscope developed for Analytical Ferrography

340°C for 90 s yields oxide film thicknesses that are in the range of the wavelengths of visible light. Reflection of light off the metal surface underlying the oxide layer produces interference effects, resulting in coloring of the particles. Different classes of alloys exhibit predictable colors. Therefore, the prior heat history of a particle may sometimes be apparent as temper colors or variations in color of the heat-treated surface [4, 7].

By quantifying the ferrographic patterns (i.e., number, shape, size, and texture) and determining the composition of different particles on the ferrogram, the origin, mechanism, and level of wear can be determined. A wear particle atlas such as in [6] can be used as a guide for data analysis.

Machines already subjected worldwide to ferrographic analysis include gearboxes, bearings, diesel engines, gas turbine engines, gasoline engines, hydraulic systems, compressors, etc. Some of the relevant industries include aircraft, naval, automotive, pulp and paper, etc. For example, Analytical Ferrography has been applied in the Israel Air Force with great success to monitor the health of helicopter gearboxes [7]. Figure 10.8 shows images at (the same) high magnification of two ferrograms—one from an assembly that experienced normal wear and the other from an assembly that experienced severe wear. The difference in the concentration of wear particles is clearly evident. Recently, the use of DR-Ferrography and Analytical Ferrography has been extended by Levi and Eliaz to condition monitoring of aircraft engines with an open-loop oil system [4].

Fig. 10.8 Ferrograms from helicopter assemblies with (**a**) severe wear, and (**b**) normal wear. The operative recommendation will be immediate disassembly or next sampling after additional 25 h of flight, respectively

Figure 10.4 compares the detection capabilities of Analytical and DR-Ferrography to those of AES and EDS. It is evident that while both Analytical Ferrography and SEM allow for determination of the shape and surface morphology of relatively large wear particles, the former is the only suitable technique for such an analysis of the smaller particles. This information is often necessary for determination of the wear mechanism. The number of wear particles can be determined by Analytical Ferrography, while the density of wear particles can be determined based on DR-Ferrography or AES. This information is required for determination of the level of wear. The exact size of wear particles can be determined by SEM (for fairly large particles only, due to the practical procedure by which the particles are sampled) and Analytical Ferrography, whereas DR-Ferrography only allows to distinguish between "large" and "small" particles. The border between these two size classifications is marked in Fig. 10.4 by a vertical dot line. It should be recalled that the size of wear particles is related to the level of damage. Chemical analysis is required for determination of the origin of damage. While SEM/EDS often allows for unambiguous identification of the material from which the worn item was made, AES provides only the distribution of elements in the mixture (including the oil itself and contaminants).

2.2.3 On-Line Ferrograph

The *On-Line Ferrograph* [21–25] is installed on the machine itself and provides its output in electrical form, which can be analyzed to automatically initiate protective actions. Thus, it is not necessary to draw fluid samples from the machine, send them to the lab, and use trained operators to evaluate them. The principle of operation is usually based on real-time cyclic separation of wear debris from a fluid system onto a surface-effect capacitive sensor so that relative quantities of small (<2 μm) and large debris, which deposit near the sensor's exit and entry, respectively, may be determined. Whenever the accumulated debris reaches a predetermined level, the on-line unit is automatically flushed and the cycle repeated. The severity of wear is therefore inversely proportional to the length of the cycle. By instrumentally measuring fluid temperature and cycle time, the volume of fluid required to deposit that quantity of debris may be computed when the viscosity is known. Thus, C_L, C_S, and WPC can be instrumentally determined and displayed [21–24]. Alternatively, the lubricating oil may flow through a transducer coil, causing small changes in the inductance of the coil as any metallic particles pass. The coil is used as the resonating inductance in an oscillator, whose frequency is compared to a fixed crystal controlled oscillator in a phase locked loop (PLL). The PLL is arranged so that it compensates automatically for any slow drift in the oscillator frequency, while transient disturbances produced by metal particles can be detected and amplified. The direction of the frequency shift allows discrimination between ferrous and nonferrous particles (ferrous particles increase the inductance of the coil, while nonferrous particles reduce it), and the signal amplitude gives a measure of debris size and concentration [25].

2.2.4 Bio-Ferrograph (BF)

Bio-Ferrography is the most recent modification of the conventional Analytical Ferrography. It was specifically developed to allow magnetic isolation of target cells or tissues [26–29]. It exhibits a unique combination of strengths compared with other techniques commonly used in biotechnology (see also Fig. 10.9), including (1) the ability to quantify biological matter and, at the same time, analyze its microscopic and chemical features; (2) preservation of the structure and morphology of the captured particles, which might otherwise not be observed if an acid dissociation step is required; (3) extremely high selectivity and sensitivity due to vertical flow; (4) a requirement for less sample manipulation relative to conventional immunomagnetic separation techniques; (5) applicability to any liquid sample, including whole blood; (6) samples as small as 1 μL and target particles as small as several nanometers (depending on their composition) can be analyzed [30]; (7) the possibility to simultaneously process up to five samples within bracketed areas (channels) on a single slide, without cross-contamination.

The BF 2100, from Guilfoyle Inc. [26–29], is a bench-top cytometry-based instrument (see Fig. 10.10a). It utilizes a magnetic field that has a maximal field strength across an interpolar gap, where the collection of magnetically susceptible particles takes place (Fig. 10.10b). Since the gradient of that field is maximal at the edges of the gap, two parallel deposition strips—primary and secondary—are formed, and a rectangular deposition band can be observed on the ferrogram even by naked eye (Fig. 10.10c). Compared to the conventional Analytical Ferrography, in Bio-Ferrography a stronger magnet is used, the flow is vertical,

Feature	*Bio-Ferrograph 2100*	Flow cytometers & sorters	High Gradient Magnetic Cell Concentration methods*(1)*	Other magnetic concentrators	Monolayer cell preparation systems*(2)*	Automated Slide Stainers
Process multiple samples simultaneously	✓				✓	
Disposable components	✓		✓	✓	✓	
Capture cells directly from whole blood	✓					
Capture cells in a defined array	✓				✓	
Stain cells	✓					✓
Wash cells while holding them in place	✓					
Capture cells by their immunological properties	✓	✓	✓	✓		
Process large numbers of cells	✓		✓	✓		
Preserve Cell Morphology	✓	✓		✓	✓	
Concentrate cells for culturing	✓	✓	✓	✓		
Separate cells from liquid samples *(3)*	✓				✓	
Process sample volumes from 10 mL to 5 ml	✓		✓	✓		
Generate a permanent record of your test	✓				✓	✓
Short fluid flow path with no obstructions	✓			✓	✓	
Moderate to High throughput process	✓		✓	✓	✓	
Easy to clean	✓			✓	✓	

Notes:

(1) Systems where the cells pass through a column filled with fibers or beads that create a localized high gradient field.

(2) Methods where cells are pulled down onto a filter using a vacuum or centrifuged onto a microscope slide.

(3) Many systems leave cells suspended in liquid in a test tube and further work is needed before analysis can be undertaken.

Fig. 10.9 Comparison between the characteristics of different cell/particle capturing techniques

and the ferrogram is very thin (thus, the deposition surface is close to the interpolar gap). A very high magnetic flux density is established at the interpolar gap, generally around 1.8 T. The vertical flow separates the vertical gravitational force from the nearly horizontal magnetic force, so that only the latter acts to retain magnetic particles moving downward through the flow chamber [26, 28]. This deposition scheme results in elimination of gradual deposition of particles according to their size, while providing higher sensitivity and an ability to capture smaller particles (with lower magnetic moment).

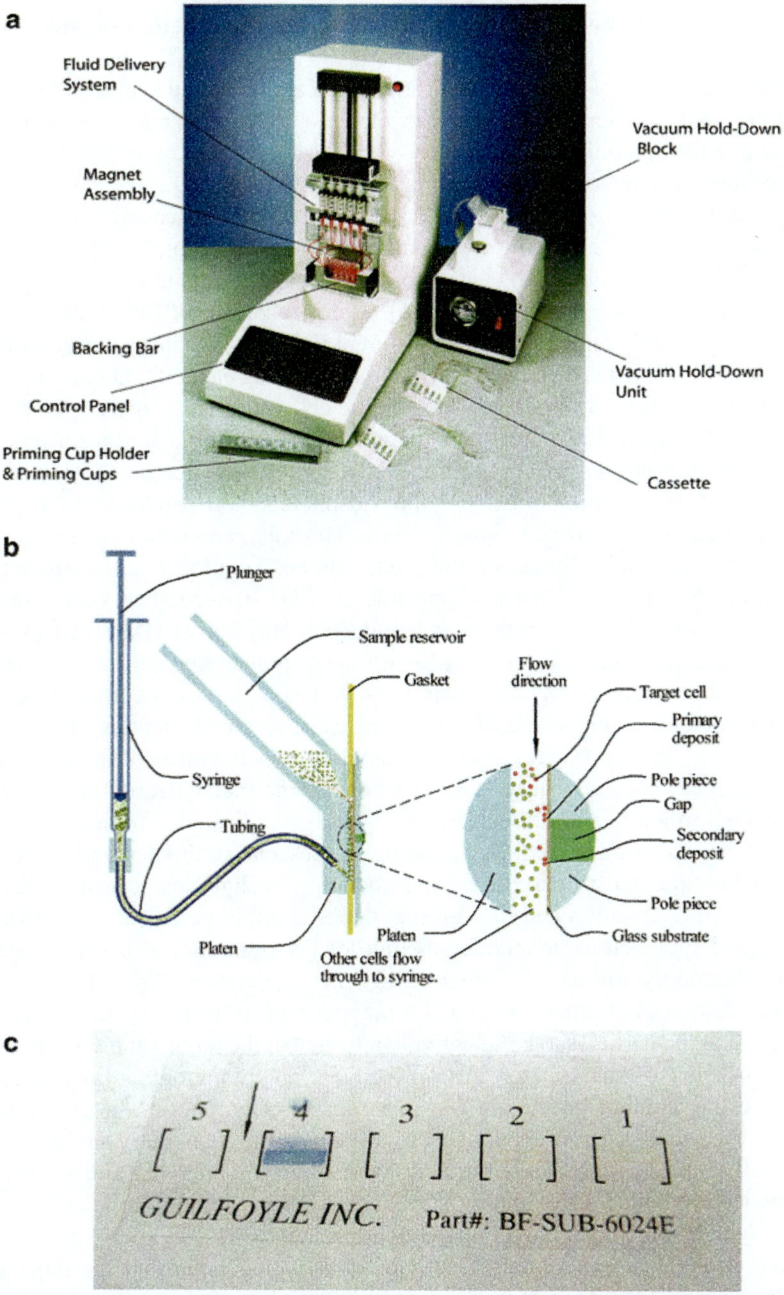

Fig. 10.10 The principles of Bio-Ferrography. (**a**) Bio-Ferrograph 2100 (Guilfoyle, Inc.) [27, 29]. (**b**) The deposition scheme of captured particles on the slide [29]. (**c**) A bio-ferrogram with rectangular deposition bands, with (channel 4) or without (channels 1 and 2) staining. The arrow on the ferrogram marks the direction of flow

2.2.5 Magnetic Labeling of Biological Matter and Synthetic Polymers

Magnetization does not naturally occur in biological matter and synthetic polymers, which are diamagnetic; hence, wear particles of such origins must be magnetized prior to any ferrographic analysis.

One approach is to use *molecular magnetic labels*, e.g., the paramagnetic lanthanide erbium cations (Er^{3+}) [31–33]. Owing to the presence of unpaired 4f-electrons, these cations have exceptionally high atomic magnetic dipole moment (9.6 μ_B). In addition, their high charge-to-volume ratio results in strong affinities to some negatively charged ligands [34, 35]. The mechanism of Er^{3+} binding to the cell surface is mostly ionic, with many different Er^{3+} binding sites, such as carboxyl groups in glycoproteins, differing in affinity and binding capacity. Other lanthanide binding sites are the Ca^{2+} receptor sites on the cell wall [31]. Cartilage binds Er^{3+} mainly via the carboxyl groups that are attached to glucuronic acid residues in the glycosaminoglycan (GAG) chondroitin sulfate. On the other hand, in bone the major sites of sequestration of Er^{3+} are the phosphate moieties in the hydroxyapatite mineral. Their interaction is extremely strong. The collagenous components of bone and cartilage may also sequester Er^{3+}, but this contribution to magnetization is small [35]. A solution of erbium chloride ($ErCl_3$, 10 mM) has been used for magnetization of ultrahigh molecular weight polyethylene (UHMWPE) [36–39], polymethylmethacrylate (PMMA) [37–39], and polycarbonate-urethane (PCU) [40, 41] wear particles from hip implants, as well as of bone and cartilage particles suspended in the synovial fluids of non-arthritic and osteoarthritic human and sheep joints [35, 37, 39, 42–50]. Yet, magnetic labeling by adsorption of the Er^{3+} cation suffers from several drawbacks (1) it is a nonspecific magnetization method, thus data analysis might be more time consuming and complicated; (2) it might interact with different staining materials, including fluorescent antibodies and dyes (e.g., Alcian Blue and Alizarin Red), thus introducing background staining [42]; (3) troublesome precipitation and background deposits might be present [45, 49].

A second approach for magnetic labeling for Ferrography and Bio-Ferrography is that commonly known as *immunomagnetic separation (IMS)* [31]. In this approach, a system of an antibody and a magnetic bead (or a magnetic colloid) is used, causing the particles to be selectively attached to the target matter when added to a suspension of that matter. An antibody is an immunoglobulin capable of specific combination with the antigen that caused its production in a susceptible animal. The small site on an antigen to which a complementary antibody may specifically bind is called an "epitope." Most often, monoclonal antibodies (i.e., homogeneous populations of antibodies that recognize a single epitope) are used for IMS, although sometimes polyclonal antibodies are also used [31, 51–54].

Eliaz et al. have developed a method for isolating bone and cartilage wear particles suspended in the synovial fluids of human hip, knee, and ankle joints by specifically labeling collagens I and II with monoclonal mouse anti-human primary antibodies coupled to goat anti-mouse IgG MACS microbeads [42, 55]. The magnetic beads (from Miltenyi Biotec.) are 50 nm in diameter, have a Dextran polymer shell, and are immobilized by secondary antibodies. In such experiments, it is very important to enable good conjugation between the target matter, primary antibodies, and secondary antibodies that are attached to the magnetic beads by properly selecting the test conditions (namely, incubation time, ratio between primary and secondary antibodies, temperature, vortexing, etc.). Based on our experience, it is recommended to first add the primary antibody to the target sample, allow incubation for at least 1 h at room temperature, add the magnetic beads to which a secondary antibody is attached, and then allow a second incubation for at least 30 min at room temperature prior to ferrographic analysis [54]. Another option is to first conjugate the primary and secondary antibodies attached to magnetic beads, and then add this cocktail to the target sample [26, 56–59].

3 Applications of Ferrography and Bio-Ferrography in Life Sciences and Medicine, Excluding Orthopedics

The success of the conventional Analytical Ferrography in condition monitoring of engineering systems triggered few feasibility studies in the fields of life sciences and medicine already in the 1980s. These included erythrocyte and white blood cell separation [46, 60–62] and bacterial tracking [20, 61, 62]. In those studies, Er^{3+} was the magnetizing agent.

Since the introduction of the Bio-Ferrograph in the late 1990s, more feasibility studies have been carried out, some of which use antibodies coupled to magnetic beads for magnetic labeling. Johnson et al. have used the BF to track *Escherichia coli* bacteria at low concentrations in natural waters [56–59, 63–66]. Several advantages of the BF compared to filtration were claimed: (1) higher recovery rates (the number of visually identified bacteria was an order of magnitude higher than the number of cfu determined from membrane filtration); (2) higher sensitivity (as low as 20 bacteria in 1 mL water), with analytical errors ranging between 10% and 20%; (3) selective collection of bacteria without collection of associated mineral debris such as clays and other mineral colloids; (4) the ability to visually identify indigenous bacterial aggregation and shape; (5) the possibility to undertake different confirmation steps (e.g., PCR or rRNA) afterwards; (6) rapid analysis; (7) small volumes of suspension are sufficient. Drake et al. [67] later examined whether pathogens are present in biofilms and whether the biofilms contain higher microbial densities compared to ships' ballast water. Bio-Ferrography was used in combination with biochemical tests and fluorescent microscopy to isolate and characterize low concentrations of *Vibrio cholerae* from the ballast water.

Bio-Ferrography has also been used to isolate rare MCF-7 breast carcinoma cells from human peripheral leukocytes [68]. The cell mixture was labeled with anti-epithelial membrane antigen antibody and a magnetic colloid. The recovery of the MCF-7 cells from the original mixture ranged from 20% to 60%; the detection limit was 10^{-6}. BF was found sensitive, relatively inexpensive, simple, and fairly quick (as it combines cell enrichment and immunocytology slide preparation in a single step, and the deposition is confined to a small area on the slide). Another advantage was the preservation of the morphology of the cancer cells during the experiment. In another work [69], Bio-Ferrography was applied on samples of CD4 cells in leukocytes and murine lymphoma cells in human peripheral blood. It was concluded that while Bio-Ferrography had good specificity with respect to cell capture, it might not be applicable to rare-event selection due to the inability to capture all available target cells. Bio-Ferrography has also been evaluated for capturing breast BT-20 tumor cells directly from whole blood [70]. It was claimed that the technique was more sensitive and allowed for identification of breast cancer at an earlier stage compared to other diagnostic techniques common in medicine.

Ishay et al. used the BF in a study aimed at determining the role of magnetic minerals embedded in the comb cells of Vespinae [71]. Ceilings and walls from the same comb were crushed separately in clean glass beakers, and suspensions were prepared from each fine powder by adding deionized water. Neither magnetizing agents nor any other chemicals or reagents were added to the suspensions during the capturing process. The number of particles captured on the ferrogram was higher than in previous publications where a BF was not used.

Parkansky et al. used the BF to isolate magnetic particles from a suspension of carbon powder in ethanol, which was produced by a pulsed arc [30]. It was found that the magnetically separated particles included chains of ~30–50 nm diameter spheres, and nanotubes and nanorods with lengths of 50–250 nm and diameters of 20–30 nm. In contrast, the residual particles which passed through the BF consisted of 1 μm and larger microparticles, and nanoparticles without any definite shape. The particles captured by Bio-Ferrography were similar in shape and dimensions to those observed by high-resolution transmission electron microscopy (HR-TEM) of the raw powder. Although the application in [30] is not in the areas of life sciences and medicine, it is included here because this seems to be the first time where the high sensitivity of Bio-Ferrography in isolation of nanoparticles was demonstrated in the literature. This may make the technique even more attractive in a variety of novel therapeutic, sensing and diagnostic modalities in nanomedicine [72–74].

4 Synovial Joints

A natural *joint* (*articulation*) is a point of contact between bones. *Synovial joints* (diarthroses) are freely movable joints. They are remarkable bearings that consist of the following structural elements [75]:

1. Joint capsule. Forms a complete casing around the ends of the bones, thereby binding them to each other.
2. *Synovial membrane* (synovium). Moist, slippery membrane that lines the inner surface of the joint capsule and attaches to the margins of the articular cartilage. It also secretes and removes *synovial fluid (SF)* from the joint

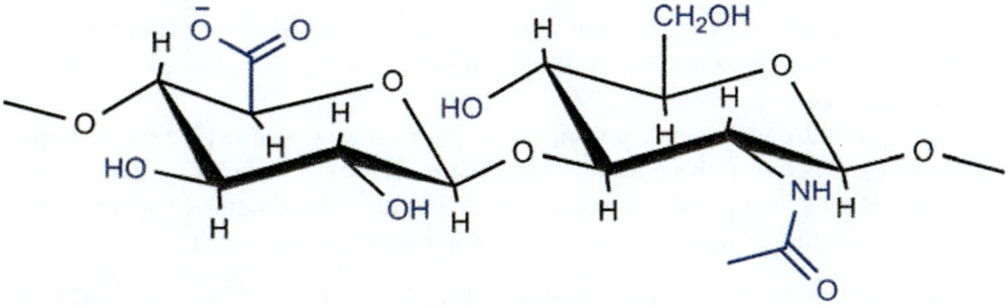

Fig. 10.11 The repeat disaccharide of hyaluronan: –D-glucuronic acid-β-1,3-*N*-acetylglucosamine-β-1,4–

space, adjusting both the volume of SF and its macromolecular composition [76]. SF is an ultra-filtrate of plasma, which has non-Newtonian properties (i.e., its viscosity increases as the shear rate decreases) and is thixotropic (i.e., its viscosity decreases with time under stress) [77]. It enables efficient movement of the joint, acting as a lubricant, supplying nutrients, and removing catabolic products. Endogenous *hyaluronan (HA)* is the main constituent of SF [78]. This polysaccharide acts as a lubricant during low impact joint movement and as a shock absorber during high impact movement [79]. It is secreted in high concentrations in the extracellular matrix (ECM) of connective tissues, although its highest concentration is in the SF (2.5–4.0 mg/mL in normal joints) [76, 80]. The molecular structure of HA is shown in Fig. 10.11. The polysaccharide chain is made of repeating disaccharide units of *N*-acetylglucosamine and glucuronic acid. The average molecular weight of healthy human HA is 5,000 kDa [80, 81]. The viscosity of the SF is affected by the length and conformation of the polymer chains, as well as by interaction between adjacent chains [76].

3. *Articular cartilage*. Thin layer of hyaline cartilage covering and cushioning the articular surfaces of bones. The deeper zones of cartilage are nourished also by blood vessels in the subchondral bone. The ECM of normal articular cartilage is composed mainly of water, collagens, and proteoglycans. Collagen II represents approximately 90% of the overall content of collagen in the articular cartilage of adults, while being present at minor concentrations in other tissues [82]. Articular cartilage acts in the natural joint as the "bearing material" and as a shock absorber. The coefficient of friction between the articulating cartilage surfaces in human hips or knees is ultra-low ($\mu < 0.002$ at pressures as high as 5 MPa) [83, 84]. The extremely efficient lubrication arises from the presence of a brush-like phase of charged macromolecules at the surface of the superficial zone. This phase forms when charged macromolecules, including lubricin and aggrecan, cross the interface between the superficial zone and the synovial cavity as they are secreted into the synovium from within the bulk of the cartilage [85].
4. Joint cavity. Small space between the articulating surfaces of the two bones of the joint.

5. Meniscus (articular disk). Pads of fibrocartilage located between the articulating ends of bones in some diarthroses (the knee joint, for example, contains two menisci).
6. Ligaments. Strong cords of dense, white fibrous tissue at most synovial joints. These grow between the bones, lashing them even more firmly together.
7. Bursa. Some synovial joints contain a closed pillow-like structure called a bursa, which is formed of synovial membrane filled with synovial fluid.

The hip and knee joints are illustrated in Fig. 10.12. The *hip joint* (Fig. 10.12a) is a *ball-and-socket joint*. Its stability derives mainly from the shapes of the head of the femur and of the acetabulum, the socket of the hip bone into which the femur head fits. This joint permits multiaxial movements: flexion (decreasing the angle between bones, e.g., by bending one on the other), extension (increasing the angle between bones, namely stretching or straightening a part to its anatomical position), abduction (moving a part away from the median plane of the body), adduction (moving a part toward the median plane), rotation (pivoting a bone on its own axis), and circumduction (moving a part so its distal end moves in a circle) [75].

The *knee joint* (Fig. 10.12b) is the largest and one of the most complex and most frequently injured joints in the body. It is both a *condyloid joint* (between the two condyles of femur and between the condyle and meniscus of tibia) and a *gliding joint* (between the patella and femur). The condyles of the femur articulate with the flat upper surface of the tibia. Counteracting forces are supplied by a joint capsule, cartilages, and numerous ligaments and muscle tendons [75].

For clarity, several terms, which are mentioned in this chapter and describe directions and planes of the body, are hereby provided. The term "superior" means "above" or "upper," while "inferior" means "below" or "lower." The term "anterior" means "front" or "in front of," while "posterior" means "back" or "in back of." The term "medial" means "toward the midline of the body," while "lateral" means "toward the side of the body, or away from its midline." The term "proximal" means "toward or nearest the trunk of the body, or nearest the point of origin of one of its parts," while "distal" means "away from or farthest from the trunk or the point of origin of a body part." All of these are directional terms. The sagittal plane runs from front to back and divides the body or any of its parts into right and left sides [75].

The natural synovial joint provides low wear over decades. This may be ascribed to the development of a fluid film, predominantly in a combination of EHL and micro-EHL (μEHL) lubrication modes [86, 87]. μEHL is a localized form of EHL, whereby pressure perturbations cause substantial flattening of asperities at the material's surfaces, increasing conformity and assisting in the maintenance of a lubricious film [88]. These two modes of full-film lubrication operate under most conditions (high loads) in healthy joints. At low loads, on the other hand, boundary lubrication prevails.

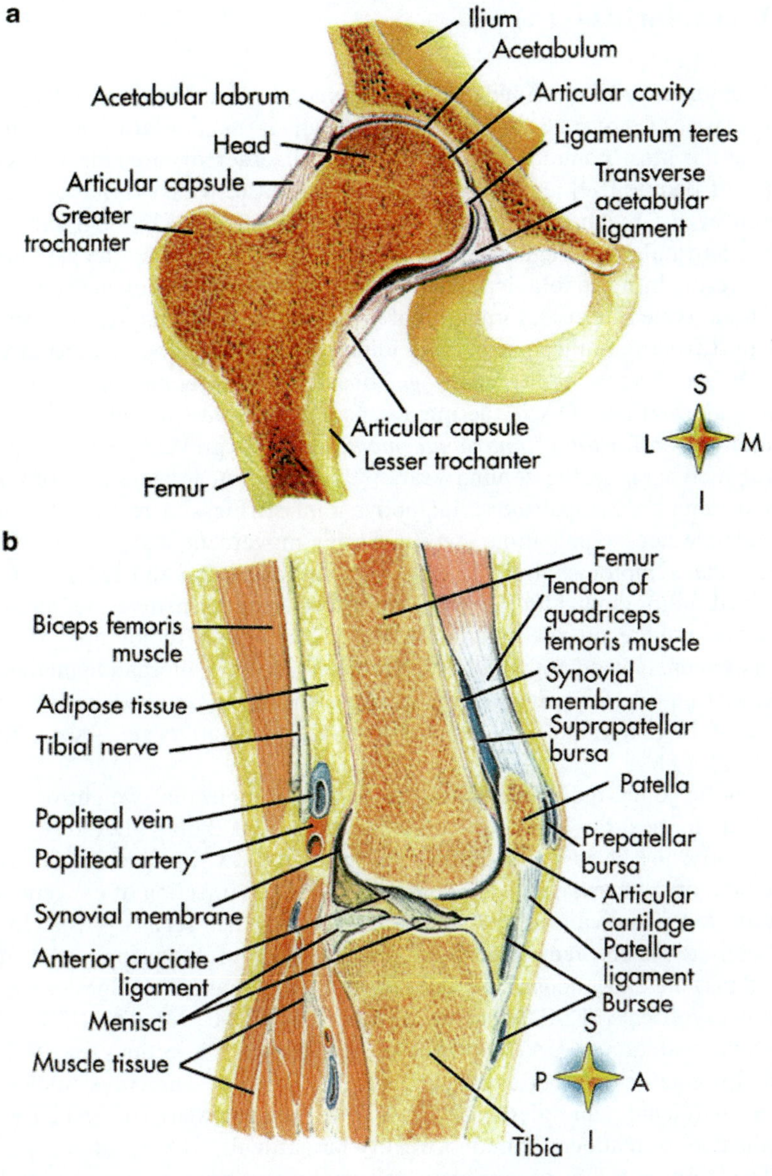

Fig. 10.12 (**a**) Frontal section through the hip joint. (**b**) Sagittal section through the knee joint. The anatomical rosettes mark the anterior (A), lateral (L), inferior (I), medial (M), posterior (P), and superior (S) directions. Thibodeau and Patton [75], Anatomy and Physiology, copyright 1999 Mosby. This material is reproduced with permission of Elsevier

5 Osteoarthritis (OA)

Joint disorders may be classified as noninflammatory joint disease or inflammatory joint disease. *Osteoarthritis (OA)*, also known as *degenerative joint disease (DJD)*, is the most common noninflammatory disorder of movable joints [75]. Millions of people [89] are stricken with this disease, which is associated with biochemical and mechanical processes and is characterized by degradation and atrophy of articular cartilage and hypertrophy of bone. If severe, the pathological changes result in morphological changes, such as narrowing of the joint space, subchondral bone sclerosis, formation of osteophytes and metaphyseal cysts [90]. OA is most common in weight-bearing joints, the knee joint being most affected. The main risk factors for OA are age, obesity, joint trauma and, sometimes, genetic abnormalities. OA is the main cause of disability of elder persons and the major reason for *total joint replacement (TJR)*. Its prevalence is expected to continue increasing in the coming years [89]. The clinical symptoms of OA are pain and functional impairment, including joint stiffness, a restricted range of motion, tenderness to palpation, crepitus during movement, and swelling. Unfortunately, the symptoms often do not correlate well with the level of OA as determined radiographically [91, 92]. More precise and sensitive measurements, such as magnetic resonance imaging (MRI) and computed tomography (CT), are still very expensive and not widely accessible. Therefore, alternative methods are needed that can detect osteoarthritic changes in the joints at an early stage of the disease in a reliable, sensitive, objective, quantitative, and cost-effective manner [42, 55].

In arthritic joints the concentration and average molecular weight of HA are lower than in healthy joints, at 0.8–2.0 mg/mL and 3,000 kDa, respectively. Consequently, the viscosity of the SF is abnormally low, its functionality is lost, and rubbing of the cartilage occurs. Intraarticular knee injection of exogenous HA (a treatment also known as *viscosupplementation* [80]) has been suggested in order to compensate for the reduction in the concentration of endogenous HA. It was assumed that the exogenous HA could restore the rheological, analgesic, and anti-inflammatory effects of SF, which are lost in OA [81, 90, 93]. The US Food and Drug Administration (FDA) approved exogenous HA injections in 1997, but classified it as a device, not as a medicine. Nevertheless, numerous studies have reported significant pain relief and a better knee function that can persist for up to 6 months due to treatment with 3–5 weekly intraarticular HA injections. As the - half-life of HA is a few days, several injections are given to provide sufficient efficacy, although some producers seek to minimize the treatment to one injection only [55].

In OA, wear particles, such as bone and cartilage fragments, are released into the SF of the affected joint [44, 94]. These particles may trigger the release of enzymes, such as collagenase [44], which results in inflammation of the synovial membrane. A variety of biomarkers have been suggested for analysis of the SF (as well as of serum and urine) [42, 55, 90, 91, 94].

6 Artificial Joints: Total Hip and Total Knee Arthroplasty

Joints that have been severely damaged by diseases such as OA or by injury may be replaced surgically with *artificial joints* (also called *prostheses*) in a procedure referred to as *arthroplasty*. The goals of arthroplasty are to relieve pain and increase the range of motion. During this procedure, the ends of the damaged bones are removed, and artificial joints made of metallic, ceramic, and plastic components are fixed in place. Although most joints in the body can be repaired by arthroplasty, the ones most commonly replaced are the hips, knees, and shoulders [95]. The numbers of hip and knee replacement surgeries increased sharply during the last decade. For example, 285,000 total hip replacements and 523,000 total knee replacements were performed in the United States alone in 2005. It is anticipated that the growth and aging of the population, together with improved diagnosis and treatment options, will result in continuous, sharp increase in these numbers. Projections to 2030 predict that these two procedures will jump to 572,000 and 3.4 million, respectively [96]. Potential complications of arthroplasty include infection, blood clots, loosening or dislocation of replacement components, and nerve injury [95].

Figure 10.13a illustrates the principles of total hip arthroplasty. While partial hip replacement involves only the femur, *total hip replacement (THR)* also involves both the acetabulum and head of the femur. For example, the damaged ball (the upper end of the femur) may be replaced by a metal ball attached to a metal stem fitted into the femur, and a plastic socket is implanted into the pelvis, replacing the damaged socket. THR was revolutionized half a century ago by Sir John Charnley, who introduced a system consisting of a stainless steel femoral stem head combined with an UHMWPE acetabular cup. This metal-on-polyethylene (MoPE) system is the gold standard treatment, which presently demonstrates a survival rate of 79% after 11 years and 51–60% after 25 years, in young patients (<55 years) [97]. Failure of this implant is mostly due to *aseptic loosening* [98, 99], which is associated with *osteolysis* and bone resorption, induced by wear particles generated from the implant [100, 101]. The MoPE artificial joint operates in a mixed-mode lubrication regime [102]. Abrasive and/or adhesive wear mechanisms can thus lead to particle generation. Traditional MoPE systems have been reported to have a wear rate of 30–100 mm^3 per year [103]. Other hard-on-hard bearings, such as metal on metal (MoM) or ceramic on ceramic (CoC), produce lower wear rates than the conventional MoPE joint due to their hardness. The short-term clinical performance of MoM and CoC is reported to be 0.3 and 0.01–0.1 mm^3 per million cycles (Mc), respectively [103]. Although a low wear volume is an important factor governing the long-term clinical outcome of THR, other factors such as particle size, morphology, material chemistry, and biological response to the wear particles must also be taken into account. MoM prostheses, for example, have not been associated with osteolysis and demonstrate a low wear volume; however, wear particles generated by MoM prostheses are significantly smaller than UHMWPE particles produced by MoPE prostheses. Consequently, a higher number of metal particles is produced

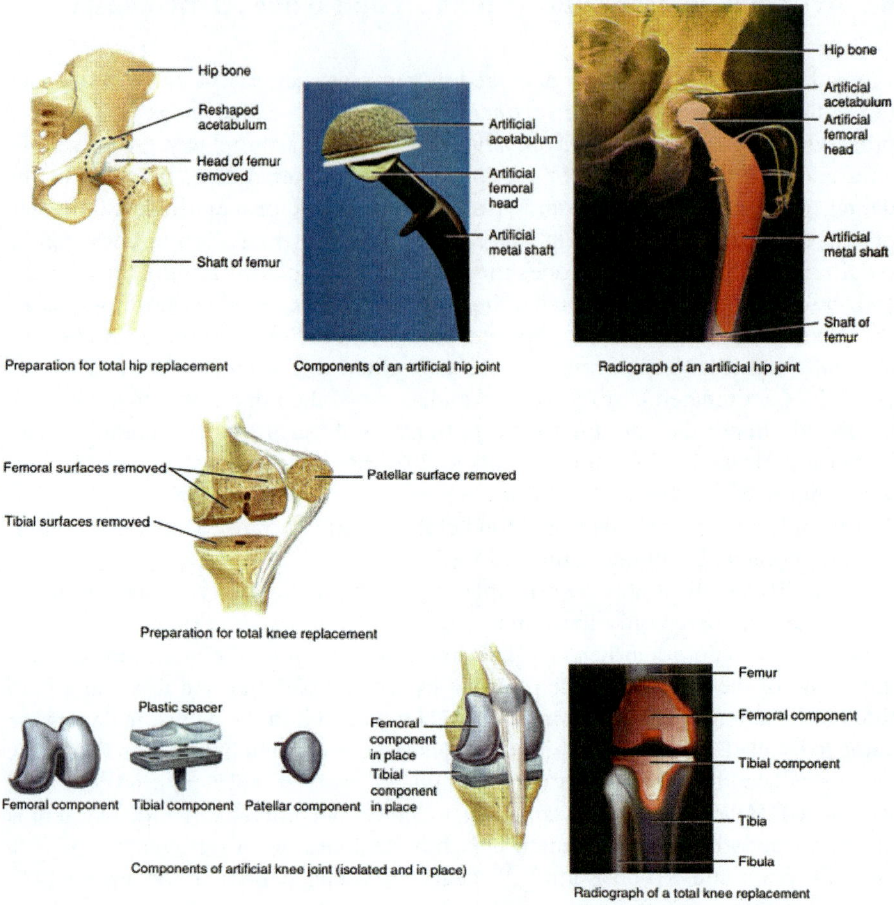

Fig. 10.13 Preparation for, components of and radiographs of total hip replacement and total knee replacement. Tortora and Derrickson [95], Principles of Anatomy and Physiology, copyright 2009 Wiley. This material is reproduced with permission of John Wiley & Sons, Inc

overall ($6.7 \times 10^{12} - 2.5 \times 10^{14}$ metal particles per year, compared to 5×10^{11} UHMWPE particles per year produced by a conventional MoPE prosthesis) [104]. Nanometer-sized metal particles have been shown to be disseminated throughout the body—in the lymph nodes, spleen, and bone marrow [105, 106]. The high activity of metallic nano-debris results in enhanced corrosion and release of metal ions to the joint. Certain metal ions have been shown to induce *hypersensitivity* and implant intolerance reactions [107, 108].

The use of compliant layer joints in artificial joints to promote EHL and μEHL and to reproduce the tribological function of the joint has gained much interest in recent years [40, 41]. Unlike rigid materials, employing a compliant hydrophilic polymer layer in the acetabular cup could restore the natural lubrication regime of the natural joint (full fluid-film lubrication) to further reduce wear. PCU, for example, has shown promising results as a candidate material for hip arthroplasty in terms of its mechanical properties, frictional behavior, and lubrication, which are similar to natural cartilage [87, 109–111].

Figure 10.13b illustrates the principles of total knee arthroplasty. In a *total knee replacement (TKR)*, the damaged cartilage is removed from the end of the femur, proximal end of the tibia, and the back of the patella (if the latter is not badly damaged, it may be left intact). The femur is shaped and fitted with a metal femoral component. The tibia is reshaped and fitted with a plastic tibia component that is cemented in place. If the back surface of the patella is badly damaged, it is replaced with a plastic component. In contrast, in a partial knee replacement, only one side of the knee joint is replaced. Once the damaged cartilage is removed from the end of the femur, the femur is reshaped and a metal component is cemented in place. Then, the damaged part from the proximal end of the tibia is removed, along with meniscus. The tibia is reshaped and fitted with a plastic component that is cemented into place [95].

7 The Use of Ferrography for Monitoring the Wear of Natural Joints

The conventional Analytical Ferrography has been used in several studies to monitor the wear of natural diarthrodial joints [35, 37, 39, 44, 45, 47–50, 112–114]. Evans et al. [39, 44] were probably the first to examine synovial fluid aspirates and saline washings of joints. Ferrograms were made from patients who presented a range of different arthritides. Wear particles, ranging in size from a few micrometers to several hundred micrometers, were clearly identifiable and fell into a limited number of discrete morphological categories. As with machines, the size of the wear particles increased with the severity of mechanical erosion of the articular surface. The ferrograms made from rheumatoid synovial fluid were quite different from those made from mechanically injured joints, indicating the diagnostic potential of ferrographic analysis of synovial fluid. In a subsequent study, Evans et al. [45] injected saline into the knee cavity of healthy, near healthy, and symptomatic patients (altogether: 43 patients) and captured wear particles by the conventional Analytical Ferrography, using $ErCl_3$ as the magnetizing agent. Different numbers of lamellae of superficial cartilage were found in the three healthy and near healthy patients (11–14 years old). No concentrations of wear particles were reported in the manuscript. Nowadays, it is unlikely that Helsinki approval to draw SF, inject saline into the knee cavity, or obtain synovial tissue from healthy donors, in order to establish a baseline for WPC, could be obtained. The results of ferrographic analysis were compared with the results of the arthroscopic examination of the same knees [45]. *Arthroscopy* is an imaging technique that is often used either as a diagnostic procedure or to perform joint surgery [75]. A narrow tube with lenses and a fiber-optic light source is inserted into the joint space through a small puncture. Isotonic saline solution is injected through a needle to expand the volume of the synovial space, thus spreading the joint structures and making the view easier. Thus, this technique allows the orthopedic surgeon to examine the internal structure of a joint without the use of an extensive surgery. Evans et al. [45] found the conventional Analytical Ferrography to be very sensitive in monitoring articular erosion, with a resolution much greater than that of arthroscopy. This was particularly apparent with knees suffering from a torn anterior cruciate ligament: while arthroscopy detected no damage to the cartilaginous surfaces, Ferrography detected a substantial level of microdamage. The higher sensitivity of Ferrography may be related to (1) the intrinsic higher sensitivity of examination of isolated wear particles compared to examination of the bulk material, and (2) the higher magnifications that can be used in ferrographic analysis compared to arthroscopy (the latter is typically used at a $\times 10$ magnification). Other findings of Evans are summarized elsewhere [35, 114]. For example, amorphous deposits with no optical activity were identified as fragments of the synovial tissue.

Mills and Hunter [47] analyzed ferrograms that represented approximately 50 patients with knee effusions. Particles ranging in size from submicrometer to approximately 2 mm were identified. A sample from what was considered to be a normal knee contained two large cartilage particles, a significantly lower number of small cartilage particles compared to symptomatic knees, and no bone particles. While the majority of particles present in the synovial fluid could be recovered, it was difficult to attribute a particular type of particle to a specific category of arthritic disease. Stachowiak et al. [48–50, 112, 113] used Analytical Ferrography, SEM, and numerical analysis to characterize the morphology of wear particles from knee joints. In one study [112], pairs of sheep knee joints were used. One of the joints was worn in a simulator over different periods of time under carefully simulated physiological forces and kinematic cycles, while the other was kept intact for comparison. Wear particles were extracted from a synovial fluid from both the worn and unworn sheep joints and then examined microscopically (by SEM). The shape of the particles was characterized by boundary fractal dimension, shape factor, and convexity. It was found that subtle changes in the shape of the particle boundary occur during the wear process. In a subsequent study [49], particles suspended in the knee joints of 30 non-arthritic and osteoarthritic patients were analyzed. A number of numerical descriptors were calculated for each one of 565 particles and correlated to the degree of OA using nonparametric tests. Significant differences were observed between the numerical descriptors calculated for wear particles from healthy and osteoarthritic joints. Many particles that were visible in the suspension were washed down the slide during Ferrography and subsequently lost for analysis. Consequently, statistical analysis was not possible for the individual patient. The shape of the wear particles was related to the distinct histological zones in articular cartilage [115]. In a third study [50], 6 sheep knee joints were worn in a joint simulator for different periods of time. The worn joints were compared to 6 unworn (control) joints to determine the changes related to wear. The surface morphology of articular cartilage was imaged using both SEM and environmental SEM (ESEM), the latter being used to avoid fixation and artifacts resulting from it. It was shown that morphological features previously attributed to articular cartilage were in fact a result of the imaging technique. Permanent wear damage to the cartilage surface was observed following even short periods of wear (e.g., 15 min/ 2,313 cycles). This damage results directly from the failure of lubrication. Analysis of the numerical characteristics indicated that the particle area decreased as wear progressed, while the boundary fractal dimension increased. In a fourth study [48], samples of synovial fluid were collected from the knee joints of 30 non-arthritic and osteoarthritic patients during diagnostic arthroscopy for the complaint of knee pain. SEM particle images were analyzed numerically. It was concluded that the particle shape can be used as an indicator of the synovial joint condition, i.e., in the diagnosis and prognosis of joint diseases. Particles obtained from osteoarthritic joints were larger in area and length than particles obtained from non-arthritic joints. In addition, particles obtained from non-arthritic joints exhibited lower boundary fractal dimensions than particles obtained from osteoarthritic joints.

8 The Use of Ferrography for Monitoring the Wear of Artificial Joints

The conventional Analytical Ferrography has also been used in several studies to monitor the wear of artificial joints [35, 37–39, 114]. Mears et al. [37, 38] analyzed specimens of synovial fluid and implants that were obtained during operation from 60 patients in whom surgical revision was necessary, after implantation periods of 2–10 years. Both Ferrography and histological analysis of the synovial tissue were employed. A bichromatic microscope and polarized light microscopy were used. It was concluded that the number and morphology of the wear particles on the ferrogram correlated with the rate and mechanisms of wear, as confirmed by examination of the implant and the adjacent synovial tissue. The metallic wear particles ranged in length from less than 0.25 μm to 1 mm for artificial joints experiencing normal or severe wear, respectively. Wear particles from stainless steel surgical tools were also found. PMMA particles from the bone cement were observed and differentiated from the UHMWPE particles. The latter, most often in the form of shredded fibers, ranged from 1 to 10 μm in diameter and up to several hundred micrometers in length [38].

Evans [35, 39] further suggested that periodic ferrographic evaluation could serve to monitor the performance of the implant and to prognosticate its future performance. Such information should prove invaluable in the design of improved prostheses with longer lifetimes and better biocompatibility.

9 The Use of Bio-ferrography for Monitoring the Wear of Natural Joints

The main weaknesses of the studies described in Sect. 7 are (1) the use of erbium cations for magnetic labeling (see mechanism of binding, advantages and drawbacks in Sect. 2.2.5); (2) the use of a conventional Analytical Ferrograph, with its limitations (see Sect. 2.2.2 and, in comparison, Sect. 2.2.4).

The only work involving Bio-Ferrography for monitoring the wear of natural joints has been carried out by Eliaz et al. [42, 55]. First, Eliaz et al. developed a method for isolating bone and cartilage wear particles suspended in the synovial fluids of human hip, knee, and ankle joints by specifically labeling collagens I and II with monoclonal mouse anti-human primary antibodies coupled to goat anti-mouse IgG MACS microbeads [42]. One of the control groups was magnetized by $ErCl_3$ solution. Synovial fluid aspirates were drawn from 14 patients, either during arthroscopy or during TJR. Several staining protocols, namely a fluorescent antibody and three dyes (Alcian Blue for cartilage, Alizarin Red and Von Kossa for bone), were found not useful for different reasons, which are explained in [42]. To eliminate charging effect of the nonconducting ferrogram inside the SEM, a thin layer of copper (Cu) was sputtered on the ferrogram. Copper coating was preferred over gold (Au) coating in order to better resolve the sulfur (S), phosphorous (P), calcium (Ca), and magnesium (Mg) peaks in EDS analysis.

Bone is composed of inorganic (mineral) and organic constituents, the former representing around 69 wt% of wet cortical bone. The primary inorganic constituent of all mammalian skeletal tissues is apatite—mainly hydroxyapatite (HAp, $Ca_5(PO_4)_3(OH)$) [116, 117]. Hence, calcium (Ca) and phosphorous (P) may be used in chemical analysis to identify bone particles ($Ca/P = 1.67$ in stoichiometric HAp, and around 1.60 in biological apatite) [35]. Due to its proteoglycan aggrecan content, cartilage normally contains appreciable amounts of sulfur (S), although these decline markedly in arthritic lesions [35, 118]. The zone of calcified cartilage forms an important interface between cartilage and bone for transmitting force, attaching cartilage to bone, and limiting diffusion from bone to the deeper layers of cartilage [118]. It contains Ca and P in addition to S [35].

The determination of the origin of wear particles was based primarily on the type of collagen they are rich in as well as on their chemical analysis, as follows:

1. Particles rich in Ca and in collagen I: subchondral bone fragments.
2. Particles rich in Ca and in collagen II: the interface between subchondral bone and calcified cartilage. It has been suggested that the biomineralization process is accompanied by a significant drop in the content of proteoglycans [118], which is expected to result in a decrease in the concentration of S compared to Ca.
3. Particles rich in S and in collagen II: articular cartilage fragments.
4. Particles rich in S and in collagen I: repaired cartilage, calcified cartilage, degenerated cartilage, synovium, or meniscus fragments (the latter is relevant to knee joints only). These particles are referred to as "cartilaginous."

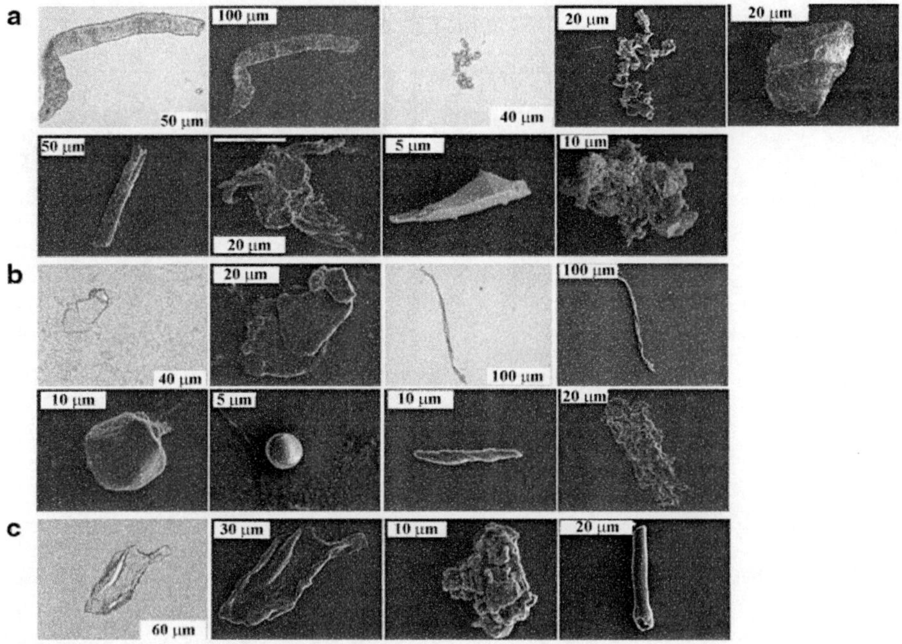

Fig. 10.14 Typical wear particles isolated from the synovial fluids of humans, as observed by SEM and light microscopy. (**a**) Bone particles rich in collagen I (9 images), (**b**) cartilage particles rich in collagen II (8 images), (**c**) cartilaginous particles rich in collagen I (4 images). Numbering from top left to bottom right: images a1, a3, b1, b3, and c1 are light microscope images. All other images are SEM images. Reprinted from [42] with permission from Elsevier

5. Particles containing neither Ca nor S (typically consisted of only Mg, P, or C): uncertain origin. Mg-rich particles may be formed during either biochemical dissolution of bone or selective precipitation from body fluids.

All particles were sorted into one of the five following shapes: chunky, platelet, rod-like, fibers, or irregular shapes. In addition, quantitative image analysis was used to characterize each particle in terms of its area A, perimeter P, length L, width W, shape factor $S.F.$, elongation Δ, and roundness R [119]. More details are provided in [42].

Figure 10.14 shows typical SEM images of bone, cartilage, and cartilaginous particles. Light microscope images of some of the particles are also shown. Bone particles were mostly angular, with either crystalline or spongy appearance; the latter may be attributed to the trabecular structure of bone. Most of them were characterized under polarized light by high optical activity, which may be related to their crystalline structure. The shape of cartilage wear particles may be related to their origin within the tissue (see Fig. 10.15) [115]. For example, it is expected that wear of the surface layer of cartilage (lamina splendens), which consists of collagen leaves with no bridging fibrils, will generate platelet particles in patients with a lower grade of OA [49]. Such particles exhibited low optical activity (bluish-silvery

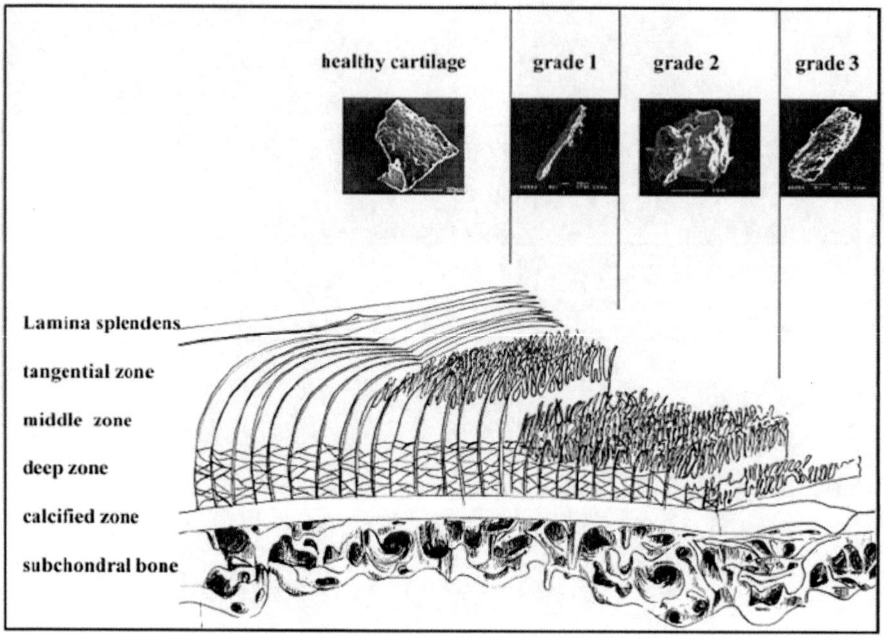

Fig. 10.15 The correlation between the shape of wear particles and their origin within the cartilage tissue (i.e., the grade of OA) [49]. By permission of Oxford University Press

color) [42]. In contrast, chunky particles will be generated in patients with a higher grade of OA from the deeper zones of cartilage, where a network of bridging fibrils exists. During advanced stages of the disease, irregular shapes are also common [49].

The Kellgren and Lawrence (K–L) scheme [120] was used to grade the severity of OA based on radiographs, from 0 (normal) to 4 (most severe). The average area, maximal area, average length, and maximal length of cartilage particles was found to increase as the level of OA increased. Ferrograms of patients that were graded as Kellgren 4 contained a significantly higher number of cartilage wear particles. For each grade of OA, there were always less bone particles than cartilage particles. The number of bone particles was significantly higher in the case of Kellgren 4 than in lower grades of OA. In the case of Kellgren 1, no bone particles were detected. The average length and average shape factor of bone particles increased with the severity of disease. Thus, the assessment of natural joint degradation by BF was usually supported by clinical diagnosis. Nevertheless, in several cases the high sensitivity of BF enabled detection of wear particles in the synovial fluid, which were not expected based on arthroscopic evaluation. This may be explained by the higher magnification at which the particles are observed on the ferrogram and by the higher sensitivity of monitoring separate wear debris compared to inspection of a bulk material.

Univariate analysis of variance (ANOVA) showed that, as the level of OA increases, bone and chunky (as well as rod-like) cartilage particles become more prominent (Fig. 10.16). It is evident that bone particles are generated at a higher

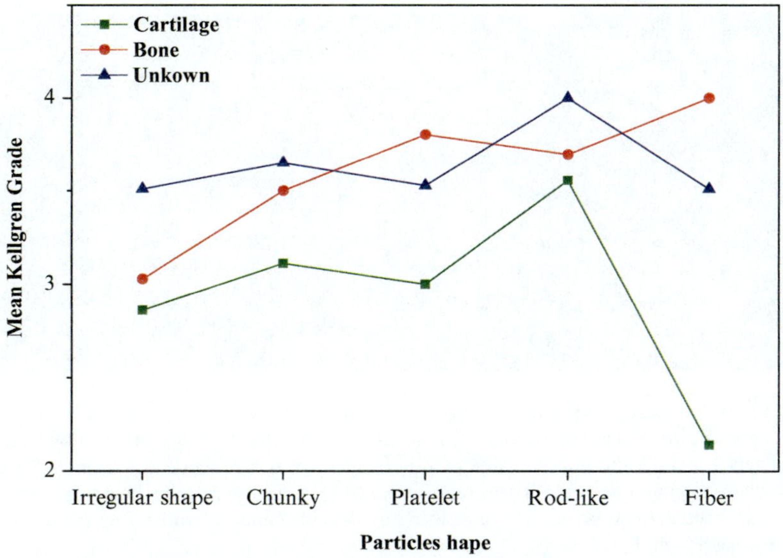

Fig. 10.16 Univariate ANOVA revealing the correlation between the chemical composition and shape of wear particles captured by Bio-Ferrography and the Kellgren and Lawrence grading of osteoarthritis. Reprinted from [42] with permission from Elsevier

grade of disease compared to cartilage particles. This is expected because at higher grades of the disease there is less cartilage present overall, hence the contacting surfaces are much less of cartilage on cartilage or even cartilage on bone, and much more of bone on bone. It is also evident that chunky cartilage particles are related to higher grades of OA compared to platelet cartilage particles, in accordance with the three-dimensional architecture of collagen [115]. The trends for particles of uncertain origin, which are more similar to those of bone particles than to those of cartilage, support the hypothesis that the source of Mg-rich particles is bone.

In a second study by Eliaz et al., Bio-Ferrography was used to capture magnetically labeled cartilage and bone debris from the synovial fluids in human knees before each of four injections of exogenous HA (Euflexxa™) [55]. The wear particles were counted and characterized microscopically and chemically. Western Ontario and McMaster Universities Osteoarthritis Index (WOMAC), Visual Analogue Scale (VAS), SF-36®, and Knee Score (KS) questionnaires indicated significant pain relief during the treatment, but suffered from inconsistency. The evaluation of joint degradation by Bio-Ferrography was found to correlate well with WOMAC assessments. Bio-Ferrography showed a reduction in the concentration of both cartilage and bone particles (Fig. 10.17), with a minimum after the third HA injection (Fig. 10.18). The HA therapy was found extremely efficient for patients with low grades of OA. It was concluded that while the exogenous HA treatment may temporarily slow the wear rate to an extent which is beyond a placebo effect, it does not prevent the joint degradation altogether. This seems to be the first ever study that demonstrates the use of Bio-Ferrography in evaluating the efficacy of a drug.

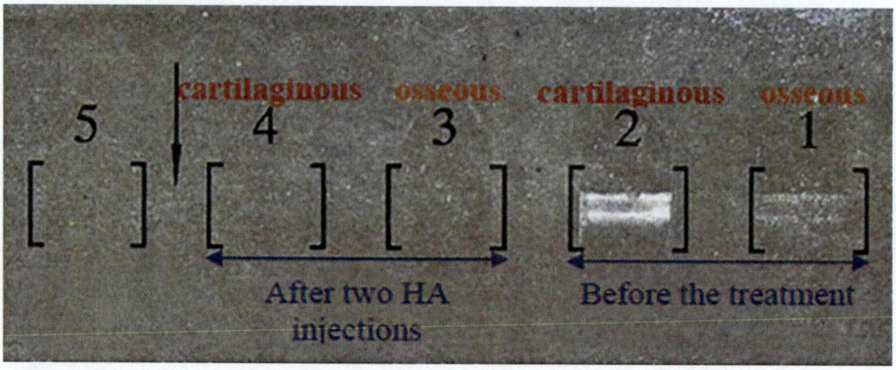

Fig. 10.17 A macroscopic view of a ferrogram of a patient, for whom the effect of HA injections on the concentration of wear particles is significant. The direction of flow is marked by an *arrow*. On channels 1 and 3 a cocktail containing anti-collagen I was run, while on channels 2 and 4 a cocktail containing anti-collagen II was run. Before the HA treatment, the rectangular deposition bands which contain the wear particles are clearly evident in channels 1 and 2. Reprinted from [55] with permission from Elsevier

10 The Use of Bio-ferrography for Monitoring the Wear of Artificial Joints

The accumulation of wear debris in the body can activate macrophages to stimulate the production of antibodies, which attack the debris, the implant, and the surrounding bone. This process may lead to the undesirable outcomes of osteolysis, aseptic loosening, and subsequent failure of the implant [98, 99]. Thus, the monitoring, isolation, and characterization of wear debris generated either *in vitro* or *in vivo* is of great importance.

Meyer et al. [36] were the first to report on the use of Bio-Ferrography for monitoring the wear of artificial implants. UHMWPE wear debris was separated from hip simulator's bovine serum lubricating fluid, and a comparison was made with conventional filtration. An enzymatic cocktail was developed and used to clean the lubricating fluid from extraneous sugars, proteins, and lipids that might interfere with the UHMWPE particle separation.

Elsner et al. [40, 41] subsequently used Bio-Ferrography and the aforementioned enzymatic cocktail in order to separate and characterize wear debris from a novel compliant *polycarbonate-urethane (PCU)* cushion form acetabular bearing for artificial hip joints during 20 million gait cycles in a wear simulator. This long-term test may be considered as equivalent to 20 years of clinical use in an average patient [121, 122], although according to few reports 10 years *in vivo* may be a better estimate [123].

It is important to simulate correctly the anatomical positioning, movement, and physiological loads that occur during normal gait, even if choosing to ignore the wider ranges of load and motion that are applied during various activities. Thus, one of the most recent advancements in hip joint simulation is the realization of the fundamental

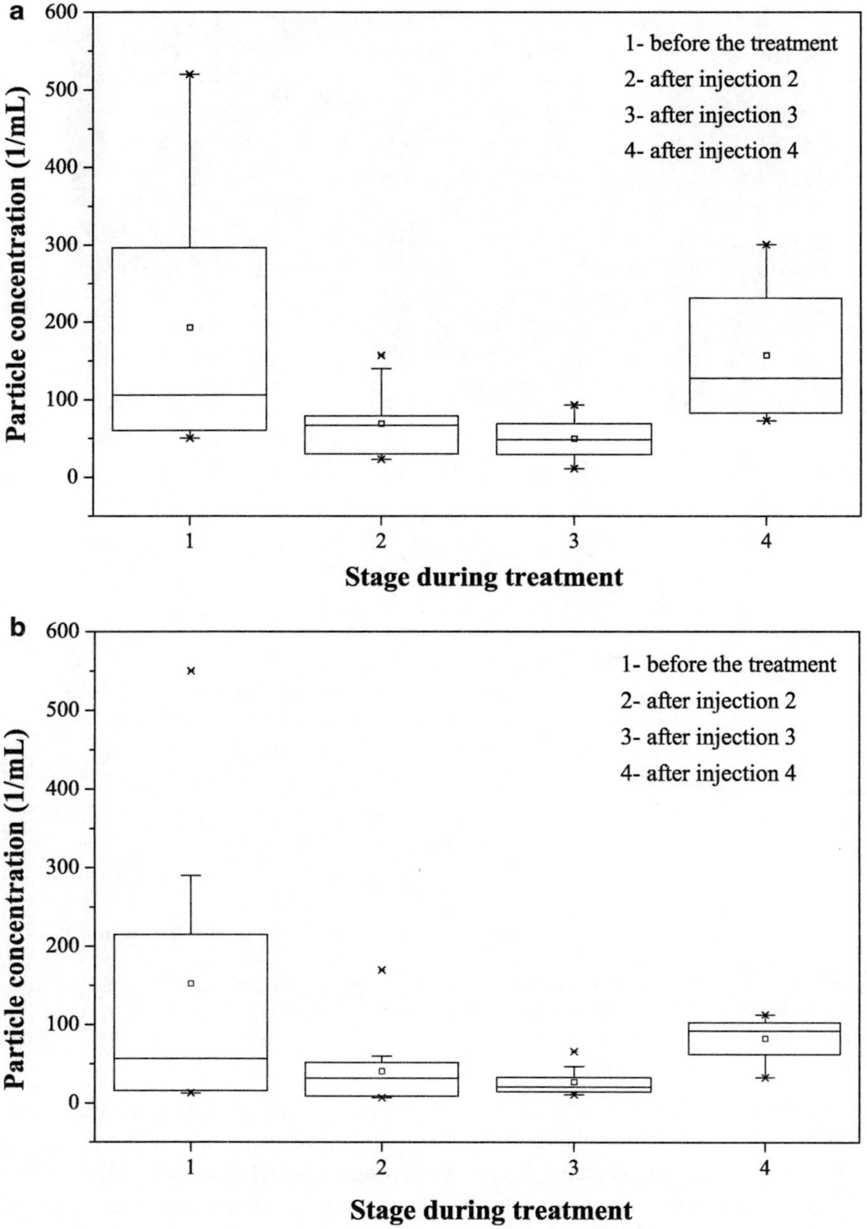

Fig. 10.18 The concentration (particles/mL) of cartilage (**a**) and osseous (**b**) wear particles as measured by Bio-Ferrography before and during the HA treatment. Reprinted from [55] with permission from Elsevier

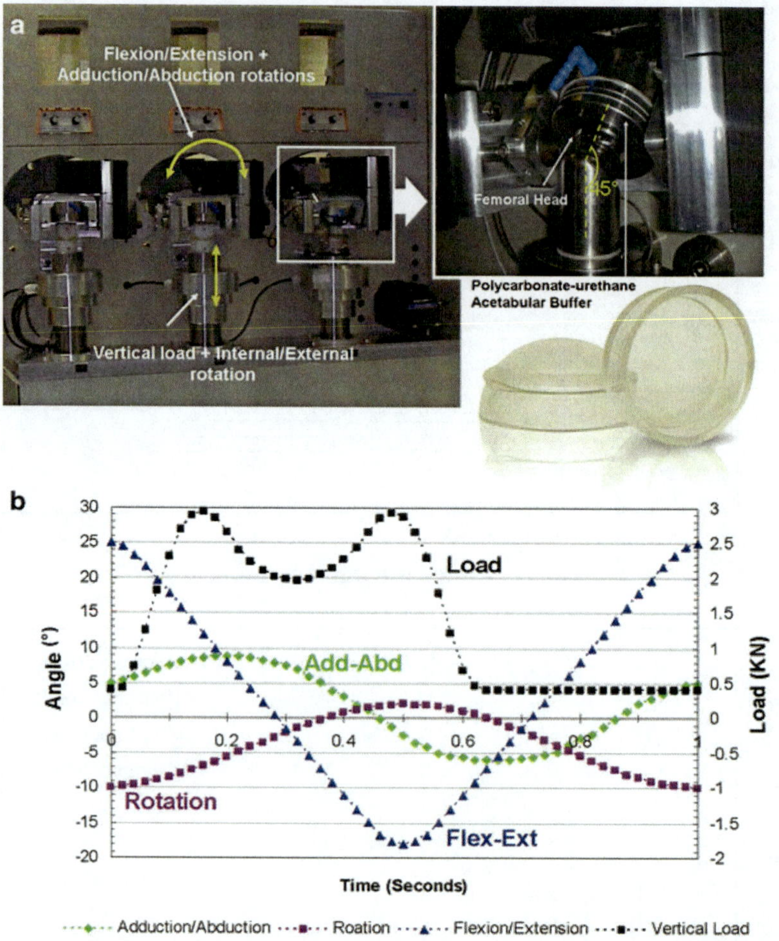

Fig. 10.19 A physiological anatomical hip joint simulator setup (**a**) and load/movement settings representative of a single load cycle, based on ISO-14242 (**b**). Reprinted from [40] with permission from Elsevier

importance of multidirectional motion to reproduce the clinically relevant wear mechanisms, wear debris, and wear magnitudes of polymer implants [124]. Figure 10.19a shows a specially designed *hip simulator* (AIC, Israel) [40]. This hip joint simulator has six articulating stations, each with three independently controlled motions: abduction–adduction, flexion–extension, and internal–external rotation, and vertical loading programmed to full gait simulation according to ISO-14242 (Fig. 10.19b). The simulator is equipped with motion and force control systems capable of generating the angular movements of the femoral component with an accuracy of $\pm 3°$ at the maxima and minima of the motion, maintaining the magnitude of the maxima and minima of the force cycle to a tolerance of $\pm 3\%$ of the maximum force value, and $\pm 1\%$ of the cycle time.

Soft bearing (compliant) materials that aim to reproduce the tribological function of the natural joint are gaining popularity as an alternative concept to conventional hard bearing materials such as polyethylene, metals, and ceramics in artificial hip and knee joints. Soft bearing materials may reduce wear by maintaining a fluid film between the articulating surfaces [40, 41]. The development of this fluid film is predominantly due to a combination of EHL and micro-elastohydrodynamic lubrication (μEHL). μEHL is a localized form of EHL, whereby pressure perturbations cause substantial flattening of asperities at the material's surfaces, increasing conformity and assisting in the maintenance of a lubricious film [40].

A commercial hip system based on a PCU acetabular liner (see Fig. 10.21a) and CoCr femoral head (Tribofit® Hip System, AIC, Memphis TN) has been available on the European market for several years. Few short-term follow-up reports on the clinical outcomes of this System have demonstrated excellent clinical results, as evident by the Harris and Oxford hip scores [125, 126].

In their first study [40], Elsner et al. measured the wear rate of the PCU bearing coupled against a CoCr femoral head over 8 million simulated gait cycles by means of gravimetry, filtration, and Bio-Ferrography. Bio-Ferrography was found to be more sensitive towards the detection of wear particles compared to the conventional filtration method and less prone to environmental fluctuations than the gravimetric method. Magnetization of the polymer particles was done via addition of $ErCl_3$. The PCU demonstrated a low particle generation rate, with the majority of wear particle mass lying above the biologically active range of polymeric particles (0.2–10 μm) that is thought to induce osteolysis [127]. Figure 10.20 shows representative SEM micrographs of PCU wear particles isolated by Bio-Ferrography (a–c) and filtration (d–f). Particles were found to have elongated (a, d), globular (b, e), or lamellar (c, f) shapes. All shapes have surface morphology similar to that of ground PCU powder (g) [40].

In their subsequent study [41], Elsner et al. extended the study to 20 million gait cycles. In addition, the wear rate at the back side of the PCU liner was monitored, and the surface of the liner was characterized microscopically, aiming to evaluate the state of the implant following service and its potential *in vivo* performance. Visual inspection of the implants did not reveal any change in their appearance (Fig. 10.21b). However, microscopic evaluation of the samples showed that the texture of the articulating surface became smoother than that prior to testing. A very low level of damage to the PCU implants' articulating surface was identified at high magnification.

Figure 10.22 shows the results in terms of cumulative mass of worn liner and particle generation rate, as measured by gravimetry, filtration, and Bio-Ferrography. Data is presented both for the articulating surface and for the backside surface. Bio-Ferrography was found to be more sensitive towards particle isolation compared to filtration, as depicted by a larger mass of wear particles captured during the initial run-in period and a slightly higher wear rate measured during the steady state phase. Backside wear was found to have a negligible contribution to the total wear mass. The PCU liner showed excellent wear characteristics in terms of its low steady-state volumetric wear rate (5.8–7.7 mm^3/million cycles) and low particles generation rate

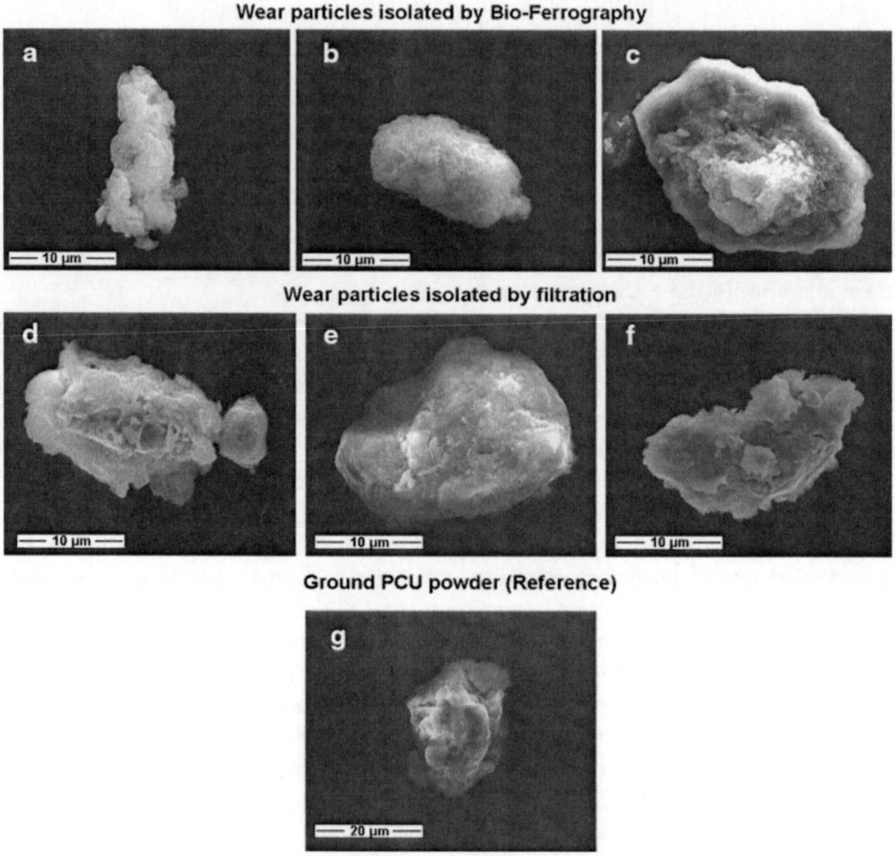

Fig. 10.20 Representative SEM micrographs of PCU wear particles isolated by Bio-Ferrography (**a–c**) and filtration (**d–f**). Particles have elongated (**a**, **d**), globular (**b**, **e**), or lamellar (**c**, **f**) shapes. All shapes have surface morphology similar to that of ground PCU powder (**g**). Reprinted from [40] with permission from Elsevier

($2–3 \times 10^6$ particles/million cycles). The former is substantially lower than that reported for the conventional UHMWPE bearings (30–100 mm^3/million cycles), and of the same order of magnitude as new-generation highly cross-linked UHMWPE (HXLPE) and MoM bearings (Fig. 10.23a). The latter is 5, 6, and 6–8 orders of magnitude lower than that of UHMWPE, HXLPE, and MoM bearings, respectively (Fig. 10.23b) [41].

It should be noted that the long-term clinical performance of a THR is affected by the volume, size, shape, and surface morphology of wear particles, as well as by the related biological response to them. Nanometer-sized metal particles, typical of MoM hard bearings, have been shown to be disseminated throughout the body—in the lymph nodes, spleen, and bone marrow [105, 106]. The high activity of metallic nano-debris results in their enhanced corrosion and release of metal ions to the joint. Certain metal ions have been shown to induce

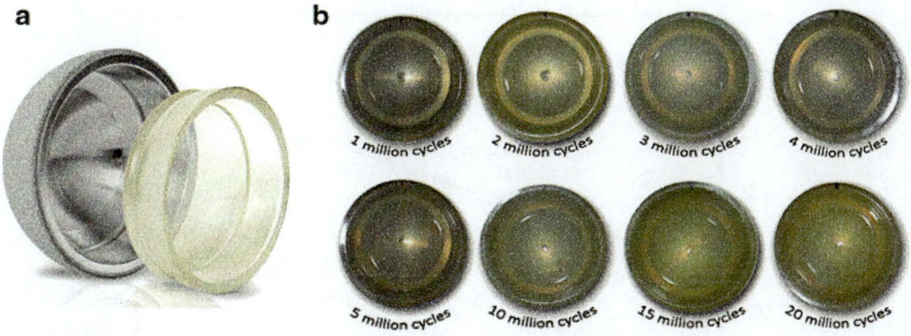

Fig. 10.21 (**a**) Side view of a PCU acetabular liner (TriboFit® Hip System, AIC, Memphis, TN). The liner component is fixed to the metal shell backing by snap-fit mechanism based on a ring shaped flange. (**b**) Photographs of the liner component, demonstrating its appearance after 1, 2, 3, 4, 5, 10, 15, and 20 million load cycles. Reprinted from [41] with permission from John Wiley and Sons. © 2011 Orthopaedic Research Society

hypersensitivity and implant intolerance reactions [107, 108]. Wear particles produced by UHMWPE bearings are larger and are therefore less likely to spread throughout the body. However, the major part of UHMWPE particles mass lies mostly in the 0.2–10 μm size range, where they have been reported to be most active, stimulating macrophages to produce high levels of the cytokine TNF-α [127]. In contrast, only 3.4% of the PCU wear particle mass was found to lie within this size range, while the majority of particle mass was associated with larger particle sizes. Another potential advantage of PCU with this respect is that, while HXLPE and UHMWPE materials share similar values of specific biological activity [128], a recent study has shown that PCU is less inflammatory to periprosthetic tissue and bone, compared to HXLPE [129].

A study of functional biological activity (FBA) of HXLPE, which took into account the wear volume and specific biological activity, has found significantly lower FBA values compared to non-cross-linked UHMWPE, due to the reduced wear volume found with the HXLPE material [128]. Thus, based on the combination of larger wear particles, less reactivity, and lower particle generation rate, it was hypothesized that the osteolytic risk of PCU is lower than that of hard bearings in THR [41].

Another long-term aspect of implants performance is their fatigue resistance. Cross-linking of UHMWPE has been shown to increase its wear resistance, but to decrease its fracture resistance. There is an increased concern for HXLPE not only due to its initially lower fatigue strength compared to un-aged UHMWPE, but also because larger head sizes with thinner PE can now be used [130]. While still uncommon, there is growing evidence on failures of HXLPE liners in clinically adverse events, specifically due to femoral neck impingement on the rim of an acetabular liner [130–132]. The behavior of the PCU liner is interesting with this respect. Its microscopic examination after 20 million cycles did not indicate on the

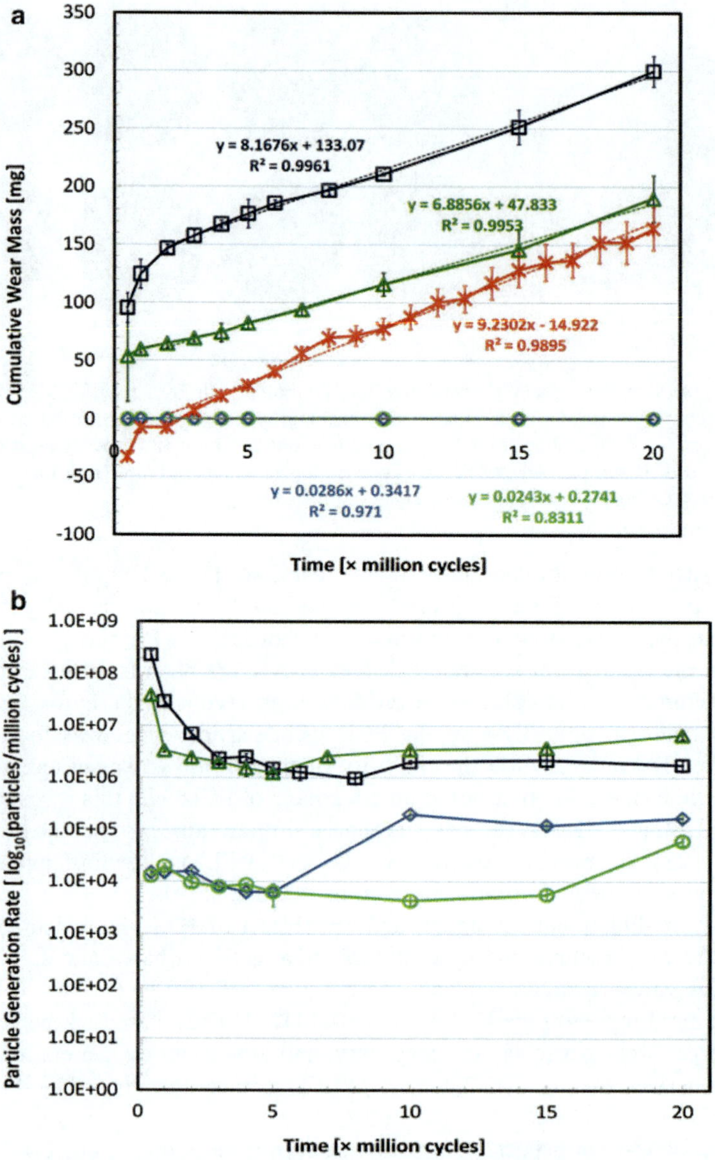

Fig. 10.22 (**a**) Cumulative mass of worn PCU during 20 M load cycles, as determined by gravimetrical measurements of the implants (*asterisks*), analysis of wear particles isolated by filtration (*open triangle*—articulating surface, *open circle*—back-side surface), or by Bio-Ferrography (*open square*—articulating surface, *open diamond*—back-side surface). (**b**) Particle generation rate during 20 million load cycles. Data is presented as mean ± standard error of the mean. Reprinted from [41] with permission from John Wiley and Sons. © 2011 Orthopaedic Research Society

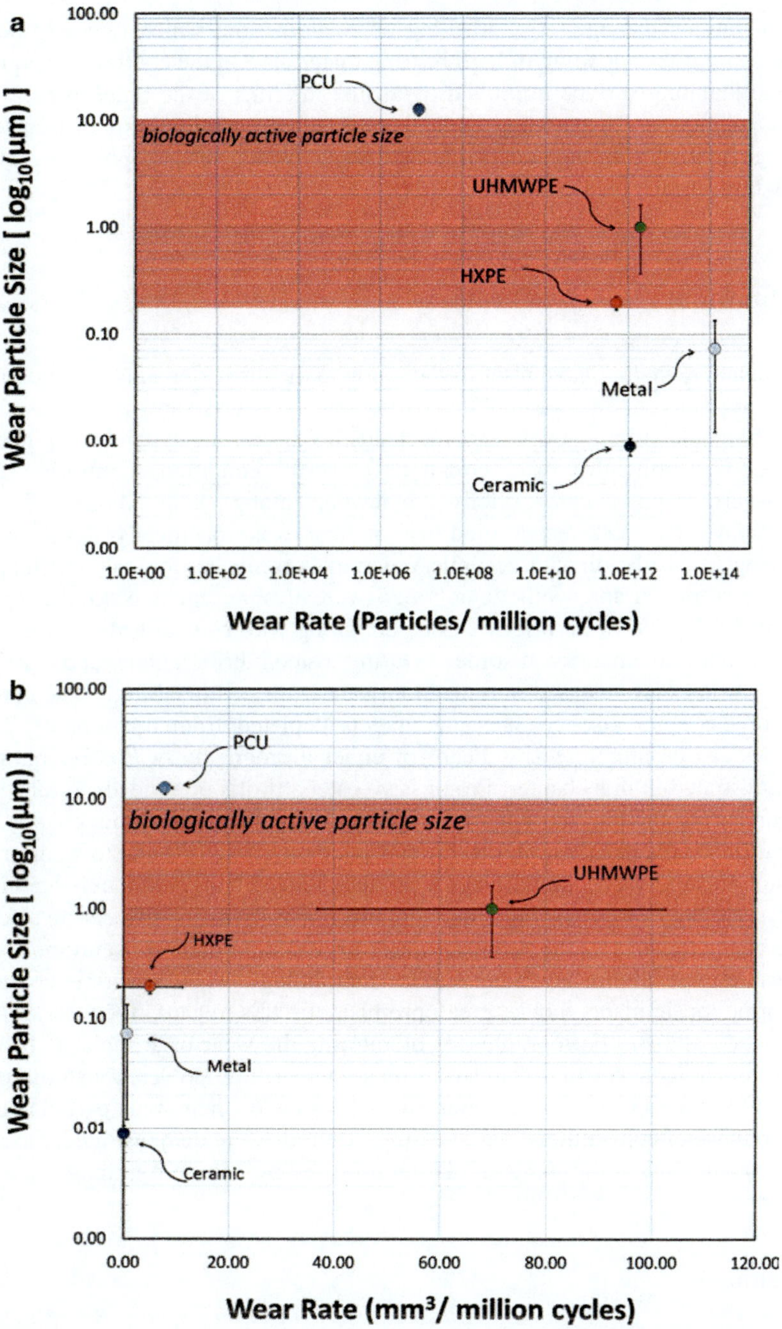

Fig. 10.23 Overview of typical wear particle sizes versus wear rates reported for various acetabular liners used in THR. Data is presented based on the volumetric wear rate (**a**) and wear particle generation rate (**b**). Reprinted from [41] with permission from John Wiley and Sons. © 2011 Orthopaedic Research Society

onset of fatigue damage, e.g., cracking or delamination, which typically occurs following long-term loading of hard bearing materials. On the contrary, the quality of the articulating surface improved over time in terms of the reduced measured surface roughness. This finding could be explained by microscopic viscoelastic rearrangement of soft segments of the polymer, which lead to polishing of the articulating surface [41].

11 Summary

In this chapter, we first gave brief introduction to the three elements of tribology—friction, lubrication, and wear. Wear particle analysis was shown to be useful in determining the origin, level, and mechanism of wear in dynamic engineering systems. The principles of a common, powerful condition monitoring technique—Ferrography—were briefly reviewed, along with those of Bio-Ferrography—its most recent modification that was specifically developed to allow magnetic isolation of target cells or tissues. Approaches for magnetic labeling of biological matter and synthetic polymers were also presented. Next, the hip and knee synovial joints were briefly discussed, along with osteoarthritis—their most common noninflammatory disorder. During osteoarthritis, bone and cartilage "wear" particles are released into the "lubricant"—the synovial fluid. The composition of the latter also changes and it gets depleted from hyaluronan. These processes can be monitored by means of either Ferrography or Bio-Ferrography, as demonstrated in this chapter. Progressive osteoarthritis often leads to total joint replacement. Thus, hip and knee arthroplasty is summarized in this chapter too. Unfortunately, the degradation of artificial joints *in vivo* (e.g., due to wear and/or corrosion) results in accumulation of wear debris in the body, ultimately leading to osteolysis, aseptic loosening, and implant failure. These adverse effects are affected by the characteristics of the wear particles, for example their concentration, size, chemical composition, shape, and surface morphology. Thus, real-time condition monitoring of implants holds great promise. In this regard, the use of both Ferrography and Bio-Ferrography for monitoring the wear of artificial joints was reviewed. The concept of soft bearing materials such as polycarbonate-urethane (PCU) was introduced, and the safer characteristics of their wear particles compared to those of conventional hard bearing materials were demonstrated, based on long-term anatomical hip joint simulation followed by Bio-Ferrography.

References

1. ASM International (2002) ASM handbook, vol 18: friction, lubrication and wear technology. ASM International, Metals Park, OH
2. ASM International (2002) ASM handbook, vol 11: failure analysis and prevention. ASM International, Metals Park, OH

3. ASTM International (1987) ASTM G40—10b: standard terminology relating to wear and erosion. ASTM International, West Conshohocken, PA
4. Levi O, Eliaz N (2009) Failure analysis and condition monitoring of an open-loop oil system using ferrography. Tribol Lett 36:17–29
5. Lakstein D, Eliaz N, Levi O, Backstein D, Kosashvili Y, Safir O, Gross AE (2011) Fracture of cementless femoral stems at the mid-stem junction in modular revision hip arthroplasty systems. J Bone Joint Surg Am 93:57–65
6. Bowen ER, Westcott VC (1976) Wear particle atlas (revised), vol. 1. Naval Air Engineering Center contract number N00156-74-C01682. Naval Air Engineering Center, Lakehurst, NJ
7. Eliaz N, Latanision RM (2007) Preventative maintenance and failure analysis of aircraft components. Corros Rev 25:107–144
8. Neale MJ (1973) Tribology handbook. Wiley, New York
9. Lockwood FE, Dalley R (1992) Lubricant analysis. In: Henry SD et al (eds) ASM handbook, vol 18. ASM International, Metals Park, OH, pp 299–312
10. Staff report (1986) Ferrography: a tool for wear-particle analysis. Hydraul Pneum November:59–61
11. Davies A (1997) Handbook of condition monitoring: techniques and methodology. Chapman & Hall, London
12. Seifert WW, Westcott VC (1972) A method for the study of wear particles in lubricating oil. Wear 21:27–42
13. Westcott VC (1977) Method and apparatus for segregating particulate matter. US Patent 4,047,814
14. Reda AA, Bowen R, Westcott VC (1975) Characteristics of particles generated at the interface between sliding steel surfaces. Wear 34:261–273
15. Scott D, Seifert WW, Westcott VC (1974) The particles of wear. Sci Am 230:88–97
16. Roylance BJ (2005) Ferrography—then and now. Tribol Int 38:857–862
17. Stecki JS, Kuhnell BT (1985) Condition monitoring of jet engines. Lubr Eng 41:485–493
18. Doyle ED, Atkin ML (1985) A review and case study of wear mechanisms and condition monitoring. In: Joint national symposium 1985: the influence of aviation on engineering and the future of aeronautics in Australia, preprints. The Institution of Engineers, Melbourne, Australia, p 23
19. Stecki JS (1980) Failure prediction using ferrographic oil analysis techniques. In: Proceedings of the conference on lubrication, friction and wear in engineering. The Institution of Engineering, Melbourne, Australia, p. 281
20. Zborowski M, Malchesky PS, Savon SR, Green R, Hall GS, Nosé Y (1991) Modification of ferrography method for analysis of lymphocytes and bacteria. Wear 142:135–149
21. Bowen ER (1979) Development of real time ferrograph, final rep. 80015. Naval Air Propulsion Center, Lakehurst, NJ
22. Merhib CP (1980) The on-line ferrograph. In: Proceedings of the conference on lubrication, friction and wear in Engineering. The Institution of Engineering, Melbourne, Australia, p 277
23. Centers PW (1983) Laboratory evaluation of the on-line ferrograph. Wear 90:1–9
24. Barwell FT (1984) The role of particle analysis—a review of ferrography. In: Dowson D, Taylor CM, Godet M, Berthe D (eds) Developments in numerical and experimental methods applied to tribology. Butterworths, London, p 3
25. Whittington HW, Flynn BW, Mills GH (1992) An on-line wear debris monitor. Meas Sci Technol 3:656–661
26. Seifert WW, Westcott VC, Desjardins JB (1998) Flow unit for ferrographic analysis. US Patent No. 5714059
27. Guilfoyle I (2000) New on the market: Bio-Ferrograph 2100. Nature 407:818
28. Desjardins JB, Seifert WW, Wenstrup RS, Westcott VC (2001) Ferrographic apparatus. US Patent No. 6303030
29. Guilfoyle I (2001) Bio-Ferrograph 2100 users manual. Belmont, MA

30. Parkansky N, Alterkop B, Boxman RL, Leitus G, Berkh O, Barkay Z, Rosenberg Yu, Eliaz N (2008) Magnetic properties of carbon nano-particles produced by a pulsed arc submerged in ethanol. Carbon 46:215–219
31. Šafařík I, Šafaříková M (1999) Use of magnetic techniques for the isolation of cells. J Chromatogr B 722:33–53
32. Evans CH, Russel AP, Westcott VC (1989) Approaches to paramagnetic separations in biology and medicine. Part Sci Technol 7:97–109
33. Evans CH (1990) Biochemistry of the lanthanides. Plenum Press, New York
34. Evans CH, Tew WP (1981) Isolation of biological materials by use of erbium (III)-induced magnetic susceptibilities. Science 213:653–654
35. Evans CH (1983) Application of ferrography to the study of wear and arthritis in human joints. Wear 90:281–292
36. Meyer DM, Tillinghast A, Hanumara NC, Franco A (2006) Bio-ferrography to capture and separate polyethylene wear debris from hip simulator fluid and compared with conventional filter method. J Tribol 128:436–441
37. Mears DC, Hanley EN, Rutkowski R, Westcott VC (1978) Ferrography: its application to the study of human joint wear. Wear 50:115–125
38. Mears DC, Hanley EN, Rutkowski R, Westcott VC (1978) Ferrographic analysis of wear particles in arthroplastic joints. J Biomed Mater Res 12:867–875
39. Evans CH, Mears DC (1981) The wear particles of synovial fluid: their ferrographic analysis and pathophysiological significance. Bulletin Prosthet Res Fall:13–26
40. Elsner JJ, Mezape Y, Hakshur K, Shemesh M, Linder-Ganz E, Shterling A, Eliaz N (2010) Wear rate evaluation of a novel polycarbonate-urethane cushion form bearing for artificial hip joints. Acta Biomater 6:4698–4707
41. Elsner JJ, Shemesh M, Mezape Y, Levenshtein M, Hakshur K, Shterling A, Linder-Ganz E, Eliaz N (2011) Long-term evaluation of a compliant cushion form acetabular bearing for hip joint replacement: a 20 million cycles wear simulation. J Orthop Res 29:1859–1866
42. Mendel K, Eliaz N, Benhar I, Hendel D, Halperin N (2010) Magnetic isolation of particles suspended in synovial fluid for diagnostics of natural joint chondropathies. Acta Biomater 6:4430–4438
43. Evans CH, Bowen ER, Bowen J, Tew WP, Westcott VC (1980) Synovial fluid analysis by ferrography. J Biochem Biophys Methods 2:11–18
44. Evans CH, Mears DC, McKnight JL (1981) A preliminary ferrographic survey of the wear particles in human synovial fluid. Arthritis Rheum 24:912–918
45. Evans CH, Mears DC, Stanitski CL (1982) Ferrographic analysis of wear in human joints: evaluation by comparison with arthroscopic examination of symptomatic knees. J Bone Joint Surg 64B:572–578
46. Hunter JA, Mills GH, Sturrock RD (1982) Ferrography: a new method for isolation of particles from biological fluids. J Clin Pathol 35:689–690
47. Mills GH, Hunter JA (1983) A preliminary use of ferrography in the study of arthritic diseases. Wear 90:107–111
48. Podsiadlo P, Kuster M, Stachowiak GW (1997) Numerical analysis of wear particles from non-arthritic and osteoarthritic human knee joints. Wear 210:318–325
49. Kuster MS, Podsiadlo P, Stachowiak GW (1998) Shape of wear particles found in human knee joints and their relationship to osteoarthritis. Br J Rheumatol 37:978–984
50. Graindorge SL, Stachowiak GW (2000) Changes occurring in the surface morphology of articular cartilage during wear. Wear 241:143–150
51. Chemicon International, Inc (1998) Introduction to Antibodies. Temecula, CA
52. Janeway CA Jr, Travers P, Walport M, Shlomchik MJ (2001) Immunobiology, 5th edn. Garland Publishing, New York
53. Goodchild S, Love T, Hopkins N, Mayers C (2005) Engineering antibodies for biosensor technologies. Adv Appl Microbiol 58C:185–226

54. Goding JE (1996) Monoclonal antibodies: principles and practice, 3rd edn. Academic, San Diego, p 141
55. Hakshur K, Benhar I, Bar-Ziv Y, Halperin N, Segal D, Eliaz N (2011) The effect of hyaluronan injections into human knees on the number of bone and cartilage wear particles captured by bio-ferrography. Acta Biomater 7:848–857
56. Zhang P, Johnson WP, Rowland R (1999) Bacterial tracking using ferrographic separation. Environ Sci Technol 33:2456–2460
57. Johnson WP, Zhang P, Fuller ME, Scheibe TD, Mailloux BJ, Onstott TC, Deflaun MF, Hubbard SS, Radtke J, Kovacik WP, Holben W (2001) Ferrographic tracking of bacterial transport in the field at the Narrow Channel Focus Area, Oyster, VA. Environ Sci Technol 35:182–191
58. Johnson WP, Zhang P, Gardner PM, Fuller ME, DeFlaun MF (2001) Evidence for detachment of indigenous bacteria from aquifer sediment in response to arrival of injected bacteria. Appl Environ Microbiol 67:4908–4913
59. Johnson WP, McIntosh OW (2003) Tracking of injected and resident (previously injected) bacterial cells in groundwater using ferrographic capture. Microbiol Methods 54:153–164
60. Graham MD, Selvin PR (1982) Separation of lanthanide binding cells. IEEE Trans Magn 18:1523–1525
61. Russell AP, Westcott VC, Demaria A, Johns M (1983) The concentration and separation of bacteria and cells by ferrography. Wear 90:159–165
62. Jones WR (1983) Wear particle analysis using the ferrograph. NASA Technical Memorandum 83422. NASA, Cleveland, OH
63. Zhang P, Johnson WP (1999) Rapid selective ferrographic enumeration of bacteria. J Magn Magn Mater 194:267–274
64. Fuller M, Mailloux B, Zhang P, Streger S, Hall J, Vainberg S, Beavis A, Johnson W, Onstott T, DeFlaun M (2001) Field-scale evaluation of CFDA/SE staining coupled with multiple detection methods for assessing the transport of bacteria in situ. FEMS Microbiol Ecol 37:55–66
65. DeFlaun MF, Fuller ME, Zhang P, Johnson WP, Mailloux BJ, Holben WE, Kovacik WP, Balkwill DL, Onstott TC (2001) Comparison of methods for monitoring bacterial transport in the subsurface. J Microbiol Methods 47:219–231
66. Zhang P, Johnson WP, Scheibe TD, Choi KH, Dobbs FC, Mailloux BJ (2001) Extended tailing of bacteria following breakthrough at the Narrow Channel Focus Area, Oyster, Virginia. Water Resour Res 37:2687–2698
67. Drake LA, Meyer AE, Forsberg RL, Baier RE, Doblin MA, Heinemann S, Johnson WP, Koch M, Rublee PA, Dobbs FC (2005) Potential invasion of microorganisms and pathogens via 'interior hull fouling': biofilms inside ballast water tanks. Biol Invasions 7:969–982
68. Fang B, Zborowski M, Moore LR (1999) Detection of rare MCF-7 breast carcinoma cells from mixture of human peripheral leukocytes by magnetic deposition analysis. Cytometry 36:294–302
69. Turpen PB (2000) Isolation of cells using bioferrography. Cytometry 42:324
70. Desjardins JB (2001) private communication. Guilfoyle Inc., Belmont, MA, April, 4
71. Ishay JS, Barkay Z, Eliaz N, Plotkin M, Volynchik S, Bergman DJ (2008) Gravity orientation in social wasp comb cells (Vespinae) and the possible role of embedded minerals. Naturwissenschaften 95:333–342
72. Zhang L, Gu FX, Chan JM, Wang AZ, Langer RS, Farokhzad OC (2008) Nanoparticles in medicine: therapeutic applications and developments. Clin Pharmacol Ther 83:761–769
73. Latorre M, Rinaldi C (2009) Applications of magnetic nanoparticles in medicine: Magnetic fluid hyperthermia. PRHSJ 28:227–238
74. Yildiz I, Shukla S, Steinmetz NF (2011) Applications of viral nanoparticles in medicine. Curr Opin Biotechnol 22:1–8
75. Thibodeau GA, Patton KT (1999) Anatomy and physiology, 4th edn. Mosby, St. Louis, MO
76. Sokoloff L (1978) The joints and synovial fluid, vol 1 & 2. Academic, London

77. Buchanan WW, Kean WF (2002) Osteoarthritis II: pathology and pathogenesis. Inflammopharmacol 10:23–52
78. Goldberg VM, Buckwalter JA (2005) Hyaluronans in the treatment of osteoarthritis of the knee: evidence for disease-modifying activity. Osteoarthritis Cartilage 13:216–224
79. Kirchner M, Marshall D (2006) A double-blind randomized controlled trial comparing alternate forms of high molecular weight hyaluronan for the treatment of osteoarthritis of the knee. Osteoarthritis Cartilage 14:154–162
80. Watterson JR, Esdaile JM (2000) Viscosupplementation: therapeutic mechanisms and clinical potential in osteoarthritis of the knee. J Am Acad Orthop Surg 8:277–284
81. The Institute for Clinical Care Council for Osteoarthritis Pain Management. http://www.clincare.org/council-for-osteoarthritis-pain-management. Accessed 14 July 2011
82. Eyre D (2002) Review: collagen of articular cartilage. Arthritis Res 4:30–35
83. Klein J (2009) Repair or replacement: a joint perspective. Science 323:47–48
84. Chen M, Briscoe WH, Armes SP, Klein J (2009) Lubrication at physiological pressures by polyzwitterionic brushes. Science 323:1698–1701
85. Klein J (2006) Molecular mechanisms of synovial joint lubrication. Proc IMechE J J Eng Tribol 220:691–710
86. Unsworth A, Dowson D, Wright V (1975) Some new evidence on human joint lubrication. Ann Rheum Dis 34:277–285
87. Scholes SC, Unsworth A, Blamey JM, Burgess IC, Jones E, Smith N (2005) Design aspects of compliant soft layer bearings for an experimental hip prosthesis. Proc Instn Mech Engrs H 219:79–87
88. Dowson D (1981) Basic tribology. In: Dowson D, Wright V (eds) An introduction to the biomechanics of joints and joint replacement. Mechanical Engineering Publications Limited, London, p 49
89. Arthritis program. Arthritis related statistics (2008) Centers for Disease Control and Prevention, Atlanta, GA. http://www.cdc.gov/arthritis/data_statistics/arthritis_related_statistics.htm. Accessed 16 July 2011
90. Wieland HA, Michaelis M, Kirschbaum BJ, Rudolphi KA (2005) Osteoarthritis—an untreatable disease. Nat Rev Drug Discov 4:331–344
91. DeGroot J, Bank RA, Tchetverikov I, Verzijl N, TeKoppele JM (2002) Molecular markers for osteoarthritis: the road ahead. Curr Opin Rheumatol 14:585–589
92. Hannan MT, Felson DT, Pincus T (2000) Analysis of the discordance between radiographic changes and knee pain in osteoarthritis of the knee. J Rheumatol 27:1513–1517
93. Moreland LW (2003) Intra-articular hyaluronan (hyaluronic acid) and hylans for the treatment of osteoarthritis: mechanisms of action. Arthritis Res Ther 5:54–67
94. Herrero-Beaumont GR, Sánchez-Pernaute O, Acebes C, Palacios I, Mas S, Rodriguez I, Egido J, Vivanco F (2001) Cartilage and bone biological markers in the synovial fluid of osteoarthritic patients after hyaluronan injections in the knee. Clin Chim Acta 308:107–115
95. Tortora GJ, Derrickson BH (2009) Principles of anatomy and physiology, vol 1, 12th edn. Wiley, Hoboken, NJ
96. Robb WJ, Healy WL, Berry DJ, Hozack WJ, Kyle RF, Lewallen DG, Trousdale RT, Jiranek WA, Iorio R (2008) Orthopaedic surgeon workforce assessment for THA/TKA in the US—preparing for an epidemic. The 75th Annual Meeting of the American Academy of Orthopaedic Surgeons (AAOS), San Francisco, CA
97. Georgiades G, Babis GC, Hartofilakidis G (2009) Charnley low-friction arthroplasty in young patients with osteoarthritis: outcomes at a minimum of twenty-two years. J Bone Joint Surg Am 91:2846–2851
98. Clohisy JC, Calvert G, Tull F, McDonald D, Maloney WJ (2004) Reasons for revision hip surgery: a retrospective review. Clin Orthop Relat Res 429:188–192
99. Ulrich SD, Seyler TM, Bennett D, Delanois RE, Saleh KJ, Thongtrangan I, Kuskowski M, Cheng EY, Sharkey PF, Parvizi J, Stiehl JB, Mont MA (2008) Total hip arthroplasties: what are the reasons for revision? Int Orthop 32:597–604

100. Howie DW, Vernon-Roberts B, Oakeshott R, Manthey B (1988) A rat model of resorption of bone at the cement-bone interface in the presence of polyethylene wear particles. J Bone Joint Surg Am 70:257–263
101. Willert HG, Semlitch M (1977) Reactions of the articular capsule to wear products of artificial joint prosthesis. J Biomed Mater Res 11:157–164
102. Jin ZM, Dowson D, Fisher J (1997) Analysis of fluid film lubrication in artificial hip joint replacements with surfaces of high elastic modulus. Proc Instn Mech Engrs H 211:247–256
103. Tipper JL, Firkins PJ, Besong AA, Barbour PSM, Nevelos J, Stone MH, Ingham E, Fisher J (2001) Characterization of wear debris from UHMWPE on zirconia ceramic, metal-on-metal and alumina ceramic-on-ceramic hip prostheses generated in a physiological anatomical hip joint simulator. Wear 250:120–128
104. Doorn PF, Campbell PA, Worrall J, Benya PD, McKellop HA, Amstutz HC (1998) Metal wear particle characterization from metal on metal total hip replacements: transmission electron microscopy study of periprosthetic tissues and isolated particles. J Biomed Mater Res 42:103–111
105. Case CP, Langkamer VG, James C, Palmer MR, Kemp AJ, Heap PF, Solomon L (1994) Widespread dissemination of metal debris from implants. J Bone Joint Surg Br 76:701–712
106. Langkamer VG, Case CP, Heap P, Taylor A, Collins C, Pearse M, Solomon L (1992) Systemic distribution of wear debris after hip replacement. A cause for concern? J Bone Joint Surg Br 74:831–839
107. Willert HG, Buchhorn GH, Fayyazi A, Flury R, Winder M, Köster G, Lohmann CH (2005) Metal-on-metal bearings and hypersensitivity in patients with artificial joints. A clinical and histomorphological study. J Bone Joint Surg Am 87:28–36
108. Korovessis P, Petsinis G, Repanti M, Repantis T (2006) Metallosis after contemporary metal-on-metal total hip arthroplasty. Five to nine year follow-up. J Bone Joint Surg Am 88:1183–1191
109. Khan I, Smith N, Jones E, Finch DS, Cameron RE (2005) Analysis and evaluation of a biomedical polycarbonate urethane tested in an in vitro study and an ovine arthroplasty model. Part I: materials selection and evaluation. Biomaterials 26:621–631
110. Khan I, Smith N, Jones E, Finch DS, Cameron RE (2005) Analysis and evaluation of a biomedical polycarbonate urethane tested in an in vitro study and an ovine arthroplasty model. Part II: in vivo investigation. Biomaterials 26:633–643
111. Scholes SC, Burgess IC, Marsden HR, Unsworth A, Jones E, Smith N (2006) Compliant layer acetabular cups: friction testing of a range of materials and designs for a new generation of prosthesis that mimics the natural joint. Proc Instn Mech Engrs H 220:583–596
112. Stachowiak GW, Podsiadlo P (1997) Analysis of wear particle boundaries found in sheep knee joints during in vitro wear tests without muscle compensation. J Biomech 30:415–419
113. Kirk TB, Stachowiak GW (1994) Computer wear particle analysis for the study of arthritic disorders. In: Engin AE, Ertas A (eds) Proceedings of the 1994 engineering systems design and analysis conference, vol 4. The American Society of Mechanical Engineers, New York, p 69
114. Evans CH (1985) Ferrographic and biochemical studies on the wear particles of human and prosthetic joints. University of Pittsburgh, PA, Kappa Delta Award
115. Jeffery AK, Blunn GW, Archer CW, Bentley G (1991) Three-dimensional collagen architecture in bovine articular cartilage. J Bone Joint Surg Br 73:795–801
116. Hughes S, Sweetnam R (1980) The basis and practice of orthopaedics. Heinemann, London, UK
117. Eliaz N (2008) Electrocrystallization of calcium phosphates. Isr J Chem 48:159–168
118. Oegema TR, Carpenter RJ, Hofmeister F, Thompson RC (1997) The interaction of the zone of calcified cartilage and subchondral bone in osteoarthritis. Microsc Res Tech 37:324–332
119. Russ JC (1990) Computer-assisted microscopy: the measurement and analysis of images. Plenum Press, New York
120. Kellgren JH, Lawrence JS (1957) Radiological assessment of osteoarthritis. Ann Rheum Dis 16:494–501

121. Wallbridge N, Dowson D (1982) The walking activity of patients with artificial hip joints. Eng Med 11:95–96
122. Sauer WL, Anthony ME (1998) Predicting the clinical wear performance of orthopaedic bearing surfaces. In: Jacobs JJ, Craig TL (eds) ASTM STP 1346: alternative bearing surfaces in total joint replacement. American Society for Testing and Materials, West Conshohoken, p 1
123. Silva M, Shepherd EF, Jackson WO, Dorey FJ, Schmalzried TP (2002) Average patient walking activity approaches 2 million cycles per year. J Arthrop 17:693–697
124. Essner A, Wang A (2009) Tribological assessment of UHMWPE in the hip. In: Kurtz SM (ed) UHMWPE biomaterials handbook, 2nd edn. Academic, Boston, p 369
125. Siebert W, Mai S, Moroni A, Chiarello E, Giannini S (2009) A two-year prospective and retrospective multi-center study of the TriboFit hip system. J Long Term Eff Med Implants 19:149–155
126. Moroni A, Nocco E, Hoque M, Diremigio E, Buffoli D, Cantù F, Catalani S, Apostoli P (2012) Cushion bearings versus large diameter head metal-on-metal bearings in total hip arthroplasty. Arch Orthop Trauma Surg 132:123–129
127. Green TR, Fisher J, Stone M, Wroblewski BM, Ingham E (1998) Polyethylene particles of a 'critical size' are necessary for the induction of cytokines by macrophages in vitro. Biomaterials 19:2297–2302
128. Galvin AL, Tipper JL, Jennings LM, Stone MH, Jin ZM, Ingham E, Fisher I (2007) Wear and biological activity of highly crosslinked polyethylene in the hip under low serum protein concentrations. Proc Inst Mech Eng Part H J Eng Med 221:1–10
129. Smith RA, Maghsoodpour A, Hallab NJ (2010) In vivo response to cross-linked polyethylene and polycarbonate-urethane particles. J Biomed Mater Res A 93:227–234
130. Oral E, Muratoglu OK (2009) Highly crosslinked UHMWPE doped with vitamin E. In: Kurtz SM (ed) UHMWPE biomaterials handbook, 2nd edn. Academic, Boston, p 221
131. Schroder DT, Kelly NH, Wright TM, Parks ML (2011) Retrieved highly crosslinked UHMWPE acetabular liners have similar wear damage as conventional UHMWPE. Clin Orthop Relat Res 469:387–394
132. Furmanski J, Anderson M, Bal S, Greenwald AS, Halley D, Penenberg B, Ries M, Pruitt L (2009) Clinical fracture of cross-linked UHMWPE acetabular liners. Biomaterials 30:5572–5582

Biography

Noam Eliaz is an Associate Professor at Tel-Aviv University, Israel, where he serves as the Head of the Biomaterials and Corrosion Laboratory. He also serves as a Chief Editor of the journal *Corrosion Reviews* (jointly with Ron Latanision). He received his B.Sc. and Ph.D. (direct track) in Materials Engineering, both *cum laude*, from Ben-Gurion University. Next, he became the first ever materials scientist to receive, simultaneously, a Fulbright postdoctoral award and a Rothschild postdoctoral fellowship and worked for two years in the H.H. Uhlig Corrosion Laboratory at M.I.T. To-date, he has contributed more than 230 journal and conference publications, including 31 plenary and invited talks, as well as 5 book chapters. In addition to editing this *Degradation of Implant Materials* book, he has edited a double volume entitled Applications of Electrochemistry and Nanotechnology in Biology and Medicine for the reputed book series *Modern Aspects of Electrochemistry* (Springer). He has garnered numerous accolades, including the T.P. Hoar Award for the best paper published in *Corrosion Science* during 2001 (on corrosion of Ti–Ag-based alloys processed by three-dimensional printing for biomedical applications), and NACE International's Herbert H. Uhlig Award (2010) and Fellow Award (2012). His main research interests include environment-induced degradation of materials, failure analysis, Bio-Ferrography, biomaterials (with focus on electrocrystallization of hydroxyapatite and other calcium phosphates), and electrochemical processing (namely, electrodeposition, electroless deposition, and electropolishing) of materials.

Keren Hakshur received her B.Sc. in Chemical & Biotechnology Engineering *cum laude* from Ariel University Center of Samaria, Israel, and her M.Sc. degree in Materials Science and Engineering *cum laude* from Tel-Aviv University (TAU), Israel. Her M.Sc. thesis was conducted under the supervision of Prof. Noam Eliaz on the application of Bio-Ferrography to evaluate the effect of hyaluronan injections into human knees. She is currently pursuing Ph.D. in Physical Electronics at the School of Electrical Engineering, TAU. Her current research interests include optical and biomedical sensors.

Chapter 11
Fatigue Failure of Materials for Medical Devices

M.S.K. Chong, Y.E. Teo, and S.H. Teoh

Abstract Medical devices such as joint prostheses are subjected to high stresses and high cycle loading. Coupled with the aggressive body environment, fatigue failure is common. A fatigue wear process causes the generation of wear debris, which invokes acute host–tissue reactions. This chapter reviews some fundamental concepts from mechanics to the methods of evaluation for biomaterials which would include wear debris morphology characterisation so as to understand the host–tissue reaction to wear debris. Case studies on hip and knee prostheses and dental restoratives are also presented. The development of fatigue fracture resistant and wear resistant biomaterials geared towards biocomposite systems with different phases to cope with the conflicting properties of fatigue fracture resistance and hard, but brittle, phases required for wear resistance and a good lubrication phase seems to provide some future direction. The ability to engineer biomaterials that have the capability to trap/isolate wear debris and promote easy removal of such wear debris remains a challenge.

1 Introduction

Medical implants are placed in harsh environments, where they are subject to severe chemical stresses. On top of that, the same implants are subject to repeated and fluctuating mechanical loads. It is a direct consequence of these cyclic loads that the implants fail as a result of what is termed fatigue failure [1]. This chapter is aimed at discussing the role of fatigue failure in medical implants. The first part will cover basic fundamentals and terminologies in fatigue mechanics. Subsequently, case studies will be covered in the second part to illustrate the role of fatigue in the failure of some common medical implants.

M.S.K. Chong • Y.E. Teo • S.H. Teoh (✉)
Division of Bioengineering, School of Chemical and Biomedical Engineering, Nanyang Technological University, 70 Nanyang Drive, Singapore
e-mail: teohsh@ntu.edu.sg

N. Eliaz (ed.), *Degradation of Implant Materials*, DOI 10.1007/978-1-4614-3942-4_11,

2 Fundamental Concepts

This section is primarily aimed at helping students who are unfamiliar with mechanics and will cover the fundamental concepts involved in fatigue and fracture mechanics.

2.1 Mechanics of Materials

Mechanics refers to the study of objects under forces. When a force is applied on a body, it is referred to as a "load". Such loads may be applied perpendicularly (often referred to as "normal") to a surface. Such normal loads may be acting to compress ("compressive loads") or stretch ("tensile loads") the object. Because force is a vector quantity, these forces are assigned directions. Accordingly, tensile forces are typically assigned positive values, and compressive forces, negative. Forces may also be applied in a direction parallel to a surface, whereby it is referred to as a "shear" force. A typical example of this is friction. In real situations, objects are likely to be subject to a combination of both. These are summarised in Fig. 11.1.

Typically, however, forces rarely act on a single point and are usually distributed over an area. Pressure is the measure of force per unit area applied on a surface. Similarly, in mechanics, stress is a measure of force per unit area acting on the surface of the object or on a plane within the object. As with forces, stresses are vector quantities and include normal and shear stresses. Many textbooks have been written on this topic; one of them which is of particular interest is "The Science and Design of Engineering Materials" by Schaffer et al. (2006).

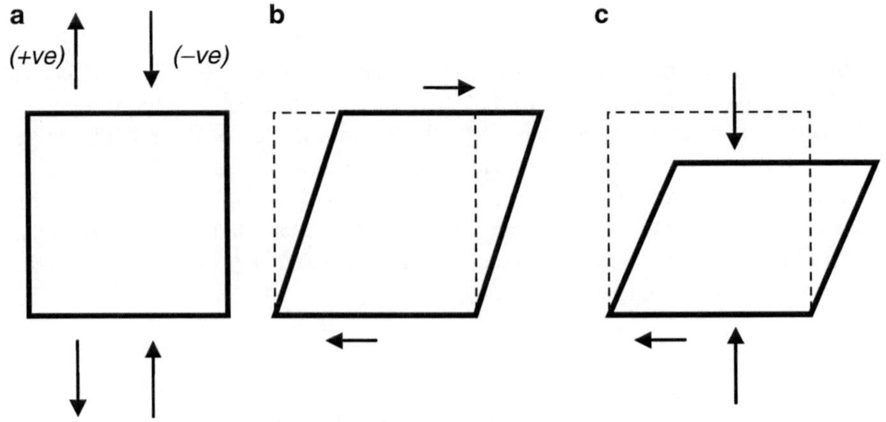

Fig. 11.1 Schematic diagram to demonstrate (**a**) normal forces, (**b**) shear forces and (**c**) compressive and shear forces acting on a body

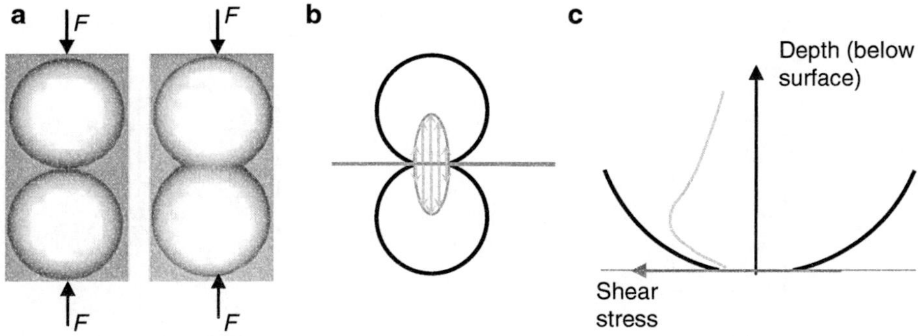

Fig. 11.2 (**a**) Deformation of bodies under normal load, (**b**) distribution of normal stress at the contact area, and (**c**) distribution of shear stress through the body

2.2 *Contact Mechanics*

Contact mechanics refers to the study of interactions of two bodies coming in contact with each other. Deformations of the bodies will typically take place, leading to alterations in stress distribution patterns, which may sometimes be counter-intuitive and unexpected. For example, consider the case of contact between the two spheres shown in Fig. 11.2a. Under a normal load, deformation takes place and the initial point contact becomes an area contact.

It can be shown that the stress distribution across the contact area is elliptical (as shown in Fig. 11.2b), with a maximum pressure, $p_{\max}$, of

$$p_{\max} = \frac{3F}{2\pi a^2} \tag{11.1}$$

occurring at the centre of the contact area, where F refers to the normal load, and a refers to the diameter of contact area.

It can further be shown that the shear stress (τ) at a depth of z below the surface is derived by the following equation:

$$\tau = p_{\max}\left[\left[1 - \frac{z}{a}\tan^{-1}\left(\frac{a}{z}\right)\right](1+\nu) - \frac{1}{2\left[1+(z/a)^2\right]}\right] - \frac{p_{\max}}{1+(z/a)^2}. \tag{11.2}$$

It can be seen from the shear stress distribution under the surface (Fig. 11.2c) that the maximum shear stress is experienced not at the surface, but at a depth under the surface. This may result in the generation of subsurface cracks, which are very difficult to detect.

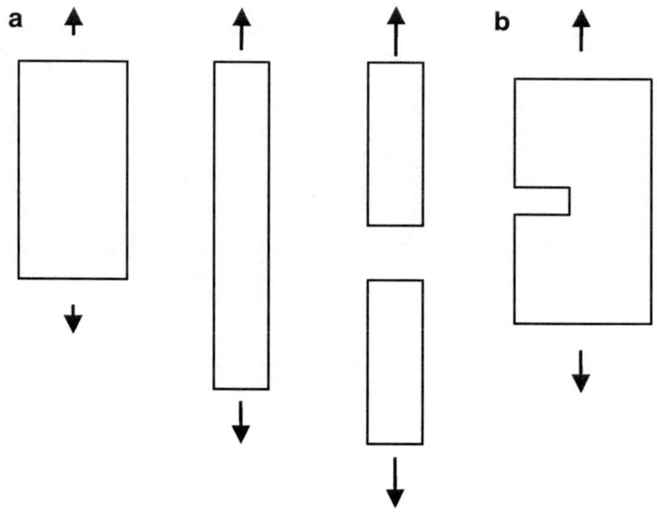

Fig. 11.3 Objects under tension (**a**) objects stretch and elongate under tensile stress and eventually fracture, (**b**) definition of crack in fracture terminology

2.3 Fracture Mechanics

Consider an object, such as a thin rubber band, being stretched. Typically, objects placed under such tensile stress will first stretch and elongate, as seen in Fig. 11.1. As the load is increased, the object will continue stretching, to a point at which it breaks. The amount of stress that it can take just at the point when it breaks is referred to as the tensile strength. On a molecular level, it can be envisioned as the point at which all the molecular forces holding the object intact on that plane are simultaneously broken, resulting in separation into two bodies, or "fracture". Thus, in a perfect object, the theoretical strength of a material should correspond to the sum of all molecular bonds within the plane of fracture. In real life, however, materials typically fail at conditions below this value. This is partly due to molecular defects and also because of the presence of cracks on the surface.

Figure 11.3b shows an object placed under tension with a crack. Briefly, the presence of pre-existing cracks results in alterations in the stress distribution profiles and concentrates stresses at the crack tip. The crack will then progress and extend (crack propagation), leading ultimately to failure. It was deduced experimentally that the tensile strength varies inversely with the square root of the crack size, i.e. $\sigma_f \sqrt{a} = C$, where σ_f refers to the tensile strength and "a" refers to the crack size. This is commonly referred to as the Griffith Criterion. C is a constant, which may be theoretically derived from mechanical properties, such as stiffness and free surface energy. The Griffith Criterion can also be rearranged to $a = (C/\sigma_f)^2$, which now suggests that, for a fixed value of applied stress, there will exist a critical size for cracks, beyond which the material will fail under this

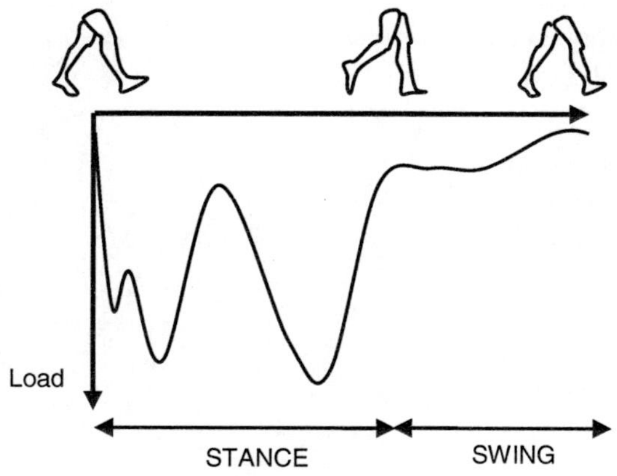

Fig. 11.4 Load experienced by the hip joint during normal gait

load. Hence, this can be used as a design criterion in which existing cracks in the material cannot exceed this critical size.

To facilitate materials selection and design, the parameter "fracture toughness" (denoted by K_{IC}) has been developed that is related to the constant C in Griffith's relation. K_{IC} is a material property indicative of the ability of a material containing a crack to resist fractures and may be obtained from materials handbooks or data sheets. The derivation of K_{IC} requires in-depth discussion of thermodynamic interactions in the crack zone and is beyond the scope of this chapter; the reader is directed to some established books written in this field [2].

2.4 Fundamentals of Fatigue

Medical prostheses are commonly placed in positions of motion, experiencing repetitive, alternating periods of loading and unloading. Examples include opening and closing of heart valve, gait-associated loading and unloading on hip joints between stance and swing phases. Materials placed under such cyclic loads may fail at stress levels significantly lower than the rated tensile strength. The process of progressive degeneration of a material by cyclic loading (and consequent failure) is referred to as fatigue. A typical example of this is seen in the repeated bending and unbending of a metal paper clip, resulting in eventual fracture. Figure 11.4 shows the fluctuating forces experienced by the hip joint during different phases of the gait cycle.

Shifting our focus now to the implant/host interface, we can see how the surface substructure will influence the fatigue failure of biomaterials [3]. Figure 11.5 illustrates the surface of a metallic implant placed in a physiological environment, under an applied load from an external body. It can be seen that there are three distinguishable layers at the surface of the metal as a result of the load: (1) the molecular adsorbed layer, (2) the passive oxide film and (3) the deformed layer. The interaction of these layers with the physiological environment during the

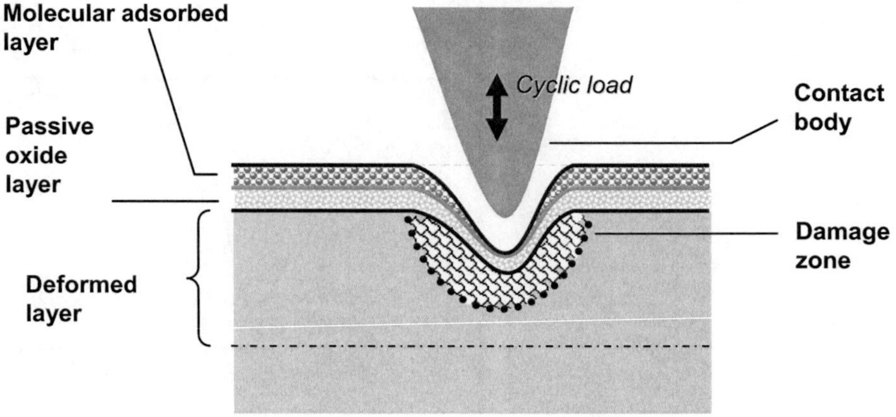

Fig. 11.5 Interactions at the surface of a deformed metallic material

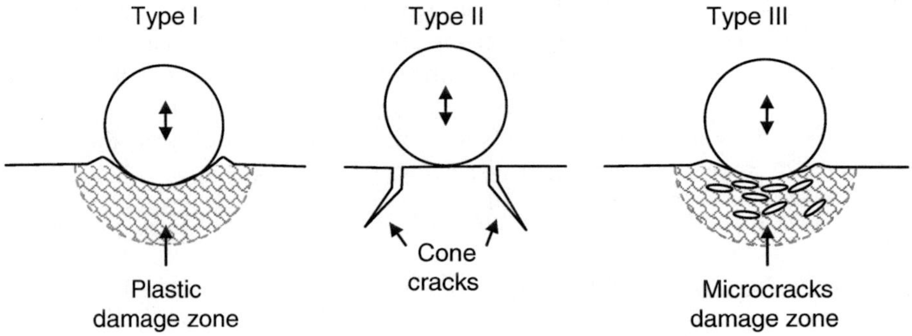

Fig. 11.6 Schematic illustration of surface and subsurface damage under cyclic loading from a spherical contact body

fatigue/wear process will have crucial influence on the outcome and service life of the medical device.

Figure 11.6 shows schematically the types of surface and subsurface damage exhibited by different materials under a spherical indentor. Type I behaviour is typical of materials having high fracture toughness and ductility, such as metals. The plastic zone (where permanent deformation has taken place) with build-up of material around the indentor is obvious. Type II behaviour shows a damage zone beneath the indentor with little deformation and cone cracks forming near the perimeter of the indentor. This is commonly encountered in brittle materials with high yield strength but low fracture toughness, such as ceramics. Type III behaviour is a mix of both and is typical of materials with moderate toughness and yield strength. Micro-cracking is often observed in the damage zone. Numerous dental restorative materials exhibit this type of behaviour under repeated impact loading.

2.4.1 Fatigue Fracture

Fatigue fracture is the breaking up of a body into two or more components as a result of an external cyclic force. Typically, the forces leading to this type of fracture are well below that of the material's tensile strength. Fatigue fractures are progressive, often starting off as small cracks which grow to eventually fracture the component.

What Are the Stages of Fatigue Fracture?

The three stages leading to fatigue fracture are initiation, propagation and final rupture.

During crack initiation, submicroscopic cracks are formed at one or more locations as a result of cyclic loading. The cyclic loads lead to distortions of the arrangement of atoms, resulting in defects on an atomic level (commonly referred to as dislocations). These defects accumulate under repeated loading, resulting in the formation of a tiny crack. Although prevention of crack initiation appears an obvious way of preventing fatigue failure from even beginning, identifying and arresting the fatigue fracture at this stage is highly challenging. The presence of debris and inclusions within the material may also act as a stress concentrator, leading to higher possibility of crack initiation, and consequently, the choice of materials with high mechanical strength and good manufacturing practices may reduce the initiation of the fatigue process.

The propagation stage involves the growth of the microcrack into the material. It is at this stage that defects may be identified, typically during routine checks and non-destructive tests. It is thus crucial to understand the progression of the crack across the cross-section. Crack propagation is highly dependent on crack geometry. Sharp, notch-shaped cracks serve to concentrate stresses at the crack tip, leading to the generation of a deformation zone ahead of the crack tip. If the energy accumulated is sufficiently high, the crack will extend further and advance across the plastic zone. Correspondingly, the crack progresses in discrete steps with each cycle of forces applied, creating "striation" marks on the crack surface of the component. Striation marks are observable microscopic features that form as a result of the advancement of the crack front. Thus, striation marks are a distinguishing hallmark of fatigue fracture and often used during failure analysis and identification.

As the fatigue crack deepens and progresses, the cross-sectional area of the component is decreased. This continues to a point whereby the crack reaches a critical size, beyond which the component fails or ruptures upon the application of a single subsequent load cycle (as similarly discussed in Sect. 2.4.1). Although progression to rupture indicates that the object has failed, the failure analyst should pay close attention to details of the final rupture area, such as geometry and location, as these provide valuable information on the actual stresses that the component was subject to during its service life.

Factors Affecting Fatigue Life

Magnitude of Stress

From the discussion of fatigue mechanism earlier, it can be seen that fracture mechanics is highly dependent on the magnitude of the applied stress:

1. During crack initiation, sufficient stress must be applied to induce atomic dislocations.
2. During crack propagation, the applied stress must be high enough to drive the progression of the crack tip
3. The magnitude of the applied stress determines the size of the critical crack at final rupture, as defined using Griffith's criterion.

It thus follows that, for an applied cyclic load of a known magnitude, there will be a corresponding number of cycles that the material can go through, before it fails by fatigue. It is also conceivable, especially for fracture-resistant material, that there will exist a minimum value of stress, below which the material will not fail, regardless of the number of cycles it undergoes. This is referred to as the "endurance limit" of the material. This relation may be experimentally derived and plotted in a stress versus number of cycles to failure (S–N) curve. This helps to predict the safe life of the component or facilitate the choice of a fatigue-tolerant material.

Design Factors

Considering the central role of cracks in fatigue, much research has focused on design to prevent crack formation. Design features such as notches, grooves, holes and geometrical discontinuities can act as initiation sites for cracks and also stress raisers, and are generally avoided by using smooth, rounded edges. Similarly, holes or slots with sharp corners are always avoided in favour of tapered edges. In addition, surface finish is critical to prevent the formation of crack initiators on the material surface. In materials selection, ductile materials which have a lower "notch sensitivity" are usually chosen.

Surface Treatments

Cracks are often initiated at the surface through the introduction of small scratches during fabrication and machining. Surface treatments and modifications can be carried out to reduce the effect of surface features and, in turn, increase the components' fatigue life.

A particular effective strategy is to impose residual compressive stress within a thin outer surface layer. Residual stresses refer to stresses that remain after the original source of stress has been removed. For example, in a process known as shot peening, a metallic surface is impacted by small and hard particles propelled at high speed. The surface will deform and spread in response to impact, while the bulk material remains unchanged. The deformation is permanent and will remain even after the particles are removed. Consequently, these deformations place the surface in compression and residual compressive stresses remain. Such compressive stresses at the surface serve to nullify tensile stresses that may result in crack initiation.

Alternative surface modifications include case hardening, where steel surfaces are exposed to carbonaceous or nitrogenous atmosphere at high temperature, to facilitate the diffusion of carbon or nitrogen atoms into the surface molecular layers, leading to chemical and physical changes in the "case". A direct relation is often seen between the outer case hardness and fatigue life.

Environmental Effects

The service environment that the material is subjected to is crucial in engineering design, particularly for devices to be implanted into the human body due to the highly adverse conditions *in vivo*. For example, fluctuating temperatures may lead to fluctuating stresses arising from thermal expansion and contraction. A design consideration is thus to incorporate room for unhindered expansion and contraction at the implant site.

More crucially, medical implants are particularly susceptible to corrosion fatigue due to the harsh chemical environment. Corrosion fatigue arises as a result of cyclic stresses compounded with chemical attack. For example, corrosion of a metallic surface may generate small notches (known as "pits") which can act as stress raisers or crack initiation sites. Corrosion also serves to exacerbate fatigue by accelerating the crack propagation process. To reduce and prevent the occurrence of environmental influenced fatigue, adequate material selection or introduction of protective coatings may be necessary.

2.4.2 Fatigue Wear

A discussion on cyclic loads is incomplete without the discussion of wear. It is useful at this point to distinguish fatigue wear from the more commonly known abrasive wear (Fig. 11.7). Abrasive wear refers to the removal of material by the cutting actions of hard particles on a softer surface. This can be caused by tiny, sharp projections (known as asperities) found on the hard surface or through third-body wear. Third-body wear is a specific form of abrasive wear in which hard particles are trapped between two articulating surfaces, and similarly resulting in the removal of material.

Fatigue wear refers specifically to the removal of small fragments of the component surface as a result of cyclic loading. This results in pits or cavities formation where the small pieces of material separate from the surface. Besides the obvious damage and physical changes to shape and size of the implant, the defect sites act as stress raisers and stress concentration points, leading to crack formation and fatigue fracture. It should also be noted that the displaced wear debris contributes towards third-body wear, and consequently, most wear scenarios involve a combination of both fatigue and abrasive wear.

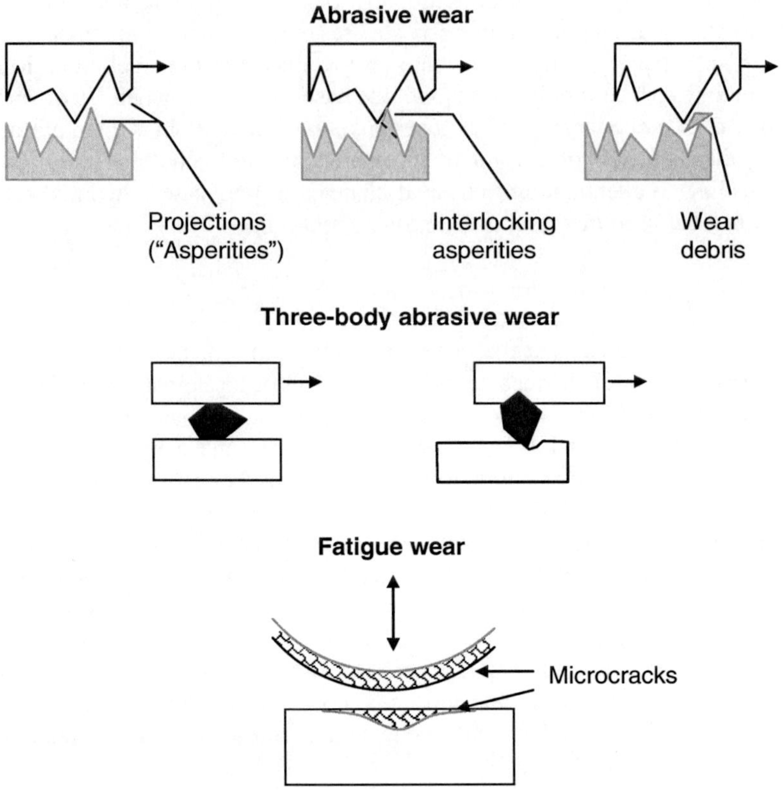

Fig. 11.7 Schematic description of the processes of abrasive, three-body and fatigue wear

What Are the Types of Crack Origin?

Fatigue wear refers to a particular case of fatigue, where micro-sized cracks and fractures occur on the surface. The fractured particles are displaced by the cyclic motion and constitute wear debris. These cracks may arise in a variety of ways.

Subsurface-Origin Fatigue

In subsurface-origin fatigue, the crack originates a short distance below the surface of the component. This is primarily due to high stress levels being concentrated at a relatively short distance below the surface. Subsurface origins are predominantly found where rolling motions are compounded with compressive stresses, to result in subsurface regions of high stress. These are typically seen in engineering systems involving rotating cams or wheels, and may be experienced in the body at the knee or ankle joints. Other causes for this phenomenon include inclusions or impurities present at the subsurface level, which act as stress concentrators. Repeated cyclic loading will cause the crack to propagate and travel towards the surface of the component or towards other crack sites, leading ultimately to microfractures and the dislodgment of material. In addition, the process leaves pits on the surface of the material, which might serve as crack sites for propagation of fatigue.

Surface-Origin Fatigue

Cracks may also originate from the surface of the component, when the maximum shear stress is experienced at the surface. In contrast to subsurface origins, surface-origin fatigue is predominantly experienced in systems involving a combination of sliding and rolling actions, and is largely influenced by sliding friction and traction forces. As a result, v-shaped cracks are initiated at the surface. As the crack expands and deepens, the wedge-shaped debris is dislodged, leaving an arrowhead-shaped pit pointing in the direction of rotation. Over several repeated cycles, however, the v-shape breaks down and is often not observed in the parts after some time.

Subcase-Origin Fatigue

Cracks are often initiated in regions of inhomogeneity, and this includes the subcase region in case-hardened materials, particularly at the interface between the case and bulk material. Consequently, subcase origins are very similar to subsurface origins, except that they can occur at much deeper regions depending on the heat treatment step of the component. Hence, larger pieces of material fallout (wear debris) are typically observed.

Cavitation Fatigue

Cavitation is a phenomenon caused by the vibration and movement in various fluids (including blood), leading to the formation of negative-pressure "bubbles" that implode onto the component after movement. These implosions on the material surface exert high compressive stresses and create pits and cavities. Thus, cavitation is largely influenced by fluid flow profiles and can be reduced by facilitating normal flow. This is achieved by increasing the smoothness of the surface and engineering design. Material selection also plays a big part, as cavitation fatigue is accelerated in corrosive environments (although it should be noted that cavitation can occur even in the absence of corrosive agents).

The Host Response to Wear Debris

Besides the obvious repercussions of material loss and geometrical changes associated with wear, wear-generated debris elicit immunological responses, with often more catastrophic effects than the actual wear itself [4]. Wear debris generated by implants is typically small, ranging from submicron to millimetre in sizes. Some wear particles may accumulate in tissue and may contribute to blackening of surrounding soft tissue. The main cause for concern over wear debris, however, appears to be the natural defence mechanism of the body. Wear debris inevitably trigger inflammatory and immunological response that recruit leukocytes, macrophages and giant cells to move in on the foreign wear particles. A series of biochemical activities occurs at this stage, including the induction of a highly acidic, harsh environment. In general, four possible ensuing scenarios may arise [3]:

1. Digestion of the debris and subsequent clearance of waste products by the liver and kidney.
2. If the foreign material cannot be digested, the body will react by excreting it.
3. If the foreign matter cannot be digested or expelled, it will be encapsulated in fibrous tissue to isolate it. This scenario may have additional implications for anchored implants, such as orthopaedic or ear implants, where the displacement and micromotions may have severe effects on prosthesis performance.
4. Finally, wear debris have been shown to elicit unexpected secondary host responses. In prosthetic joints, for example, recruited macrophages participate in osteolysis and produce inflammatory mediators which may activate osteoclasts, resulting in massive bone loss in the surgical site.

Factors Affecting Wear

Wear System

Wear systems typically involve the relative motion of two articulating surfaces under the application of a normal load. Wear processes are thus highly dependent on the type of motion, which determines the manner articulating surfaces interact. For example, human joints involve a forward–reverse motion (i.e. reciprocation), which can result in up to fivefold increase in wear, as compared to unilateral motions. It is the pattern of motion, however, that has a dominating effect, while other parameters, such as speed, may have limited and indirect implications on wear. In polymers, for example, the sliding speed of moving components may result in local heating due to friction, and subsequent softening due to temperature changes, that may have an indirect influence on wear.

Wear processes are also dependent on the magnitude of applied loads: increased normal loads, for example, result in increased friction and increased wear. Increased normal loads may also accelerate the rate at which three-body abrasive wear takes place. Under the application of a continuous load, permanent changes in geometry of the component may also arise due to the applied load (a phenomenon known as creep). Such changes may also have an indirect influence on wear.

Wear Debris

Besides being a direct by-product of wear and material loss, wear debris contribute directly by participating in third-body abrasion and accelerates the degradation process. Thus, the shape of the debris is known to influence wear characteristics.

It also follows that conforming surfaces that retain wear debris may be subject to higher wear rates. Thus, the use of some ceramic coatings, such as those commonly used in hip implants, may generate rounder wear particles and consequently reduce wear rates.

Surface Roughness

The surface finish of the component is vital, as very rough surfaces increase the tendency for wear. A highly polished surface results in greater conformity between the two surfaces, increasing the contact area and correspondingly reducing the contact stresses. In contrast, rough surfaces typically result in interlocking of asperities, which increases the abrasive wear rate. The surface roughness can be controlled by fabrication processes, post-treatments and the environment at which it is being used in. It should be noted that, when used in a corrosive environment, the material may be subjected to surface corrosion, which causes an increase in surface roughness.

Fabrication and Post-treatments

The fabrication processes and any post-treatment can affect the wear rates of the material as they may change the material's bulk properties and surface roughness. For example, when ultrahigh molecular weight polyethylene (UHMWPE), a commonly used material for implant applications, is moulded between 190 and 200 °C, antioxidants may be added to improve wear resistance. On the other hand, moulding at higher pressures may decrease the wear resistance of UHMWPE. In addition, sterilisation of polymeric implants by gamma irradiation can also cause a decrease in fatigue threshold and crack propagation resistance. These highlight the need to consider the influences of secondary processes on wear properties.

3 Fatigue in a Clinical Setting

Hip, knee and dental implants as well as heart valves are examples of prostheses that experience extensive cyclic loads. The heart valve, for example, functions as a check valve and opens or closes to regulate the direction of blood flow. Consequently, the open–close cycle occurs some 40 million times a year, or 1.2 billion times over the expected life span of the valve. Fatigue is thus a very real problem for medical prostheses in service, and prosthetic heart valves have now been designed to endure fatigue lifetimes in excess of 10^9 cycles in blood-analogue environments. Damage-tolerant designs, in conjunction with the advent of modern materials, such as pyrolytic carbon, have very effectively improved the service life of the device. In addition, non-invasive methods of detecting and monitoring fatigue crack propagation have markedly reduced the chance of catastrophic failure. Taken together, these measures serve to highlight how the knowledge of fatigue mechanisms can

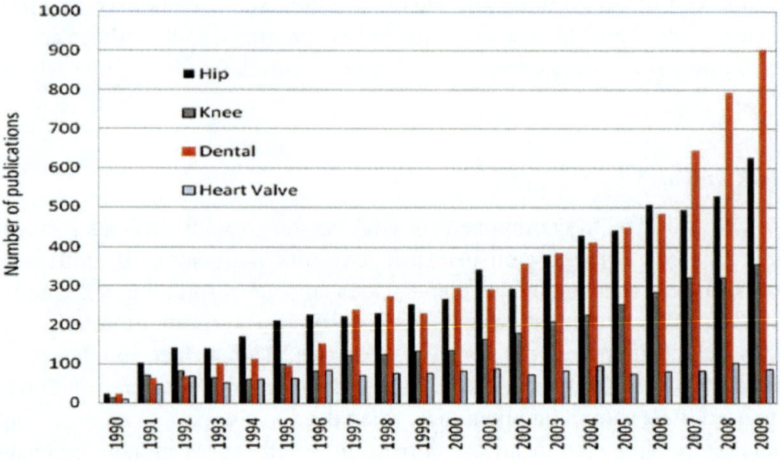

Fig. 11.8 Research trends in prosthesis failure (1990–2009). Data compiled from Thomson Reuters, Web of Science 2010

apply to manage fatigue-associated problems, even under such extreme conditions. An illustration of research trends on prostheses failure can be found in Fig. 11.8. It can be seen that, since 1990, research interest in the failure of hip, knee and dental implants has increased exponentially, highlighting the critical need to better understand the associated problems. The fatigue-related problems will be discussed in the next section to help the reader appreciate and apply concepts of fatigue in the clinical setting.

4 Case Study: Hip Prostheses

The hip prosthesis (Fig. 11.9) is the most common artificial joint in human beings. A typical total hip replacement (THR) implant consists of a cup type acetabular component and a femoral component whose head is designed to fit into the acetabular cup and thus enable joint articulations. The femoral stem is tapered so that it can be fixed into a reamed medullary canal of the femur. In current designs, the femoral head is typically a ceramic ball that can be fitted onto the stem just prior to implantation. The acetabular component comprises a metal cup with a smooth plastic polyethylene. This is placed into the pelvis, creating a socket for insertion of the femoral head. Over time, the implant integrates with the surrounding bone and effectively replaces the hip joint. Implants come in a wide range of sizes to provide a customised match for the unique bone geometry, weight and size of each patient.

Conventional hip implants use metals for the femoral shaft and neck. In modern design, the head piece is separate, and typically constructed of ceramic. Steel alloys were used in earlier designs of acetabular cups, which were found to constitute high

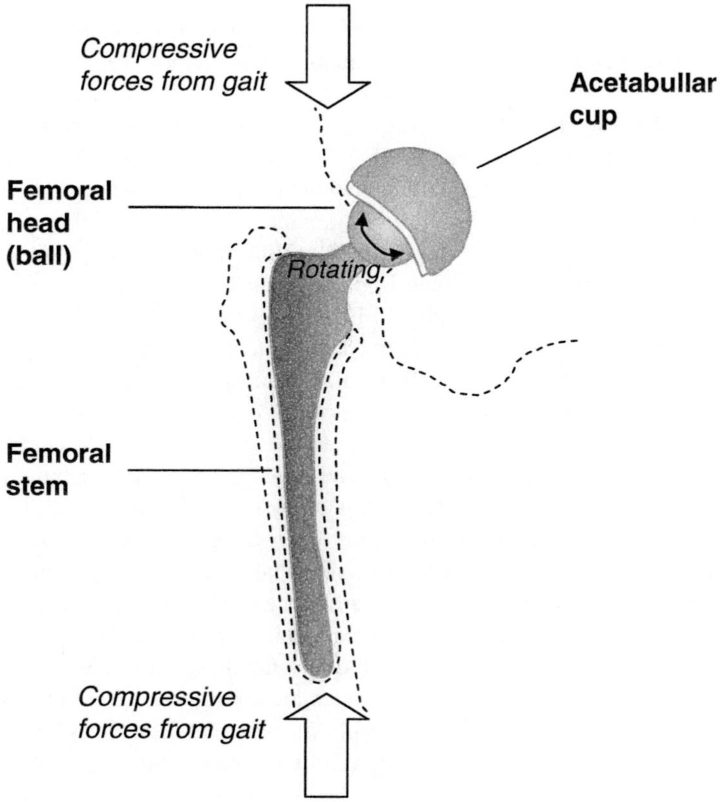

Fig. 11.9 Schematic of hip prosthesis

friction as well as metallic wear particles. These were replaced by polymers in the early 1960s, leading to the modern version employing UHMWPE cups.

It can be seen that the prostheses experiences two separate sources of cyclic loading as the patient walks. Firstly, the hip is placed under fluctuating compressive loads during normal gait, which translates to bending at the femoral neck. Next cyclic shear stresses are experienced at the articular interface at the ball–socket joint.

4.1 Fatigue-Related Issues of the Hip Prosthesis

4.1.1 Fracture

Fatigue fracture was not uncommon in early designs of hip prostheses, occurring in up to 6% of patients. It was later recognised that these failures carried many distinguishing signs associated with fatigue, originating from micro-defects within the implants arising from inadequate manufacturing techniques. With the improvement of engineering techniques, fatigue fractures have almost been eliminated by the early 1980s. This includes the use of advanced manufacturing technology such as hot-isostatic pressing (HIP) and forging, rather than standard casting as previously used. In particular, the use of HIP generated implants with extremely fine microstructures, which are able to reduce the rate of crack propagation, and are thus extremely resistant to fatigue. In addition, greater efforts were made to eliminate impurities and inclusions, which may otherwise contribute to crack initiation.

Outside of the implant, the bone-implant surface has received considerable attention [5]. Bone cement, typically PMMA, is commonly used as an adhesive to bond the implant to surrounding bone. The cement, however, is itself prone to deterioration over time, and typically results in loosening. This might lead to a situation where there is proximal loosening of the femoral stem, with a firm distal fixation, and resulting in a cyclic, cantilever-like action that may contribute to fatigue stresses on the stem, as well as accelerated breakdown of surrounding bone cement. Such loosening is directly associated with pain and immobility of the patient. Cement-free approaches have thus been developed, where the surfaces are coated with porous metal or polymeric materials, which promote the ingrowth of bone. Hydroxyapatite (HAp) coatings are some of the most commonly used in the clinic, which can serve as a base of osteogenic ingrowth and osteointegration. Ironically, the use of porous coatings introduces notches and crack initiation sites on the surface, particularly where the particles are strongly bonded to the underlying alloy surface. Thus, the coatings are only introduced to low-stress areas on the implant, which can be determined through finite-element modelling and analysis.

4.1.2 Wear

Advances in the materials, designs of the prosthetic components and surgical techniques have greatly improved the long-term mechanical fixation of components to the skeleton. However, wear of the articulating surfaces (Fig. 11.10) and the associated biological reactions to the wear debris subsequently emerged as the major problems in total joint replacements that present the greatest challenge to the long-term success of joint replacements.

While the mechanical consequences of wear can be minimised by design and surgical techniques, the biological consequences of wear debris remain a major unsolved problem. Even low wear rates can generate large numbers of wear particles, which might enter the periprosthetic tissue. Many studies have shown that the liberation of submicrometer wear particles into the tissues leads to foreign-body macrophage and giant-cell reactions, contributing to periprosthetic osteolysis. Such bone loss may be sufficient for component loosening or even fracture. Unfortunately, revision surgery in such cases is particularly difficult and often requires the use of special components and massive bone grafts. In some cases, the extent of bone loss is so massive as to preclude revision surgery altogether. Osteolysis is thus the most common complication of total hip arthroplasties,

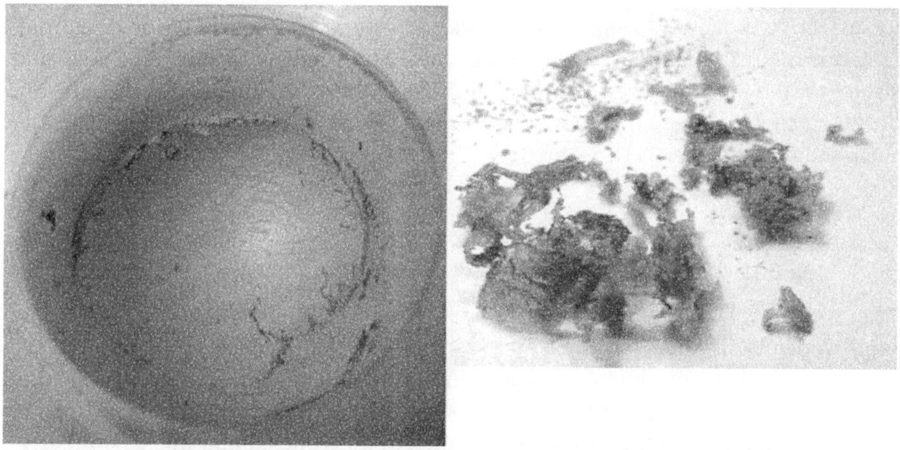

Fig. 11.10 Photo of UHMWPE cup subjected to *in vitro* testing after 700,000 cycles. *Inset* shows a close-up photograph of wear particles, measuring from submicrometre to millimetre size

occurring in 7–10 % of retrievals at 10–15 years post-operation and is always caused by implant wear. Significant debate exists over "acceptable" clinical wear rates, and it is generally recognised that wear rates of 30 mm^3 per year are acceptable, while rates above 150 mm^3 per year are problematic.

4.1.3 Stress Shielding

While cyclic loads are associated with fatigue failure of synthetic material, the removal of cyclic stresses from biological tissue has conversely been associated with tissue deterioration [6]. Bony healing and osteointegration involves three steps (1) bony ingrowth into the wound site, (2) adaptation to increased loading by reinforcement with lameallar bone and (3) remodelling by resorption and functional adaptation. Step (3) has been shown to critically influence the density and architecture of the healed bone, where, in accordance with Wolff's law, bone remodels itself to be denser and stronger when placed under increased loads. Correspondingly, when loading is removed, osteopenia takes place as the bone remodels to become weaker, with accompanying mass loss.

Stress shielding is thus the phenomenon in which osteopenia takes place when normal stresses are transferred from the bone to the implant. In the case of the hip implant, this may take place in the area surrounding the femoral stem, where the prosthesis itself serves as the main load-bearing member, and relieves the surrounding bone from compressive stresses [5]. In some designs which incorporate a metallic backing of the acetabular cup, stress shielding may also be experienced in the pelvic socket. Both processes result in bone loss, which may directly lead to loosening of the implant. Due consideration must thus be paid to effective load transmission to the surrounding matrix. For example, non-cemented implants are designed to have sharp edges to transfer stress and strain to the surrounding interlocked bone, whereas cemented implants are designed to be round, in order to reduce the peak stresses experienced in the surrounding cement.

5 Case Study: Knee Prostheses

Improvements in knee prostheses have made knee arthroplasty operations very successful and increasingly common in recent years. Knee arthroplasty may involve just replacement of the damaged half of the knee (i.e. unicompartmental knee replacement), if the arthritis is restricted to only one compartment. Similarly, for severely damaged patella-femoral joints, a patella-femoral knee replacement may be adequate. In the most severe cases, the whole knee joint is removed and replaced.

The modern prosthetic knee design consists of three main components (Fig. 11.11): a metallic femoral component, a polymeric tibial component and a patella component. The femoral component is designed to wrap around the end of

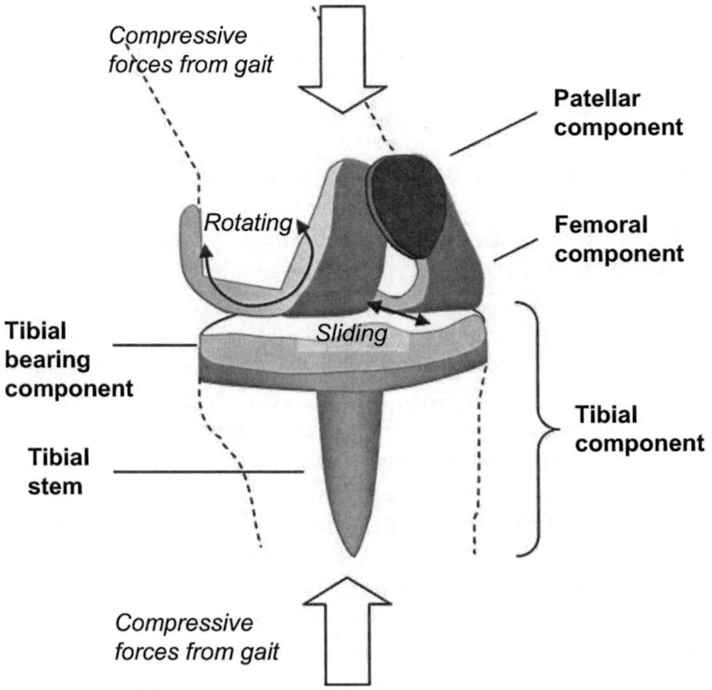

Fig. 11.11 Schematic of a typical knee prosthesis

the femur and has a groove which facilitates the movement of the patella smoothly against the bone as the knee flexes. The design is contoured to allow articulation on the corresponding surface of the tibial bearing component. The tibial component is typically a flat metal platform with a polyethylene cushion, which has a raised surface with a centre cam (posterior stabilised). It is anchored by a tibial stem, and bone cement is often used to hold the component in place. Finally, the patellar component is usually a dome-shaped piece of polyethylene shaped to resemble the patella.

It can be seen that the prosthesis experiences three separate sources of cyclic loading as the patient walks. Firstly, the knee is placed under fluctuating compressive loads during normal gait. Next, flexion at the knee is achieved by a mixture of rolling and sliding. These lead to cyclic loads being experienced at two locations: the patellar–femoral interface and the femoral–tibial interface.

5.1 Fatigue-Related Issues with the Knee Prosthesis

5.1.1 Fatigue Loosening

Fatigue fracture of the individual components is rarely seen and is largely avoided through the use of materials with high fracture toughness, such as stainless steel or other metallic alloys. Loosening is more common and occurs as a result of the weakening of bond between the bone and the prosthesis over time. Mechanical factors which cause loosening are primarily cyclic stresses brought about by everyday activities, which are transferred to bond between the prosthesis and the bone. Thus, younger patients are more susceptible as they tend to lead more active lifestyles which generate increased cyclic stresses. This issue is exacerbated by the inability of the artificial knee to fully recapitulate natural motion, which results in jarring motions which further stress the interface. To address this issue, mobile-bearing designs have been developed.

Like fixed-bearing replacements, mobile-bearing knees use three components to provide a relatively natural and even interface. The difference between them is in the bearing surface. In a mobile-bearing knee, the femoral component and tibia tray slide across a polyethylene insert to create a dual-surface articulation. This reduces the restriction to natural movement, and consequently, lessens stress transmission to the bone/prosthesis interface. Mobile-bearing knees are also designed to allow greater rotation of the knee. By promoting natural motion, congruent contact between femoral and tibial components and conformity between moving parts are maximised throughout the range of knee joint movement, thus serving to reduce wear. As with all engineering designs, however, there is a trade-off between freedom of motion and stability. Consequently, the mobile bearing limb is subject to a greater risk of dislocation and costs significantly more than traditional fixed-bearing implants.

5.1.2 Wear

Wear is the predominant cause of implant damage and failure in the total knee implant [7]. The generated wear debris contributes towards third-body abrasion and further accelerates removal of surface material. More important, these wear particles are implicated in osteolysis, as discussed with the hip implants. It is an extremely serious problem faced by patients that undergo total knee replacements, where a common consequence of osteolysis is loosening of the knee prosthesis. In severe cases, large sections of the patient's skeleton can be destroyed. It is thus unsurprising that much research focus has focused on reducing wear.

Components in the knee prosthesis are designed so that metal always articulates against plastic, which provides smooth movement. As a consequence, the polymeric material (typically, polyethylene) is often subject to severe wear, and UHMWPEs are often used (just as in hip implants). UHMWPE is a very tough and impact-resistant material. It is very dense, and consequently, highly resistant to moisture absorption and corrosion. Crucially, it has a very low coefficient of friction, and consequently, good wear characteristics. Additional cross-linking may be used to minimise surface orientation and improve the wear characteristics further. Excessive cross-linking, however, has negative effects on the fracture toughness and impact resistance of the material. This, in turn, increases the rate of crack propagation in fatigue, and care should be taken to avoid fatigue fracture.

It should be noted that the wear process involves two articulating surfaces. Consequently, the femoral component should be considered as well. Studies have shown that the introduction of just a few small scratches onto the metallic surface can drastically increase the wear rate on the UHMWPE counterface, by up to 70-fold. Despite advances in manufacturing to produce increasingly smooth mirror finishes, implants often become scratched or damaged during service. These scratches may arise due to (1) surgical instruments, (2) abrasion from a third body, such as those arising from inadequately contained bone cements and (3) surface scratching. Ceramics such as alumina, which have increased scratch resistance, have thus been the subject of much study. Not only does this approach reduce the rate of polyethylene wear, generated wear particles are rounder and reduce inflammatory responses. To avoid issues associated with inherent brittleness, these ceramics are typically employed as coatings, where they are sintered onto the underlying metallic surface.

5.1.3 Metallosis

Another major problem is metallosis [8]. In a well-performed knee replacement with the ligaments balanced and the bone cuts accurately aligning the leg, the polyethylene implant should last 10 to 15 years. This time period will depend in part on the patient's weight and level of activity. When the polyethylene implant is completely worn out, however, contact between the metallic components occurs, resulting in the release of metal debris within the joint and subsequent reactive synovitis, known as metallosis. Chronic metallosis may also contribute to osteolysis around the implant, leading to loosening of the prosthesis. Little can be done to prevent this, although replacing the polyethylene pad is a relatively straightforward operation that can be preformed without changing the femoral or tibia implants, if they are still well fixed to bone.

6 Case Study: Dental Restoratives

Dental restoratives, commonly referred to as dental fillers, are used to restore the integrity and morphology of compromised tooth structures. Amalgam has been the material of choice for dental restorations, but has recently faced severe controversy in use, in particular over potential mercury toxicity. Health advisories and laws limiting the use of mercury-based amalgams have been in place in some European countries since 1993, leading to eventual discontinuation of use. Thus, despite the hitherto excellent clinical record of amalgams in dental reconstruction, health concerns and aesthetic reasons have contributed to a shift towards composite resins as dental restoratives.

Composites refer to materials comprising two or more materials, often with very different and complementary properties. Composite resins have been developed and researched on for dental restoration since the 1960s, and typically comprise a tough, wear resistant resin matrix and ceramic fillers. Commonly used matrix resins contain dimethacrylate monomers that are hardened as a result of chemical reactions between the resin monomers (or "cured") *in situ*. Thus, the mechanical properties of the resin are largely influenced by the composition of monomers used and the degree of curing. The cured resin serves as a rigid network surrounding the filler materials and defines the shape of the reconstruction. The filler materials, on the other hand, serve to improve the mechanical properties of the composite, including increased strength, reduced polymer contraction and reduced thermal expansion. In addition, the fillers are used to adjust colour and improve aesthetics. Typical fillers used are made from ceramics such as quartz or borosilicate glasses.

6.1 Fatigue-Related Issues with Dental Restorations

As the teeth, together with any restorations, move in contact against each other, wear inevitably takes place. In addition, wear results from any abrasive particles or debris sandwiched between the teeth during the mastication process. Furthermore, the teeth are subject to chemical attack from dietary or regurgitation acids. The effects of these sources of wear may be described as attrition, abrasion and erosion.

Attrition refers to wear at the sites of direct contact, and manifests in the flattening of cusp tips, with associated wear facets on the occlusal surfaces [9]. When attrition takes place at a restoration site, enamel and dentine may be preferentially worn away, resulting in "cupping" of the occlusal surface. In contrast to attrition, abrasion refers to wear that takes place at non-contacting sites. It is prevalently caused by wear with objects other than another tooth, such as in toothbrush abrasion. Finally, erosion takes place as a result of fluid flow [10]. It refers to surface loss resulting from softening of the tissue by acids, and eventual rubbing off by movements of soft tissues or food slurry. These acids may be derived from dietary sources, such as citric acid, or from regurgitative sources, such as gastric reflux [11]. During the mastication cycle, a combination of attrition, abrasion and erosion can be seen.

Masticatory wear can be studied in two phases. In the initial phase, the teeth are brought from the open position to a position of near contact (open phase). This is followed by the second part of the cycle, where a force is applied to the food bolus, and shredding of the food takes place by lateral sliding of the teeth (closed phase). The effect of abrasive particles in the diet differs between the phases. In the open phase, particles are suspended in the food bolus, which thus acts as a slurry. There is free movement of the abrasive particles within the suspension, leading to abrasion of the surface of the restored tooth. During the closed phase, the teeth are brought close, and the abrasive particles become trapped between the surfaces. Consequently, they no longer act as a slurry but, instead, cause scratching and scoring as a third-body abrasive during sliding of the teeth.

The wear of restorative materials can be understood by considering the relationship between sliding wear and slurry wear. Slurries tend to preferentially abrade the softer resin material, leaving the harder filler material protruding from the surface. During sliding contact wear, these particles may transmit the sliding force to the surrounding matrix, resulting in microcracking at the filler/matrix interface and eventual dislodgement of the filler. Certain composites, such as porcelain, comprise two phases of similar hardness. Consequently, slurry wear does not cause protrusions in these materials. However, due to the loading pattern, the material may be vulnerable to subsurface cracking and consequent fatigue wear [12].

Considering the wear process, it follows that wear resistance of the composite restorative may be improved by decreasing the interparticle spacing [13]. This may reduce the effect of slurry wear, particularly if the spacing is smaller than the size of the abrasive particles. Wear resistance may also be improved by the use of smaller filler particles, which will result in smaller surface protrusions and reduce the extent of filler plucking and surface degradation during sliding wear.

7 Conclusions

Numerous medical devices such as joint prostheses are subjected to high stresses and high cycle loading. Coupled with the aggressive body environment, this might lead to fatigue failure. A fatigue wear process causes the generation of wear debris, which invokes acute host–tissue reactions, which tend to aggravate the fatigue problems of the biomaterial by producing enzymes and chemicals that are highly corrosive.

The methods of evaluation for biomaterials must include wear debris morphology characterisation so as to understand the host-tissue reaction to wear debris and simulate as close as possible the imposed stress–strain and environmental conditions *in vivo*. The development of fatigue fracture resistant and wear resistant biomaterials is still in its infancy. Research geared towards biocomposite systems with different phases to cope with the conflicting properties of fatigue fracture resistance and hard, but brittle, phases required for wear resistance and a good lubrication phase seems to provide some future direction. The morphology of wear debris that forms must be readily acceptable by the body. The ability to engineer biomaterials that have the capability to trap/isolate wear debris and promote easy removal of such wear debris remains a challenge.

References

1. Teoh S (2003) Failure in biomaterials. In: Mai Y, Teoh S (eds) Comprehesive structural integrity series. Elsevier, Oxford, UK, pp 1–33
2. Anderson TL (2005) Fracture mechanics: fundamentals and applications, 3rd edn. CRC Press, Boca Raton, FL
3. Teoh SH (2000) Fatigue of biomaterials: a review. Int J Fatigue 22(10):825–837
4. Goodman S (2005) Wear particulate and osteolysis. Orthop Clin North Am 36(1):41–48, vi
5. Morscher EW (2003) Failures and successes in total hip replacement—why good ideas may not work. Scand J Surg 92(2):113–120
6. Chen JH et al (2010) Boning up on Wolff's Law: mechanical regulation of the cells that make and maintain bone. J Biomech 43(1):108–118
7. Paling IH (2000) Wear in total hip and knee replacements. J Bone Joint Surg Am 82-A (12):1806
8. Matsushita I et al (2007) Severe bone defects and reduced mineralization caused by massive metallosis after total knee arthroplasty: histopathologic and bone morphometric findings. Mod Rheumatol 17(6):507–510
9. Shafer W, Hine M, Levy B (1983) Regressive alterations in teeth. In: Shafer W, Hine M, Levy B (eds) A textbook of oral pathology. Saunders, Philadelphia, pp 318–338
10. Powers JM, Fan PL (1980) Erosion of composite resins. J Dent Res 59(5):815–819
11. Smith BG, Knight JK (1984) A comparison of patterns of tooth wear with aetiological factors. Br Dent J 157(1):16–19
12. Mair LH et al (1996) Wear: mechanisms, manifestations and measurement. Report of a workshop. J Dent 24(1–2):141–148
13. Yap AU, Teoh SH, Chew CL (2002) Effects of cyclic loading on occlusal contact area wear of composite restoratives. Dent Mater 18(2):149–158

Biography

Dr Mark Chong received his B.Eng (Hons) in Mechanical Engineering and PhD in Bioengineering from the National University of Singapore. He is currently a lecturer in the School of Chemical and Biomedical Engineering, and research fellow at the National University Health System. Dr Chong's main field of research is in tissue engineering, with particular emphasis on engineering biomaterial and stem cell responses. Dr Chong's other commitments towards research and education include serving on the editorial board of the journal ISRN Tissue Engineering, as well as being the current vice-chair of the Tissue Engineering & Regenerative Medicine International Society (TERMIS) – Student & Young Investigator Section.

Professor S.H. Teoh received his B.Eng (1st Honors) and PhD from the Department of Materials Engineering at Monash University, Australia, in 1978 and 1982, respectively. He is presently Professor and Director of the Centre of Biomedical Materials & Applications (BIOMAT) at the Department of Mechanical Engineering, Executive Member of the NUS Tissue Engineering Program, National University of Singapore (NUS). His main field of research is biomaterials, especially 3D bioresorbable scaffolds for bone and ultrathin membrane tissue engineering. He is a member of the editorial boards of the journals Tissue Engineering, Journal of Tissue Engineering and Regenerative Medicine, Journal of Mechanical Behaviour of Biomedical Materials, Journal of Oral & Maxillofacial Research, and Proceedings of the Institution of Mechanical Engineers Part H: Journal of Engineering in Medicine. He has served in numerous international committees, including ASTM (USA) F04 on Surgical Implant Materials, the American Society of Artificial Internal Organs (ASIAO), and the Biomedical Engineering Society (Singapore).

Dr. Erin Teo received her B.Eng (with Honors) and PhD from the Department of Mechanical Engineering, National University of Singapore, in 2005 and 2011, respectively. She is currently a postdoctoral research fellow at the National University of Singapore, working on layer-by-layer technology for skeletal muscle tissue engineering and anti-adhesion barrier membrane technology for skin tissue engineering. Her PhD work involved the development of an antibiotic delivery system for wound healing and tissue engineering applications. She also serves as the treasurer of the Tissue Engineering International & Regenerative Medicine Society (TERMIS)—Student & Young Investigator section.

Chapter 12
Hypersensitivity to Implant Debris

Nadim J. Hallab

Abstract Total Joint Replacements (TJR) have been extremely successful over the past 50 years, restoring mobility and function to millions of people each year. However, over time these implants need replacing for a number of reasons, such as infection or increasing immune reactivity to implant debris. All implant metals degrade *in vivo*, and the released products biologically interact locally and systemically. Local and/or systemic immune reactivity to implant debris may become excessive, specifically to one or more of the materials (metals) used in the implant alloys. When this excessive reactivity to implant debris involves the adaptive immune system where lymphocytes respond to specific stimuli it can be characterized as a sensitivity or hypersensitivity response. Dermal hypersensitivity reactions to metals (such as Nickel) are common and affect approximately 10–15% of the population in USA and Europe. In its extreme form, metal sensitivity exists as a relatively rare complication in only a few highly susceptible patients with joint replacements (i.e., less than 1% of joint replacement recipients).

However, the role of implant-related metal sensitivity in implant performance is likely underreported due to the scarcity of diagnostic testing. The person-dependent mechanism(s) by which metal sensitivity occurs in some people and not others has not been completely explored with current hypersensitivity testing techniques. This issue is becoming increasing popular due to recent failures of metal-on-metal hips and increasing numbers of joint replacement procedures, worldwide. Better materials and availability of appropriate immunologic testing (e.g., LTT) will likely enhance future assessment of patients susceptible to hypersensitivity responses.

N.J. Hallab (✉)
Department of Orthopedic Surgery, Rush Medical College, 1653 W. Congress Parkway, Chicago, IL 60612, USA
e-mail: nhallab@rush.edu

N. Eliaz (ed.), *Degradation of Implant Materials*, DOI 10.1007/978-1-4614-3942-4_12,

1 Introduction

It is surprising to learn that almost all of the >1,000,000 total joint replacements/year that are performed in the USA are expected to eventually fail if their recipients live long enough. This eventual failure is due to implant debris-induced inflammatory reactions [1]. This is particularly horrible for millions of elderly people who may need a revision of their current implants in their last decades of life where the incidence of death caused by major surgery can be as high as 13% (vs. <1% in patients <75 years of age) [2]. Implant loosening due to debris-induced aseptic osteolysis accounts for over 75% of all total joint arthroplasty (TJA) implant revisions and is the predominant factor limiting the longevity of current total joint arthroplasties; the other reasons include infection (7%), recurrent dislocation (6%), periprosthetic fracture (5%), and surgical error (3%) [3].

Excessive biologic reactivity to implant debris can be defined as any toxicologic or immunologic reactivity to implant debris that causes, or is in the process of causing, implant failure prematurely, where "prematurely" is generally considered less than 5–7 years. Particle-induced osteolysis, or "particle disease," generally refers to the slow process of peri-implant osteolysis, where implant loosening and inflammation are due to implant particulate debris interacting with innate immune system cells (i.e., tissue macrophages termed histiocytes) where modest inflammation persists for many years. This innate inflammatory response is typically associated with the eventual failure of metal-on-polymer THAs. Excessive reactivity to implant debris, or hypersensitivity to implant debris, is slightly different and typically involves the adaptive immune system where conditioned lymphocytes respond to specific stimuli. The degree to which excessive innate response factors into this hypersensitivity immune response is not known. Implant hypersensitivity has been predominantly characterized as specific, and of the delayed type hypersensitivity (DTH) response, and has been more associated with metal-on-metal bearing implants.

Implant surfaces are not known to cause hypersensitivity. Implant debris is the known cause of immune responses to implants. This distinction is important, because when debris is minimized, the chances of hypersensitivity decrease. But what is hypersensitivity to implants? In its broadest definition hypersensitivity to implants is any aseptic (nonbacterial) material-driven excessive immune response that causes peri-implant pathology, such as decreased bone homeostasis or massive local inflammation of T-cells or B-cells or macrophages. When an implant fails prematurely (<7 years) due to an exuberant immune response to a normal (and typically very tolerable) amount of implant debris, that is what can be categorized as "metal-allergy," "implant-allergy," "implant sensitivity," or "hypersensitivity." These terms have been used interchangeably in scientific studies. Implant-related metal sensitivity has been well reported in case and group studies; however, there is still much we do not know about this phenomenon [4–6]. All implant metals corrode and/or wear *in vivo* [7, 8], and the released products (particles and ion) interact with plasma proteins, and then with local and systemic cells including those of the immune system. Materials such as polymers that are less chemically promiscuous (less able to alter proteins at the molecular level) are also less likely to activate the immune system and have not been as widely implicated in causing an implant allergic response. Dermal hypersensitivity reactions to metals have been reported to cause immune reactions, which most typically manifest as skin hives, eczema, redness, and itching, affecting approximately 10–15% of the US and Europe's population [4, 5, 9–12]. Implant materials have been optimized over time to eliminate materials that demonstrate adverse host responses. Some subtle adverse responses, such as *in vivo* metal hypersensitivity or hypersensitivity-like reactivity to metallic biomaterials, are difficult to characterize before and after surgery. Hypersensitivity is caused by the debris of metallic biomaterials that include particulate wear debris, colloidal organometallic complexes (specifically or nonspecifically bound), free metallic ions, inorganic metal salts/oxides, and precipitated organometallic storage forms (basically, particles and ions).

All metals begin to degrade via corrosion when in contact with their biological environment [7, 13]. The released metal ions can activate the immune system by forming complexes with native proteins [12, 14–16]. These metal–protein complexes are what elicit hypersensitivity responses. Plastic (polymeric) biomaterials used in orthopedics are not easily chemically degraded *in vivo* and have not been intensely investigated or found to be involved in case or group studies as sources of hypersensitivity-type immune responses. This is in part due to the relatively large degradation products (>50 nm) associated with polymeric debris *in vivo*, which do not readily form polymer–protein haptenic complexes with human antibodies. However, there have been reports of immunogenic reactions associated with polymethylmethacrylate (PMMA) [17], but these may be due to an unreacted PMMA monomer which is highly toxic and can elicit an immune response at very low levels.

Metals that are well known to cause hypersensitivity reactions are beryllium [18], nickel [9–11, 18], cobalt [18], and chromium [18] and to a lesser degree tantalum [19], titanium [20, 21], and vanadium [19]. The incidence of metal sensitivity among the general US population is approximately 10–15% (Fig. 12.1), where nickel hypersensitivity is the most prevalent (approximately 14%) [4], followed by cobalt and chromium [4, 12]. The amounts of these metals found in medical grade alloys are presented in Table 12.1.

2 Metal Sensitivity

Hypersensitivity can take one of two central forms (1) an immediate (within minutes) humoral response (initiated by antibody–antigen complexes of types I, II, and III reactions), or (2) a delayed (hours to days) cell-mediated response [22, 23]. Implant-related hypersensitivity reactions are generally associated with delayed-type responses and have been categorized as type IV DTH.

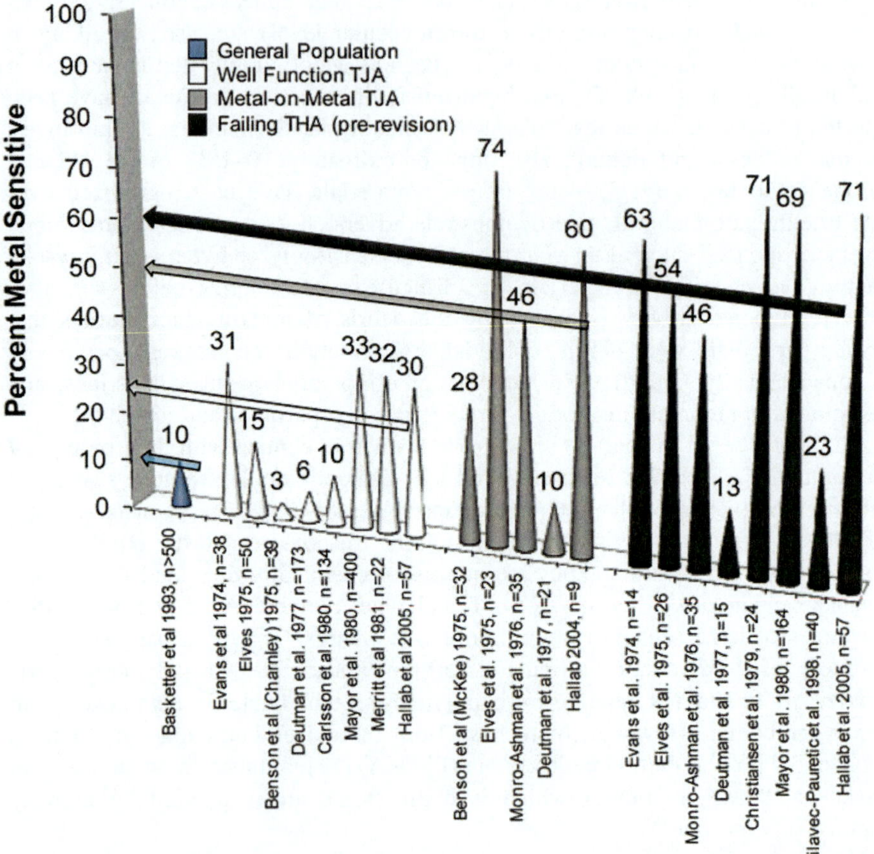

Fig. 12.1 A compilation of averaged incidence percentages of metal sensitivity (nickel, cobalt, or chromium) among different groups: (1) the general population, (2) patients after receiving a metal-containing TJA, (3) patients with metal-on-metal bearing arthroplasty, and (4) patient populations with significant osteolysis or due to be revised. Note: Studies by Hallab et al. used Lymphocyte Transformation Testing to measure hypersensitivity, all other used dermal patch testing (picture courtesy of Orthopedic Analysis Inc)

Cell-mediated DTH is an adaptive immune response and is characterized by antigen activation of sensitized T-helper lymphocytes releasing various cytokines, which result in the recruitment and activation of macrophages. These $T_{H\text{-}1}$ cells are characterized by the cytokines they release, including interferon-γ (IFN-γ), tumor necrosis factor-α (TNF-α), interleukin-1 (IL-1), and interleukin-2 (IL-2). $T_{H\text{-}1}$ cells are generally associated with responses to intracellular pathogens and autoimmune diseases. Metal-activated T-cells in conjunction with activated antigen presenting cells (APCs) can secrete a variety of cytokines that recruit and activate other innate immune cells, i.e., macrophages, monocytes, and neutrophils. These cytokines include IFN-γ and TNF-β which produce a number of effects on local endothelial

Table 12.1 Approximate weight percent of different metals within popular orthopedic alloys

Alloy	Ni	N	Co	Cr	Ti	Mo	Al	Fe	Mn	Cu	W	C	Si	V
Stainless Steel														
(ASTM F138)	10–15.5	<0.5	[a]	17–19	[a]	2–4	[a]	61–68	[a]	<0.5	<2.0	<0.06	<1.0	[a]
CoCrMo Alloys														
(ASTM F75)	<2.0	[a]	61–66	27–30	[a]	4.5–7.0	[a]	<1.5	<1.0	[a]	[a]	<0.35	<1.0	[a]
(ASTM F90)	9–11	[a]	46–51	19–20	[a]	[a]	[a]	<3.0	<2.5	[a]	14–16	<0.15	<1.0	[a]
(ASTM F562)	33–37	[a]	35	19–21	<1	9.0–11	[a]	<1	<0.15	[a]	[a]	[a]	<0.15	[a]
Ti Alloys														
CP-Ti (ASTM F67)	[a]	[a]	[a]	[a]	99	[a]	[a]	0.2–0.5	[a]	[a]	[a]	<0.1	[a]	[a]
Ti-6Al-4V (ASTM F136)	[a]	[a]	[a]	[a]	89–91	[a]	5.5–6.5	[a]	[a]	[a]	[a]	<0.08	[a]	3.5–4.5
45TiNi	55	[a]	[a]	[a]	45	[a]	[a]	[a]	[a]	[a]	[a]	[a]	[a]	[a]
Zr Alloy														
(95% Zr, 5% Nb)	[a]	[a]	[a]	[a]	[a]	[a]	[a]	[a]	[a]	[a]	[a]	[a]	[a]	[a]

[a]Indicates less than 0.05%

Note: Alloy compositions are standardized by the American Society for Testing and Materials (ASTM vol. 13.01) cells facilitating infiltration and release of migration inhibitory factor (MIF), which inhibits the migration of macrophages away from the site of a DTH reaction. Therefore, in the final phases of a DTH response there is infiltration, activation, and eventual migration inhibition of innate immune cells such as macrophages, controlled by adaptive immune cells (T-cells). These locally attracted and activated macrophages have the increased ability to phagocytose, process, and present metal–protein complexes in class II MHC complexes and IL-1, which trigger the activation of more T cells, which in turn activates more macrophages, which activate more T-cells, in a vicious cycle. This DTH self-perpetuation response can create extensive tissue damage. Current research efforts to use immunosuppressive therapy in people to temporarily stop this cycle are currently underway by us and others.

There is much about implant debris sensitivity that remains incompletely understood, including all the reactive lymphocyte subpopulations and the cellular mechanisms of recognition and activation, as well as specific antigenic metal-protein determinants. The Langerhans cells are well characterized as the primary antigen presenting cells (APCs) associated with skin hypersensitivity. While APCs in the periprosthetic region include macrophages, endothelial cells, lymphocytes, Langerhans cells, dendritic cells and, to lesser extent, parenchymal tissue cells, tissue macrophages (histiocytes) are presumed to be the primary peri-implant APCs. The T-cell receptor (TCR) has been widely acknowledged as involved in metal-induced activation [24–26]. Metals have also been shown to act as both typical antigen (such as tetanus toxin), but have also been shown to cause nontypical activation of T-cells by cross-linking receptors (e.g., VB17 of CDR1 T-cell receptor) to create what is called "superantigen"-like enhancement of T-cell receptor–protein contact [24, 27]. Despite reports of nontypical DTH-like metal-induced lymphocyte activation, the traditional DTH response (where there is one clonally specific group of lymphocytes specific to a single lock-and-key type T-cell receptor mechanism of activation) remains the dominant mechanism associated with implant-related hypersensitivity responses [28–30].

3 Testing for Metal Sensitivity

Testing for metal allergy can be accomplished by skin testing (i.e., so-called patch testing or intradermal testing) or by lymphocyte transformation testing (LTT). There are commercial kits that contain some of the metals in orthopedic implants [22, 31] (e.g., TrueTest™, Glaxo Dermatology, Research Triangle Park, NC). However, there is concern about the applicability of skin testing to diagnose immune responses around implants [14–16, 32, 33]. Patch testing involves antigens (e.g., 1% aqueous nickel sulfate) in a carrier, such as petrolatum, that are placed on dermal tissue for approximately 48–96 h after which reactions are graded on a scale of 1 (mild or absent response) to 4 (severe red rash that can contain small encrusted weeping blisters). One problem with this is that the immunogenic potential of metals on the dermis is likely quite different from the closed periprosthetic *in vivo* environment. On the skin dermal Langerhans cells are the primary hypersensitivity effector cells and around implants macrophages are the primary antigen presenting cells, which are not nearly as efficient as Langerhans dendritic cells [23, 34]. Other general concerns associated with patch testing include the possible induction of hypersensitivity in a previously insensitive patient [35].

In vitro metal allergy testing, called lymphocyte proliferation testing (also known as LTT), involves measuring the proliferative response of lymphocytes after they are activated by an antigen. A radioactive marker is used to precisely measure the amount of cell division over a set time period by measuring the amount of radioactive [H^3]-thymidine that is incorporated into the cellular DNA upon cell division after 4–6 days of exposure to antigen. [^3H]-thymidine uptake is measured using liquid scintillation, and the amount of immune response (proliferation factor or stimulation index) is calculated using measured radiation counts per minute (cpm):

$$\text{Proliferation factor} = (\text{cpm with treatment})/(\text{cpm without treatment}).$$

The use of proliferation testing to measure metal allergy and in the assessment of general drug sensitivity has been well established as a method of testing in a variety of clinical settings [36–40]. There is a growing use of LTT testing for implant-related metal sensitivity that has been shown to have diagnostic efficacy particularly in the area of metal-on-metal implants, which have led to higher rates of metal sensitivity responses [41–43]. Several investigations indicate that metal allergy can be more easily detected by LTT than by dermal patch testing [41, 44, 45]. Thus, given the growing number of studies utilizing the highly quantitative nature of LTT testing in orthopedics it is likely better suited for the testing of implant-related sensitivity than dermal patch testing [36–41, 46, 47].

4 Case Studies in Metal Implant-Related Metal Sensitivity

There have been many reports over the past 30 years of implants that have elicited an allergy or sensitivity type responses. In these reports implant degradation products have been shown to be temporally linked with specific responses such as severe dermatitis, urticaria, vasculitis [48–53], and/or nonspecific immune suppression [54–58].

One of the first correlations of eczema reaction to metallic orthopedic implants was made in 1966 by Foussereau and Laugier [59], where a nickel-containing implant was associated with hypersensitivity reactions. There have been growing numbers of case reports over the past 40 years that link immune responses with adverse performance of metallic cardiovascular [48, 60, 61], orthopedic [5, 49, 50, 52, 53, 62], plastic surgical [63], and dental [64–70] implants. In some cases immunological reactions have necessitated device removal, which then results in stopping the immune reactions [48–53]. Some of these severe skin reactions [48, 51, 53, 60–62, 71, 72] have been associated with the relatively more general phenomena of metallosis (dark metallic staining of tissue due to excessive implant debris), excessive periprosthetic fibrosis, and muscular necrosis [50, 73, 74].

In one of the earliest case studies implicating an orthopedic implant as a source of metal sensitivity [49], a 20-year-old woman was examined for extensive rashes on her chest and back, 5 months after she had stainless steel screws put in to treat chronic patellar dislocation. Topical steroids helped her condition for 1 year. Then, it worsened with further generalized dermal rashes. "Out of sheer desperation," [49] the stainless steel screws were removed, and in less than 72 h her eczema completely disappeared. "The orthopedist still doubted that the steel screws could be the cause of her dermatitis and applied a stainless steel screw to the skin of her back. In a period of 4 h, generalized puritus and erythema developed" [49]. Metal allergy patch testing showed reactions to nickel, nickel sulfate, and the steel screw. This satisfies Koch's postulate as a causative agent, that when the suspected cause is removed the symptoms abate, and when it is returned the symptoms also return, thus metal allergy associated with implant was solidified nearly 40 years ago as a real phenomenon.

In another example, a 50-year-old woman suffered from persistent abdominal pain and urticaria postoperatively following a cholecystectomy using tantalum metal clips that could only get symptom relief after a plasma exchange, but not with corticosteroids or antihistamines. After removal, the tantalum clips showed visible signs of corrosion, indicating that it was likely the corrosion debris that she was reactive to [5, 50, 52, 53, 63]. There are a number of case studies that show similar temporal and physical evidence of immune reactivity to orthopedic implants [12]. It is these cases of severe metal sensitivity that present the greatest problems.

There are more cases of stainless steel and cobalt alloy implant-induced immune responses than there are to titanium alloy components [5, 12, 51–53, 61, 62, 71, 75, 76]. A case report of cobalt hypersensitivity implicated in the poor performance of cobalt alloy plates and screws used in the fracture fixation of 45-year-old woman's left radius and ulna[43] indicated the induction of periprosthetic fibrosis, patchy muscular necrosis, and chronic inflammatory changes peripherally, 7 years after implantation. However, after the implant was removed and the symptoms (swelling) disappeared, the patient remained reactive to cobalt, as indicated by patch testing [50].

5 Cohort Studies of Implant-Related Metal Sensitivity

Cohort studies generally indicate a correlation between the presence of a metal implant and metal sensitivity [5, 31, 35, 77–85]. When these studies are compared (Fig. 12.1), the incidence of metal sensitivity among patients with well-functioning implants is approximately 25%, roughly twice as high as that of the general population [31, 35, 76, 77, 79, 80, 84, 86, 87]. The average prevalence of metal sensitivity among patients with a failed or poorly functioning implant (as judged by a variety of criteria) is approximately 60% [35, 76, 80, 86, 87], where the prevalence of metal sensitivity in people with failed or failing implants is approximately 6 times that of the general population, and approximately 2–3 times that of all patients with well-functioning implants. It is unknown to what extent this association is reflective of a causal link between immunogenicity to implant debris and poor implant outcome.

Cohort studies showing sensitivity to polymeric materials among patients with well-functioning implants have not been well established, although it has been reported [17, 88], where a 50% incidence of PMMA hypersensitivity of patients ($n = 26$ subjects) was associated with loose total hip prostheses using patch testing and mononuclear cell subset analysis [17]. Other studies have refuted this, where people with well-functioning implants showed no hypersensitivity reactions to PMMA, as determined by patch testing ($n = 112$ subjects) [89].

Generally, investigations of immune responses to implant debris suggest one of three possible outcomes hypotheses (1) metal degradation products are immunogenic [26, 28, 29, 90–92], (2) metal degradation products are immunosuppressive [93–95], or (3) metal degradation products are immunoneutral (i.e., nonbioreactive) [96, 97]. While all three possibilities have been shown to occur, the type of reaction that will occur in any one individual is dependent on the individual, the environment, and the type of implant.

Specific types of implants that release more metallic debris *in vivo* are more likely to induce metal sensitivity. Total hip prostheses with metal-on-metal bearing surfaces have been associated with metal sensitivity to a greater extent than similar designs with metal-on-ultrahigh molecular weight polyethylene bearing surfaces [77, 86]. New generations of metal-on-metal total hip replacement have advantages like larger head sizes that decrease the rates of dislocation after surgery, but have greater reports of failures attributable to excessive inflammatory reactions to metal debris characterized as hypersensitivity-like responses. Some reports show that 76–100% of the people with these metal-on-metal implants which have aseptic implant failures requiring revision also have evidence of histological inflammation accompanied by extensive lymphocyte infiltrates. This is characteristic of delayed-type hypersensitivity responses that are characterized by infiltrations of lymphocytes not normally seen in the peri-implant tissue [98, 99]. These reports have shown that in people with metal-on-metal bearing implant with aseptic loosening, all those being revised have been shown to have extensive lymphocytic infiltrates around the metal debris, indicative of unwanted adaptive immune system reactivity. The rates of sensitivity of these people with well-performing metal-on-metal implants are shown in Fig. 12.1 and are nearly twice that of people with well-performing implants.

However, whether metal sensitivity is causal or not may be beside the point once sensitivity has been established as a negative feedback to implant performance. It is very likely that metal-stimulated lymphocytes participate in the pathogenesis of aseptic osteolysis given that activated lymphocytes release powerful cytokines such as IL-2, IFN-γ, and RANKL (receptor-activated NF-KB ligand), which can directly and indirectly affect bone resorption and turnover by promoting

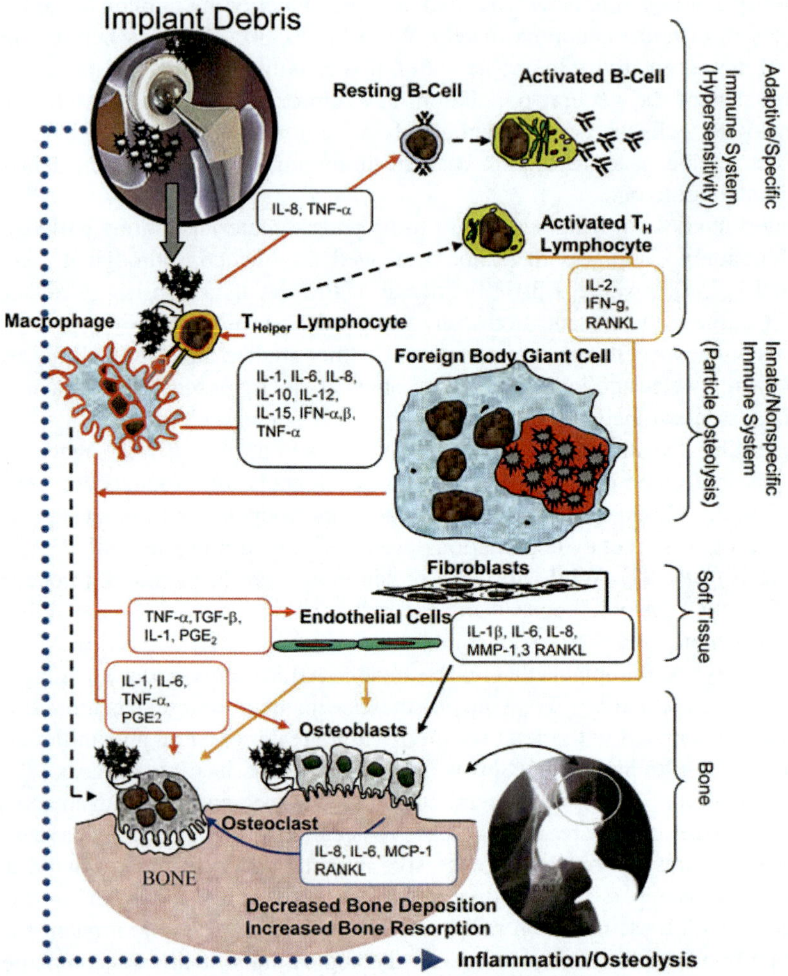

Fig. 12.2 The biologic reactivity of implant debris causes local immune responses primarily mediated by macrophages, which produce reactive oxygen intermediates and pro-inflammatory cytokines that affect a host of local cell types and induce a widening zone of soft-tissue damage and inflammation (picture courtesy of BioEngineering Solutions Inc) [100]

osteoclast activity (bone resorption) and inhibiting osteoblast activity (bone deposition) in a disruptive set of events that result in vicious cycle of inflammation and bone erosion, Fig. 12.2 [101].

Thus, it seems that, although metal-on-metal implants produce less overall debris than metal-on-polymer articulations do, the rates of revision/failure of metal-on-metal THAs may ultimately prove to be as high as metal-on-polymer bearings, and that these failures may be entirely due to exuberant biological responses to implant debris, be they toxic or hypersensitivity-like immune responses. It is likely that, if metal sensitivity could be effectively screened out preoperatively, revision rates of metal-on-metal THAs would drop.

6 Conclusions

It is unclear whether hypersensitivity responses to metallic biomaterials affect implant performance in other than a few highly predisposed or implant-debris sensitized people [12, 22, 102]. It is clear that some patients experience excessive immune reactions directly associated with implanted metallic materials [5, 49, 50, 52, 53, 62]. Metal sensitivity exists as an extreme complication in only a few highly susceptible patients (i.e., less than 1% of joint replacement recipients), and a more common subtle contributor to implant failure. In addition to direct immunogenic responses, metal degradation products may mediate indirect immunologic effects due to cell toxicity, where immune reactions are secondary to this person-dependent toxicity responses. It is likely that implant-related metal sensitivity is underreported due to the scarcity of diagnostic testing. The person-dependent mechanisms by which *in vivo* metal sensitivity occurs in some people and not is others have not been identified, making it difficult to know whether a known condition of metal hypersensitivity will elicit an overaggressive immune response [22, 102]. Increasing popularity of immunologic testing will likely enhance future assessment of patients susceptible to hypersensitivity responses. In the event of temporally related signs of allergic response to implant placement, metal sensitivity should be considered, after infection has been ruled out. To appropriately weigh optimum treatment of patients presenting with signs of an allergic reactions, evaluation for sensitivity should be conducted. Removal of the device that has served its function should be considered, since removal may alleviate the symptoms that may produce other immune-related disorders. Patients who have allergic reactions to cheap jewelry are more likely to have reactions to orthopedic implants. It is important to note that there is increasing awareness of the phenomenon of metal hypersensitivity and many surgeons now take this into account when planning which implant is optimal for each patient. The importance of further study, diagnostic testing, and treatment for sensitivity to implant debris is paramount, as almost 1 in 2 people will eventually require an orthopedic implant, and revision surgery over the age of 75 can result in >13% mortality [2, 103].

References

1. Willert HG, Semlitsch M (1977) Reactions of the articular capsule to wear products of artificial joint prostheses. J Biomed Mater Res 11:157–164
2. Radcliffe GS, Tomichan MC, Andrews M, Stone MH (1999) Revision hip surgery in the elderly: is it worthwhile? J Arthroplasty 14:38–44

3. Holt G, Murnaghan C, Reilly J, Meek RM (2007) The biology of aseptic osteolysis. Clin Orthop Relat Res 460:240–252
4. Basketter DA, Briatico-Vangosa G, Kaestner W, Lally C, Bontinck WJ (1993) Nickel, cobalt and chromium in consumer products: a role in allergic contact dermatitis? Contact Dermatitis 28:15–25
5. Cramers M, Lucht U (1977) Metal sensitivity in patients treated for tibial fractures with plates of stainless steel. Acta Orthop Scand 48:245–249
6. Fisher AA (1977) Allergic dermatitis presumably due to metallic foreign bodies containing nickel or cobalt. Current Contact News 19:285–295
7. Black J (1984) Systemic effects of biomaterials. Biomaterials 5:12–17
8. Jacobs JJ, Gilbert JL, Urban RM (1994) Corrosion of metallic implants. In: Stauffer RN (ed) Advances in orthopaedic surgery, vol 2. Mosby, St. Louis, pp 279–319
9. Gawkrodger DJ (1993) Nickel sensitivity and the implantation of orthopaedic prostheses. Contact Dermatitis 28:257–259
10. Haudrechy P, Foussereau J, Mantout B, Baroux B (1994) Nickel release from nickel-plated metals and stainless steels. Contact Dermatitis 31:249–255
11. Kanerva L, Sipilainen-Malm T, Estlander T, Zitting A, Jolanki R, Tarvainen K (1994) Nickel release from metals, and a case of allergic contact dermatitis from stainless steel. Contact Dermatitis 31:299–303
12. Merritt K, Rodrigo JJ (1996) Immune response to synthetic materials. Sensitization of patients receiving orthopaedic implants. Clin Orthop Relat Res 326:71–79
13. Jacobs JJ, Shanbhag A, Glant TT, Black J, Galante JO (1994) Wear debris in total joint replacements. J Am Acad Orthop Surg 2:212–220
14. Yang J, Black J (1994) Competitive binding of chromium cobalt and nickel to serum proteins. Biomaterials 15:262–268
15. Yang J, Merritt K (1994) Detection of antibodies against corrosion products in patients after Co-Cr total joint replacements. J Biomed Mater Res 28:1249–1258
16. Yang J, Merritt K (1996) Production of monoclonal antibodies to study corrosion of Co-Cr biomaterials. J Biomed Mater Res 31:71–80
17. Gil-Albarova J, Lacleriga A, Barrios C, Canadell J (1992) Lymphocyte response to polymethylmethacrylate in loose total hip prostheses. J Bone Joint Surg [Br] 74:825–830
18. Liden C, Maibach HI, Howard I, Wahlberg JE (1995) Skin. In: Goyer RA, Klaasen CD, Waalkes M (eds) Metal toxicology. Academic, New York, pp 447–464
19. Angle C (1995) Organ-specific therapeutic intervention. In: Goyer RA, Klaassen CD, Waalkes MP (eds) Metal toxicology. Academic, New York, pp 71–110
20. Lalor PA, Revell PA, Gray AB, Wright S, Railton GT, Freeman MA (1991) Sensitivity to titanium. A cause of implant failure. J Bone Joint Surg 73-B:25–28
21. Parker DC (1993) T cell-dependent B cell activation. Annu Rev Immunol 11:331–360
22. Hensten-Pettersen A (1993) Allergy and hypersensitivity. In: Morrey BF (ed) Biological, material, and mechanical considerations of joint replacements. Raven Press, New York, pp 353–360
23. Kuby J (1991) Immunology, 2nd edn. W.H. Freeman and Company, New York
24. Moulon C, Vollmer J, Weltzien HU (1995) Characterization of processing requirements and metal cross-reactivities in T cell clones from patients with allergic contact dermatitis to nickel. Eur J Immunol 25:3308–3315
25. Saito K (1996) Analysis of a genetic factor of metal allergy–polymorphism of HLA-DR, -DQ gene. Kokubyo Gakkai Zasshi 63:53–69
26. Silvennoinen-Kassinen S, Poikonen K, Ikaheimo I (1991) Characterization of nickel-specific T cell clones. Scand J Immunol 33:429–434
27. Gamerdinger K, Moulon C, Karp DR, Van BJ, Koning F, Wild D, Pflugfelder U, Weltzien HU (2003) A new type of metal recognition by human T cells: contact residues for peptide-independent bridging of T cell receptor and major histocompatibility complex by nickel. J Exp Med 197:1345–1353

28. Griem P, Gleichmann E (1995) Metal ion induced autoimmunity. Curr Opin Immunol 7:831–838
29. Griem P, von Vultee C, Panthel K, Best SL, Sadler PJ, Shaw CF (1998) T cell cross-reactivity to heavy metals: identical cryptic peptides may be presented from protein exposed to different metals. Eur J Immunol 28:1941–1947
30. Kubicka-Muranyi M, Griem P, Lubben B, Rottmann N, Luhrmann R, Gleichmann E (1995) Mercuric-chloride-induced autoimmunity in mice involves up-regulated presentation by spleen cells of altered and unaltered nucleolar self antigen. Int Arch Allergy Immunol 108:1–10
31. Rooker GD, Wilkinson JD (1980) Metal sensitivity in patients undergoing hip replacement. A prospective study. J Bone Joint Surg 62-B:502–505
32. Hallab N (2001) Metal sensitivity in patients with orthopedic implants. J Clin Rheumatol 7:215–218
33. Woodman JL, Black J, Jiminez SA (1984) Isolation of serum protein organometallic corrosion products from 316LSS and HS-21 in vitro and in vivo. J Biomed Mater Res 18:99–114
34. Korenblat PE (1992) Contact Dermatitis, 2nd edn. W.B. Saunders Company, Philidelphia
35. Mayor MB, Merritt K, Brown SA (1980) Metal allergy and the surgical patient. Am J Surg 139:477–479
36. Everness KM, Gawkrodger DJ, Botham PA, Hunter JA (1990) The discrimination between nickel-sensitive and non-nickel-sensitive subjects by an in vitro lymphocyte transformation test. Br J Dermatol 122:293–298
37. Secher L, Svejgaard E, Hansen GS (1977) T and B lymphocytes in contact and atopic dermatitis. Br J Dermatol 97:537–541
38. Svejgaard E, Morling N, Svejgaard A, Veien NK (1978) Lymphocyte transformation induced by nickel sulphate: an in vitro study of subjects with and without a positive nickel patch test. Acta Derm Venereol 58:245–250
39. Veien NK, Svejgaard E (1978) Lymphocyte transformation in patients with cobalt dermatitis. Br J Dermatol 99:191–196
40. Veien NK, Svejgaard E, Menne T (1979) In vitro lymphocyte transformation to nickel: a study of nickel-sensitive patients before and after epicutaneous and oral challenge with nickel. Acta Derm Venereol 59:447–451
41. Carando S, Cannas M, Rossi P, Portigliatti-Barbos M (1985) The lymphocytic transformation test (L.T.T.) in the evaluation of intolerance in prosthetic implants. Ital. J Orthop Traumatol 11:475–481
42. Granchi D, Ciapetti G, Stea S, Savarino L, Filippini F, Sudanese A, Zinghi G, Montanaro L (1999) Cytokine release in mononuclear cells of patients with Co-Cr hip prosthesis. Biomaterials 20:1079–1086
43. Granchi D, Verri E, Ciapetti G, Stea S, Savarino L, Sudanese A, Mieti M, Rotini R, Dallari D, Zinghi G, Montanaro L (1998) Bone-resorbing cytokines in serum of patients with aseptic loosening of hip prostheses. J Bone Joint Surg [Br] 80:912–917
44. Granchi D, Ciapetti G, Savarino L, Cavedagna D, Donati ME, Pizzoferrato A (1996) Assessment of metal extract toxicity on human lymphocytes cultured in vitro. J Biomed Mater Res 31:183–191
45. Granchi D, Verri E, Ciapetti G, Savarino L, Cenni E, Gori A, Pizzoferrato A (1998) Effects of chromium extract on cytokine release by mononuclear cells. Biomaterials 19:283–291
46. Granchi D, Ciapetti G, Stea S, Cavedagna D, Bettini N, Bianco T, Fontanesi G, Pizzoferrato A (1995) Evaluation of several immunological parameters in patients with aseptic loosening of hip arthroplasty. Chir Organi Mov 80:399–408
47. Granchi D, Savarino L, Ciapetti G, Cenni E, Rotini R, Mieti M, Baldini N, Giunti A (2003) Immunological changes in patients with primary osteoarthritis of the hip after total joint replacement. J Bone Joint Surg Br 85:758–764
48. Abdallah HI, Balsara RK, O'Riordan AC (1994) Pacemaker contact sensitivity: clinical recognition and management. Ann Thorac Surg 57:1017–1018

49. Barranco VP, Solloman H (1972) Eczematous dermatitis from nickel. JAMA 220:1244
50. Halpin DS (1975) An unusual reaction in muscle in association with a vitallium plate: a report of possible metal hypersensitivity. J Bone Joint Surg 57-B:451–453
51. King J, Fransway A, Adkins RB (1993) Chronic urticaria due to surgical clips. N Engl J Med 329:1583–1584
52. Merle C, Vigan M, Devred D, Girardin P, Adessi B, Laurent R (1992) Generalized eczema from vitallium osteosynthesis material. Contact Dermatitis 27:257–258
53. Thomas RH, Rademaker M, Goddard NJ, Munro DD (1987) Severe eczema of the hands due to an orthopaedic plate made of Vitallium. Br Med J 294:106–107
54. Bravo I, Carvalho GS, Barbosa MA, de Sousa M (1990) Differential effects of eight metal ions on lymphocyte differentiation antigens in vitro. J Biomed Mater Res 24:1059–1068
55. Gillespie WJ, Frampton CM, Henderson RJ, Ryan PM (1988) The incidence of cancer following total hip replacement. J Bone Joint Surg [Br] 70:539–542
56. Merritt K, Brown SA (1985) Biological effects of corrosion products from metal. American Society for Testing and Materials, Philadelphia
57. Poss R, Thornhill TS, Ewald FC, Thomas WH, Batte NJ, Sledge CB (1984) Factors influencing the incidence and outcome of infection following total joint arthoplasty. Clin Orthop Relat Res (182):117–126
58. Wang JY, Wicklund BH, Gustilo RB, Tsukayama DT (1997) Prosthetic metals interfere with the functions of human osteoblast cells in vitro. Clin Orthop Relat Res (339):216–226
59. Foussereau J, Laugier (1966) Allergic eczemas from metallic foreign bodies. Trans St John's Hosp Derm Soc 52:220–225
60. Buchet S, Blanc D, Humbert P, Girardin P, Vigan M, Anguenot T, Agache P (1992) Pacemaker dermatitis. Contact Dermatitis 26:46–47
61. Peters MS, Schroeter AL, Hale HM, Broadbent JC (1984) Pacemaker contact sensitivity. Contact Dermatitis 11:214–218
62. Rostoker G, Robin J, Binet O, Blamutier J, Paupe J, Lessana-Liebowitch M, Bedouelle J, Sonneck JM, Garrel JB, Millet P (1987) Dermatitis due to orthopaedic implants. A review of the literature and report of three cases. J Bone Joint Surg 69-A:1408–1412
63. Holgers KM, Roupi G, Tjellstrom A, Bjurstem LM (1992) Clinical, immunological and bacteriological evaluation of adverse reactions to skin-penetrating titanium implants in the head and neck region. Contact Dermatitis 27:1–7
64. Bruze E, Bjorkner BB, Moller H (1994) Clinical relevance of contact allergy to gold sodium. J Am Acad Dermatol 31:579–583
65. Guimaraens D, Gonzalez MA, Conde-Salazar L (1994) Systemic contact dermatitis from dental crowns. Contact Dermatitis 30:124–125
66. Helton J, Storrs F (1994) The burning mouth syndrome: lack of a role for contact urticaria. J Am Acad Dermatol 31:205
67. Hubler WJ, Hubler WS (1983) Dermatitis from a chromium dental plate. Contact Dermatitis 9:377–383
68. Laeijendecker R, vanJoost T (1994) Oral manifestations of gold allergy. J Am Acad Dermatol 30:205–209
69. Spiechowitz E, Glantz PO, Axell T, Chmieleweski W (1984) Oral exposure to a nickel-containing dental alloy of persons with hypersensitive skin reactions to nickel. Contact Dermatitis 10:206–211
70. Vilaplana J, Rmoaguera C, Cornellana F (1994) Contact dermatitis and adverse oral mucus membrane reactions related to the use of dental prosthesis. Cantact Dermatitis 30:80–84
71. Gordon PM, White MI, Scotland TR (1994) Generalized sensitivity from an implanted orthopaedic antibiotic minichain containing nickel. Contact Dermatitis 30:181–182
72. Parker AW, Drez-Jr D, Jacobs JJ (1993) Titanium dermatitis after failure of a metal-backed patellas. Am J Knee Surg 6:129–131

73. Black J, Sherk H, Bonini J, Rostoker WR, Schajowicz F, Galante JO (1990) Metallosis associated with a stable titanium-alloy femoral component in total hip replacement. J Bone Joint Surg 72A:126–130
74. Nakamura S, Yasunaga Y, Ikutu Y, Shimogaki K, Hamada N, Takata N (1997) Autoantibodies to red cells associated with metallosis-a case report. Acta Orthop Scand 68:495–496
75. Black J (1988) Orthopaedic Biomaterials in Research and Practice. Churchill Livingstone, New York
76. Elves MW, Wilson JN, Scales JT, Kemp HB (1975) Incidence of metal sensitivity in patients with total joint replacements. Br Med J 4:376–378
77. Benson MK, Goodwin PG, Brostoff J (1975) Metal sensitivity in patients with joint replacement arthroplasties. Br Med J 4:374–375
78. Brown GC, Lockshin MD, Salvati EA, Bullough PG (1977) Sensitivity to metal as a possible cause of sterile loosening after cobalt-chromium total hip-replacement arthroplasty. J Bone Joint Surg Am 59-A:164–168
79. Carlsson AS, Macnusson B, Moller H (1980) Metal sensitivity in patients with metal-to-plastic total hip arthroplasties. Acta Orthop Scand 51:57–62
80. Deutman R, Mulder TH, Brian R, Nater JP (1977) Metal sensitivity before and after total hip arthroplasty. J Bone Joint Surg [Am] 59-A:862–865
81. Fischer T, Rystedt I, Safwenberg J, Egle I (1984) HLA -A, -B, -C and -DR antigens in individuals with sensitivity to cobalt. Acta Derm Venereol (Stockh) 64:121–124
82. Kubba R, Taylor JS, Marks KE (1981) Cutaneous complications of orthopedic implants. A two-year prospective study. Arch Dermatol 117:554–560
83. Merritt K (1984) Role of medical materials, both in implant and surface applications, in immune response and in resistance to infection. Biomaterials 5:53–57
84. Merritt K, Brown S (1981) Metal sensitivity reactions to orthopedic implants. Int J Dermatol 20:89–94
85. Pinkston JA, Finch SC (1979) A method for the differentiation of T and B lymphocytes and monocytes migrating under agarose. Stain Technol 54:233–239
86. Evans EM, Freeman MA, Miller AJ, Vernon-Roberts B (1974) Metal sensitivity as a cause of bone necrosis and loosening of the prosthesis in total joint replacement. J Bone Joint Surg 56-B:626–642
87. Munro-Ashman D, Miller AJ (1976) Rejection of metal to metal prosthesis and skin sensitivity to cobalt. Contact Dermatitis 2:65
88. Wooley PH, Fitzgerald RHJ, Song Z, Davis P, Whalen JD, Trumble S, Nasser S (1999) Proteins bound to polyethylene components in patients who have aseptic loosening after total joint arthroplasty. A preliminary report [see comments]. J Bone Joint Surg [Am] 81:616–623
89. Carlsson A, Moller H (1989) Implantation of orthopaedic devices in patients with metal allergy. Acta Derm Venereol (Stockh) 69:62–66
90. Ungersbock A, Pohler O, Perren SM (1994) Evaluation of the soft tissue interface at titanium implants with different surface treatments: experimental study on rabbits. Biomed Mater Eng 4:317–325
91. Warner GL, Lawrence DA (1986) Cell surface and cell cycle analysis of metal-induced murine T cell proliferation. Eur J Immunol 16:1337–1342
92. Warner GL, Lawrence DA (1988) The effect of metals on IL-2-related lymphocyte proliferation. Int J Immunopharmacol 10:629–637
93. Kubicka-Muranyi M, Goebels R, Goebel C, Uetrecht J, Gleichmann E (1993) T lymphocytes ignore procainamide, but respond to its reactive metabolites in peritoneal cells: demonstration by the adoptive transfer popliteal lymph node assay. Toxicol Appl Pharmacol 122:88–94
94. Thompson GJ, Puleo DA (1995) Effects of sublethal metal ion concentrations on osteogenic cells derived from bone marrow stromal cells. J Appl Biomater 6:249–258

95. Wang JY, Tsukayama DT, Wicklund BH, Gustilo RB (1996) Inhibition of T and B cell proliferation by titanium, cobalt, and chromium: role of IL-2 and IL-6. J Biomed Mater Res 32:655–661
96. Kohilas K, Lyons M, Lofthouse R, Frondoza CG, Jinnah R, Hungerford DS (1999) Effect of prosthetic titanium wear debris on mitogen-induced monocyte and lymphoid activation. J Biomed Mater Res 47:95–103
97. Vollmer J, Fritz M, Dormoy A, Weltzien HU, Moulon C (1997) Dominance of the BV17 element in nickel-specific human T cell receptors relates to severity of contact sensitivity. Eur J Immunol 27:1865–1874
98. Korovessis P, Petsinis G, Repanti M, Repantis T (2006) Metallosis after contemporary metal-on-metal total hip arthroplasty. Five to nine-year follow-up. J Bone Joint Surg Am 88:1183–1191
99. Milosev I, Trebse R, Kovac S, Cor A, Pisot V (2006) Survivorship and retrieval analysis of Sikomet metal-on-metal total hip replacements at a mean of seven years. J Bone Joint Surg Am 88:1173–1182
100. Gristina AG (1994) Implant failure and the immuno-incompetent fibro-inflammatory zone. Clin Orthop Relat Res (298):106–118
101. Hallab NJ, Vermes C, Messina C, Roebuck KA, Glant TT, Jacobs JJ (2002) Concentration- and composition-dependent effects of metal ions on human MG-63 osteoblasts. J Biomed Mater Res 60:420–433
102. Boyan BD (1993) Discussion of Toxicity and Allergy. Raven Press, New York
103. Jacobs JJ, Skipor AK, Doorn PF, Campbell P, Schmalzried TP, Black J, Amstutz HC (1996) Cobalt and chromium concentrations in patients with metal on metal total hip replacements. Clin Orthop Relat Res (329 Suppl):S256–S263

Biography

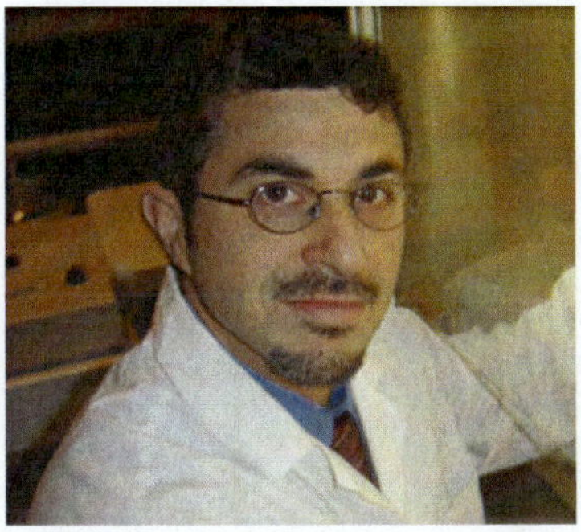

Dr. Nadim James Hallab was born in Baton Rouge, LA, USA, received his BS and MS in Mechanical Engineering at Texas A&M University, and his PhD in Biomedical Engineering at Tulane University in 1996. He was a postdoctoral fellow at Rush University Medical Center in the Orthopedic Department for 2 years, studying implant corrosion, metal-protein binding, and the biologic effects of soluble metallic implant debris. He is currently an Associate Professor in Orthopedic Surgery at Rush University Medical Center where he has been a researcher for the past 13 years, investigating implant debris generation and the biologic reactivity to implant debris. He is an adjunct faculty member in the Department of Cell Biology/Anatomy, Department of Immunology at Rush, and the Dept. of Biomedical Engineering at University of Illinois at Chicago. He is also currently the CEO of an implant debris analysis company (BioEngineering Solutions Inc.) and a metal allergy testing company (Orthopedic Analysis Inc.), which have been in business since 2004. He has over 65 peer reviewed publications and 15 book chapters in the field of implant degradation and immunologic reactivity to implant debris. Current awards include the 2002 Kappa Delta Award (Team Member) and the 2009 Harris Award for excellence in Orthopedic Research for the discovery that danger signaling mechanisms (inflammasome pathways) are associated with inflammation and hypersensitivity reactions to soluble and particulate implant debris. While occasionally still conducting corrosion and biomechanical testing research, the bulk of his current research involves innate and adaptive immunologic reactivity to implant debris.

Chapter 13
Implant Infections and Infection-Resistant Materials

Davide Campoccia, Lucio Montanaro, and Carla Renata Arciola

Abstract Infection is broadly recognized as one of the most critical and devastating complications associated with the use of biomaterials, particularly in orthopaedic prosthesis surgery.

The idea of combining biomaterials with antimicrobial substances for prophylaxis of infections through local delivery was first developed in dentistry and then the combination of antibiotic cement was proposed also in orthopaedic surgery. Various prophylaxis measures or antimicrobial treatments are analysed and their different efficacies are considered. Different strategies in designing and using infection-resistant and anti-infective biomaterials are presented and discussed.

1 Implant Infections and Historical Approach to Their Prevention

The event of infection is broadly recognized as one of the most critical and devastating complications associated with the use of biomaterials in medicine [1–3]. Over the last decades, improvements in the practice of asepsis and perioperative prophylaxis provided only partial success in reducing the rate of implant-related infections and, anyway, the limited achievements have been counterbalanced by the explosive diffusion of prosthetic materials in all sectors of medicine. In particular, the failure to obtain a total control over contamination of the surgical field, totally

D. Campoccia
Research Unit on Implant Infections, Rizzoli Orthopaedic Institute, Bologna, Italy

L. Montanaro • C.R. Arciola (✉)
Research Unit on Implant Infections, Rizzoli Orthopaedic Institute, Bologna, Italy

Experimental Pathology Department, University of Bologna, Via S. Giacomo 14, Bologna 40126, Italy
e-mail: carlarenata.arciola@ior.it

N. Eliaz (ed.), *Degradation of Implant Materials*, DOI 10.1007/978-1-4614-3942-4_13,

eliminating the chance of bacterial seeding on materials surfaces, and the spreading of bacterial strains resistant to most antibiotics used in standard protocols of perioperative prophylaxis, has progressively led to an increased awareness that the tactic to deal with this complex phenomenon should rely on multiple strategies. The attempt to preventively interfere with and reduce the chances of bacterial seeding on biomaterials surfaces has led to very modest advancements, mostly observable with surgical implantation of totally internal devices. However, many other indwelling devices remain percutaneous (e.g. external orthopaedic fixation systems), totally internal but with periodic external access (e.g. arteriovenous graft for haemodialysis), or totally external though in contact with vulnerable mucosal membranes (e.g. urinary catheters). Under these circumstances, the exposure to microbial contamination is not confined just to the short time window of surgical implantation or to the remote risk of haematogenous spreading from a distant site of infection thereafter. The continuity of the exposure to microbial contamination and colonization makes this type of medical devices particularly prone to infections, which can easily turn from superficial to deep and invasive, no matter the insertion was performed with best aseptic techniques avoiding any source of initial contamination. Perioperative prophylaxis is of no value or even not applicable in such conditions.

General awareness of implant-related infections and their huge cost in individual, social, and economic terms has been progressively growing with the explosion in the use of biomaterials in all medical sectors especially observed in the last 30–40 years. Nevertheless, the development of new adjuvant preventive strategies, not based on systemic perioperative use of antibiotics, is dated very long ago. Indeed, adaptation of clinical procedures to prevent or treat medical post-surgical infections was introduced in surgery at a very early stage [4], even before the age of antibiotics.

In the era preceding the discovery of microorganisms, the selection of implantable devices was at times empirically driven towards the choice of materials intrinsically bactericidal. For instance, the use of silver-wire ligatures was practiced before the year 1849 [5], and as early as 1855 silver was the recommended material for indwelling catheters in gynaecology [6]. However, only by the end of the nineteenth century, with the discovery of inhibitive and bactericidal action of various metals [4], dressings based on silver foil found an expressed indication as biomaterials selectively endowed with bacterio-inhibitive power and selective action with respect to host tissue toxicity.

This debut of a dressing with claimed antiseptic properties can probably be considered one of the earliest conscious moves in the development of bioactive biomaterials, and a primary pioneer cultural step in the path of establishing strategies to prevent or combat infections at a local level.

The discovery of antibiotics, substances with a highly specific activity against prokaryotic cells, represented an important step towards the management of infections, not just in clinical treatment but even in the perioperative prophylaxis of postsurgical infections. In fact, first reports on perioperative antibiotic prophylaxis date back to the early 1950s [7], following a pair of decades of growing evidences of the efficacy of antibiotic use for clinical treatment in different infective pathologies. The introduction of antibiotic-based prophylaxis has certainly to be regarded as a major advancement in the fight of surgical infections and, among them, implant-related infections, although best achievements were obtained only with the definition of appropriate protocols of perioperative antibiotic administration.

The idea of combining biomaterials with antibiotic substances for prophylaxis of infections through local delivery was first developed in dentistry just after the introduction of systemic perioperative prophylaxis in surgical practice [8]. However, it took nearly 2 decades before the combination of antibiotic cement was proposed also in orthopaedic surgery [9, 10].

Analogously to what described for acrylic cements, even the possibility of locally eluting antibacterial drugs by means of resorbable materials was initially explored in dentistry in a rather early phase [11, 12]. Degradable biomaterials certainly possess great potential for local delivery as they can control the kinetics of drug release and, at least in some cases, they can completely degrade without leaving residual particles which could offer a lasting support for biofilm growth (Fig. 13.1) and microbial survival in the tissues. However, despite the points of strength behind the rational of using degradable biomaterials for in-site antibiotic release, a number of recent papers raise doubts on the *in vivo* performance of antibiotic adsorbed in natural polymers such as collagen [13, 14].

With the years, clinical use of biomaterials possessing antibacterial properties has broadly been extended to a variety of surgical and non-surgical procedures in different medical fields. The introduction in clinical use took place often in the absence of a sound evidence of *in vivo* efficacy. Although the reasons that lead surgeons to clinical use of local antibiotics driven by their good sense and care for the patient are perfectly understandable, every effort should be made to have clear epidemiological confirmations of the correctness of the medical protocols applied. In this regard, especially in the case of devices releasing subinhibitory amounts of antibiotics in orthopaedics, the overall benefits for the patient as well as for the community have often been debated, appearing questionable both for the non-proved prophylactic activity in the patient and for the potential of favouring the emergence of antibiotic-resistant strains [15].

2 A Solution Requiring More Than a Single Strategy

Large clinical studies have shown, as in the prevention of implant infections, that the best results are achieved when antibiotic prophylaxis is given both systemically and in the bone cement [16]. While systemic treatments act effectively to reduce viable bacteria present in distant vascularized tissues surrounding the implant (dispersed in the extracellular space or within host cells phagosomes), antibiotics eluted from bioactive materials more actively interfere with the process of surface adhesion and colonization, performing their action where needed, i.e. in close proximity to the implant surface. Furthermore, the delivery at the site of

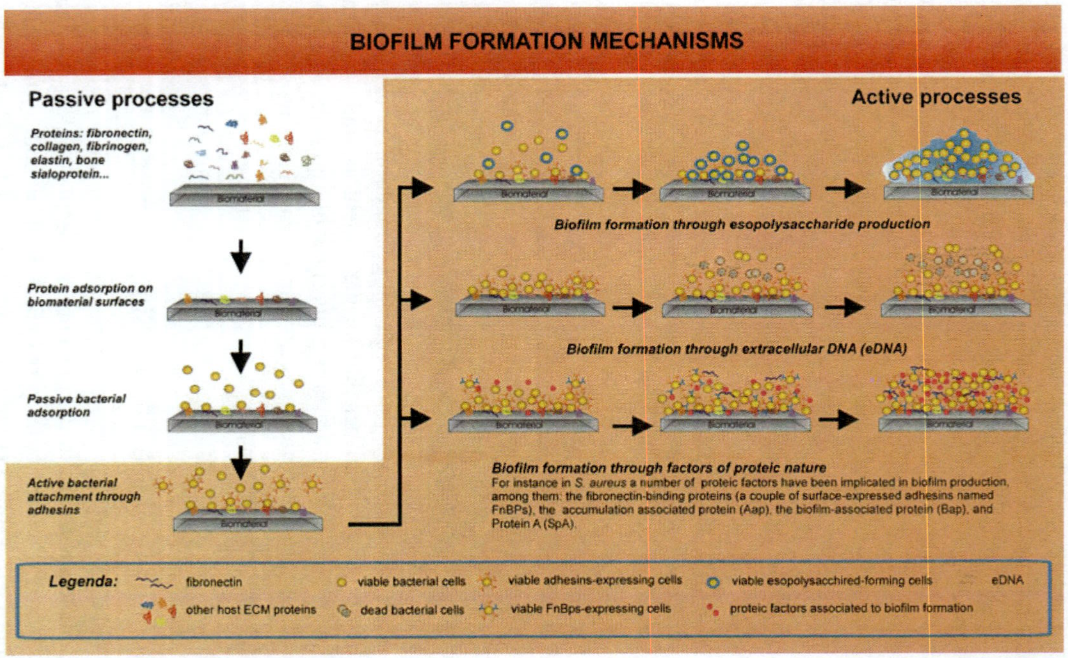

Fig. 13.1 The presence of a biomaterial favours bacterial adhesion and colonization. Scheme representing the complex process leading to microbial biofilm formation starting from the initial phases of passive adsorption of bacterial cells on the biomaterial surface. Following the adhesion of the cells mediated by receptorial proteins termed adhesins, capable to selectively bind host extracellular matrix (ECM) proteins, a number of mechanisms of biofilm formation can be actuated. Different bacterial clones can be variously endowed in their virulon with genes enabling specific mechanisms associated with bacterial accumulation and biofilm production. At times, more than one mechanism can coexist within the same strain, being used contemporarily or in distinct phases of cell aggregation and biofilm accumulation under the influence of cell density or of the niche and the interstitial milieu [25]

implantation permits the diffusion of antibiotic substances even in ischemic regions that cannot be reached by systemic therapies. Similarly, in the case of some percutaneous external fixation devices in use in orthopaedic surgery, anti-infective surfaces have demonstrated to be capable of reducing the occurrence of superficial infections deriving from external colonization of the implant. Interestingly, in the same model, deep infections were better prevented by systemic antibiotics. The efficacy of various prophylactic measures is illustrated in Fig. 13.2.

Altogether, these findings strongly suggest that to adequately deal with implant infections the approach has to rely on more than just a single strategy [2, 17]. A range of different expedients have been proposed to prevent and treat implant-related infections, varying from systemic (e.g. systemic antibiotics, vaccines targeting capsular components, virulence determinants, or even DNA sequences specific of pathogens implicated in the pathogenesis of implant-related infection, and so on) to local systems usually based on bioactive biomaterials.

3 Infection-Resistant Materials

Focussing just on the local strategies centred on biomaterials we should first distinguish those biomaterials designed with the main scope to exert an antibacterial effect at the site of application, from implant biomaterials that combine a specific anatomic or physiological function, that of native implant, with an adjuvant function conferring to the device resistance to microbial colonization. The former biomaterials can be either viscous, jelly, or solid and they are principally aimed at locally delivering antibiotic or antibacterial drugs in wounds of surgical and non-surgical origin at high risk of infection. In this category we may include medicated bioactive dressings, when the application is superficial, or even biomaterials performing also a cavity filling function, when the application is deep and invasive. For instance, antibiotic-loaded poly-methyl-methacrylate (PMMA) beads have been used for long time to locally treat infected tissues after the removal for septic failure of orthopaedic implants. In light of the high risk of recidive in these circumstances, their use was introduced to drastically abate tissue contamination, combining systemic antibiotic therapies with local delivery. More recent developments have brought to dismiss the use of beads in favour of antibiotic loaded temporary spacers capable additionally to perform the anatomic function of the removed implant.

The latter category of biomaterials has been largely broadening over the last decades, following the unmet success in completely controlling the phenomenon of implant-related infection through the practice of asepsis and perioperative antibiotics. The achievement of implants with anti-infective properties can be pursued through different tactics. The first approach to reduce the rate of infection on biomaterial surfaces is to interfere with the early steps of colonization, i.e. the initial adhesion of bacterial cells. Surfaces characterized by low bacterial adhesion can be obtained through various polymeric coatings and surface conditioning

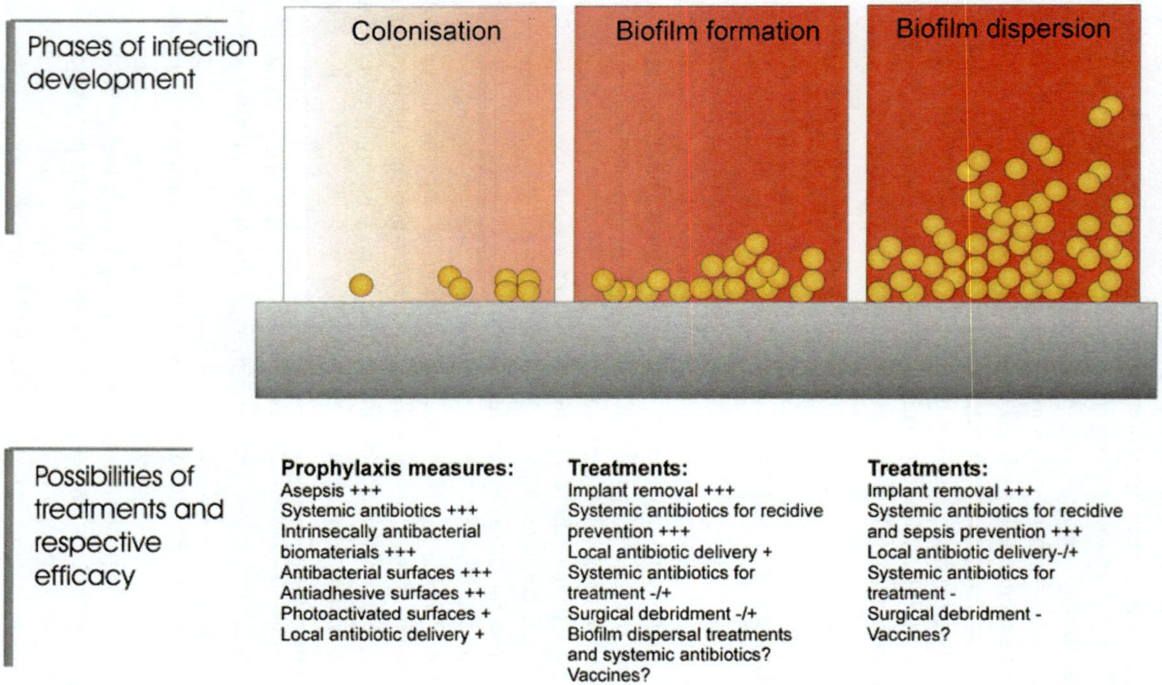

Fig. 13.2 Different efficacy of various prophylactic measures. This scheme attempts to enlist the different efficacy of various prophylactic measures or antimicrobial treatments based on the phase of development of an implant infection, from the early adhesion phenomena to the production of a mature biofilm, and finally to its latest dispersion. What reported has to be regarded as very indicative to illustrate the different impact of various strategies depending on the stage of the infection. In fact, the effects of the different treatments greatly change with the type of medical device and the site of implantation. Moreover, for many treatments there is still large controversy and lack of definitive results on the level of efficacy based on clinical evidence

treatments, including the use of heparinized hydrophilic polymer [18], cross-linked poly(ethylene glycol)-based polymer [19], hydrophilic polymer brushes [20], as well as low-surface energy polymers (e.g. poly(methylpropenoxyfluoroalkylsiloxane)s or poly(perfluoroacrylate)s) [21]. However, the efficacy of low adhesion surfaces is significantly affected by the conditions of use and the anatomic location of the device and, at times, the efficiency of the system is highly affected simply by the presence of protein-rich physiologic fluids. Furthermore, surface energy and charge of the bacterial surface varies greatly in different microbial species when not even in different strains of the same species, affecting the dynamics of passive microbial absorption and the efficacy of the surface treatment.

Another approach to prevent bacterial adhesion acting at the superficial level is based on engineering the biomaterial surface at a nanoscale, creating appropriate nanotopographies [22]. Although the idea of acting on the topology rather than on the chemistry of the biomaterial surface could certainly have a huge potential, evidences of clinically relevant effects are still not documented, and some findings seem to suggest that in physiological conditions the simple presence of extracellular proteins could have profound effects by far more significant [23].

For both chemically modified/conditioned surfaces and nanostructured surfaces it is crucial that the alteration introduced does not translate into a modified tissue response that could compromise compatibility and integration of the implant. Interesting studies have demonstrated the possibility to reduce bacterial adhesion while promoting eukaryotic cell attachment in a selective manner [24].

Alternative, more potent, strategies to achieve infection-resistant biomaterials are oriented to the design of implants made of biomaterials intrinsically bactericidal. In this case, not just the biomaterial surface but even the bulk is both capable of actively contrasting microbial infection thanks to their microbicidal action. Even when passive adhesion of bacteria takes place, microbial survival and growth are actively hampered. The number of biomaterials at the same time compatible with human tissues and possessing bactericidal properties is anyway restricted to very few. These include chitosan, quaternary polymers (e.g. copolymers of 4-vinylpyridine and poly(ethylene glycol) methyl ether methacrylate), silver compounds, photoactivated TiO_2 among them. Strong bactericidal properties of materials are rarely conjugated with a high specificity and absence of any tissue toxicity.

Just the objective of developing biomaterials retaining the desired biocompatibility while exerting a selective action against microbial cells has characterized the most traditional approach consisting of combining well-established biomaterials with antimicrobial substances already pharmacologically tested such as antibiotics. In these circumstances, the way to direct clinical use as well as the industrial registration of a new biomedical device is much simplified with respect to the development of totally new materials, which need to be investigated all the way through. However, as anticipated earlier, the utilization of antibiotics-doped materials for prophylaxis of infections is still being debated. Preclinical and clinical studies have enlightened the possible risk of emergence of antibiotic-resistant strains and concern has also been expressed by epidemiologists, especially on the indiscriminate use of important antibiotic substances such as vancomycin. Other,

Table 13.1 Strategies for infection-resistant materials

Reduction of bacterial adhesion	Finishing the material surface with a repellent coat
	Modulating surface hydrophilicity
	Modifying artificial surfaces with surfactants
	Hydroxyapatite coating
	Coating with biologic molecules such as heparin
Release of bactericidal substances	TiO_2 coating, which acts with a photocatalytic bactericidal mechanism
	Hydrogels containing Cu^{2+} and Ag^{+} complexes
	Sol-gel coatings releasing nitric oxide
	Chlorhexidine-releasing methacrylate composite materials and bone cements
	Antibiotic-loaded biomaterials
Permanent self-sterilizing materials	Hydrophobic polycations containing antimicrobial monomers such as quaternary ammonium compounds
	Silicone catheters coated with Ag and TiO_2 nanocomposite thin film

less documented and more remote concerns regard the possibility of sensitization or even anaphylactic reaction associated with the use of antibiotics-laden biomaterials [15]. In view of all these aspects, ideal bioactive biomaterials should avoid the use of conventional antibiotic molecules and possibly rely on alternative bactericidal substances that can be delivered locally. Table 13.1 summarizes principal strategies for infection-resistant materials.

Certainly, the story of antimicrobial biomaterials spans over many decades and up to now has produced relatively little fruits. The versatile adaptation of microbial genome enables growth even under extreme conditions and constantly challenges all the efforts made in biomedicine to find a permanent solution to all infective pathologies. Nevertheless, the design of bioactive implants with ideal self-sterilizing properties remains one of the most ambitious challenges for future research in the biomaterials area.

References

1. Darouiche RO (2004) Treatment of infections associated with surgical implants. N Engl J Med 350(14):1422–1429
2. Schierholz JM, Beuth J (2001) Implant infections: a haven for opportunistic bacteria. J Hosp Infect 49(2):87–93
3. Campoccia D, Montanaro L, Arciola CR (2006) The significance of infection related to orthopedic devices and issues of antibiotic resistance. Biomaterials 27(11):2331–2339
4. Lewis JS (1909) IX. Note on silver foil in surgery. Ann Surg 50(4):793–796
5. Sims JM (1869) Ovariotomy: pedicle secured by silver-wire ligatures: cure. Br Med J 1(432):326
6. Ockerblad NF, Carlson HE (1939) Vesico-vaginal fistula. South Med J 32(6):653–657
7. Zondek B, Sadowsky A, Brzezinski A (1950) Intra-abdominal routine prophylaxis by penicillin in gynaecological operations and caesarean sections. Harefuah 39(10):115–117
8. Colton MB, Ehrlich E (1953) Bactericidal effect obtained by addition of antibiotics to dental cements and direct filling resins. J Am Dent Assoc 47(5):524–531

9. Hessert GR (1971) Tensile strength and structure of bone cement Palacos mixed with gentamicin sulfate. Arch Orthop Unfallchir 69(4):289–299
10. Wahlig H, Buchholz HW (1972) Experimental and clinical studies on the release of gentamicin from bone cement. Chirurg 43(10):441–445
11. Alexander AD (1951) Bacitracin and gelfoam. Combined use in dentistry. US Armed Forces Med J 2(8):1247–1250
12. Goldman DR, Kilgore DS, Panzer JD, Atkinson WH (1973) Prevention of dry socket by local application of lincomycin in Gelfoam. Oral Surg Oral Med Oral Pathol 35(4):472–474
13. Bennett-Guerrero E, Pappas TN, Koltun WA, Fleshman JW, Lin M, Garg J, Mark DB, Marcet JE, Remzi FH, George VV, Newland K, Corey GR, SWIPE 2 Trial Group (2010) Gentamicin-collagen sponge for infection prophylaxis in colorectal surgery. N Engl J Med 363(11):1038–1049
14. Bennett-Guerrero E, Ferguson TB Jr, Lin M, Garg J, Mark DB, Scavo VA Jr, Kouchoukos N, Richardson JB Jr, Pridgen RL, Corey GR, SWIPE-1 Trial Group (2010) Effect of an implantable gentamicin-collagen sponge on sternal wound infections following cardiac surgery: a randomized trial. JAMA 304(7):755–762
15. Campoccia D, Montanaro L, Speziale P, Arciola CR (2010) Antibiotic-loaded biomaterials and the risks for the spread of antibiotic resistance following their prophylactic and therapeutic clinical use. Biomaterials 31(25):6363–6377
16. Engesaeter LB, Lie SA, Espehaug B, Furnes O, Vollset SE, Havelin LI (2003) Antibiotic prophylaxis in total hip arthroplasty: effects of antibiotic prophylaxis systemically and in bone cement on the revision rate of 22,170 primary hip replacements followed 0–14 years in the Norwegian Arthroplasty Register. Acta Orthop Scand 74(6):644–651
17. Arciola CR, Alvi FI, An YH, Campoccia D, Montanaro L (2005) Implant infection and infection resistant materials: a mini review. Int J Artif Organs 28(11):1119–1125
18. Nagaoka S, Kawakami H (1995) Inhibition of bacterial adhesion and biofilm formation by a heparinized hydrophilic polymer. ASAIO J 41(3):M365–M368
19. Saldarriaga Fernández IC, van der Mei HC, Lochhead MJ, Grainger DW, Busscher HJ (2007) The inhibition of the adhesion of clinically isolated bacterial strains on multi-component cross-linked poly(ethylene glycol)-based polymer coatings. Biomaterials 28(28):4105–4112
20. Roosjen A, Boks NP, van der Mei HC, Busscher HJ, Norde W (2005) Influence of shear on microbial adhesion to PEO-brushes and glass by convective-diffusion and sedimentation in a parallel plate flow chamber. Colloids Surf B Biointerfaces 46(1):1–6
21. Tsibouklis J, Stone M, Thorpe AA, Graham P, Peters V, Heerlien R, Smith JR, Green KL, Nevell TG (1999) Preventing bacterial adhesion onto surfaces: the low-surface-energy approach. Biomaterials 20(13):1229–1235
22. Anselme K, Davidson P, Popa AM, Giazzon M, Liley M, Ploux L (2010) The interaction of cells and bacteria with surfaces structured at the nanometre scale. Acta Biomater 6(10):3824–3846
23. Campoccia D, Montanaro L, Agheli H, Sutherland DS, Pirini V, Donati ME, Arciola CR (2006) Study of *Staphylococcus aureus* adhesion on a novel nanostructured surface by chemiluminometry. Int J Artif Organs 29(6):622–629
24. Maddikeri RR, Tosatti S, Schuler M, Chessari S, Textor M, Richards RG, Harris LG (2008) Reduced medical infection related bacterial strains adhesion on bioactive RGD modified titanium surfaces: a first step toward cell selective surfaces. J Biomed Mater Res A 84 (2):425–435
25. Vergara-Irigaray M, Valle J, Merino N, Latasa C, García B, Ruiz de Los Mozos I, Solano C, Toledo-Arana A, Penadés JR, Lasa I (2009) Relevant role of fibronectin-binding proteins in *Staphylococcus aureus* biofilm-associated foreign-body infections. Infect Immun 77(9):3978–3991

Biography

Prof. Arciola, MD, PhD in Experimental Pathology, is Professor of General Pathology at the Medical Faculty of the University of Bologna. Member of the Academic Board of the PhD Course in Oncology and Experimental Pathology of the Bologna University. She is Specialist Doctor in Clinical Pathology for Laboratory Management, in Bioengineering, and in Ophthalmology. She currently is the Head of the Research Unit on Implant Infections at the Rizzoli Orthopaedic Institute of Bologna. From 1989 to 2005 she was Assistant-Director, vice-Director, and Director of the Rizzoli Research Laboratory on Biocompatibility of Implant Materials. Prof. Arciola is Associate Editor of the journal *Biomaterials* for the clinical outcome "Implant Infections", Section Editor for "Biocompatibility and Graft Infections" and permanent Guest Editor of the annual special issue "Focus on Implant Infections" of the *International Journal of Artificial Organs*. She is a member of international scientific evaluation boards as France's National Research Agency, Swiss National Science Foundation, Research Grants Council of Hong Kong and China, Laboratories of Excellence of the French Research Funding Agency. She is reviewer of over forty journal quoted by ISI.

Dr. Campoccia, MSc, PhD in Clinical Engineering. He currently holds a position of Senior Biology Research Manager at the Research Unit on Implant Infections of the Rizzoli Orthopaedic Institute of Bologna. Formerly, he was a Research Biologist in the Rizzoli Research Laboratory on Biocompatibility of Implant Materials. He had been a Guest Researcher at the former Institute of Medical & Dental Bioengineering now Department of Clinical Engineering, Royal Liverpool Hospital, University of Liverpool, UK. He is a member of the Editorial Board of the journal *Biomaterials*. He is a member of the European Society for Biomaterials and of Italian Association for Cell Culture. He is reviewer of numerous scientific journals.

Prof. Montanaro, MD, PhD, is Full Professor of General Pathology at the Medical Faculty of the University of Bologna. Formerly, he was Director of the Experimental Pathology Department, of the University of Bologna; Coordinator of the PhD Course on Experimental Pathology of the University of Bologna; Director of the Medical Specialization School in Clinical Pathology of the University of Bologna; Director of the Rizzoli *Research Laboratory on Biocompatibility of Implant Materials*; and Director of the First Department of Rizzoli Institute. Currently, he is Project Liaison Manager at the Research Unit on Implant Infections and appointed as Advisor of the Scientific Director of the Rizzoli Orthopaedic Institute of Bologna.

Chapter 14
Biomaterial Calcification: Mechanisms and Prevention

Amy Munnelly, Frederick Schoen, and Naren Vyavahare

Abstract Pathologic calcification of implants is a detrimental condition that can severely impact device performance and ultimately lead to implant failure. Although calcification affects a wide variety of medical implants, both synthetic and biologically derived, the pathogenesis of the disease is not well understood. In biologically derived implants, such as bioprosthetic heart valves, the major mechanisms of and factors contributing to calcification are chemical crosslinking, cellular damage, extracellular matrix composition, patient factors, and mechanical stress on the device. However, in synthetic-material implants, for example, intraocular lenses, pacemakers, and vascular replacements, calcification is largely due to surface conditions, such as material porosity, surface defects, and protein adsorption. This chapter investigates the mechanisms of implant calcification and discusses anti-calcification strategies, using specific medical devices as examples.

1 Overview

Deposition of calcium phosphates on a surface or within a tissue or material is called mineralization or calcification. The deposits are generally composed of poorly crystalline hydroxyapatite, the mineral found in bone [1]. Although mineralization is a normal physiological process and is vital for bone and teeth formation and maintenance, calcification can also occur as a disease state in tissues or

A. Munnelly • N. Vyavahare (✉)
Department of Bioengineering, Clemson University, 501 Rhodes Hall,
Clemson, SC 29634, USA
e-mail: narenv@clemson.edu

F. Schoen
Department of Pathology, Brigham & Women's Hospital, Harvard Medical School,
Boston, MA, USA

N. Eliaz (ed.), *Degradation of Implant Materials*, DOI 10.1007/978-1-4614-3942-4_14,

implants that are not intended to calcify. For example, pathologic calcification of soft tissue, known as *ectopic* calcification (i.e., outside of a normal calcified site), can alter tissue structure and lead to necrosis in the kidneys, blood vessels, skin, and tendons [2]. Calcification of implants (other than hard tissue implants such as bones and teeth) is undesirable, and the deposition of calcific deposits and their subsequent aggregation into gross nodules can have severe consequences on device performance. Calcification of implants can lead to stiffening, structural instability, and ultimately device failure.

Pathologic calcification is generally classified according to the nature and condition of the substrate and the host biochemical environment. *Dystrophic* calcification occurs in individuals with normal calcium metabolism and is marked by the deposition of calcium phosphate in association with implants or diseased tissue. *Metastatic* calcification, in contrast, occurs in individuals with abnormal calcium metabolism due to diseases, such as diabetes or renal disease that leads to high serum calcium and/or phosphorus. In metastatic calcification, calcium salts are deposited in previously healthy tissue. Pathologic calcification may be classified further by its location. *Intrinsic* calcification occurs deep within the implant material, whereas extrinsic calcification is found at the material surface. *Extrinsic* calcification is generally associated with cells, proteins, thrombi, or vegetations that have attached to the implant surface [3].

Host factors often play a large role in the timing and extent of pathologic calcification, and different individuals receiving identical implants can have wide ranges of calcification, varying from almost none to massive deposits that cause device failure. It has been long established that age at implantation is strongly related to calcification, as calcium metabolism is enhanced in juveniles and adolescents [4]. This is why juvenile mice, rats and adolescent sheep are used to model accelerated calcification and why devices fail due to calcification much sooner in young patients than in the elderly. Diseases such as renal disease or diabetes that cause elevated blood calcium levels or enhanced calcium deposition also predispose patients to enhanced implant calcification [5]. Additionally, substrate factors and biochemical factors can act synergistically, for example, calcification of implants is enhanced in both young individuals and patients with abnormal calcium metabolism [1, 6]. The implantation site can have an effect, as certain body fluids contain more calcium and certain areas are more metabolically active than others [7]. Additionally, the host's foreign body response can lead to the formation of a fibrous capsule, which has the potential to calcify. Capsular calcification is especially of concern for breast implants and intraocular devices [7, 8].

Although calcification is such a prevalent condition in biomedical implants, the pathogenesis of the disease is still not fully understood, and no therapies have been successful in completely preventing this disease process. This chapter will discuss mechanisms and prevention strategies of calcification in the two classes of implants: biologically derived and synthetic.

2 Introduction to Biological Material Calcification

Many implants are derived from biological sources such as tissue transplants from human donors, cryogenically preserved cadaver tissue, or animal tissue. While early researchers were mostly concerned about the immunogenicity of these structural implants (as with whole organ transplants such as kidney or heart), it turned out that mitigating the immune response was often more straightforward than managing calcification. Biological tissue contains cells and extracellular matrix components, which are the principal nucleation sites for pathologic calcification [1, 3, 6]. Furthermore, crosslinking chemicals such as glutaraldehyde, which may be used to preserve bioprosthetic tissue, potentiate calcification by devitalizing cells and eliminating cellular calcium regulatory mechanisms. Additionally, certain host factors are specific to biologically derived implants. For example, pathologic calcification of native coronary arteries and aortic valves occurs in disease states such as atherosclerosis and aortic stenosis. Atherosclerotic calcification is highly related to blood cholesterol levels, and cholesterol-lowering statin therapy has proven to reduce the progression of coronary calcification. Similarly, hypercholesterolemia has been identified as a risk factor for calcification of cardiovascular bioprosthetic implants. It is hypothesized that lipids and cholesterol esters absorb into the bioprosthetic tissue and serve as nidi for calcium deposition [9]. Mechanical stress also plays a role, as we will see below [10].

Two of the most recent examples of biological implant calcification, the bioprosthetic heart valve and biologically derived tissue-engineering scaffolds, are discussed below.

3 Bioprosthetic Heart Valves

Heart valve disease necessitates valve replacement surgery in over 300,000 people worldwide each year. Replacement valves generally fall into one of the two categories: mechanical or bioprosthetic. Mechanical valves are commonly made from pyrolytic carbon and are highly durable. However, they also can be thrombogenic, and patients who receive these types of implants must be on lifetime anti-coagulation therapy [11, 12]. Thus, patients with contraindication for anticoagulation receive bioprosthetic valve replacements [13]. The majority of clinically available bioprosthetic heart valves are composed of stent-mounted, glutaraldehyde-fixed porcine aortic valve, or bovine pericardial tissue [13]. Glutaraldehyde fixation effectively crosslinks the collagen in the tissue, and to a great extent eliminates the immunogenicity and thrombogenicty of the implant. However, within 15 years of implantation, over 50% of bioprosthetic heart valves fail [14]. One of the major factors contributing to failure is pathologic calcification of the valve causing tissue stiffening and/or tearing.

Bioprosthetic materials are devitalized tissue that cannot repair damage which occurs during function. Calcification is marked by the deposition of poorly

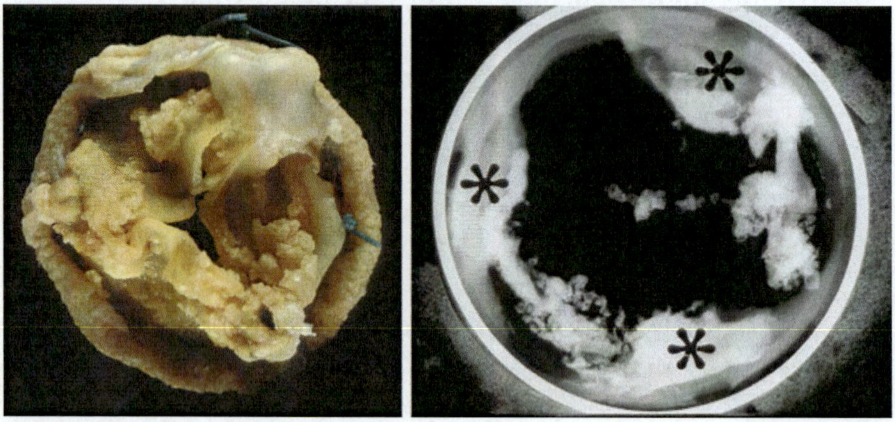

Fig. 14.1 Calcified clinical bioprosthetic valves. (**a**) Photograph of explanted bovine pericardial bioprosthetic valve following several years function. (**b**) Radiograph of another calcified bioprosthetic valve. *Arrow* points to calcific deposit; *asterisks* designate commissures (i.e., the places where the cusps come together)

crystalline hydroxyapatite in the bioprosthetic valve tissue [15]. Calcification of the cusps causes stiffening (with stenosis) or tears (Fig. 14.1), with regurgitation. Calcification of both the cusps and the aortic wall can cause stenosis of the valve. It has been observed that calcification is accelerated in young patients and in areas of high tissue stress [14, 16–18]. Because calcification of heart valve prostheses is such a widespread problem, many anti-calcification strategies have been evaluated, and several are utilized in clinically available bioprosthetic valves [3, 19]. This section will detail the mechanisms of bioprosthetic heart valve calcification and highlight methods to reduce this problem.

3.1 Mechanisms

Five main factors—glutaraldehyde pretreatment, cellular injury, matrix composition, recipient biochemistry, and mechanical stress—act in concert to determine the extent and kinetics of bioprosthetic heart valve calcification [1, 3, 15]. A schematic of the hypothesized calcification pathway is shown in Fig. 14.2.

3.1.1 Role of Glutaraldehyde

Glutaraldehyde has been the preferred crosslinker for bioprosthetic heart valves since the 1970s, since it renders the tissue nonimmunogenic and nondegradable, while thoroughly crosslinking collagen [20]. However, calcification of bioprosthetic

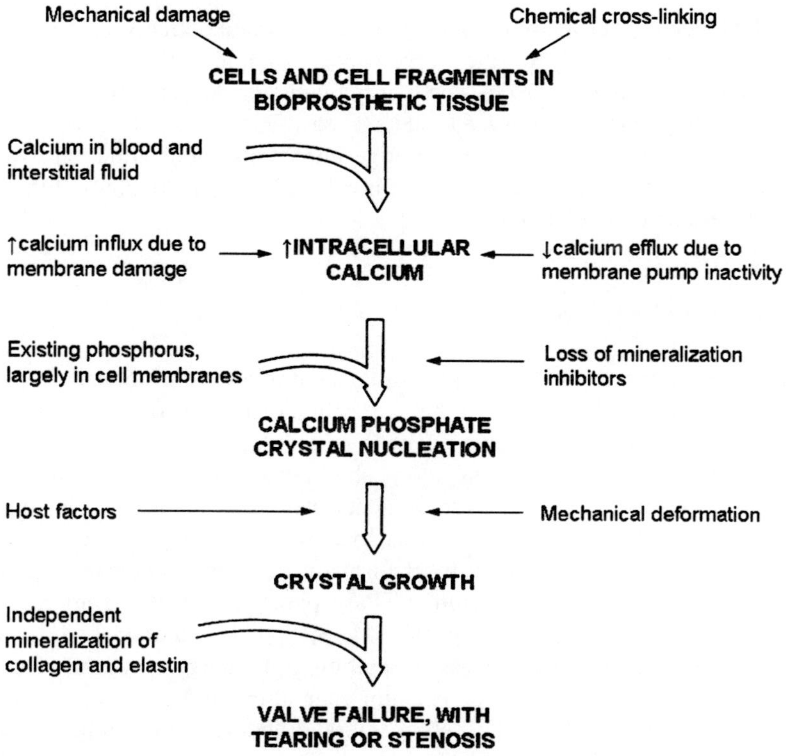

Fig. 14.2 Proposed model of calcification in bioprosthetic heart valve tissue. Reprinted with permission from [15]

heart valves is highly dependent on the extent of glutaraldehyde crosslinking. A study by Gershon Golomb showed that tissues fixed with low levels of glutaraldehyde had little calcification but exhibited peripheral inflammation, whereas tissues fixed with higher glutaraldehyde concentrations had no immune or inflammatory response but extensive calcification [21]. Paradoxically, however, a critical level of glutaraldehyde is needed to stabilize the tissue and crosslink the collagen to provide durability. A non-glutaraldehyde-based crosslinker has not yet achieved successful clinical application.

Another key factor in bioprosthetic valve calcification is devitalization of the tissue and the associated interference with cellular calcium regulation [22]. Cellular debris and the phospholipid membranes of the devitalized cells can serve as nucleation sites for calcification [1, 3, 20]. Additionally, unreacted aldehyde groups on glutaraldehyde-fixed implants can leach into the surrounding tissue and cause necrosis and calcification [23]. Furthermore, the chemical processes that occur during glutaraldehyde fixation and valve storage are not completely understood. The reaction products that are produced when glutaraldehyde reacts with valvular proteins to form crosslinks have been only partially characterized [24]. Variations in the glutaraldehyde crosslinking procedure have variable effects on the stability and calcification potential of the valves. For example, valves fixed in high concentrations of glutaraldehyde, stored in glutaraldehyde long term, or fixed at high temperatures exhibit decreased tendency to calcify [25, 26]. Obviously, the glutaraldehyde fixation chemistry has a large impact on calcification, but without understanding all of the chemical reactions involved in crosslinking, it is difficult to predict mineralization outcomes.

3.1.2 Role of Devitalized Cells

In healthy cells, the intracellular calcium concentration is maintained at a level 1,000 to 10,000 times below that of the extracellular calcium concentration due to active calcium pumps. After glutaraldehyde crosslinking, devitalized cells in the tissue do not possess active processes to push out calcium. Thus, devitalized cells have been shown to be the predominant sites of calcification in clinical biopsosthetic valves [1, 3, 6, 15]. It was hypothesized that preparation induces interruption of a key homeostatic mechanism in cellular calcium regulation by active processes such as the calcium-ATPase pump at the cell membrane and calcium binding proteins inside the cell [1, 3, 6, 15]. Glutaraldehyde fixation devitalizes the cells in the valve tissue, inhibiting all energy-requiring processes and eliminating the possibility of calcium regulation. It has been shown that glutaraldehyde permeabilizes the cell membrane and that intracellular calcium levels increase in a dose-dependent manner immediately upon glutaraldehyde exposure [21]. It was hypothesized that glutaraldehyde binds to the ATP-binding sites of calcium pumps, rendering them nonfunctional [21]. Since the cell cannot extrude the extra calcium, intracellular calcium concentrations increase dramatically. In contrast, The phosphorous concentration is normally greater inside the cell than outside. The intracellular concentrations of calcium and phosphorous eventually reach levels sufficient to cause nucleation of hydroxyapatite crystals [27]. The plasma membrane and organelles are nucleation sites for calcification because they are rich in phospholipids. Eventually, the calcium deposits grow large enough to converge into nodules that stiffen and weaken the tissue, ultimately leading to prosthesis failure [1, 27]. Microscopic features of clinical and experimental bioprosthetic valve calcification are shown in Fig. 14.3.

3.1.3 Role of Extracellular Matrix Components

While cell devitalization is a major contributing factor in valve calcification, extracellular matrix components have also been implicated as secondary calcification sites [1]. Porcine aortic valve bioprostheses are composed of valve leaflets and aortic wall, which have varying amounts of collagen, elastin, and glycosaminoglycans (GAGs) in the extracellular matrix. The leaflets have substantial collagen (60% dry weight) with 10% elastin and 4% GAGs, whereas the aortic

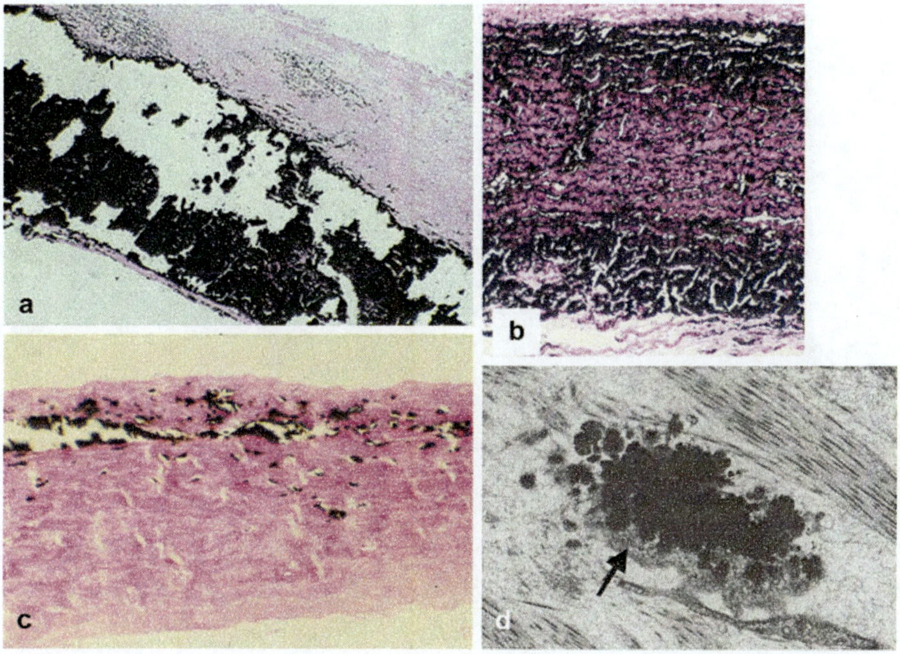

Fig. 14.3 Microscopic features of clinical and experimental bioprosthetic valve calcification. (**a**) Large calcific nodule in explanted clinical pericardial valve. (**b**) Early (3 days) calcification of pericardium in the subcutaneous rat model. (**c**) Late (21 days) calcification of pericardium in the subcutaneous rat model. (**d**) Transmission electron microscopy of calcified residual cell in pericardial tissue implanted into the subcutaneous rat model. *Arrow* designates cell with dark calcific deposit. (**a–c**) stained with von Kossa stain (calcium phosphates black). (**d**) reprinted by permission from [6]

wall is dominated by elastin (50%) with 15% collagen [28]. Calcification of collagen and elastin in glutaraldehyde-fixed tissue has been demonstrated by several studies [29–32]. However, non-glutaraldehyde-treated tissue and non-glutaraldehyde-treated Type I collagen sponges do not calcify *in vivo*, which suggests that glutaraldehyde is responsible for matrix calcification as well [33]. It is proposed that when glutaraldehyde devitalizes the tissue, it removes certain cellular components and exposes matrix sites for hydroxyapatite nucleation [6]. For example, glutaraldehyde does not fix GAGs, and when the GAGs degrade, they may expose calcifiable sites in the extracellular matrix [34]. Furthermore, GAG-degrading enzymes and matrix metalloproteinases (MMPs) remain active in the fixed tissue and may also degrade the tissue to expose sites for mineralization [35].

Elastin has also been shown to calcify, and glutaraldehyde does not stabilize this ECM component [29, 31, 36]. Degeneration of elastin fibers is a feature of the calcification of native arteries [37–39]. Elastin in bioprosthetic valves may be prone to degradation by endogenous enzymes or MMPs from the host

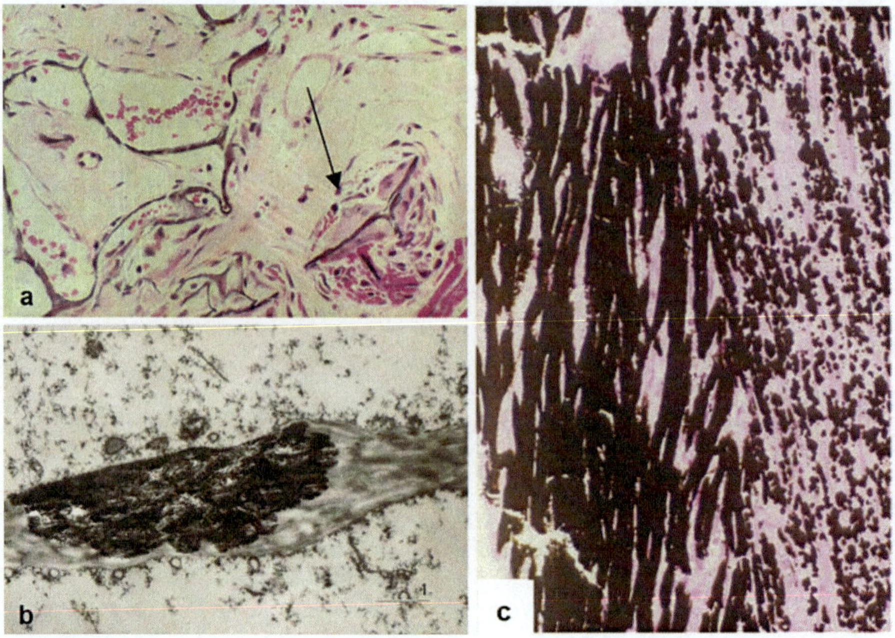

Fig. 14.4 Calcification of extracellular matrix. (**a**, **b**) Calcification of subcutaneously implanted type collagen sponge; (**a**) Light microscopy (*arrow* indicates calcified collagen); (**b**) Transmission electron microscopy. (**a**) Stained by hematoxylin and eosin (H&E); (**c**) Calcified elastin (stained by von Kossa stain, calcium black). Reprinted with permission from [33]

environment, which infiltrate the tissue [35]. It has been shown that MMP inhibitors can significantly reduce elastin-specific calcification *in vivo* [36]. Additionally, elastin mineralization is especially prominent in the aortic wall, since it contains large amounts of elastin [28, 40]. Calcification of the aortic wall is observed in both stented valves, where the wall is largely covered by prosthetic material and stentless valves, where the aortic wall is directly exposed to blood flow. Stented valves are constrained by the rigid stent that encircles the aortic wall. Stentless valves allow for more movement and expansion of the wall during function, leading to improved hemodynamics [41, 42]. However, since such a large area of the aortic wall is exposed to blood flow, calcification of aortic wall in stentless valves may be especially detrimental and eliminate the benefits that these valves have over stented prostheses. Such calcification could stiffen the valve root, alter the hemodynamic function, and in extreme cases could lead to wall rupture (see Figure 14.5) [3]. Anti-calcification strategies are generally aimed at reducing cell-mediated calcification, and are thus ineffective at preventing elastin-specific aortic wall mineralization [43, 44]. A treatment to reduce elastin-mediated calcification is likely the best option for minimizing aortic wall mineralization, as described later in this review. Calcification of extracellular matrix in biomaterial implants is depicted in Figs. 14.4 and 14.5.

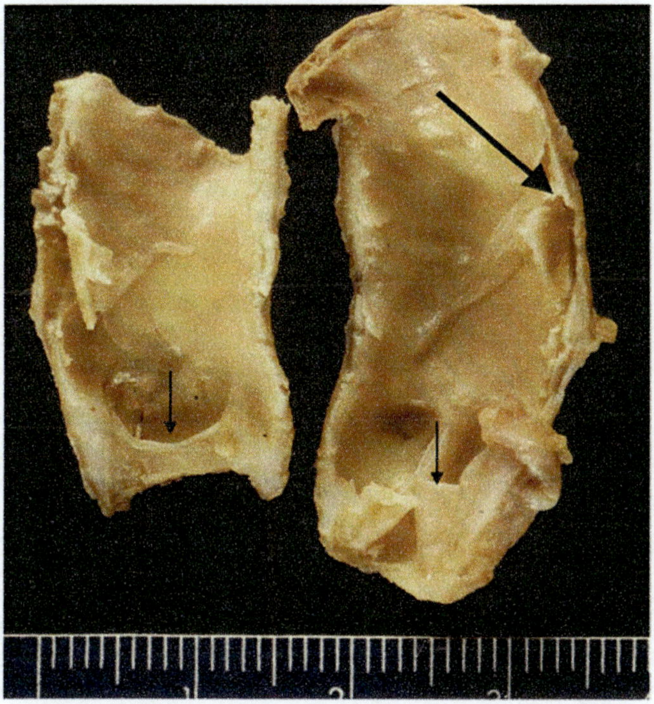

Fig. 14.5 Removed clinical stentless porcine aortic valve bioprosthesis. Aortic wall (*thick arrow*) is calcified but cusps (*thin arrows*) are not. Reprinted with permission from [130]

3.1.4 Mechanical and Inflammatory/Immune Factors

Calcification of bioprosthetic valves is generally enhanced in areas of high mechanical stress [16–18, 45]. Stressed areas may disrupt structural integrity and allow for greater influx of plasma ions and proteins, or cyclic deformations might cause calcium deposits to migrate and aggregate in deformed areas [46]. Calcific damage and structural stresses have synergistic effects on tissue deterioration, increasing the likelihood of tissue failure in highly stressed regions.

A hypothetical role of inflammation and an immune response in calcification is plausible and has been highly debated, but research suggests that neither of these pathophysiologic processes impact pathologic mineralization of bioprosthetic valves *in vivo*. While antibodies against valve tissue and mononuclear inflammatory cells have been found around failed heart valves and in the plasma, there has been no definitive research to suggest that the inflammatory or immune response directly influences calcific valve degeneration [47–50]. Indeed, non-crosslinked tissue evokes a strong immune response, but does not calcify, whereas fixed tissue shows a minimal immune response, but may calcify heavily [21]. However, it is generally recognized that physiological molecules play a regulatory role in calcification. Promineralization proteins such as osteopontin, tenascin-C, and TGF-α have been found in calcified bioprosthetic valves, suggesting that they may mediate pathologic calcification [51–54].

Table 14.1 Anti-calcification methods investigated in heart valve bioprostheses (adapted from [3])

Alteration of glutaraldehyde fixation method
Use of alternative fixatives
Removal of residual aldehyde groups
Alteration of tissue charge
Heat treatment (Thermafix)
Cell/phospholipid targeted treatment
Surfactants
Ethanol
Decellurization of the tissue
Amino-oleic acid (AOA)
Extracellular matrix targeted treatment
Ethanol
Trivalent metal ions
Inhibition of hydroxyapatite formation
EHBP
EDTA

3.2 Calcification Prevention Strategies

A wide array of anti-calcification strategies has been investigated for preventing calcification of bioprosthetic heart valves. Finding an ideal treatment has been challenging, as the technique not only has to sufficiently prevent calcification, but also must not compromise device safety, efficacy, durability, and mechanical properties. The tissue treatment itself should not induce an inflammatory or immune response, thrombosis, toxicity, or tissue degradation [1, 3, 55]. Moreover, the mode of anti-calcification must not adversely affect the recipient. All of the anti-calcification methods tested fall into one of the three general approaches: systemic therapy, local drug delivery, or material modification. Of these three approaches material modification by, post-fixation modification of the bioprosthetic tissue has shown the most promise to be translated to the clinical environment and has been the most widely researched method in recent years. Modified biomaterials are attractive, as they eliminate the concern of undesirable side effects from drugs delivered systemically or locally. Common strategies of material modification (Table 14.1) include alteration of the glutaraldehyde fixation method, removal or alteration of calcium nucleation sites, extracellular matrix-targeted inhibition methods, and inhibition of hydroxyapatite formation [3].

3.2.1 Glutaraldehyde-Targeted Anti-calcification Methods

Since glutaraldehyde fixation is a major contributing factor in bioprosthetic heart valve mineralization, modification of the fixation procedure has been investigated. In fact, elimination of glutaraldehyde altogether has been considered. Non-glutaraldehyde crosslinking agents such as epoxies, acyl azide, diisocyanates, carbodiimides, and naturally derived agents such as genipin have shown to reduce calcification in animal models, but have not been adopted clinically [56–62]. Photo methods such as UV light and dye-mediated photo-oxidation have also been proposed [63–65]. They showed excellent results in animal studies [64, 65], but clinical use of photooxidation was problematic in a pericardial design tissue with suboptimal mechanical properties, which led to accelerated degeneration and mechanical failure of the valve [66]. One alternative crosslinking agent has shown to improve biocompatibility and biomechanical properties while reducing calcification. Triglycidyl amine (TGA) is a water-soluble, poly-epoxide crosslinker, whose epoxide groups are highly reactive with the amino groups of collagen, leading to the formation of stable collagen crosslinks [56]. It has been shown that TGA-crosslinked aortic valve cusps remained more compliant than GLUT-fixed tissues and also exhibited reduced calcium levels after subdermal implantation [56, 67]. TGA significantly reduced calcification in 21-day implants by altering the extracellular matrix signaling activity. However, calcification was observed in longer term implants, and TGA alone did not reduce aortic wall calcification [56]. Thus, a secondary anti-calcification strategy was investigated. Tissue calcium levels were further reduced when TGA-fixed tissues were treated with 2-Mercaptoethylidene-1,1-bisphosphonic acid (MABP), a bisphosphonate that binds calcium to prevent crystal formation and also reduces alkaline phosphotase activity [67]. Moreover, combined TGA-MABP treatment prevented calcification of both the aortic valve cusps and aortic wall over a 90-day implant period [67]. Biomechanical analysis by Sacks et al. confirmed that TGA-MABP did not negatively affect tissue mechanics, as the tissue samples retained compliance and stability [68]. Thus, TGA-MABP may be a feasible alternative to glutaraldehyde for fixing bioprosthetic heart valves. Much more research is needed before an alternative crosslinker is ready for clinical use.

Chemical modification of glutaraldehyde-fixed tissue has also been attempted. Cytotoxic residual aldehyde groups, which are shown to cause calcification, can be blocked with L-glutamic acid, lysine, homocysteic acid, or diamines [22, 23, 69]. Tissue reactivity can be reduced through treatment with reducing agents like borohydride or cyanoborohydride, or the charge on the tissue can be altered to repel metal ions [70]. However, methods solely related to glutaraldehyde neutralization have not shown to fully prevent calcification in a manner that would be clinically useful.

Heat treatment of the glutaraldehyde-fixed cusps was also considered as a non-chemical anti-calcification method. Studies have shown that bioprosthetic valves heated in glutaraldehyde solution at 50 °C for at least 2 months resist calcification in the rat subdermal implant model for 4 months and in the sheep mitral valve replacement model [25]. It is hypothesized that heating may extract or denature the proteins and phospholipids that serve as nuclei for calcification. Since the exact procedure is proprietary, the mechanism for calcification reduction is not completely understood. Heat may denature proteins, but the relation between denaturation and calcification has not been demonstrated. Furthermore, it is unclear how glutaraldehyde is affected by 2 months of storage at 50 °C. The anti-calcification property is only seen after 2 months of heating, and not sooner [25].

It has been suggested that heat alters how glutaraldehyde reacts with tissue components in such a way as to render the valves resistant to calcification. However, why this happens only after 2 months of heating or whether it could occur at room temperature has not been elucidated. Additionally, heat treatment has only proven effective in reducing leaflet calcification, not elastin-associated aortic wall calcification, suggesting that the mechanism of action may be due to the effect on devitalized cells.

3.2.2 Cell and Phospholipid-Targeted Anti-calcification Methods

Phospholipids in and around devitalized cells are the most prominent calcification nucleation sites; thus, removal of these tissue components has been proposed to reduce mineralization [1, 3]. Studies have shown these to be effective calcification prevention strategies. Surfactants such as sodium dodecyl sulfate are used clinically to extract phospholipids from the cells, thus reducing membrane calcification [71], and an FDA-approved valve utilizing this approach is available. Polysorbate 80 and Triton-X, as well as organic solvents like ethanol, ether, and chloroform/methanol have similarly been used [72]. One FDA approved bioprosthetic valve utilizes ethanol treatment to prevent calcification. Pretreatment with 80% ethanol extracts phospholipids from the tissue while also causing a change in collagen conformation that increases the valve's resistance to collagenase [32, 44, 73]. The ethanol treatment is reported to extract almost all phospholipids and cholesterols from the valve, thus eliminating calcification of the devitalized cells [44]. Additionally, ethanol treatment also prevents adsorption of exogenous phospholipids and cholesterol from the host environment [44]. However, long-term studies are needed to show that this effect is permanent and will prevent calcification indefinitely.

Another FDA-approved valve uses a pretreatment with 2-α-amino-oleic acid (AOA), which has been shown to inhibit calcium flux through the tissue of bioprosthetic heart valve cusps [43, 69]. It is also proposed that AOA may remove phospholipids and slow calcium diffusion though the tissue, and that bound AOA may chelate calcium and prevent crystallization [43]. However, AOA pretreatment only prevents calcification of the valve cusp tissue, and not the aortic wall. Aortic wall calcification could cause valve stiffening or stenosis that alters hemodynamic properties. Thus, a combination of treatments may be needed to effectively inhibit calcification of the entire valve (cusp and wall), especially in stentless valves.

Rather than just removing phospholipids from the cell, removing the cells entirely from the tissue, leaving only the extracellular matrix, may be beneficial [74]. Recent studies with decellularized aortic valve allografts in sheep showed that the allograft valves exhibited minimal calcification while control fresh valve conduits showed significant thickening and calcification [75]. Additionally, a clinical series of traditional cryopreserved pulmonary valves and decellularized cryopreserved valves revealed that the decellularized valves displayed less stenosis and valvular insufficiency [76]. Thus, removing the cellular component of the valve tissue may reduce calcification by eliminating the main nucleation sites of mineral deposition. Recently, studies have also been reported where decellularized valves were crosslinked with agents such as procyanidins and quercetin as alternatives for glutaraldehyde crosslinking [77, 78].

3.2.3 Extracellular Matrix-Targeted Anti-calcification Methods

Most commercially available anti-calcification treatments aimed at preventing cell-mediated calcification do not specifically target calcification of the extracellular matrix components. The exception is ethanol treatment, which may mitigate collagen-associated calcification [32]. Ethanol treatment, either prior to or following glutaraldehyde fixation, alters the collagen conformation such that the collagen becomes resistant to calcification [32]. However, ethanol only partially prevents calcification of elastin fibers. Thus, calcification of the aortic wall, which is dominated by elastin, is not effectively reduced by ethanol treatments [28].

Since elastin is a major component of the extracellular matrix, especially in the aortic wall, a successful calcification prevention strategy will have to target elastic fibers as well as devitalized cells. One system includes a proprietary combination of ethanol treatment for the valve leaflets and aluminum chloride treatment (biLinx) for the aortic wall [79]. Trivalent metal ions, such as aluminum and iron, have been extensively studied and used in $FeCl_3$ and $AlCl_3$ pretreatments of glutaraldehyde-fixed tissue [80–82]. Aluminum tightly binds to elastin to induce a conformational change that reduces elastin's affinity for binding calcium ions [31]. The metal ions react with the phosphate anions to form a complex and prevent formation of calcium phosphate [83]. Additionally, metal ions also interfere with MMPs and alkaline phosphatase, an enzyme involved in bone formation, and thus prevent calcification by this pathway as well [81, 84]. Aluminum-pretreated aortic wall also exhibits altered gene expression of MMPs in the local cellular environment; this altered gene expression may limit MMP-induced degradation of elastin, further reducing elastin calcification and improving the durability of the implant [85].

3.2.4 Inhibition of Hydroxyapatite Formation

Calcium phosphate crystal formation leads to the accumulation of hydroxyapatite nodules that can damage cuspal tissue, leading to stiffening and tearing. Retarding or preventing growth of these crystals can be accomplished with bisphosphonates [86]. Ethane-1-hydroxy-1,1 bisphosphonate (EHBP) is a calcium chelator that, when delivered systemically, has shown to prevent calcification of implants in animal models. However, this method produces systemic side effects, such as poor bone growth [87]. Local delivery of EHBP through locally implanted polymers has also shown some success in animal models, but would be difficult to implement clinically and therefore has not been investigated in humans [88]. To circumvent systemic side effects, bioprosthetic heart valves may be pretreated with EHBP, which is hypothesized to poison the growth of calcific crystals [81, 89].

Similarly, ethylenediaminetetraacetic acid (EDTA) is a calcium chelator that has been shown to inhibit mineralization of bovine pericardial tissue [90]. It is proposed that EDTA binds to calcium ions on the outer shell of hydroxyapatite crystals. Chelation of these ions removes material from the crystals, causing them to shrink, thus demineralizing the sample. Pretreatment of tissues with EDTA slows the progression of calcification by binding calcium before it can react to form hydroxyapatite. Studies show that EDTA treatment alone reduced calcium levels in bovine pericardium by approximately 50%; thus, it would best be utilized in conjunction with another anti-calcification agent in order to fully prevent mineralization [90].

In summary, calcification of bioprosthetic heart valves is a clinically significant problem that contributes to implant failure. Pathologic calcification is generally attributed to cell devitalization by glutaraldehyde crosslinking, which induces calcium influx and calcium phosphate nucleation in the cell membrane. Extracellular matrix components such as collagen, GAGs, and elastin are also involved in valve calcification. The early hydroxyapatite deposits can eventually aggregate to form nodules that contribute to leaflet stiffening, tearing and mechanical failure. However, all of the mechanisms of pathologic calcification are not completely understood. There is a strong age effect, with younger individuals having accelerated calcification. Thus, host factors also play a large role. However, immunological/inflammatory mechanisms are likely not important. While a number of anti-calcification strategies have been implemented clinically over the last few years, and many more are under experimental investigation, a long-term clinical data showing complete elimination of pathologic calcification of bioprosthetic valve implants is yet to be published.

4 Calcification of Collagen and Elastin Tissue Engineering Scaffolds

Collagen and elastin scaffolds have widely been studied for use in tissue engineering applications because of their high tensile strength and elasticity. However, collagen and elastin have both been noted as sites of calcification nucleation in bioprosthetic tissue, as described above, and elastin films have an especially strong calcific response *in vivo* [31].

4.1 Mechanisms of Collagen and Elastin Mineralization

Elastin microfibrils contain fibrillin, which has EGF domains that can bind calcium [91]. Calcification of elastin generally begins in the microfibrils surrounding the elastic fiber, suggesting that the microfibrils play a major role in mineralization [92]. A study by Daaman et al. examined the contribution of microfibrils to elastin and collagen scaffold calcification [93]. Scaffolds of purified collagen, collagen +

purified elastin, collagen + purified fibril-containing elastin, and collagen + elastic ligament were prepared using carbodiimide crosslinking and implanted subdermally into young rats. The highly purified collagen used in this study did not calcify, whereas other studies using less well-defined collagen matrices did observe calcification, indicating that contaminants may be the source of calcification of collagen scaffolds [94]. Furthermore, Tedder et al. [95] reported that crosslinked collagen scaffolds made from decellularized and purified porcine pericardium did not calcify *in vivo*, whereas glutaraldehyde-treated scaffolds had significant mineralization. Thus, glutaraldehyde may be largely responsible for collagen calcification. Contaminants may also contribute to elastin mineralization, as the non-pure elastin ligament in the Daaman study calcified to a greater extent than the purified elastin matrices [93]. However, the removal of microfibrils from the elastin did not result in reduced calcification, suggesting that microfibrils are not the source of elastin calcification. Another possibility is that matrix degradation and elastin fragmentation may create nucleation sites for calcification. However, the purified elastin in this study remained intact, and the exact mechanism of elastin calcification could not be determined [93]. Thus, the mechanism of elastin calcification *in vivo* remains uncertain.

4.2 Calcification Prevention Strategies

Pure collagen does not calcify on its own, but non-crosslinked collagen scaffolds experience rapid and severe enzymatic degradation [95]. Thus, to create a viable collagen scaffold for tissue engineering, fixation is necessary. Glutaraldehyde, however, has been cited as a cause of collagen calcification [33, 94, 95], leading researchers to consider alternative crosslinking methods. One successful method has been carbodiimide chemistry using *N*-(3-dimethylaminopropyl)-*N*′-ethylcarbodiimide (EDC) and *N*-hydroxysuccinimide (NHS) [96]. Collagen scaffolds crosslinked with EDC/NHS show improved resistance to *in vivo* calcification as compared to glutaraldehyde controls, possibly because EDC/NHS crosslinking does not leave deleterious residual chemicals in the scaffold [97].

Prevention of calcification of elastin scaffolds is not as simple as altering the crosslinking method, as pure, untreated elastin also calcifies *in vivo* [93]. Mineralization of elastin scaffolds can be prevented by pretreatment with trivalent metal ions, such as aluminum chloride [31, 97].

5 Introduction to Synthetic Materials Calcification

While calcification of heart valve and other tissue-derived materials is the most widely documented and studied form of pathologic mineralization observed clinically, calcification of synthetic materials, such as polymers and to some extent metals, has also been observed. Because synthetic materials are free of cellular and other debris (except that adherent to the surface, as discussed below) and extracellular matrix fibers, calcification is generally less severe than that of biologically derived materials. However, mineralization of synthetic devices can still be extensive enough to cause device malfunction or failure. Calcification of synthetic materials is generally present in one of two ways: mineralization of the material itself, or calcification of the fibrous capsule or adherent biologic material. Material-specific mineralization most often begins at the surface of the implant, and deposits may impinge into the material as calcification progresses. Surface defects, material porosity, and protein adsorption are the major determinants of material calcification [98]. Capsular calcification occurs in the fibrous capsule, which can create a hypoxic environment that causes necrosis and the accumulation of cellar debris [8]. Calcific nodules can form around the cell fragments and at the interface of the surface and fibrous capsule [99].

Both material-specific and capsular calcification can interfere with device performance and accelerate degeneration of the implant. For example, calcific nodules can stiffen and tear polymeric heart valve leaflets [5], cause stenosis of vascular grafts [98], interfere with pacemaker function [99], or lead to opacification of ocular implants [7] and stiffening of breast implants [8]. Vascular access grafts for hemodialysis and synthetic vascular replacements composed of Dacron or expanded polytetrafluoroethylene (e-PTFE) also calcify in some patients [100, 101]. In some cases, the disease can progress to the point of necessitating device removal or replacement. This section details specific mechanisms of both surface and capsular calcification of synthetic implants and describes anti-calcification methods that have been investigated.

6 Polyurethane Calcification

6.1 Mechanisms

Polyurethane is a polymer commonly used in cardiovascular implants such as trileaflet heart valves, ventricular assist devices, and total heart systems. Calcification of polyurethane valve implant is shown in Fig. 14.6. Similarities in the calcification of these devices have been noted. For example, the early calcium deposits are irregularly shaped and form elevated plaques on the surface of the device [102, 103]. The mechanisms of mineralization of polyurethane are not completely understood, but two hypotheses have been proposed. Calcium and phosphorus may directly adsorb to the surface to form crystals, eventually leading to subsurface crystallization [5, 102]. This would likely occur by metal ion (calcium) complexation to the polyurethane soft segment [104]. It has also been suggested that degraded cellular components from erythrocytes and platelets may adhere to the polymer surface and serve as nucleation sites for

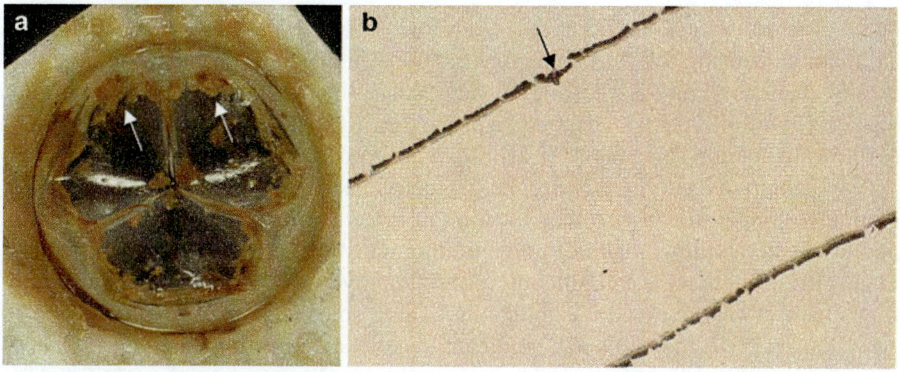

Fig. 14.6 Calcification of polyurethane. (**a**) Removed polyurethane valve implanted for 5 months in a sheep mitral valve model. *Arrows* show calcified thrombotic deposits. (**b**) Polymer sheet implanted 21 days in subcutaneous rat model. Calcium deposits are at the surface of the polymer, without thrombus, in some areas in continuity with subsurface deposits (*arrow*). (**b**) Stained with von Kossa stain (calcium phosphates black). Reprinted with permission from [104]

calcification [103, 104], especially at defects in the material. Experimental results support the second of these hypotheses, as calcification of ventricular-assist devices and total heart systems has been associated with devitalized cells and cellular fragments. Some polyurethane blood-contacting surfaces have been textured or roughened with a polyurethane-Dacron composite to promote the formation of a protective pseudointima; however, cellular elements can become trapped in the resulting thrombus, leading to calcification [103]. Furthermore, calcification is localized to areas of high stress, which may be caused by strong hemodynamic forces that injure cells and lead to increased cell and platelet deposition [5, 102, 105, 106]. For instance, calcification of trileaflet polyurethane heart valves occurs at the subsurface and at areas of high stress, such as the cuspal commissures and flexure lines (see also Figure 14.6) [107].

6.2 Examples of Polyurethane Device Calcification

Polyurethane devices that are not subject to high stresses generally exhibit cell-related calcification in adjacent tissues. For example, polyurethane is used as an insulator for cardiac pacemaker leads, which are also prone to mineralization. Mineralization occurs in the collagenous capsule at the distal ends of the leads, where they embed in cardiac muscle. As the capsule grows thicker, the blood supply to the area is diminished, decreasing local oxygen tension. The hypoxic environment causes cell death and an accumulation of cellular debris that can nucleate mineral growth [99]. While encapsulation of the leads is a normal phenomenon, excessive tissue growth and calcification can have serious impacts on device performance. Calcific nodules can form barriers that interfere with pacing stimulation, causing loss of pacing or loss of cardiac signal sensing. Additionally, calcific damage to the lead insulation can result in inappropriate pacing or current leakage [99]. Since patients depend on proper pacing to maintain active circulation, severe calcification of pacemaker leads can have dire consequences.

Calcification of polymeric vascular grafts has also been noted. A study by Park et al. investigated the *in vitro* calcification of three common polymers used in this application [98]. Films made of polytetrafluoroethylene (PTFE, Dacron), polyurethane (PU, Pellethane), and silicone (poly dimethyl siloxane) were mounted in a flow chamber that pumped a calcium-containing solution. After 21 days of exposure, the PTFE films showed significantly more calcification than the PU and silicone films. The authors hypothesize that the rough surface of the PTFE led to higher levels of calcification. PTFE is composed of nodes and fibers that render the material porous and allow for the formation of a pseudointima. However, the porous surface may also lead to a greater influx of calcium and phosphorous ions that can nucleate to form hydroxyapatite deposits. PU and silicone showed markedly less calcification than the PTFE, possibly because the smooth, lubricated surface of these materials discourages ion deposition. The results suggest that polymer surface characteristics largely determine the extent of calcification, but the specific correlations are yet uncertain. Thus, while polyurethane does have the potential to calcify, controlling its surface properties potentially may help to reduce pathologic mineralization.

6.3 Calcification Prevention Strategies

Surface modification of polyurethanes has been shown to reduce calcification. Polyurethanes coated with EHBP, heparin, or both have been fabricated by covalently binding the modifier to the hydroxy terminus of the polyurethane chains [104, 108]. Rat subdermal implant studies showed that the modified polyurethanes did not calcify significantly during the 60-day implant period, while unmodified polyurethane did calcify [104]. Further studies investigated polyurethanes that were first bromoalkylated, and then treated with bisphosphonate, which bonded to the bromine site [109]. The bisphosphonate-derivatized trileaflet valves exhibited significantly reduced calcification as compared to unmodified control valves in both rat subdermal and sheep circulatory studies. It was hypothesized that the EHBP or bisphosphonate-modified surface reduces calcium diffusion, while also inhibiting calcium phosphate nucleation and crystal growth [104]. These studies suggest that surface modification of polyurethanes may successfully block hydroxyapatite formation and prevent pathologic calcification of polyurethane implants.

Another method for reducing calcification of polyurethanes is fabrication of composite materials. A composite of polyurethane and biaxially drawn, ultra high molecular weight polyethylene (BDUHMWPE) was shown to be more stable and resistant to oxidation than polyurethane alone [110]. When the composite material was compared to the controls of plain BDUHMWPE and polyurethane, the plain BDUHMWPE calcified the least, while plain polyurethane calcified the most. The composite material was more resistant to calcification than plain polyurethane. The authors hypothesized that the hydrophobicity of the BDUHMWPE altered the diffusion and chelation of the calcium ions, reducing the tendency of the material to calcify. FTIR spectra showed that the calcification was extrinsic, occurring on the surface of the material and leaving the bulk unaltered. The authors conclude that this surface calcification is likely due to microstructure, such as pits or defects, rather than material chemistry. Thus, the calcification of the material is dependent on the surface structure and physical properties of the material.

In summary, the mineralization of polymers such as polyurethane is only somewhat dependent on bulk material chemistry. The major determinants of polymer calcification are physical properties such as surface smoothness, porosity, and surface charge; modifying these properties can reduce the calcification potential of the material. Additionally, surface coatings can mask calcifiable sites on the polymer surface to prevent calcium nucleation and hydroxyapatite formation.

7 Silicone Breast Implant Calcification

Calcification of silicone breast implants has been implicated as a possible cause of implant rupture in older implants [8]. Breast implant mineralization can be classified as capsular calcification, as a fibrous capsule grows around almost all breast implants. Cellular debris in the capsule (and perhaps silicone gel that has diffused through the implant envelope and into the fibrous capsule), as well as irregularities on the silicone surface, can nucleate calcium phosphate crystallization. Calcific deposits in explanted breast implant specimens are generally of two types: globular nodules at the capsule–surface interface, and bone formation within the fibrous capsule tissue [111]. In either case, hard deposits can weaken the implant wall and damage the shell, and plaques with sharp edges can cause perforation of the implant. Tears and perforations in the breast implant can cause the gel to leak from the implant shell. As the gel spreads and covers the implant surface and capsule, calcification accelerates [8], as shown in Fig. 14.7. Large tears and leakage can eventually lead to deformations and necessitate implant removal. Furthermore, large calcific nodules can interfere with imaging and tumor detection [111]. However, calcification of breast implants is more common in older implants, and due to advances in material science and manufacturing techniques, may be less clinically relevant in newer implants.

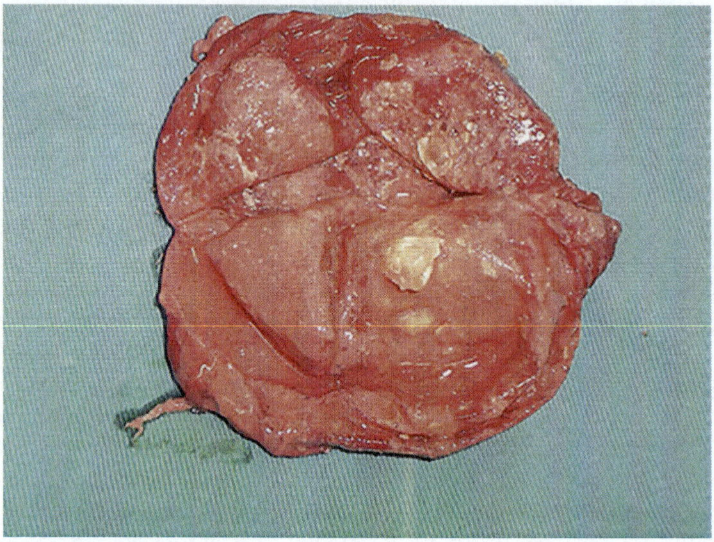

Fig. 14.7 Mineralization of breast implant capsule. The large white nodules are calcific deposits. Reprinted with permission from [8]

8 Hydrogels and Ophthalmic Implants

8.1 Hydrogels

Hydrogels are another class of polymer widely used in medical implants. Hydrogels, especially those consisting of poly hydroxyethyl methacrylate (PHEMA), have been used in cardiovascular implants, breast implants, ocular implants, dental applications, drug delivery, and nasal reconstruction [112]. This material has been extensively investigated and utilized because it is highly biocompatible, and its properties can be tailored to the application by altering the polymerization method or copolymerizing the polymer. However, calcification of PHEMA hydrogels has been observed in corneal implants [113], breast implants [114], and contact lenses [115]. A study by Lou et al. examined the effects of topography and porosity of hydrogels on their tendency to calcify *in vitro* [112]. PHEMA hydrogels of varying porosity, as well as poly HEMA-co-EEMA hydrogels were investigated. Results showed that calcification begins across the entire polymer surface, especially at surface flaws, which serve as nucleation sites for crystallization. The specimens with the greatest amount of surface flaws also exhibited the most intense calcification, as observed through Alizarin red staining. The more porous hydrogels had enhanced subsurface calcification and also deeper calcium deposits, due to the ability of the ions to diffuse through pores and channels in the material. However, calcification was not observed deep into the center of a hydrogel-containing nanopores, suggesting that the tiny pores were too small for effective calcium diffusion. It was also observed that the degree of crosslinking influenced calcification because crosslinks reduce the ability of the polymer to swell, and thus reduce water infiltration and diffusion of ions. Additionally, there was no evidence that copolymerization of PHEMA and EEMA influenced calcification tendency. The authors concluded that surface calcification of hydrogels is mainly due to flaws and voids, while bulk calcification is highly dependent on porosity. Material chemistry and composition likely play a role as well, but topography and morphology appear to be the main determinants of the degree of mineralization of hydrogels.

8.2 Ocular Implants

Hydrogels are popular materials for ocular implants, as they allow for proper hydration and oxygen permeation to the ocular environment. Calcification of ophthalmic implants is less common than calcification of blood-contacting devices, partly because the ocular environment is less metabolically active than other anatomical areas [7]. Additionally, the aqueous humor has a lower calcium concentration and fewer cellular components than the plasma [116]. Indeed, intraocular implants take longer to calcify than subcutaneous and intramuscular implants [7]. However, clinically relevant calcification of a number of acrylic, hydrophilic intraocular lenses (IOLs) has been reported [117, 118]. IOLs are implanted after the removal of cataracts; however, calcification of these implants can cause lens opacification and cloudy vision [119]. Calcification of IOLs has recently been classified into the following three categories: primary, secondary, and false-positive [118].

8.2.1 Classification of Ocular Calcification

Primary calcification is due to the material itself and is likely caused by manufacturing problems that introduce flaws into the lenses. For example, silicone particles from machines or packaging may contaminate the lenses during manufacturing; these small particles can act as nucleation sites for mineralization [120]. Energy dispersive X-ray analysis has demonstrated that elemental silicon is present in higher levels in calcium deposits than in adjacent uncalcified areas [121]. Hydrophobic silicone compounds adsorbed to IOLs interact with fatty acids in the aqueous humor to promote formation of a layer of polar, hydrophilic fatty acids on the lens surface that can nucleate calcium and phosphate [121, 122]. Since primary calcification is often associated with surface flaws, contaminants, or adsorbed substances, it generally begins at the surface and then migrates into the bulk material.

Secondary calcification describes calcification of the material surface, but is due to the ocular environment and pre-existing conditions or diseases that increase calcium deposition. The local concentrations of calcium, phosphate, and albumin are key factors in calcification, as diffusion of calcium and phosphate into the lens leads to the development of salts [119, 123]. Increased levels of calcium and

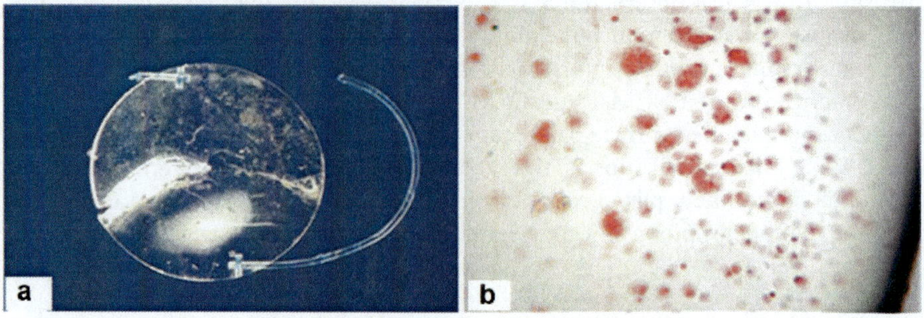

Fig. 14.8 Secondary intraocular lens opacification; (**a**) Gross morphology and (**b**) light microscopy showing positive Alizarin red staining for calcium. Reprinted with permission from [118]

phosphorous have been observed when residual cataract material is left after surgery, indicating that surgical technique also influences calcification tendency [119]. Furthermore, the adsorption of proteins such as albumin onto the lens can also encourage mineralization nucleation [124]. An example of an intraocular lens exhibiting calcific opacification is shown in Fig. 14.8a. The presence of calcium deposits is verified with Alizarin red staining (Fig. 14.8b). False-positive calcification refers to improperly diagnosed calcification of IOLs, such as residues being mistaken for calcific deposits [118].

8.2.2 Determinants of Ocular Calcification

Calcification of IOLs shares morphologic similarities with calcification of other ocular implants, such as silicone implants and hydrogel contact lens. The implants have grainy, white deposits, nodules, and films, and the surface may additionally be pitted [7]. It is hypothesized that calcification begins at the surface, but calcific damage leads to pits which allow the calcification to migrate into the bulk material [7]. The acrylate polymer surface can complex with calcium ions, thus providing nucleation sites [119]. For example, PHEMA has a greater tendency toward calcification than other acrylates, possibly due to the presence of hydroxyl groups, which can ionize or complex with phosphate to initiate surface nucleation. Additionally, water content of the lenses plays an important role, as enhanced hydration leads to more ionized groups on the surface, thus promoting calcification [125]. Since most intraocular hydrogels are porous, the uptake of proteins and lipids is also an important factor in material calcification [126]. Differences in porosity lead to differences in cell penetration, and more porous materials generally show greater calcification due to the increased infiltration of calcification precursors [7, 127]. Extrinsic calcification (cell related) appears to occur first, with calcium plaques around the edge of the implant apparent after 30 days. Intrinsic (material related) calcification is secondary, as nodules appear in patches across the implant surface. Intrinsic calcification is thought to be largely diffusion dependent [7].

9 Intrauterine Device Calcification

Intrauterine devices (IUDs) have been used as contraceptives for decades. IUDs are plastic or metal devices that are implanted into a woman's uterus to prevent pregnancy. Calcification of these implants was first noted in the 1980s. As calcium deposition progresses, the IUD surface can become incrusted. Incrustation increases the risk of complications such as abdominal cramps, infection, and inflammation, and can interfere with the normal function of the device [128]. For example, incrustation can prevent the effective release of pregnancy-preventing agents, like copper ions or hormones, and could lead to unwanted pregnancy. Plastic IUDs are more inert than copper ones, as copper corrosion and dissolution can lead to the precipitation of carbonate and calcium ions which interact to form calcium carbonate deposits on the IUD surface [129]. Calcium carbonate accounts for 75% of incrustation materials, the other 25% being composed of hydroxyapatite (5%) and matrix materials (20%) [128]. Copper corrosion, however, is not a requirement for IUD calcification, as chemically inert plastic IUDs can become incrusted as well. The calcification of plastic IUDs is highly dependent on the metabolic environment and the particular surface quality of the implant. In both plastic and copper IUDs, if inflammation is present, incrustation occurs at an accelerated pace, likely due to increased metabolism and excretion of crust-forming compounds by inflammatory cells [128]. Incrusted IUDs can be detected via ultrasound and often have to be removed, as they can cause the patient significant pain or render the device ineffective.

To prevent IUD incrustation, researchers are considering new materials that may be less reactive with the in utero environment. A team of researchers in China has developed a copper/low-density polyethylene (LDPE) composite that slows the release of irritative copper ions and prevents burst release, which has been shown to increase the risk of severe side effects [130]. Samples of the composite, as well as plain copper controls, were incubated in a simulated uterine solution for 5 months, after which the surfaces were analyzed with scanning electron microscopy. Results showed that the composite surface had no deposits while the control surface had deposits composed of calcium, phosphorus, chloride, copper, and oxygen ions. Copper and its oxide are highly reactive and can adsorb ions from the uterine solution and precipitate deposits. LDPE, however, is nonpolar, has a low surface energy, and is hydrophobic. It was hypothesized that these features prevent precipitates from adhering to the composite surface. Material modification, therefore, may be a promising method for preventing calcification and incrustation of IUDs.

10 Conclusions

Pathologic calcification affects a wide array of medical implants, including both biologic and synthetic materials. In some cases, calcification is a clinically significant problem, as it limits implant durability and can lead to device failure. The use of *in vivo* material and device testing, for example, particularly in rat and sheep models, has led to a better understanding of the mechanisms of pathologic mineralization. Bioprosthetic heart valves undergo the most frequent clinically important implant calcification. In bioprosthetic tissues, devitalized cells, residual crosslinking chemicals, and extracellular matrix components are the main culprits, whereas in synthetic materials such as polymers, surface properties and physical characteristics are the main determinants of calcification. Accordingly, anti-calcification techniques in bioprosthetic implants are aimed at eliminating cellular debris, phospholipids, and residual chemicals from the tissue or reducing the tissue affinity for calcium binding and hydroxyapatite nucleation. For polymers, on the other hand, anti-calcification strategies include minimizing surface defects, eliminating contaminants, chemically or physically modifying the surface, or controlling physical properties such as porosity and roughness. While none of these strategies can completely prevent calcification, many have successfully reduced the extent and severity of implant mineralization. Combinations of these methods may produce synergistic effects and result in improved outcomes. However, until a highly effective prevention strategy is developed, pathologic calcification will remain a limiting factor in implant performance.

References

1. Schoen FJ, Levy R (1999) Founder's Award, 25th annual meeting of the Society For Biomaterials, Providence, RI, April 28–May 2, 1999. Tissue heart valves: current challenges and future research perspectives. J Biomed Mater Res 47:439–465
2. Giachelli CM (1999) Ectopic calcification: gathering hard facts about soft tissue mineralization. Am J Pathol 154:671–675
3. Schoen FJ, Levy RJ (2005) Calcification of tissue heart valve substitutes: progress toward understanding and prevention. Ann Thorac Surg 79:1072–1080
4. Flameng W, Meuris B, Yperman J, De Visscher G, Herijgers P, Verbeken E (2006) Factors influencing calcification of cardiac bioprostheses in adolescent sheep. J Thorac Cardiovasc Surg 132:89–98
5. Levy RJ, Schoen FJ, Anderson HC, Harasaki H, Koch TH, Brown W, Lian JB, Cumming R, Gavin JB (1991) Cardiovascular implant calcification: a survey and update. Biomaterials 12:707–714
6. Schoen FJ, Levy RJ, Nelson AC, Bernhard WF, Nashef A, Hawley M (1985) Onset and progression of experimental bioprosthetic heart valve calcification. Lab Invest 52:523–532
7. Buchen SY, Cunanan CM, Gwon A, Weinschenk JI 3rd, Gruber L, Knight PM (2001) Assessing intraocular lens calcification in an animal model. J Cataract Refract Surg 27:1473–1484
8. Gumus N (2009) Capsular calcification may be an important factor for the failure of breast implant. J Plast Reconstr Aesthet Surg 62:e606–e608
9. Farivar RS, Cohn LH (2003) Hypercholesterolemia is a risk factor for bioprosthetic valve calcification and explantation. J Thorac Cardiovasc Surg 126:969–975
10. Sacks MS, David Merryman W, Schmidt DE (2009) On the biomechanics of heart valve function. J Biomech 42:1804–1824
11. Zilla P, Brink J, Human P, Bezuidenhout D (2008) Prosthetic heart valves: catering for the few. Biomaterials 29:385–406

12. Koerfer R, Reiss N, Koertke H (2009) International normalized ratio patient self-management for mechanical valves: is it safe enough? Curr Opin Cardiol 24:130–135
13. Silberman S, Oren A, Dotan M, Merin O, Fink D, Deeb M, Bitran D (2008) Aortic valve replacement: choice between mechanical valves and bioprostheses. J Card Surg 23:299–306
14. Rahimtoola SH (2003) Choice of prosthetic heart valve for adult patients. J Am Coll Cardiol 41:893–904
15. Schoen FJ, Levy RJ (1992) Heart valve bioprostheses: antimineralization. Eur J Cardiothorac Surg 6(Suppl 1):S91–S93, discussion S94
16. Sabbah HN, Hamid MS, Stein PD (1989) Mechanical factors in the degeneration of porcine bioprosthetic valves: an overview. J Card Surg 4:302–309
17. Gabbay S, Bortolotti U, Wasserman F, Factor S, Strom J, Frater RW (1984) Fatigue-induced failure of the Ionescu-Shiley pericardial xenograft in the mitral position In vivo and in vitro correlation and a proposed classification. J Thorac Cardiovasc Surg 87:836–844
18. Thubrikar MJ, Deck JD, Aouad J, Nolan SP (1983) Role of mechanical stress in calcification of aortic bioprosthetic valves. J Thorac Cardiovasc Surg 86:115–125
19. Simionescu DT (2004) Prevention of calcification in bioprosthetic heart valves: challenges and perspectives. Expert Opin Biol Ther 4:1971–1985
20. Kim KM, Herrera GA, Battarbee HD (1999) Role of glutaraldehyde in calcification of porcine aortic valve fibroblasts. Am J Pathol 154:843–852
21. Golomb G, Schoen FJ, Smith MS, Linden J, Dixon M, Levy RJ (1987) The role of glutaraldehyde-induced cross-links in calcification of bovine pericardium used in cardiac valve bioprostheses. Am J Pathol 127:122–130
22. Stacchino C, Bona G, Bonetti F, Rinaldi S, Della Ciana L, Grignani A (1998) Detoxification process for glutaraldehyde-treated bovine pericardium: biological, chemical and mechanical characterization. J Heart Valve Dis 7:190–194
23. Weissenstein C, Human P, Bezuidenhout D, Zilla P (2000) Glutaraldehyde detoxification in addition to enhanced amine cross-linking dramatically reduces bioprosthetic tissue calcification in the rat model. J Heart Valve Dis 9:230–240
24. Cheung DT, Nimni ME (1982) Mechanism of crosslinking of proteins by glutaraldehyde I: reaction with model compounds. Connect Tissue Res 10:187–199
25. Carpentier SM, Chen L, Shen M, Fornes P, Martinet B, Quintero LJ, Witzel TH, Carpentier AF (1998) Heat treatment mitigates calcification of valvular bioprostheses. Ann Thorac Surg 66:S264–S266
26. Zilla P, Weissenstein C, Bracher M, Zhang Y, Koen W, Human P, von Oppell U (1997) High glutaraldehyde concentrations reduce rather than increase the calcification of aortic wall tissue. J Heart Valve Dis 6:502–509
27. Schoen FJ, Levy RJ (1992) Bioprosthetic heart valve calcification: membrane-mediated events and alkaline phosphatase. Bone Miner 17:129–133
28. Lee CH, Vyavahare N, Zand R, Kruth H, Schoen FJ, Bianco R, Levy RJ (1998) Inhibition of aortic wall calcification in bioprosthetic heart valves by ethanol pretreatment: biochemical and biophysical mechanisms. J Biomed Mater Res 42:30–37
29. Singla A, Lee CH (2002) Effect of elastin on the calcification rate of collagen-elastin matrix systems. J Biomed Mater Res 60:368–374
30. Singla A, Lee CH (2003) Inhibition of CEM calcification by the sequential pretreatment with ethanol and EDTA. J Biomed Mater Res A 64:706–713
31. Vyavahare N, Ogle M, Schoen FJ, Levy RJ (1999) Elastin calcification and its prevention with aluminum chloride pretreatment. Am J Pathol 155:973–982
32. Vyavahare NR, Hirsch D, Lerner E, Baskin JZ, Zand R, Schoen FJ, Levy RJ (1998) Prevention of calcification of glutaraldehyde-crosslinked porcine aortic cusps by ethanol preincubation: mechanistic studies of protein structure and water-biomaterial relationships. J Biomed Mater Res 40:577–585

33. Levy RJ, Schoen FJ, Sherman FS, Nichols J, Hawley MA, Lund SA (1986) Calcification of subcutaneously implanted type I collagen sponges. Effects of formaldehyde and glutaraldehyde pretreatments. Am J Pathol 122:71–82
34. Lovekamp JJ, Simionescu DT, Mercuri JJ, Zubiate B, Sacks MS, Vyavahare NR (2006) Stability and function of glycosaminoglycans in porcine bioprosthetic heart valves. Biomaterials 27:1507–1518
35. Simionescu DT, Lovekamp JJ, Vyavahare NR (2003) Extracellular matrix degrading enzymes are active in porcine stentless aortic bioprosthetic heart valves. J Biomed Mater Res A 66:755–763
36. Vyavahare N, Jones PL, Tallapragada S, Levy RJ (2000) Inhibition of matrix metalloproteinase activity attenuates tenascin-C production and calcification of implanted purified elastin in rats. Am J Pathol 157:885–893
37. Aikawa E, Aikawa M, Libby P, Figueiredo JL, Rusanescu G, Iwamoto Y, Fukuda D, Kohler RH, Shi GP, Jaffer FA, Weissleder R (2009) Arterial and aortic valve calcification abolished by elastolytic cathepsin S deficiency in chronic renal disease. Circulation 119:1785–1794
38. Price PA, Chan WS, Jolson DM, Williamson MK (2006) The elastic lamellae of devitalized arteries calcify when incubated in serum: evidence for a serum calcification factor. Arterioscler Thromb Vasc Biol 26:1079–1085
39. Bobryshev YV (2005) Calcification of elastic fibers in human atherosclerotic plaque. Atherosclerosis 180:293–303
40. Walther T, Falk V, Diegeler A, Rauch T, Weigl C, Gummert J, Autschbach R, Mohr FW (1999) Effectiveness of different anti-calcification treatments for stentless aortic bioprostheses. Thorac Cardiovasc Surg 47:23–25
41. Borger MA, Carson SM, Ivanov J, Rao V, Scully HE, Feindel CM, David TE (2005) Stentless aortic valves are hemodynamically superior to stented valves during mid-term follow-up: a large retrospective study. Ann Thorac Surg 80:2180–2185
42. Jasinski MJ, Hayton J, Kadziola Z, Wos S, Sosnowski AW (2002) Hemodynamic performance after stented vs stentless aortic valve replacement. J Cardiovasc Surg (Torino) 43:313–317
43. Chen W, Kim JD, Schoen FJ, Levy RJ (1994) Effect of 2-amino oleic acid exposure conditions on the inhibition of calcification of glutaraldehyde cross-linked porcine aortic valves. J Biomed Mater Res 28:1485–1495
44. Vyavahare N, Hirsch D, Lerner E, Baskin JZ, Schoen FJ, Bianco R, Kruth HS, Zand R, Levy RJ (1997) Prevention of bioprosthetic heart valve calcification by ethanol preincubation. Efficacy and mechanisms. Circulation 95:479–488
45. Liao KK, Li X, John R, Amatya DM, Joyce LD, Park SJ, Bianco R, Bolman RM 3rd (2008) Mechanical stress: an independent determinant of early bioprosthetic calcification in humans. Ann Thorac Surg 86:491–495
46. Deiwick M, Glasmacher B, Baba HA, Roeder N, Reul H, von Bally G, Scheld HH (1998) In vitro testing of bioprostheses: influence of mechanical stresses and lipids on calcification. Ann Thorac Surg 66:S206–S211
47. Human P, Zilla P (2001) Inflammatory and immune processes: the neglected villain of bioprosthetic degeneration? J Long Term Eff Med Implants 11:199–220
48. Human P, Zilla P (2001) Characterization of the immune response to valve bioprostheses and its role in primary tissue failure. Ann Thorac Surg 71:S385–S388
49. Gong G, Seifter E, Lyman WD, Factor SM, Blau S, Frater RW (1993) Bioprosthetic cardiac valve degeneration: role of inflammatory and immune reactions. J Heart Valve Dis 2:684–693
50. Levy RJ, Schoen FJ, Howard SL (1983) Mechanism of calcification of porcine bioprosthetic aortic valve cusps: role of T-lymphocytes. Am J Cardiol 52:629–631
51. Giachelli CM (2005) Inducers and inhibitors of biomineralization: lessons from pathological calcification. Orthod Craniofac Res 8:229–231

52. Giachelli CM, Steitz S (2000) Osteopontin: a versatile regulator of inflammation and biomineralization. Matrix Biol 19:615–622
53. Jian B, Narula N, Li QY, Mohler ER 3rd, Levy RJ (2003) Progression of aortic valve stenosis: TGF-beta1 is present in calcified aortic valve cusps and promotes aortic valve interstitial cell calcification via apoptosis. Ann Thorac Surg 75:457–465, discussion 465–456
54. Srivatsa SS, Harrity PJ, Maercklein PB, Kleppe L, Veinot J, Edwards WD, Johnson CM, Fitzpatrick LA (1997) Increased cellular expression of matrix proteins that regulate mineralization is associated with calcification of native human and porcine xenograft bioprosthetic heart valves. J Clin Invest 99:996–1009
55. Schoen FJ, Levy RJ, Hilbert SL, Bianco RW (1992) Antimineralization treatments for bioprosthetic heart valves. Assessment of efficacy and safety. J Thorac Cardiovasc Surg 104:1285–1288
56. Connolly JM, Alferiev I, Clark-Gruel JN, Eidelman N, Sacks M, Palmatory E, Kronsteiner A, Defelice S, Xu J, Ohri R, Narula N, Vyavahare N, Levy RJ (2005) Triglycidylamine crosslinking of porcine aortic valve cusps or bovine pericardium results in improved biocompatibility, biomechanics, and calcification resistance: chemical and biological mechanisms. Am J Pathol 166:1–13
57. Girardot JM, Girardot MN (1996) Amide cross-linking: an alternative to glutaraldehyde fixation. J Heart Valve Dis 5:518–525
58. Xi T, Ma J, Tian W, Lei X, Long S, Xi B (1992) Prevention of tissue calcification on bioprosthetic heart valve by using epoxy compounds: a study of calcification tests in vitro and in vivo. J Biomed Mater Res 26:1241–1251
59. Zilla P, Bezuidenhout D, Torrianni M, Hendriks M, Human P (2005) Diamine-extended glutaraldehyde- and carbodiimide crosslinks act synergistically in mitigating bioprosthetic aortic wall calcification. J Heart Valve Dis 14:538–545
60. Vasudev SC, Chandy T, Sharma CP (2000) The anti-calcification effect of polyethylene glycol-immobilized on hexamethylene diisocyanate treated pericardium. Artif Cells Blood Substit Immobil Biotechnol 28:79–94
61. Somers P, De Somer F, Cornelissen M, Bouchez S, Gasthuys F, Narine K, Cox E, Van Nooten G (2008) Genipin blues: an alternative non-toxic crosslinker for heart valves? J Heart Valve Dis 17:682–688
62. Sung HW, Chang Y, Chiu CT, Chen CN, Liang HC (1999) Mechanical properties of a porcine aortic valve fixed with a naturally occurring crosslinking agent. Biomaterials 20:1759–1772
63. Bianco RW, Phillips R, Mrachek J, Witson J (1996) Feasibility evaluation of a new pericardial bioprosthesis with dye mediated photo-oxidized bovine pericardial tissue. J Heart Valve Dis 5:317–322
64. Suh H, Hwang YS, Park JC, Cho BK (2000) Calcification of leaflets from porcine aortic valves crosslinked by ultraviolet irradiation. Artif Organs 24:555–563
65. Flameng W, Ozaki S, Meuris B, Herijgers P, Yperman J, Van Lommel A, Verbeken E (2001) Antimineralization treatments in stentless porcine bioprostheses: an experimental study. J Heart Valve Dis 10:489–494
66. Schoen FJ (1998) Pathologic findings in explanted clinical bioprosthetic valves fabricated from photooxidized bovine pericardium. J Heart Valve Dis 7:174–179
67. Rapoport HS, Connolly JM, Fulmer J, Dai N, Murti BH, Gorman RC, Gorman JH, Alferiev I, Levy RJ (2007) Mechanisms of the in vivo inhibition of calcification of bioprosthetic porcine aortic valve cusps and aortic wall with triglycidylamine/mercapto bisphosphonate. Biomaterials 28:690–699
68. Sacks MS, Hamamoto H, Connolly JM, Gorman RC, Gorman JH 3rd, Levy RJ (2007) In vivo biomechanical assessment of triglycidylamine crosslinked pericardium. Biomaterials 28:5390–5398

69. Girardot MN, Torrianni M, Girardot JM (1994) Effect of AOA on glutaraldehyde-fixed bioprosthetic heart valve cusps and walls: binding and calcification studies. Int J Artif Organs 17:76–82
70. Trantina-Yates AE, Human P, Zilla P (2003) Detoxification on top of enhanced, diamine-extended glutaraldehyde fixation significantly reduces bioprosthetic root calcification in the sheep model. J Heart Valve Dis 12:93–100, discussion 100–101
71. Hirsch D, Drader J, Thomas TJ, Schoen FJ, Levy JT, Levy RJ (1993) Inhibition of calcification of glutaraldehyde pretreated porcine aortic valve cusps with sodium dodecyl sulfate: preincubation and controlled release studies. J Biomed Mater Res 27:1477–1484
72. Jones M, Eidbo EE, Hilbert SL, Ferrans VJ, Clark RE (1988) The effects of anti-calcification treatments on bioprosthetic heart valves implanted in sheep. ASAIO Trans 34:1027–1030
73. Vyavahare NR, Jones PL, Hirsch D, Schoen FJ, Levy RJ (2000) Prevention of glutaraldehyde-fixed bioprosthetic heart valve calcification by alcohol pretreatment: further mechanistic studies. J Heart Valve Dis 9:561–566
74. Courtman DW, Pereira CA, Kashef V, McComb D, Lee JM, Wilson GJ (1994) Development of a pericardial acellular matrix biomaterial: biochemical and mechanical effects of cell extraction. J Biomed Mater Res 28:655–666
75. Baraki H, Tudorache I, Braun M, Hoffler K, Gorler A, Lichtenberg A, Bara C, Calistru A, Brandes G, Hewicker-Trautwein M, Hilfiker A, Haverich A, Cebotari S (2009) Orthotopic replacement of the aortic valve with decellularized allograft in a sheep model. Biomaterials 30:6240–6246
76. Konuma T, Devaney EJ, Bove EL, Gelehrter S, Hirsch JC, Tavakkol Z, Ohye RG (2009) Performance of CryoValve SG decellularized pulmonary allografts compared with standard cryopreserved allografts. Ann Thorac Surg 88:849–854, discussion 554–845
77. Zhai W, Chang J, Lu X, Wang Z (2009) Procyanidins-crosslinked heart valve matrix: anti-calcification effect. J Biomed Mater Res B Appl Biomater 90:913–921
78. Zhai W, Lu X, Chang J, Zhou Y, Zhang H (2010) Quercetin-crosslinked porcine heart valve matrix: mechanical properties, stability, anti-calcification and cytocompatibility. Acta Biomater 6:389–395
79. Walther T, Falk V, Autschbach R, Diegeler A, Rauch T, Weigl C, Gunther B, van Son JA, Mohr FW (1998) Comparison of different anti-calcification treatments for stentless bioprostheses. Ann Thorac Surg 66:S249–S254
80. Levy RJ, Vyavahare N, Ogle M, Ashworth P, Bianco R, Schoen FJ (2003) Inhibition of 966 cusp and aortic wall calcification in ethanol- and aluminum-treated bioprosthetic heart 967 valves in sheep: background, mechanisms, and synergism. J Heart Valve Dis 12:209–216, 968 discussion 216
81. Hirsch D, Schoen FJ, Levy RJ (1993) Effects of metallic ions and diphosphonates on inhibition of pericardial bioprosthetic tissue calcification and associated alkaline phosphatase activity. Biomaterials 14:371–377
82. Hirsch D, Drader J, Pathak YV, Yee R, Schoen FJ, Levy RJ (1993) Synergistic inhibition of 959 the calcification of glutaraldehyde pretreated bovine pericardium in a rat subdermal model by 960 FeCl3 and ethanehydroxydiphosphonate: preincubation and polymeric controlled release 961 studies. Biomaterials 14:705–711
83. Webb CL, Schoen FJ, Flowers WE, Alfrey AC, Horton C, Levy RJ (1991) Inhibition of mineralization of glutaraldehyde-pretreated bovine pericardium by AlCl3. Mechanisms and comparisons with FeCl3, LaCl3, and Ga(NO3)3 in rat subdermal model studies. Am J Pathol 138:971–981
84. Bailey M, Xiao H, Ogle M, Vyavahare N (2001) Aluminum chloride pretreatment of elastin inhibits elastolysis by matrix metalloproteinases and leads to inhibition of elastin-oriented calcification. Am J Pathol 159:1981–1986
85. Bailey M, Pillarisetti S, Jones P, Xiao H, Simionescu D, Vyavahare N (2004) Involvement of matrix metalloproteinases and tenascin-C in elastin calcification. Cardiovasc Pathol 13:146–155

86. Golomb G, Schlossman A, Eitan Y, Saadeh H, Van Gelder JM, Breuer E (1992) In vitro and in vivo anti-calcification effects of novel bishydroxyiminophosphonates. J Pharm Sci 81:1004–1007
87. Levy RJ, Schoen FJ, Lund SA, Smith MS (1987) Prevention of leaflet calcification of bioprosthetic heart valves with diphosphonate injection therapy. Experimental studies of optimal dosages and therapeutic durations. J Thorac Cardiovasc Surg 94:551–557
88. Levy RJ, Wolfrum J, Schoen FJ, Hawley MA, Lund SA, Langer R (1985) Inhibition of calcification of bioprosthetic heart valves by local controlled-release diphosphonate. Science 228:190–192
89. Webb CL, Benedict JJ, Schoen FJ, Linden JA, Levy RJ (1988) Inhibition of bioprosthetic heart valve calcification with aminodiphosphonate covalently bound to residual aldehyde groups. Ann Thorac Surg. 46:309–316
90. Sucu N, Karaca K, Yilmaz N, Comelekoglu U, Aytacoglu BN, Tamer L, Ozeren M, Dondas HA, Oguz Y, Ogenler O, Dikmengil M (2006) Two stage EDTA anti-calcification method for bioprosthetic heart valve materials. Med Sci Monit 12:MT33–MT38
91. Handford PA (2000) Fibrillin-1, a calcium binding protein of extracellular matrix. Biochim Biophys Acta 1498:84–90
92. Fartasch M, Haneke E, Hornstein OP (1990) Mineralization of collagen and elastic fibers in superficial dystrophic cutaneous calcification: an ultrastructural study. Dermatologica 181:187–192
93. Daamen WF, Nillesen ST, Hafmans T, Veerkamp JH, van Luyn MJ, van Kuppevelt TH (2005) Tissue response of defined collagen-elastin scaffolds in young and adult rats with special attention to calcification. Biomaterials 26:81–92
94. Nimni ME, Myers D, Ertl D, Han B (1997) Factors which affect the calcification of tissue-derived bioprostheses. J Biomed Mater Res 35:531–537
95. Tedder ME, Liao J, Weed B, Stabler C, Zhang H, Simionescu A, Simionescu DT (2009) Stabilized collagen scaffolds for heart valve tissue engineering. Tissue Eng Part A 15:1257–1268
96. van Wachem PB, Plantinga JA, Wissink MJ, Beernink R, Poot AA, Engbers GH, Beugeling T, van Aken WG, Feijen J, van Luyn MJ (2001) In vivo biocompatibility of carbodiimide-crosslinked collagen matrices: effects of crosslink density, heparin immobilization, and bFGF loading. J Biomed Mater Res 55:368–378
97. Lu Q, Ganesan K, Simionescu DT, Vyavahare NR (2004) Novel porous aortic elastin and collagen scaffolds for tissue engineering. Biomaterials 25:5227–5237
98. Park JC, Song MJ, Hwang YS, Suh H (2001) Calcification comparison of polymers for vascular graft. Yonsei Med J 42:304–310
99. Stokes K, Anderson HC, McVenes R, McClay C (1995) The encapsulation of polyurethane-insulated transvenous cardiac pacemaker leads. Cardiovasc Pathol 4:163–171
100. Schlieper G, Kruger T, Djuric Z, Damjanovic T, Markovic N, Schurgers LJ, Brandenburg VM, Westenfeld R, Dimkovic S, Ketteler M, Grootendorst DC, Dekker FW, Floege J, Dimkovic N (2008) Vascular access calcification predicts mortality in hemodialysis patients. Kidney Int 74:1582–1587
101. Tomizawa Y, Takanashi Y, Noishiki Y, Nishida H, Endo M, Koyanagi H (1998) Evaluation of small caliber vascular prostheses implanted in small children: activated angiogenesis and accelerated calcification. ASAIO J 44:M496–M500
102. Coleman DL (1981) Mineralization of blood pump bladders. Trans Am Soc Artif Intern Organs 27:708–713
103. Harasaki H, Kambic H, Whalen R, Murray J, Snow J, Murabayashi S, Hillegass D, Ozawa K, Kiraly R, Nose Y (1980) Comparative study of flocked vs biolized surface for long-term assist pumps. Trans Am Soc Artif Intern Organs 26:470–474
104. Joshi RR, Underwood T, Frautschi JR, Phillips RE Jr, Schoen FJ, Levy RJ (1996) Calcification of polyurethanes implanted subdermally in rats is enhanced by calciphylaxis. J Biomed Mater Res 31:201–207

105. Harasaki H et al (1980) Comparative study of flocked vs. biolized surface for long-term assist pumps. Trans Am Soc Artif Intern Organs 26:470–474
106. Wisman CB, Pierce WS, Donachy JH, Pae WE, Myers JL, Prophet GA (1982) A polyurethane trileaflet cardiac valve prosthesis: in vitro and in vivo studies. Trans Am Soc Artif Intern Organs 28:164–168
107. Alferiev I, Vyavahare N, Song C, Connolly J, Hinson JT, Lu Z, Tallapragada S, Bianco R, Levy R (2001) Bisphosphonate derivatized polyurethanes resist calcification. Biomaterials 22:2683–2693
108. Alferiev IS, Connolly JM, Stachelek SJ, Ottey A, Rauova L, Levy RJ (2006) Surface heparinization of polyurethane via bromoalkylation of hard segment nitrogens. Biomacromolecules 7:317–322
109. Tang ZG, Teoh SH, McFarlane W, Poole-Warren LA, Umezu M (2002) In vitro calcification of UHMWPE/PU composite membrane. Mat Sci Eng C 20:149–152
110. Peters W, Pritzker K, Smith D, Fornasier V, Holmyard D, Lugowski S, Kamel M, Visram F (1998) Capsular calcification associated with silicone breast implants: Incidence, determinants, and characterization. Ann Plast Surg 41:348–360
111. Lou X, Vijayasekaran S, Sugiharti R, Robertson T (2005) Morphological and topographic effects on calcification tendency of pHEMA hydrogels. Biomaterials 26:5808–5817
112. Vijayasekaran S, Hicks CR, Chirila TV, Fitton JH, Clayton AB, Lou X, Platten S, Crawford GJ, Constable IJ (1997) Histologic evaluation during healing of hydrogel core-and-skirt keratoprostheses in the rabbit eye. Cornea 16:352–359
113. Calnan JS, Pflug JJ, Chhabra AS, Raghupati N (1971) Clinical and experimental studies of polyhydroxyethylmethacrylate gel ("Hydron") for reconstructive surgery. Br J Plast Surg 24:113–124
114. Levy B (1984) Calcium deposits on glyceryl methyl methacrylate and hydroxyethyl methacrylate contact lenses. Am J Optom Physiol Opt 61:605–607
115. Davson H (ed) (1980) Physiology of the eye, vol 18, 4th edn. Academic, New York
116. Bucher PJ, Buchi ER, Daicker BC (1995) Dystrophic calcification of an implanted hydroxyethylmethacrylate intraocular lens. Arch Ophthalmol 113:1431–1435
117. Neuhann IM, Kleinmann G, Apple DJ (2008) A new classification of calcification of intraocular lenses. Ophthalmology 115:73–79
118. Gartaganis SP, Kanellopoulou DG, Mela EK, Panteli VS, Koutsoukos PG (2008) Opacification of hydrophilic acrylic intraocular lens attributable to calcification: investigation on mechanism. Am J Ophthalmol 146:395–403
119. Dorey MW, Brownstein S, Hill VE, Mathew B, Botton G, Kertes PJ, El-Defrawy S (2003) Proposed pathogenesis for the delayed postoperative opacification of the hydroview hydrogel intraocular lens. Am J Ophthalmol 135:591–598
120. Werner L, Hunter B, Stevens S, Chew JJ, Mamalis N (2006) Role of silicon contamination on calcification of hydrophilic acrylic intraocular lenses. Am J Ophthalmol 141:35–43
121. Guan X, Tang R, Nancollas GH (2004) The potential calcification of octacalcium phosphate on intraocular lens surfaces. J Biomed Mater Res A 71:488–496
122. Werner L (2008) Calcification of hydrophilic acrylic intraocular lenses. Am J Ophthalmol 146:341–343
123. Nakanome S, Watanabe H, Tanaka K, Tochikubo T (2008) Calcification of Hydroview H60M intraocular lenses: aqueous humor analysis and comparisons with other intraocular lens materials. J Cataract Refract Surg 34:80–86
124. Dalas E, Kallitsis JK, Koutsoukos PG (1991) Crystallization of hydroxyapatite on polymers. Langmuir 7:1822–1826
125. Imai Y (1985) Effect of age on calcification of poly(hydroxyethyl methacrylate) in animals. Artif Organs 9:255–258
126. Sprinel L, Kopecek J, Lim D (1973) Effect of the structure of poly(glycol monomethacrylate) gel on the calcification of implants. Calcif Tissue Res 13:63–72

127. Patai K, Berenyi M, Sipos M, Noszal B (1998) Characterization of calcified deposits on contraceptive intrauterine devices. Contraception 58:305–308
128. Kosonen A (1981) Factors influencing the dissolution of copper in utero. Contracept Deliv Syst 2:77–85
129. Yang ZH, Xie CS, Cai SZ, Xia XP (2008) Effects of LDPE film on the properties of copper/LDPE composites for intrauterine contraceptive device. Mater Lett 62:4226–4228
130. Fyfe B, Schoen F (1999) Pathological analysis of nonstented freestyle aortic root bioprostheses treated with amino oleic acid. Semin Thorac Cardiovasc Surg 11:151–156

Biography

In 2009, Amy Munnelly earned her Bachelor of Science degree in Biomedical Engineering from Duke University, graduating *magna cum laude* and with distinction. While at Duke, she completed a Pratt Undergraduate Research Fellowship project which focused on determining the mechanisms of adhesion between cells in a co-culture model of a tissue-engineered blood vessel. Amy continued her education at Clemson University, where she worked with Dr. Narendra Vyavahare to complete her Master's thesis project, entitled "Vena Cava as an Alternative to Pericardium in Bioprosthetic Percutaneous Heart Valves." Currently, Amy resides in Costa Mesa, California and is employed at a biomedical device company.

Schoen is Professor of Pathology and Health Sciences and Technology, Harvard Medical School, and Executive Vice-Chairman in the Department of Pathology at the Brigham and Women's Hospital (BWH) in Boston. He is Director of the BWH Biomedical Research Institute (BRI) Technology Innovation Program and BWH liaison to the Center for Integration of Medicine and Innovative Technology

(CIMIT). Schoen's research contributions have been in host–biomaterial interactions, structure–function–pathology correlations in heart valve substitutes and other cardiovascular prostheses, calcification of bioprosthetic tissues, and cardiovascular tissue engineering. He is the author or co-author of over 425 manuscripts in journals and books. He authored Interventional and Surgical Cardiovascular Pathology: Clinical Correlations and Basic Principles (1989); and was Co-Editor of Biomaterials Science: An Introduction to Materials in Medicine (1st Ed. 1996, 2nd Ed. 2004; 3rd Ed. in preparation), and Silver's Cardiovascular Pathology, 3rd Ed. (2001). He is Past-President of the Society for Biomaterials (SFB), the Society for Cardiovascular Pathology (SCVP), and the International Society for Applied Cardiovascular Biology (ISACB), and was a Founding Fellow of the American Institute of Medical and Biological Engineering. He has received the HST London Teaching Award (1992), the 2003 Harvard Medical School Prize for Excellence in Teaching (Years 1 & 2), the SFB Clemson Award for Applied Biomaterials Research (1990), the SFB Founders Award (1999), and the Society for Cardiovascular Pathology Distinguished (Lifetime) Achievement Award in 2009. Dr. Schoen received a B.S.E. (Materials and Metallurgical Engineering) from the University of Michigan (1966), a Ph.D. in Materials Science from Cornell University (1970) and an M.D. from the University of Miami School of Medicine (1974). He completed an internship in Surgery at Jackson Memorial Hospital in Miami, and a residency in Anatomic Pathology and a fellowship in Thoracic and Cardiovascular Pathology at the University of Florida.

Dr. Vyavahare is the Hunter endowed Chair and Professor in the Department of Bioengineering at Clemson University. He is also director of NIH funded COBRE Center of Biomaterials for Tissue Regeneration (SCBIOMAT). He is serving as a Co-Chair for Cell and Organ Therapies special interest group in the Society for Biomaterials. He is also a fellow of the American Institute of Medical and Biological Engineering (AIMBE). His research has focused on cardiovascular pathology and implants for last 15 years, and he has published more than 100 scientific articles in this area. His awards include McQueen Quattlebaum faculty achievement award for exemplary leadership in the engineering profession from College of Engineering and Science, Board of Trustees research awards, and Alumni Award for outstanding achievement in research from Clemson University. Vyavahare received his B.S. and M.S. degrees in chemistry in 1985 and Ph.D. degree in biomaterials in 1990 from Pune University, India. He received further postdoctoral training at Rutgers University and the University of Michigan Medical School. Before coming to Clemson in 1999, he held faculty positions at the University of Michigan Medical School, and at the University of Pennsylvania School of Medicine. The research in his laboratory is sponsored by the National Institutes of Health (NIH).

Chapter 15
Orthopedic Implant Retrieval and Failure Analysis

Lynne C. Jones, Audrey K. Tsao, and L.D. Timmie Topoleski

Abstract Orthopedic devices make up a major percentage of the medical devices being implanted today. Success rates exceeding 95% have been reported. However, their longevity is compromised by fatigue, wear, corrosion, and other degradative mechanisms. Adverse biologic responses to products from these different mechanisms can lead to significant bone loss and may even compromise subsequent surgeries. It is imperative that we fully understand what mechanisms are in play and carefully evaluate strategies to minimize, if not eliminate, these degradation processes. Devices retrieved at revision surgery and at autopsy have provided us with important information regarding the performance of orthopedic implants. Surface characterization of the implant, biomechanical analyses, and histological analysis of the tissues surrounding the implant can provide us with information not available through other sources. This chapter will review the significance of retrieved implants to the long-term survival of these implants.

L.C. Jones (✉)
Department of Orthopaedic Surgery, Johns Hopkins University School of Medicine, Baltimore, MD, USA

Johns Hopkins Orthopaedics at The Good Samaritan Hospital, Suite 201 Smyth Building, 5601 Loch Raven Boulevard, Baltimore, MD 21239, USA
e-mail: ljones3@jhmi.edu

A.K. Tsao
Sun Valley Orthopaedic Surgeons, 12361 W. Bola Drive, Suite 100, Surprise, AZ 85374, USA
e-mail: aktsao@svos.md

L.T. Topoleski
Department of Mechanical Engineering, University of Maryland, Baltimore County, Baltimore, MD, USA

N. Eliaz (ed.), *Degradation of Implant Materials*, DOI 10.1007/978-1-4614-3942-4_15,

1 Introduction

Devices have played a fundamental role in the success of orthopedic procedures. The ultimate goal, as with any biomedical implant, is to relieve pain and restore function. For example, total joint arthroplasty is reported to be one of the most successful orthopedic procedures, with success rates exceeding 95% [1, 2]. However, this is usually framed in terms of short-term or mid-term results. The longevity of implants is impacted by material and device design, surgical technique, and patient-related factors. The previous chapters have outlined the various mechanisms that may be involved in the degradation of implant biomaterials. This chapter provides an overview of how these mechanisms may affect the longevity of orthopedic implants.

Orthopedic devices encompass the relatively simple designs of cerclage wires, rods, and screws to the complex designs of total joint replacements and electrical stimulation devices such as those used in DC electrical stimulation or with expiratory muscle activation. The materials and designs utilized depend on the location and function of the tissue being replaced. Orthopedic devices *in situ* and at revision have been evaluated to determine the limitations of the implants and to improve the longevity and success of these implants. While more difficult to obtain, postmortem retrievals may provide additional information concerning well-functioning implants.

The outcome of orthopedic procedures using devices has long been the focus of surgeons and clinician-scientists and studies have ranged from characterization of implant fatigue (fracture) to the biological responses toward metal ions and wear particles. While the FDA, since its establishment, has watched over the use of medical devices, they were empowered with more oversight responsibility in 1976 when Congress passed the Medical Device Regulation Act. Devices are required to undergo preclinical testing to determine their safety profile. Proof-of-concept studies are also performed to determine whether the device is capable of performing as hypothesized. These studies are particularly useful in screening out certain materials and devices that will not function as intended. That is, if it does not work in an *in vitro* or *in vivo* model, it is not likely to work clinically.

However, the behavior(s) of a specific device within the human body with its specific biological and mechanical constraints cannot be ascertained until it has been implanted in patients in large clinical trials. Failure of devices and surgical procedures occur despite the best efforts of research and development teams. Examination of specimens retrieved at revision or autopsy is needed to thoroughly examine the implant and the implant–host interface. In order to fully comprehend the failure mechanisms of orthopedic devices and the biologic responses to these failure mechanisms, one must first appreciate the anatomy and physiology of the tissues being replaced as well as the design and mechanics of these devices.

1.1 The Musculoskeletal System

The musculoskeletal system is comprised of hundreds of bones, muscles and tendons, ligaments, cartilage, and other connective tissues (meniscus, synovium, etc.). Medical devices have been designed to restore the function of most of these tissues. An understanding of the anatomy and physiology of these tissues is needed in order to replace or support these tissues within the mechanical and biological milieu in which they are located.

1.1.1 Bone

Bone consists of long bones, short bones, flat bones, and irregular bones. They are grouped into two major categories: the axial and appendicular skeleton. The structure of a specific bone is related to its function and location. Two primary functions of bone include mobility and protection. Bones such as the cranium and ribs protect soft tissues such as the brain and spinal cord, the lungs, heart, etc. Bones such as the humerus and tibia act as structural supports and levers. The skeleton is also a major source of calcium—an element required for blood clotting, chemical transmitter release, muscle contraction, and normal heartbeat. While it has long been recognized that the musculoskeletal system plays a primary role in mobility, protection of internal organs, and a source of calcium, recent research regarding cell-based therapies has demonstrated the importance of bone marrow as a source of progenitor cells for hematopoiesis and immunity [3–5].

Choosing materials for orthopedic implants usually focuses on the material and structural properties of the bone to be replaced. Bones experience mechanical compressive and tensile stresses and shear. They are subjected to loading under a variety of conditions depending on the position of the joint and bone with respect to the torso: axial, off-axial, rotation, and varus-valgus. Bone is considered anisotropic because it responds differently if forces are applied in different directions [6]. The mechanical requirements for implant materials depend on whether the implant will be directly weight bearing or not. Clearly, the mechanical strength of an implant must be significantly higher for an implant such as a femoral component of a total hip prosthesis as compared to bone graft used as a void or space filler.

1.1.2 Joints

A coordinated effort between motor neurons and muscles is required to permit movement of individuals within their environment. Many types of movement including flexion, extension, rotation, but also abduction, adduction, supination, pronation, dorsiflexion, plantar flexion, inversion, and eversion movements can cause various loading situations, i.e., distraction, compression, shear, rotation, and a combination at the same time within the same joint. There are different types of joints: synarthroses, amphiarthroses, and diarthoses, underlying different levels of complexity from the less complex pubic symphysis to the very complex ankle. Understanding the kinematics of these joints is vital to designing a joint replacement. In general, the long-term outcome of a joint replacement correlates with the degree of complexity: the more complex, the worse the long-term outcome and vice versa.

Implants have been designed for diaphyseal joints. The original joint contains articular cartilage laid over bone, ligaments to provide constraints to motion about the joint, synovium and synovial fluid to provide lubrication, and fibrocartilage (e.g., meniscus, labrum) to act as a stress distributor. When the knee is loaded, the meniscus distributes stress over a large area of the articular cartilage. While many joint replacements may contain different components with different materials, there are no implants that model directly the softer cartilage over harder bone. The use of metal-on-polyethylene attempts to mimic some of the tribology. The use of mobile bearing implants was introduced to decrease strain and enable some of the motions within the joint.

1.1.3 Effect of Trauma, Disease, and Aging on the Musculoskeletal System

Trauma, disease, and aging can affect the quality of the bone and other musculoskeletal structures. This can have an impact on the outcome of the surgical procedure. For example, it has long been recognized that soft osteoporotic bone or overstretched lax ligaments following trauma can negatively influence fixation and function of total knee prostheses. However, during the R&D phase, most implant designs are tested in "healthy" bones and joints. The effect of the unhealthy environment on the outcome of the interface is not known. Relevant models should be utilized, particularly if the implant being designed has direct applications to a specific condition (e.g., locking screws for osteoporotic bone).

1.2 Orthopedic Implants

Orthopedic devices have been created to replace or support the many different tissues of the musculoskeletal system. Table 15.1 summarizes some of the orthopedic implants that have been used.

Historically, surgical ingenuity has resulted in the use of a wide variety of materials as orthopedic biomaterials. The first orthopedic implants were comprised of wood, ivory, iron, gold, silver, copper, and bones from different animals [7]. Despite the likelihood of challenges regarding biocompatibility and material properties, these early procedures were likely compromised by issues relating to the length of surgery, anesthesia, and infection. A list of some of the materials utilized in orthopedic implants is given in Table 15.2.

Table 15.1 Orthopedic implants for repair and reconstruction

1. Fracture fixation
a. Wires, pins, and screws
b. Plates and screws
c. Rods and other internal fixation devices
d. Spinal fixation devices
2. Ligament repair
a. Suture and screws
b. Ligament and tendon allografts and substitutes
3. Joint replacement
a. Hip
i. Total hip
ii. Resurfacing
1. Total
2. Hemiarthroplasty
iii. Femoral
b. Knee
i. Total knee
ii. Unicondylar
iii. Unispacer
c. Ankle
d. Shoulder
e. Elbow
f. Finger
g. Wrist
4. Spine
a. Cages
b. Rods
5. Bone graft and bone graft substitutes
a. Autograft
b. Allograft: cortical bone, cancellous bone, ligaments, tendons, meniscus, cartilage
c. Bone graft substitutes
i. Tricalcium phosphate
ii. Calcium sulfate
iii. Hydroxyapatite
6. Biologics/tissue engineered constructs
a. Drug eluting materials
b. Cell-based combination products
7. Prosthetics

Metal implants were then introduced, primarily because of their strength and the ability to shape them into different structures. Metals include stainless steel, cobalt–chromium alloys, titanium and titanium alloys, and tantalum. Polymers were introduced and are still used for articulating components and the repair of soft tissues. Polymers that have been used include polyethylene, silicone, PET, ePTFR (ACL), PLA-C fiber, Dacron®, and Teflon®. Polymethylmethacrylate (PMMA) has been used for implant fixation as well as for direct injection to support

Table 15.2 Materials used in the manufacture of orthopedic implants

1. Metals
   a. Stainless steel
   b. Cobalt–chromium alloys
   c. Titanium alloys
   d. Tantalum
2. Polymers
   a. Polymethylmethacrylate (PMMA)
   b. Polyethylenes
   c. Silicone
   d. Polyethylene terephthalate (PET)
   e. Expanded polytetrafluoroethylene (ePTFR)
   f. Polylactic acid/carbon (PLA-C) fiber
   g. Dacron®
   h. Teflon®
3. Ceramics
   a. Bone graft substitutes
      i. Tricalcium phosphates
      ii. Calcium sulfate
      iii. Hydroxyapatite
   b. Arthroplasty
      i. Alumina
      ii. Bioglass
      iii. Zirconia
4. Composites
5. Biologics

bony structures (such as in Kyphoplasty). Ceramics have been used for support structures (e.g., femoral stems), articulating surfaces, and as bone substitutes. Ceramics utilized have included alumina, zirconia, calcium phosphates/hydroxyapatite, and bioactive glasses. There has been considerable interest in the use of composites. These have included carbon fiber reinforced UHMWPE, polylactic acid and polyglycolic acid composites for biodegradable pins and plates and as scaffolds, poly ether ether ketone (PEEK), aramid and dacron for artificial tendons and ligaments [8]. Composites have also been explored in order to create products that harness the strengths of different materials. Recently, biologics and tissue engineered constructs are being developed to replace worn or damaged cartilage, bone, and soft tissues.

Although not traditionally included under orthopedic implants, bone grafts are also a biologic implant used extensively in orthopedics. Originally, bone grafts primarily encompassed autografts, allografts, and xenografts; today, bone graft substitutes such as tricalcium phosphate, tricalcium sulfate, and hydroxyapatite products are frequently used in those cases where the bony architecture must be reconstructed. Grafts may now include cells with bone graft (e.g., Trinity Evolution®, Osteocel®) or for cartilage reconstruction (Carticel®). The future of biologics is likely to include various combination products including tissue-engineered constructs and drug-eluting implants.

Many factors govern material selection, including material properties: bulk properties, surface properties, biocompatibility; mechanical properties, long-term structural integrity; wear, corrosion, and fatigue. Articulating surfaces have been particularly troublesome, as will be discussed later.

1.3 Failure

Device failure has been defined by the FDA as the "failure of the device to function or perform as intended" [9]. However, it is important to recognize that failure may be a consequence of the materials or design of the implant, mechanical environment, inadequate surgical technique, or host-related factors. Material failures occur when there is a degradation of the material either due to manufacturing defect, corrosion, or exceeding the material properties of the materials used in the device. Mechanical failures may be related to material failures but may also include failures relating to malposition (e.g., varus/valgus, rotation, anterior/posterior) of the implants or inadequate fixation. There are also a number of factors related to the health status of the patient. This may include the presence of infection or sepsis, the bone quality, and whether there is evidence of osteoporosis, disease, and other comorbidities. Table 15.3 summarizes some of the failure mechanisms associated with orthopedic implants.

In previous chapters, laboratory tests have been described which are conducted during the preclinical development of medical devices, including those used for orthopedic procedures. In order to determine the safety and efficacy of the implants, preclinical testing is completed prior to implantation of a device in patients and before any clinical trials are initiated. Safety is determined by characterization of implant integrity as well as by which biological responses are activated in cell culture assays and animal models (absence of acute or chronic inflammation, tumorigenesis, hypersensitivity, toxicity). While preclinical testing is useful for screening and predicting how medical devices will behave *in situ*, medical implants retrieved from revision surgery or from autopsy specimens provide the definitive proof of how the human body responds to the implant and how the implant responds to the human environment. While a majority of the implants removed during revision surgery are removed for failure, autopsy specimens may also include those that have been characterized as well functioning. A more holistic approach to the study of medical devices requires that both types of implants be evaluated.

1.4 R&D: Preclinical Testing

During the development stage, implants undergo rigorous testing regimens to evaluate mechanical properties, biocompatibility, safety, and proof of concept. Implants for orthopedics and other disciplines are expected to pass standards

Table 15.3 Factors which may contribute to failure of the orthopaedic procedure

1. Host factors
   a. Age and gender
   b. Disease—osteoporosis, rheumatoid, CDH
   c. Trauma
   d. Health
   e. Smoking
   f. Biological responses to materials or wear
      i. Inflammation: acute and chronic
      ii. Hypersensitivity reactions
      iii. Carcinogenesis
2. Surgical techniques
   a. Poor cement technique
   b. Malpositioning with respect to alignment, rotation, varus/valgus, etc.
   c. Iatrogenic damage of the implant surface
3. Implant
   a. Fracture of a component (e.g., femoral stem)
   b. Delamination of UHMWPE (material failure)
   c. Wear—adhesion, abrasion, transfer, fatigue, third body
      i. Material selection
      ii. Malalignment
      iii. Poor implant design
      iv. Impingement
   d. Corrosion—uniform, galvanic, crevice, pitting, intergranular, stress
   e. Failure due to the biological process: nonunion of fracture, loosening, infection, allergic reaction, plastic deformation, reinjury, pathological process i.e., cancer, Paget's osteoporosis
   f. Retrieval of well-functioning implants: young age with fracture fixation, autopsy retrieval of long-term and short-term implants

regarding these properties (ANSI, ISO, ASTM). These tests are used to screen materials and implant designs to eliminate those which are more likely to fail. Various benchtop and animal tests are conducted in order to predict the limitations of a material or implant in order to determine the likelihood of failure under certain conditions. However, they are unable to predict all types of responses. It is not possible to predict 100% of the time what conditions the material or implant will be subjected to. The behavior of the implant *in situ* provides the definitive answer to whether the implant functioned as intended. Therefore, while laboratory testing is useful in predicting failure mechanisms, the performance of an implant and its material components is best evaluated following explantation from patients.

1.5 Clinical Testing

Clinical series and outcome studies are necessary to determine whether pain is relieved and how the tissue under study performs during activity, including activities of daily living as well as work and leisure activities. Clinical studies involve different types of outcomes based research (relief of pain and restoration of function). Regarding pain relief, there is no objective way to determine pain relief, so patient-based and physician-based questionnaires and measurement of impact on function are used to indirectly assess pain relief. Various imaging modalities can be used to evaluate the tissue–implant interface. However, there are limitations to this type of analysis. With orthopedic implants, by the time that the bony architecture has changed sufficiently to be seen in radiography, significant damage has been done. Another limitation is that techniques are not readily available to determine ingrowth into a device whether it is intended (porous ingrowth prostheses) or not (adhesion tissue).

2 Implant Retrieval Programs

Several retrieval programs have been established at academic institutions throughout the country [10–18]. While many of these programs have focused on characterization of implants retrieved at revision surgery, some programs collect autopsy specimens from patients who have consented to explantation at the time of death. Retrieved devices are not a model or simulation; they permit the assessment of the actual performance of the device [19]. Both retrievals obtained during revision surgery and at autopsy provide us with direct information about the implant itself as well as the host responses to the implant.

2.1 Retrievals from Revision Procedures

As one of the primary goals of an implant retrieval program is to study failure mechanisms in order to develop strategies to improve the longevity of implants, specimens obtained at revision are invaluable. These studies primarily evaluate implant surfaces in order to identify mechanisms of material failure, wear, or corrosion. Isolated retrievals provide evidence of known parameters such as cold flow or creep and wear of known factors that are patient specific to their clinical situation, but broad generalizations cannot be done [20]. As will be described later, implants retrieved at revision were instrumental in determining that delamination of the polyethylene tibial inserts was occurring with the heat-pressed polyethylene used with the PCA total knee prosthesis [21, 22]. Revision retrievals have also demonstrated that mismatched metals could lead to increased corrosion at head–neck junctions in total hip prostheses or with screws and plates [23, 24]. Not all revisions are for failure of the implant or fixation. Implants may also be removed for infection or for undersized components. Important information can also be gleaned from these implants, particularly the biological responses to these situations.

However, there are limitations to studies of revision implants. As it is important to conserve as much of the host tissue as possible during a revision, there is minimal tissue from the host–implant interface to analyze. Frequently, there are artifacts introduced to the surface of the implant during removal that may hinder the surface characterization. Additionally, the implant may not have functioned properly for some time, making it difficult to determine whether a specific feature of the implant is due to a specific mechanism [25]. For example, it may be impossible for a given implant to determine which came first: the implant failure or the loosening.

2.2 Autopsy Retrievals

Fewer studies have been conducted of autopsy specimens. This is partially due to the regulations and amount of effort required to retrieve specimens from the deceased [11]. Like revision specimens, surface characterization can be performed on autopsy specimens, and it is also possible for the surface of the implant to be damaged during the removal of the implant. A major advantage of postmortem specimens is that they provide an opportunity to fully characterize the biologic response of the surrounding tissues to the device. Autopsy specimens allow for a block to be removed containing the explanted device as well as the surrounding tissue left as undisturbed as possible. This type of specimen is ideal for studying implant fixation of arthroplasty devices as well as ligaments and other soft tissue. Furthermore, autopsy specimens can also be used to evaluate the mechanical behavior of the implant under various loading conditions [26–29]. For implants with biologic fixation, autopsy specimens permit the evaluation of biologic ingrowth into or onto the porous surface. This type of determination is not possible *in situ* using standard imaging methodologies. As radiolucencies around orthopedic implants can mean different things [25], the tissue must be examined histologically to determine whether the tissue is fibrous or full of histocytes and whether there are any particles and, if so, what they may be.

2.3 Clinical Context

Regardless of whether the sample was obtained at revision or autopsy, there is a need for clinical context. For example, it cannot be assumed that all autopsy-retrieved implants are well functioning or well fixed [25]. This demonstrates the importance of having knowledge of the patient's last follow-up and how the implant was functioning just prior to the death of the patient. Furthermore, the patients may have been relatively inactive or sedentary years before their death, thereby reducing the amount of mechanical loading of an implant [30]. What was the health of the patient? How was the implant positioned? Were there complications during or after surgery? Did the patient fall? Did they have osteoporosis? The answers to these questions may be easier to obtain from patients undergoing revision. Generally, we may know very little about the *in vivo* service conditions of a particular device. There may be no information related to the activity level of the patient/donor, whether some trauma occurred years before the retrieval, or there may be no radiographic evidence of material failure [11, 31, 32].

While retrieval studies—both revision and autopsy—have added to our understanding of how devices function, there are still limitations to our ability to determine the biological and functional history of the implant. Each case represents an individual's experience; no two cases are absolutely identical. Each case is a snapshot: one moment over the course of the entire time. Different factors may have contributed to the final appearance of the implant and the investigator is like a forensic scientist in trying to determine the major factors.

2.4 Assessments

Both revised and autopsy implants have been used to study the biological response to the implant. Due to the nature of revision, the samples taken to study the host response are usually soft tissue samples taken from the interface. Samples of periprosthetic tissues have been instrumental to the study of wear debris found at the host–bone interface of total joint replacements.

Specimens that have been retrieved after *in vivo* service are unique sources of data that can give singular insight into the service performance of a device. The condition of the retrieved implants depends on the service environment and length of time *in vivo*, but can also be influenced by the manner of retrieval, handling, and storage [11]. Changes to the implanted materials likely begin immediately after implantation. Consider, for example, that a component of an artificial joint goes from a sterile package into a human biological environment, which is wet, full of proteins, and will impose mechanical and chemical loads on the material. In cases of retrieved specimens, the condition of the material is that immediately prior to the retrieval, and we may not be able to determine the evolution of the changes from the immediately implanted state to the retrieved state. For example, when analyzing the worn surface of an articulating component (e.g., an UHMWPE tibial tray or a CoCrMo femoral head), the surface is characteristic of the state of the material just before it was removed. By close examination of the scratch patterns, it may be possible to determine which scratches are older because we may be able to detect or observe whether one scratch lies on top of another. In contrast, it may not be possible to determine whether severe damage occurred over a period of days, weeks, months, or years prior to the retrieval [33, 34]. Each implant surface holds the story of its history; however, we may not be able to speak the correct language to properly understand that story.

If we are fortunate, the retrieved specimens may contain areas that are not (or apparently not) damaged, and thus may act as an internal control or self-reference in contrast to the damaged areas. If not so fortunate, then retrievals can be compared only to either other retrieved specimens or never-implanted laboratory controls. In some studies, specimens retrieved postmortem (during autopsy) were compared to similar specimens retrieved during revision surgery. In such cases, there is a unique opportunity to investigate the characteristics and behaviors of specimens that failed *in vivo* and those that did not fail *in vivo*. In other studies, retrieved specimens may be tested and the results compared to the results of the same tests performed on never-implanted specimens, as controls.

It is difficult to control or randomize the specimens available for retrieval studies as for prospective studies; it is not possible to predict whether a device will fail, thereby providing a specimen at revision, or whether a patient will pass away, leaving a postmortem specimen available. For retrieval studies, we get what we get. Researchers are therefore careful in selection of specimens and the types of data they attempt to discern from the population of specimens available. Regardless of the care and skill of the researchers, because the available specimens, retrieval techniques, and methods of analysis may differ greatly between investigators, it may be difficult to form general conclusions from the results of the many investigators. Recognizing this, there are investigators who have taken on the task of developing consistent methods to support standard protocols to create retrieval programs [11, 35].

In considering an overview of retrieval studies, one question is how to organize the many studies that are available in order to uncover specific relationships. In orthopedic biomaterials, depending on the interest of the reader, we may wish to differentiate between studies of different devices, like artificial hips vs. artificial knees. However, there may be different studies of retrieved artificial hips that may not be related, for example a study of the surface roughness in acetabular components of metal-on-metal joints and a study on the initial stability of cemented vs. non-cemented femoral components. Any categorization may be best left to the interest of the researcher. That said, it is important to understand the different types of analyses that are available, and what those different analyses can teach us about the device.

There are two broad categories of analysis to consider: specimens that are characterized through nondestructive observation (e.g., scanning electron microscopy (SEM) or surface profilometry) and specimens where specific properties are determined (e.g., through mechanical testing or chemical analysis). In the use of either approach, we are interested in the changes that occurred to the device *in vivo*, and those changes may be referenced against never-implanted, virgin materials (representing the initial conditions of the material), well-functioning retrieved materials (retrieved at autopsy), failed specimens (retrieved during revision surgery), or specimens, either retrieved or never-implanted, subjected to laboratory testing.

It may thus be difficult to reconcile or cross-reference the results from a study that used SEM analysis of specimens retrieved during revision surgery with the results from a study that tested the Young's modulus of specimens created from material retrieved at autopsy. Nonetheless, the many researchers who have examined retrieved specimens have advanced our knowledge of the *in vivo* behavior of orthopedic biomaterials and have made enormous contributions to improving orthopedic devices.

For the sake of organization, it will help to guide this discussion to examine some of the results of retrieval analysis by looking at materials, location (hips vs. knees), and within each device category, exploring the differences in knowledge obtained from autopsy retrievals, revision retrievals, never-implanted specimens, and perhaps to focus on the information gained by the retrieval study that could not have been obtained through any other method.

3 Key Examples Over the Years

The evaluation of retrieved orthopedic implants and the surrounding tissues have led to several significant findings regarding the implants themselves as well as the biological responses to them. The remainder of the chapter will highlight some of the key examples. As most orthopedic retrieval programs have focused on total joint replacement, most of the examples relate to this type of implant.

There are many factors which may contribute to the long-term success of an orthopedic implant, including those relating to the implant, the surgical technique, and the patient. Most of this chapter will focus on the implant and the biological response to the implant.

3.1 What Have We Learned About Patient and Surgical Technique?

Much of what is known about the influence of surgical technique and patient-related factors on outcomes is not based on retrievals. Therefore, the reader is referred to several well-written reviews on these topics [31, 36–38]; a summary is provided below.

3.1.1 Clinical Factors

The importance of the proper execution of the surgical technique has been underscored by the poor clinical outcomes of malpositioned or malaligned implants [39–42], inadequate fixation into poor bone stock [43], or inadequate repair or balancing of compromised tendons or ligaments [44, 45]. Experience of the surgeon and learning curves for new techniques have also had a significant impact. Instrumentation has been shown to be important to increase the reproducibility of the techniques involved. One source of damage to articulating surfaces may occur if the surgeon nicks or scratches the surface during implantation. Thorough irrigation is required to limit the possibility of third-body wear that might have been introduced during implantation.

3.1.2 Patient Factors

Patient variables may also affect the lifespan of the procedure. It is important to appreciate that the patient's anatomy and health are not static but ever-changing. For most orthopedic procedures, bone quality is of paramount importance. Osteopetrosis or heterotopic ossification may lead to improper seating of an implant or screw fixation [35, 46], while osteoporosis may lead to unbalanced loading of the supporting host bone [47, 48]. Anatomic diversity as well as deformities may result in the patient lying outside the bell curve used to determine size of implant [31, 49]. Other patient factors which may contribute to variable results include activity level [50], preoperative discrepancies in length of bony or soft tissue structures [51], and ligamentous laxity [52]. Prior surgery, including whether the index procedure is a revision procedure, may have an effect on surgical exposure, bone quality, adhesion tissue, and wound healing.

The general health of the patient is also likely to have an effect on healing. Patient comorbidities have been shown to have an effect on the long-term outcome of orthopedic devices. These include but are not limited to obesity, osteoporosis, rheumatoid arthritis, trauma, tumors, infection, and diabetes. Smoking has been demonstrated to hinder bone healing, thus smoking abstinence before total joint replacement is recommended [53, 54].

3.2 What Have We Learned About Material Selection?

With the modern era of devices, implant components have been manufactured from metals, polymers, ceramics, and composites. There has been a significant learning curve regarding the impact of these materials on the performance of the implant as well as the biologic responses to these materials.

3.2.1 Metals

Fracture of an implant can occur through fatigue of the metal, as shown in Fig. 15.1. This has been observed with internal fracture fixation devices and screws [35, 55], as well as with hip and knee prostheses [56–58]. Fractured hip implants and internal fixation devices, when they occur, are often associated with bending [56, 59, 60]. Several factors may contribute to the fracture of a tibial tray component in total knee replacement, including implant design, alignment, and the quality of the subchondral bone [61–63].

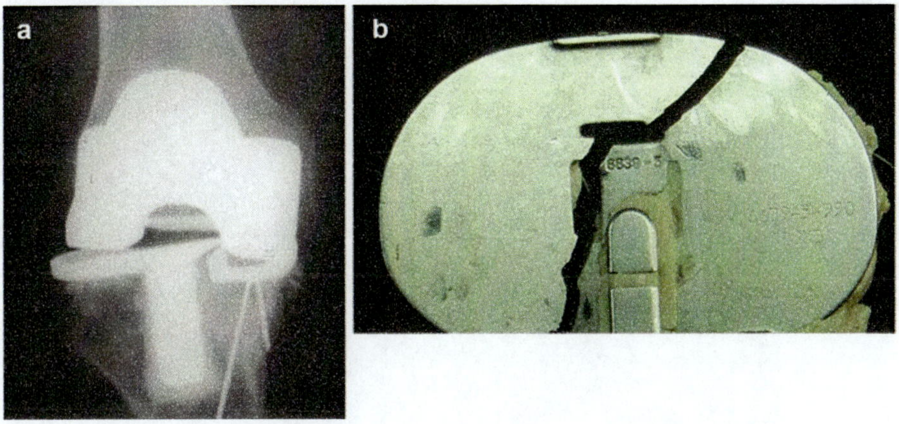

Fig. 15.1 Catastrophic failure of tibial metal baseplate through stress area through slot near with well-fixed stem. (**a**) AP radiograph; (**b**) photograph of tibial base plate

Corrosion

Corrosion can lead to improper function of an implant as well as to biological consequences relating to introduction of metal particles and ions into the periprosthetic tissues. While laboratory tests have been designed to evaluate materials regarding their potential to corrode (ASTM), it may not be possible to create a truly inert—noncorrosive metallic orthopedic implant—especially those with articulating surfaces. There are several mechanisms of corrosion including uniform, galvanic, crevice, pitting, intergranular, and stress. Corrosion resulting from each of these mechanisms has been observed on a number of different orthopedic implants as well as on the surfaces of particulate debris [32, 64–78].

Retrieval studies have focused on the interfaces between the head–neck junction of femoral components of hip systems as well as the screw–plate junction with internal fixation devices. These areas are prone to corrosion resulting from crevice and galvanic mechanisms. This may be of particular concern when the components are made by different manufacturers or of dissimilar metals as is seen with cobalt–chromium heads on titanium stems [23, 64, 71, 79, 80].

Wear

Wear is not a single phenomenon, and wear damage can be caused by several different mechanisms. Wear depends on the nature of the articulating materials and any material existing in between. Wear can be divided into two main types: two-body and three-body wear. Sandpaper on wood is an example of two (or second) body wear. One surface is abrasive toward the other. Polishing cream or cleanser are examples of three (or third) body wear. Abrasive particles (the third bodies) are free to move between the two surfaces and cause damage. In either case, material

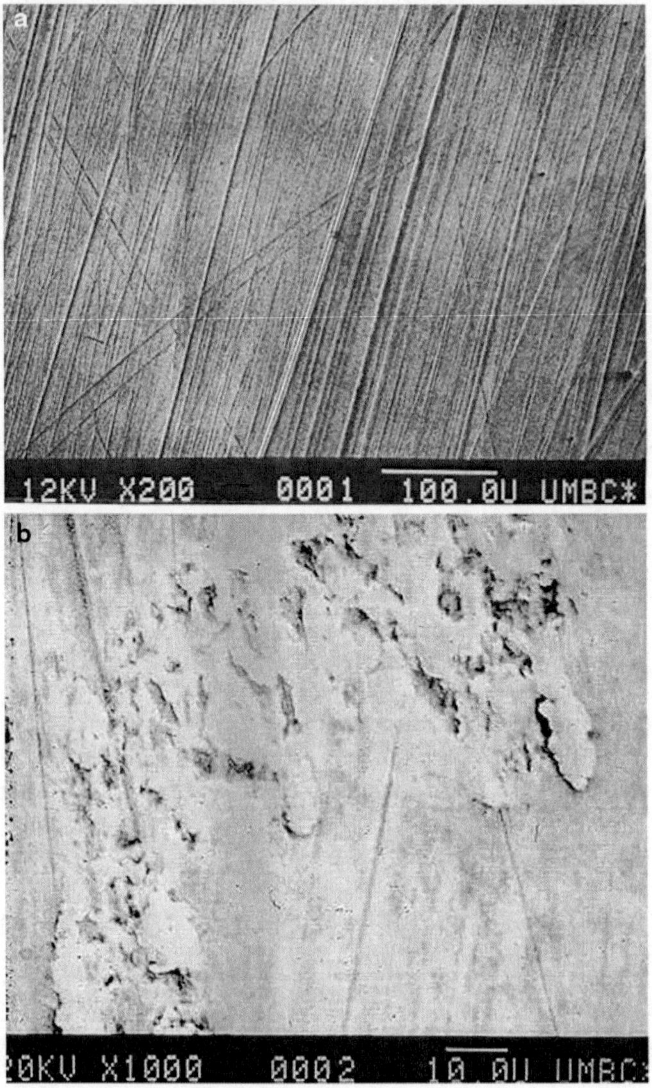

Fig. 15.2 (**a**) Scratches from abrasive wear on a retrieved CoCrMo artificial joint surface. Such scratches are usually made by abrasive particles. (**b**) Material that has been displaced on a retrieved CoCrMo artificial joint surface. The material appears to be smeared across the surface. It may be that the material was initially removed or displaced from the surface by adhesion (to the articulating counter-surface), then plastically deformed as it was repositioned on the original surface

may be removed or rearranged on a surface by different mechanisms (Fig. 15.2). Abrasive wear is caused by hard particles plowing furrows in a material. Adhesive wear is caused by high stress contact at microasperities on a surface, which essentially leads to welding of the two surfaces, and subsequent fracture and

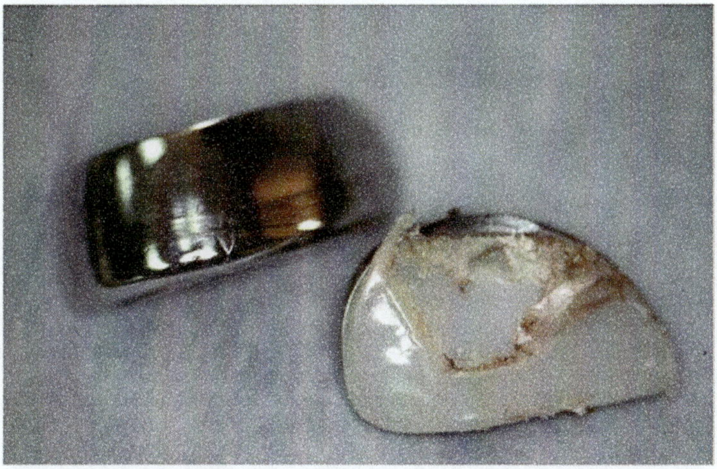

Fig. 15.3 Edge loading of polyethylene with failure to tibial tray and consequent burnishing of femoral component related to malalignment of unicompartmental knee replacement

material transfer. Particles may form on surfaces through surface fatigue (e.g., metal carbides may become dislodged through repeated loading). Erosion is a type of wear caused by an impingement of abrasive particles on a surface. These wear mechanisms can lead to either a rougher surface, or smoother surface, depending on the nature of the surfaces and the type of wear. Fretting wear occurs when small oscillating motion exists between two surfaces and creates a self-sustaining process where the oxide layer is removed and reformed.

While the focus has primarily been on polymer wear, especially with total joint prostheses, metal wear can also pose a significant challenge. Concern has focused primarily on the mechanisms of abrasive, corrosive, and fatigue wear [64]. Metal-on-metal wear can occur in metal-on-polyethylene devices with impingement or malalignment (Fig. 15.3). Mintz et al. reported substantial femoral component abrasions present in areas where metal-to-metal contact occurred in total knee prosthesis that had a worn metal-backed tibial insert [22].

Wear of the metal component, whether or not it has worn through the UHMWPE, presents at least two major problems in the implant system. First, wear of the metal results in the release of metal particles, and the exposure of new, non-oxidized, metal surface to the host. Both the metal particles, with high surface area-to-volume ratios, and the newly scratched metal surface are sources for enhanced corrosion and ion release. Second, the scratched metal surface becomes an abrasive against the polyethylene, potentially accelerating the degradation and wear of the polyethylene surface [81]. The chemical composition of the scratched surface can be quite different from that of the initial surface of the implant through *in vivo* oxidation and diffusion through the near surface [34, 82].

In metal-on-metal, ceramic-on-metal, or ceramic-on-ceramic, the type of wear debris and consequences of wear can be different. One common challenge to all of

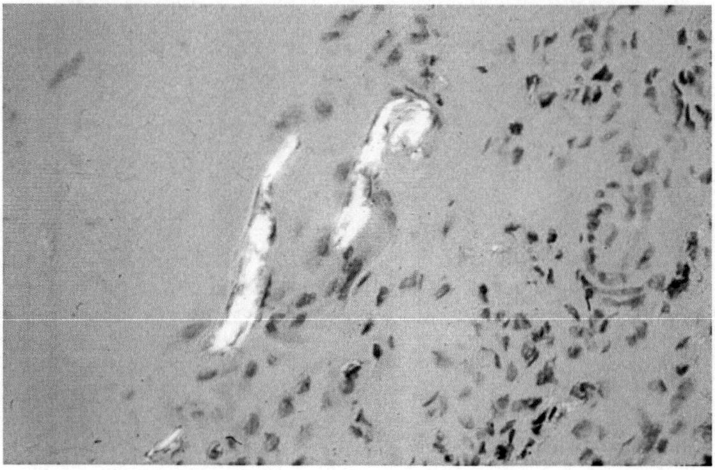

Fig. 15.4 Polyethylene debris of different shapes and sizes within the periprosthetic membrane from around a failed acetabulum. The debris is identified using polarized light

orthopedic biomaterials is that of wear of these materials leads to particulate debris and the body does not respond favorable to wear debris. For example, alumina particles are a highly aggressive abrasive against polyethylene, and UHMWPE particles are believed to be the primary mediators of osteolysis in contemporary implants.

3.2.2 Polymers

Numerous degradation mechanisms have been documented regarding the polymers used with orthopedic devices. They are influenced by the quality of the material before and after processing, device type and design, and with articulations—the other material that they may be coupled with.

Polyethylene

In 1962, Charnley transitioned from Teflon® cups to high-density polyethylene cups [83]. Ultrahigh molecular weight polyethylene (UHMWPE) was introduced because of its improved mechanical properties. In 1998, cross-linked polyethylene was introduced for use in total hip implants because of its higher wear resistance. Concern with the use of polyethylene has focused on the processing methods as well as the generation of wear debris (Fig. 15.4). Much that has been written to date about the biological responses to polyethylene debris in retrievals is related to the traditional UHMWPE.

Significant implant failure was observed with components that were manufactured using machined ram-extruded polyethylene [84–89]. Berzins et al.

proposed that net-shaped molded components may have demonstrated superior performance relating to the type of resin, the absence of calcium stearate, the consolidation method, or the method of final geometry shaping [84].

The method of sterilization was also found to affect the long-term durability of UHMWPE. Gamma sterilization is associated with oxidation of the subsurface of polyethylene implants [90, 91]. Failure of the components is associated with delamination [92, 93]. Alternative sterilization methods have included gamma irradiation in inert environments (such as argon or nitrogen) or non-irradiation methods (such as gas plasma or ethylene oxide). As irradiation is associated with the development of free radicals [91, 94–96], Vitamin E which reduces the amount of free radical available has also been investigated [97, 98].

Wear is evident on most, if not all, UHMWPE articulating surfaces. Wear rates in excess of 140 mm^3/year have been reported [99, 100]. As one would expect, there is an increase in wear with increasing times of implantation [30, 101]. There appears to be an initial "bedding in" period that is not related to wear per se [102–104]. Patterns of wear can be polar or edge loaded with high and low areas [84, 105]. There is a risk of fracture with increased polyethylene wear in the critical weight bearing area [101, 106]. A minimal thickness of the UHMWPE is of 6–8 mm is recommended [107].

Different processing approaches have been used to improve the wear rate of polyethylene. One approach involved the use of a proprietary process to change the crystalline structure to an extended chain (Hyalamer®); it was anticipated that this change would result in a UHMWPE with superior material properties (e.g., yield strength and creep resistance) [108]. However, the clinical experience with this material demonstrated a high failure rate at early stage (<5 years) [109, 110]. Another approach was to cross-link the polyethylene. Highly cross-linked polyethylene is characterized by resistance to abrasive and adhesive wear, oxidation, and plastic flow and lamellae alignment [111–114]. However, the highly cross-linked polyethylene had inferior fatigue, crack propagation resistance, and mechanical properties [115, 116]. Because of differences in the wear mechanisms between knees and hips, cross-linked polyethylene was first introduced for use with polyethylene liners for total hip replacement [116]. Improvements have been made in the processing of highly cross-linked polyethylene—the so-called second generation [115]. Highly cross-linked polyethylene tibial tray inserts are now available with higher wear and fatigue resistance than for conventional polyethylene [117]. However, the findings of elevated oxidation at the rim of annealed liners and wear debris in periprosthetic tissues indicate that additional research and development are necessary [118, 119].

PMMA

Bone cement (PMMA) used for implant fixation has two surfaces—the bone/cement interface and the cement/implant interface. The long-term stability of an orthopedic implant requires that the mechanical interface between cement and bone

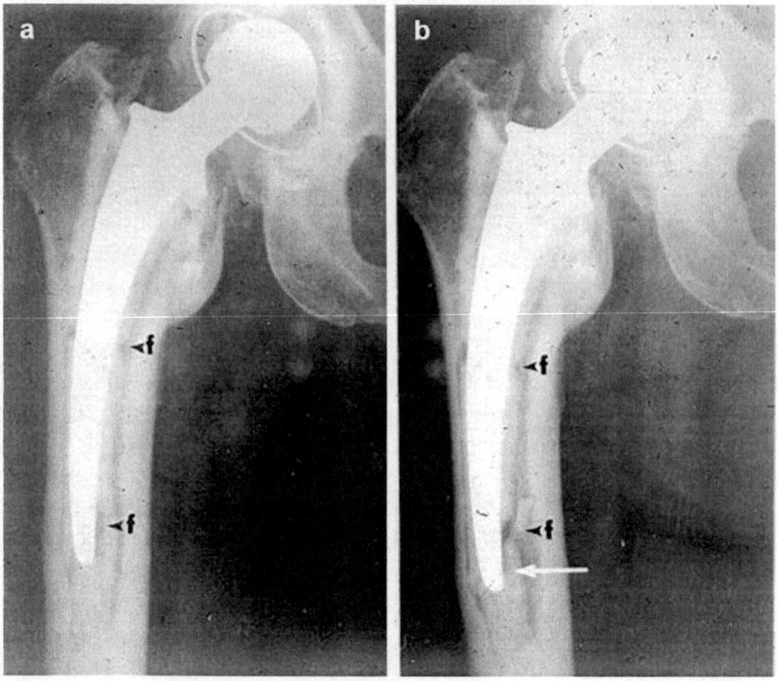

Fig. 15.5 Radiographs of a cemented total artificial hip. In (**a**), fractures of the cement mantle are evident (labeled "f"); however, the patient did not exhibit any adverse symptoms. In (**b**), taken approximately 3 years after (**a**), the cement fracture had clearly progressed. A piece of the cement has even dislodged from the mantle and is intruding into the bone. (**b**) was taken just prior to revision surgery. Figure reprinted with permission from John Wiley and Sons [120]

to be stable [29]. Inadequate cement technique can have an adverse effect on either interface. Several studies have characterized the bone/cement interface of stable and unstable orthopedic implants (see below). Fewer studies of revised and autopsy retrievals have examined the cement mantle itself. Topoleski et al. investigated specimens of retrieved, failed bone cement from the cement mantles of artificial hips (Fig. 15.5) [120]. They also created controlled fractures under different loading conditions in the laboratory. Using scanning electron microscopy (SEM) to identify unique features of the failed surfaces (a process known as fractography) and by comparing the surface features from both the retrieved and controlled laboratory conditions, they proved that fatigue failure was the predominant *in vivo* failure mode. Their retrieval study established that to improve the material performance of bone cement, the fatigue behavior must be addressed. Maloney et al., using SEM to examine retrieved femoral stems, observed debonding and fracture initiation and noted that debonding usually began either proximally or distally around the distal tip [121]. Of interest was the finding by Jasty et al. of debonding and radial fractures of the cement at the corners of a total hip prosthesis that had no clinical evidence of loosening [122]. They also found evidence of incomplete polymerization of the

PMMA. In a study of 214 cemented stems, Bishop et al reported voids in 22% and cracks and blood in <10% of the cement mantles analyzed [123]. They also observed debonding in 21% of the stems. In an analysis of 14 cemented polyethylene acetabular cups, Schmalzried et al. reported that "cement fractures were absent from regions of predicted high stresses such as at the periphery of the components" [124]. In a study of 129 total hip patients with greater than 10 year follow-up, Hernigou et al. found no association between survival rate and presence of voids [125]. However, it is generally accepted that voids affect the mechanical integrity of the cement mantle and should be reduced as much as possible [120, 126]. Contemporary cement technique may now include an intramedullary plug, pulsatile lavage, vacuum mixing, a cement gun to introduce the cement in retrograde fashion, pressurization, a proximal rubber seal, and a distal centralizer [127, 128]. An evenly distributed cement mantle of greater than 2–3 mm is recommended [129, 130].

3.3 *What Have We Learned About Implant Design?*

As one would expect, implant design has always had an effect on the long-term outcome of orthopedic implants. The medical literature is replete with examples of innovations gone wrong. Consequently, the following is an overview of some of the major factors regarding implant design of total knee and hip implants. The implication of these outcomes has an effect on all types of total joint prostheses.

Total hip and total knee arthroplasty are considered to be one of the most successful orthopedic procedures [131, 132]. Recent technological advances are likely to extend the longevity of the prostheses used in these procedures. However, as many of the currently implanted prostheses are of older designs and surgical procedures, it is also very likely that the number of revision arthroplasties will increase significantly over the next 2 decades [133]. Understanding the mechanisms of failure will allow us to make further improvements in material selection, implant design, and surgical procedures.

Over the past few decades, the major design considerations for both knee and hip prosthesis have included method of fixation, the materials to use for the articulating surfaces, and the degree to which the implant models the human anatomy.

3.3.1 Fixation

Orthopedic devices have been designed as press-fit devices, devices that are fixed with PMMA (bone cement), devices which permit biological fixation, and combinations thereof. There are advantages and disadvantages of each type of implant fixation.

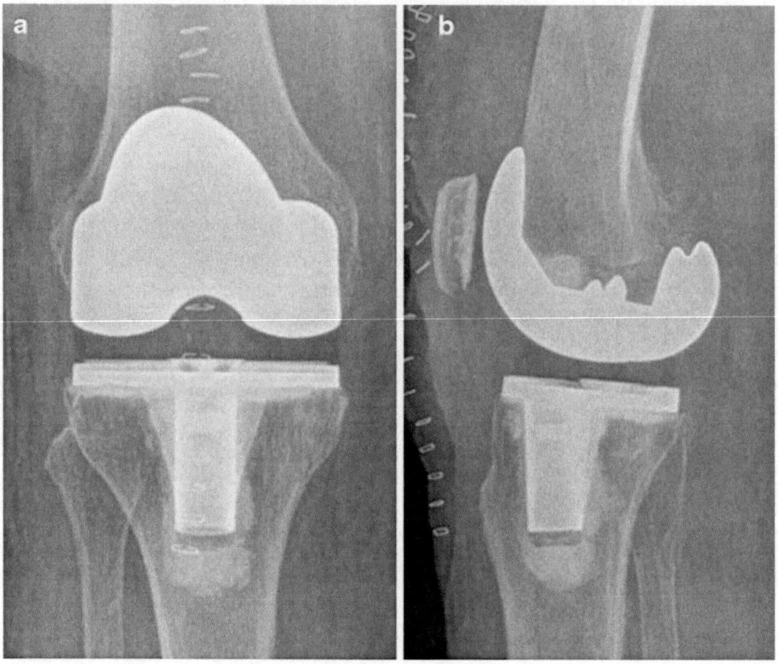

Fig. 15.6 AP and lateral radiographs of well-aligned total knee replacement with cement fixation, using contemporary cement technique

Cement

Sir Walter Charnley implanted the early total hip replacement with cement. The outcome of these procedures was superior to other procedures performed at that time for degenerative joint disease. In fact, more recent long-term clinical studies have demonstrated success rates of 77–85% in patients with Charnley Total Hip Prostheses for follow-ups of over 20 years. These results underscore the challenge of trying to make small improvements in success rates with innovative techniques.

In the late 1970s, William Harris and others published reports of loosening and significant bone loss associated with cemented implants [134–137]. Several investigators suggested that this adverse response was a consequence of the histocytic response to particles of cement debris [122, 138–141]. Inadequate mixing of the bone cement used to fix implants to bone may result in voids (air bubbles), undissolved PMMA powder, and nonuniform distribution of the cement mantle [122]. Once the fixation is compromised, the implant or implant–cement construct can move within the bony envelop creating a hostile environment associated with significant osteolysis. Once the limitations were better understood, the cement technique was improved with the use of cement spacers and centrifugation [127, 128]. While most studies have indicated improved outcomes with current cement techniques (Fig. 15.6) [127, 142, 143], others have not [144].

Engh et al. have reported cemented acetabular failure rates of 18.5% [145]. While it has been proposed that the key factor contributing to acetabular failure is mechanical loosening [146], others believe that it is biological [124]. Based upon testing of retrieved implants, Ohlin et al. demonstrated that less torque was required to failure than for newly cemented acetabulae [28]. A fibrous tissue membrane containing macrophages and inflammatory cytokines is commonly seen around a failed cemented acetabulum [124, 138]. This is associated with particulate debris and osteolysis that progresses from the perimeter and any screw holes that may be present [25, 121, 124].

With respect to the femoral component, retrieval studies demonstrate that a stable interface is attainable and that bone remodeling against the cement may occur [12]. The pattern of bone remodeling is with endosteal remodeling leading to an inner and outer cortex of bone and the formation of a secondary medullary canal with trabecular struts then lead to long-term implant stability [121, 147].

The integrity of the cement mantle is essential to the stability of the cemented components. Concerns with cement technique have revolved around the finding of an uneven cement mantle with areas devoid of cement fixation, voids within the cement mantle, and debonding of the cement from the metal stem. Cement fracture and subsequent mechanical failure have been observed in mantles that are less than 1 mm thick [121]. In thin cement mantles, cement fractures may be associated with wear at the fracture surfaces and voids within the cement mantle [122, 148]. Voids with fatigue may also be observed in well-cemented THR [27, 147]. Debonding of the cement prosthetic interface is a significant concern [121, 123, 149]. The pistoning of the metal implant within the cement mantle may create increased stresses and generate particulate debris. As with all cemented components, the long-term stability requires the mechanical interface between cement and bone to be stable [29]. Rotational micromotion is a primary concern [121].

Biologic Ingrowth Prostheses

Introduced as an alternative to cement fixation, porous-surfaced implants allowed for biologic fixation of the implant (Fig. 15.7). Initial clinical series demonstrated good to excellent results [145, 150, 151]. Engh reported that the failure rate, defined by loosening/revision, was 10.8% for uncemented femur in early implants which improved to 1.8% with modern implants [145]. Longer term studies highlighted specific issues regarding the outcome of uncemented prostheses, especially the importance of implant stability. Rapid ingrowth and the associated mechanical stability that is achieved are considered necessary [152, 153].

Retrieval studies have demonstrated the capacity for bone to grow into and onto the porous surfaces, although several constraints have been noted. While bone ingrowth into porous-surfaced implants was believed to be necessary, it was unclear as to how extensive the bony ingrowth must be to achieve a stable interface [153]. Reports of the extent of bony ingrowth have varied depending on type and design of

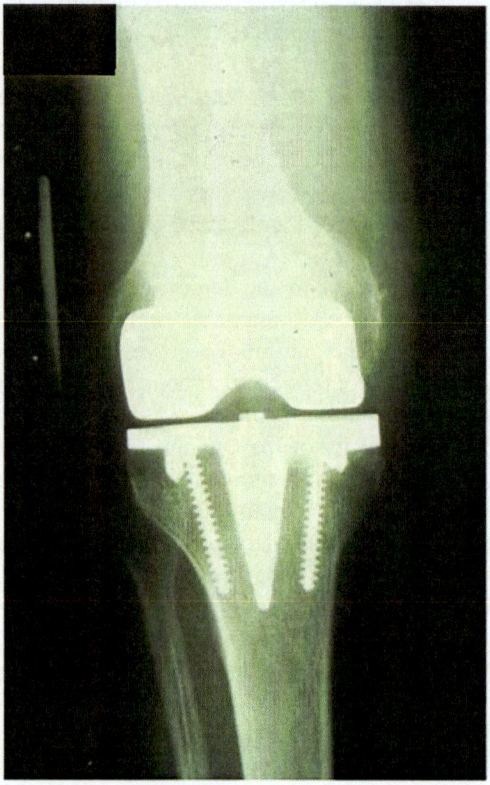

Fig. 15.7 Well-aligned uncemented total knee components with adjunctive screw fixation to provide initial stability until bone ingrowth occurred

implant and whether the implant was obtained following revision surgery or autopsy [153–158].

Failure related to micromotion and implant bone interface if the displacement was greater than 150 μm [145]. Collier et al. have proposed that, over time, remodeling with bone ingrowth and appositional growth may be the primary method of some ingrowth, and that remodeling is a long-term process potential 5 years of so [12]. With some implant designs, stress shielding was observed, especially with extensively porous coated stems or with larger, more rigid stems [48, 159, 160].

One concern raised with porous-surfaced implants was whether there was any evidence of a loss of integrity of the surface itself [161, 162]. While this could have an impact on implant stability, it may also lead to third-body wear if particles migrated to the articulations [163].

Metal-backed polyethylene components were introduced as a way to distribute stresses better and as a way to provide for cementless fixation. Modular components also permitted the exchange of polyethylene for damaged or worn components. Retrieval studies have documented several issues regarding metal-backed components, including micromotion between the polyethylene and metal backing, having too thin of a polyethylene liner or tray, and metal-on-metal wear resulting

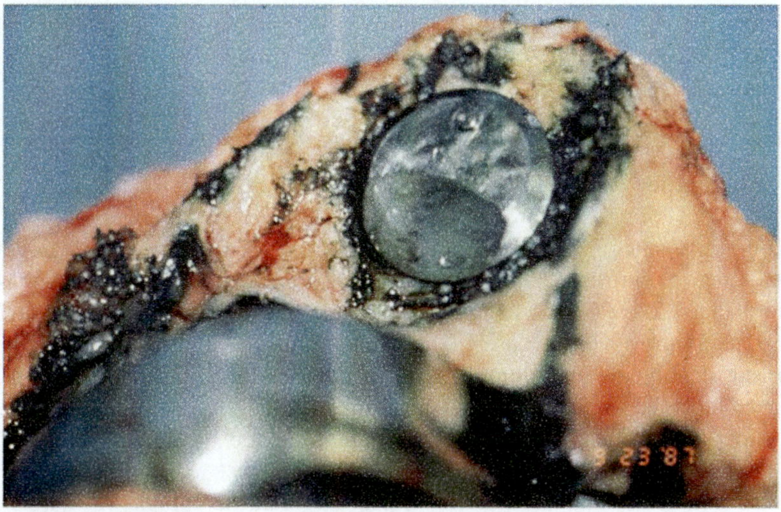

Fig. 15.8 Metal backed patellar component with edge loading and failure of polyethylene and complete dissociation of patellar component and metal wear resulting in synovial metallosis in surrounding tissues

from impingement or wear-through (Fig. 15.8). Another issue is that screw holes which permit the use of screws for increased fixation of the implant to bone may also be a site that provides access of wear debris to the bone interface [164].

As stated previously, initial implant stability is important to the success of biologic ingrowth prostheses. Hydroxyapatite (HAp) coatings were introduced to the surfaces of hip and knee implants in order to create a more intimate bond between the bone and the implant. Retrieval of HAp coated implants demonstrated that the ongrowth of bone is related to the mechanical stability of the implant from geometry and loading [165, 166]. Aebli et al. noted intimate host bone–femoral stem contact where the HAp coating had disappeared completely [167]. Geesink proposes that osteoclastic resorption of HAp is associated with the replacement of the HAp coating with bone, thereby preserving the stability of the interface [168].

3.3.2 Articulations

While metal-on-UHMWPE showed improved biocompatibility as compared to Teflon®, its longevity has been compromised by processing technique (ram-extruded and then machined vs. compression molded) or sterilization method (gamma irradiation in air vs. sterilization with gas plasma). The development of highly cross-linked polyethylene has demonstrated improved wear rates as compared to traditional UHMWPE, yet the highly cross-linked polyethylene has inferior fatigue, crack propagation resistance, and mechanical properties. Due to superior wear resistance, ceramics have been used with other materials (ceramics, metal) for articulation couples. With improved manufacturing technology and the ability to produce implants within precise specifications, metal-on-metal couples have been re-introduced.

3.3.3 Total Knee Replacement

The long-term outcome of total knee replacement has been outstanding with survival rates of up to 100% at 10 years [169, 170] and reports of implants lasting for more than 20 years [171, 172]. Most implants are metal-on-polyethylene, although ceramics have been used. The patellar and the tibial components can be comprised of polyethylene alone or with metal backing. Like total hip implants, fixation of most TKR are either cemented, uncemented, or a combination of both.

Consider what we can learn from studies that include specimens obtained both from autopsy and revision surgery. We will assume, at least initially, that any specimen obtained at autopsy was functional, and that the patient was happy with the performance; a successful implant. We can also assume that specimens obtained from revision surgery were not functional. Thus, we can study differences between the two groups' devices to investigate potential failure mechanisms. As an excellent illustration, Parks et al. determined that there was significant relative motion between the UHMWPE insert and the supporting tibial tray [173]. Further, they noted that the type and magnitude of relative motions were design/manufacturer dependent. From that test, however, it was not possible to predict the *in vivo* performance of a specific design. Engh et al. expanded on Park et al.'s work with another study of the relative motion between the UHMWPE insert and the supporting tibial tray using never-implanted controls, specimens retrieved from revision surgery, and specimens retrieved at autopsy [174]. Both types of used specimens, revision and autopsy, showed significantly greater motion than the controls. Interestingly, there was no difference between the relative motion in specimens retrieved during revision surgery and those retrieved at autopsy. These results raise an interesting question, and one that we must ask as we review any retrieval study: why is there a difference, or in this case, why is there no difference, between the two specimen populations? One possibility is that the observed measurement was not related to the failure or success of the device. Here, that would imply that the motion between the tibial insert and tibial tray did not significantly contribute to wear-related failures. Another equally possible explanation is that people will react differently to the same conditions. Thus, the consequences of relative motion between the UHMWPE tibial insert and metal tray may lead to osteolysis in one patient, but be well tolerated in another. In the case of the study by Engh et al., they showed that there was a significant change in the relative motion when the device saw *in vivo* service, but it is not clear why some devices failed and others did not [174].

Other studies that include both revision and autopsy retrievals use different measures to assess reasons for failure. Harman et al. studied the correlation between areas of articular contact and UHMWPE damage [175]. Not surprisingly, they determined that where the contact was greatest, the UHMWPE damage was also greatest in either specimen group. However, like Engh et al. [174], Harman et al. detected no difference between the revision specimens and autopsy specimens [175]. Rohrbach et al. also studied both revision and autopsy specimens, and used an elaborate scoring scheme to measure wear and damage [176]. Both retrieved groups showed fatigue and wear damage. In contrast to the studies previously discussed, Rohrbach et al. found that the damage was significantly less in the autopsy group. These results appeal to our intuition, since it makes sense for the well-functioning specimens (the autopsy retrievals) to be less damaged because the general consensus is that UHMWPE wear is a mediator of implant failure.

If a study examines only one type of specimen, either autopsy or revision retrieved, is it possible to conclude anything about the reasons for longevity or failure of the devices from those data? Consider the study by Lavernia et al. [50] Which evaluated the wear and creep in specimens from autopsy only. They investigated the relationships between the mechanical damage and the time implanted. Since there is, according to the investigators, no radiographic measure of wear in artificial knees as there is for artificial hips, wear rates were taken as an overall average over the life of the implant. There was an increase in creep with increasing time implanted; however, and surprisingly, there was a decrease in wear rate with increasing time implanted. Weight was the most important predictor of volumetric wear. Why would wear rate decrease the longer the implant was functional? There may be several answers. The authors postulate that the older the patients become, the less active they become, and thus the wear rate decreases. It also may be that there is an initially higher wear rate (possibly from the initial "wear-in" phase of articulation), which then settles into a steady-state wear. Thus, when averaged over the entire life of the prosthesis, the wear rate appears to decrease over time.

Both Wright et al. [177] and Garcia et al. [178] are examples of studies that analyzed specimens retrieved at revision only. In each study, the investigators showed that the reason for revision was severe damage to the UHMWPE. In Wright et al.'s study, they measured an increase in density of the UHMWPE. They argued that a change in the material's density would change the mechanical properties (for example, the Young's modulus would increase). If the Young's modulus increases, then so does the stress in the material, which results in an increased risk for further wear. In their study, Wright et al. not only examined the wear of the surfaces, but discovered a possible mechanism for the observed wear. In Garcia et al.'s work, they noted many different UHMWPE damage modes, but, like Lavernia et al. [50] showed that there were variables that changed as a function of implant time. The UHMWPE damage increased with longer implant time, but the surprise in their data was the damage decrease with increasing patient weight. They postulated the decrease in damage was a consequence of reduced activity by the heavier patients.

Cook et al. also examined specimens retrieved at revision only, but included both hips and knees to investigate bone ingrowth [179]. They found no relationship between the amount of ingrowth and the time implanted.

The first generation tibial components were all polyethylene and fixed with PMMA. The primary mechanism of failure was loosening. In a study of 48 total-condylar-type knees, Hood et al. observed scratching (90%), pitting (81%), burnishing (75%), surface deformation (cold flow) (62%), and abrasive wear (42%), but delamination was uncommon (2 cases) [180]. Landy and Walker, in a study of 90 total knee replacements (various designs) with up to 10 years of implantation, noted that the wear of the tibial implants was greater than that reported for hips [181]. They noted abrasion from cement debris and delamination as characterized by "the complete breakup of material in flakes and particles." They reported that the delamination appeared to be associated with "intergranular material defects and propagated by the excessive subsurface stresses beneath the contact zone." In contrast, delamination was not observed in specimens without evidence of material defects.

Biologic ingrowth total knee systems were first introduced in the late 1970s and early 1980s. While short-term outcomes were favorable, concern was raised whether significant levels of bone ingrowth were achievable. Several studies confirmed that bony ingrowth was evident, although generally at lower levels than had first been anticipated [156, 182]. Yet, reports of early failure related to porous coating and thinner implants and metal tibial tray at areas of stress risers, poor bone ingrowth, and unsupported tibial base plates led to catastrophic mechanical failures (Fig. 15.9) [16, 183]. Fibrous ingrowth has been described and found to be stable. However to optimize bone ingrowth, appropriate mechanical loading is required, especially in TKR. Enhancement of fixation with screws, slanted pegs, etc. to allow more compression on initial fixation enhances bone ingrowth. Time *in situ* did not change first generation ingrowth [179]. However, time *in situ* directly correlated with increased wear especially for fatigue failure in PCA and Duracon prostheses [176]. Kinematics showed wear predominantly in those areas of highest contact femoral tibial articulation on motion fluoroscopics, stair climbing, and gait. This also translated to the striation pattern on the articular surface found [175].

In 1991, Mintz et al. reported extensive wear and delamination in a study of 487 PCA implants [22]. They suggested a relationship between the wear and patient factors such as size, age, and activity. The design features of the knee prosthesis—nonconforming femorotibial articular surfaces, thin PE, heat-pressed PE, inadequate mechanical attachment of tibial insert to metal baseplate—were also believed to contribute to the wear process. In 1992, the same group examined 12 retrieved PCA total knee implants and found evidence of pitting, scratching, burnishing, delamination, surface deformation, and abrasion [177]. They suggested that the heat pressing, time of implantation, and body weight were concerns. Further examination of retrieved implants demonstrated extensive delamination and cracks from the periphery to the center of the condyles (Fig. 15.10) [106]. Bloebaum et al. observed that retrieved and unused heat-pressed inserts showed increased crystallinity in the surface and middle regions [21].

Engh et al. examined 86 polyethylene inserts from total and unicompartmental knee prostheses of various designs [101]. They noted "severe wear, with delamination or deformation, with longer periods of time *in situ*, lack of congruency, thin

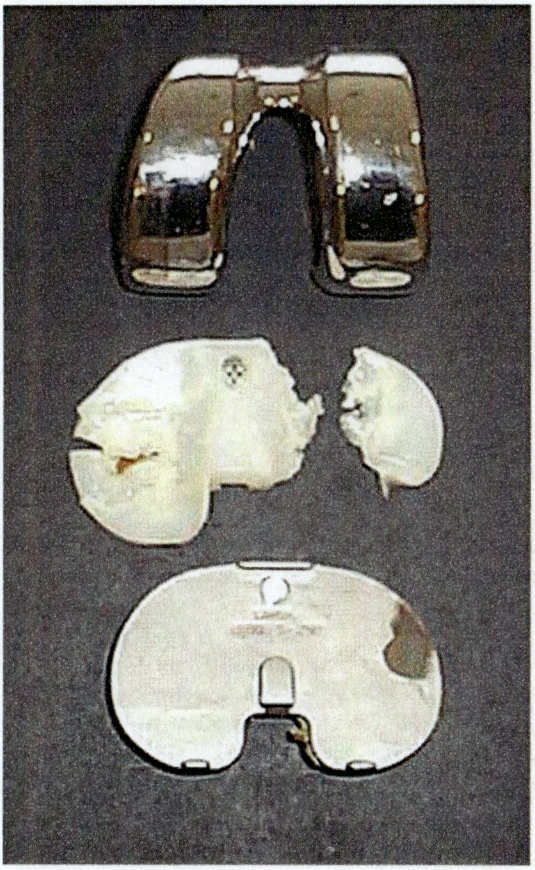

Fig. 15.9 Total artificial knee joint, removed after revision, showing catastrophic delamination and wear-through of the polyethylene component. Complete wear-through of the polyethylene component occurred some time before revision, as indicated by the wear-burnished surface of the underlying tibial tray. In this case, both polyethylene and metal particles were produced

polyethylene, third-body wear, and heat-pressed poly." Delamination was associated with significant under-surface cold flow in areas of unsupported polyethylene. As they observed significantly less delamination in thick inserts as compared to thin inserts, they recommended the use of thicker inserts. They also noted that the presence and location of screw holes for fixation were a problem.

Harman et al. examined eight knees in six patients fluoroscopically during a stair-rise and descent activity and treadmill gait and then following removal at revision surgery [175]. They found a significant correlation between where there was contact during activity and the location of the polyethylene damage. The femoral contact and PE damage occurred predominantly on the posterior half of the tibial articular surface. Asymmetric damage patterns correlated with the tibial rotations seen *in vivo*; the medial damage (38%) greater than lateral damage (32%). The most frequently observed types of damage were burnishing, scratching, and tractive striations; pits were observed on two inserts and delamination on one. Seven of eight inserts had plastic deformation on the sides of the tibial eminence. There was no correlation between damage size and weight.

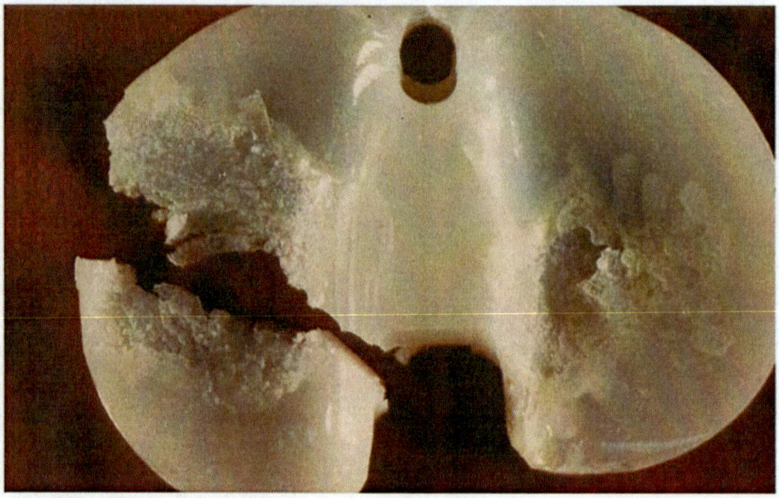

Fig. 15.10 Heat polished polyethylene tibial components exhibiting delamination, cracking, pitting, and ultimate fatigue failure of the component

There has been considerable debate about wear and conforming/nonconforming articular surfaces. Wright, Tsao, and others have observed significant wear with a nonconforming design [106, 177]. Blunn et al. found that the polyethylene experienced more torque and shear and that delamination could still occur [184]. The kinematics of the knee joint is complex with different reactions (e.g. rotation) that can occur in response to the variable forces that can be applied to the joint [185–188]. High stress load in low conforming designs can lead to polyethylene overload. This may be accelerated clinically with poor undercorrection of varus/valgus, which correlated to sliding distance on the bearing surface [176]. Implants are designed to maximize the conformity while permitting the natural motions of the knee. A less conforming joint articulation requires appropriate surgical technique to balance ligaments and achieve appropriate alignment to maintain proper mechanical loading and kinematics. This is necessary to keep functional motion and loading within the tolerances of both the material properties and the implant design [106].

The thickness of the tibial insert has an effect on the integrity of the implant. There are numerous reports of wear-through with metal-backed components [101, 106, 177, 189, 190]. Based on these findings, Wright et al. recommend that the minimum thickness of conventional polyethylene should be 6–8 mm [177]. However, it may be possible for this thickness to be smaller with highly cross-linked polyethylene and more conforming designs.

The rationale for metal backing was to permit a more even distribution of stress within polyethylene as well as to allow for modularity [191–193]. Retrievals of tibial components have shown abrasive and backside wear along with delamination [194]. Micromotion has been observed between the polyethylene insert and the metal tray [164, 173]. In fact, Parks et al. determined that micromotion between the insert and the tray was evident in nine different designs of modular implants even at the lowest loading level of 100 N [173]. In 2002, Surace et al. evaluated 25 posterior-retaining total knee implants [195]. They found that damage to the back surface was limited with polishing detected on 84% of the inserts and abrasive wear on 20%. This was primarily on the periphery of the component and around the area corresponding to the site of the screw holes on the tibial plate. Polishing was evident on 36% of the molded components and 100% of the machined inserts. Abrasive wear was detected on the back surface of 45% of the molded and none of the machined. Pitting was present in 84% of the components, but was relatively limited (less than 1% of the area in all but one components). Pitting was higher in the molded (100%) than in the machined (71%). Delamination and cracking were not observed. There was a concave deformation in the AP direction on the back surface in 96% of the cases. Failure of the locking assembly may also cause disassembly of the component or abnormal articular or backside wear [174, 196–199].

Rotating platform knees were introduced in the late 1970s to increase conformity and decrease the stresses experienced by the bearing surfaces with rotation. In a study of 40 retrieved rotating platform knees, Garcia et al. observed evidence of all damage modes on superior surface: pitting (100%), scratching (100%), embedded particulate debris (24/40), permanent deformation (16/40), and surface delamination (1/40) [178]. Backside wear is a problem; in rotating hinge with highly conforming articular surface, articular and backside wear is an issue and correlated with increased weight and time *in vivo*. On the inferior surface, they found pitting and curvilinear scratching (100%), embedded debris (1/40), and burnishing (5/40). In an evaluation of 100 LCS-RP mobile bearing total knee prostheses (DePuy), Atwood et al. observed damage on flexion and rotation surfaces of all of the polyethylene inserts and scratches on 97% of the metal trays [200]. While rotating platform total knee replacements reduce cross shear, concern remains for increasing wear with longer term implantations [178, 201].

The patellar implant remains a difficult and unsolved problem (Fig. 15.11). Clinically, adverse responses have included anterior knee pain, fracture, malalignment, and subluxation/dislocation. This implant is subject to very high stress [202, 203] and does overload the parameters for material design. Currently, the most common design is a dome-shaped configuration with a thinner bony substrate, which we know as a less conforming surface, being subject to eccentric loading and high shear [204] and may be prone to failure [205]. Metal-backed patellar components are also subject to failure resulting from poor design, thin polyethylene, thin metal backing, where subtle malalignment may cause edge loading, fatigue failure, third-body wear, and catastrophic failure [191]. A customized implant may decrease stress [203].

Alumina ceramics have also been used as an articulating surface in TKR as an alternative to metal-on-polyethylene. However, Hashiguchi reported evidence of wear debris causing scratching of the poly insert on retrieval of autopsy between 13 and 49 months [206]. Debris was found in the tissues as well.

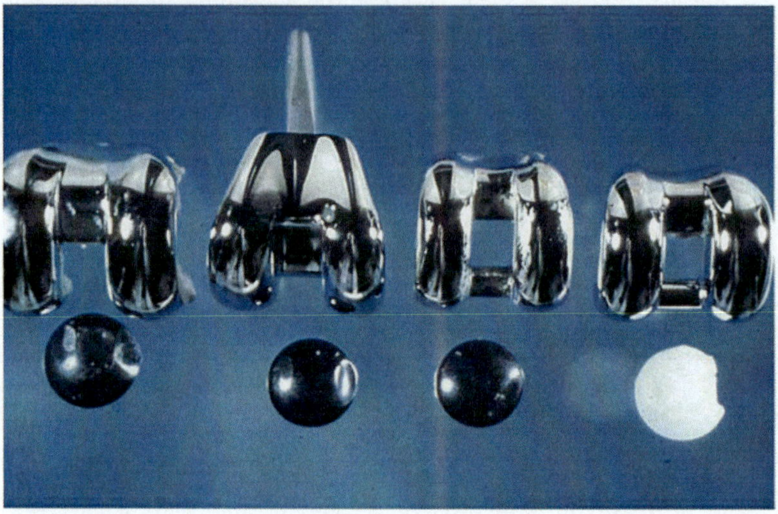

Fig. 15.11 Plastic deformation and failure of thin all-polyethylene patellar buttons against the trochlear groves of the femoral components

The above studies involved analyzing a series of implants. It is, however, equally important to discuss another source of information of device failure: the case study. Case studies tend to be reports of one or two specimens with the purpose of instructing the reader about a specific observation or unusual type of failure. A case study may be some of the first information to appear about a specific type of implant. For example, Ranawat et al. reported on two cases of porous coated implant failures and showed that metal beads from the porous coating became detached and migrated into the surrounding tissue [16].

As a summary of this section, the retrieval analyses provided information, in each case, that could not have been obtained any other way. For example, in the case of Engh et al., there would be no way of reproducing the failed devices obtained at revision and the functional devices obtained at autopsy in the lab [174]. We do not know all of the intricacies of how to create a specimen that is likely to fail vs. one that will remain functional. We can certainly create adverse conditions for the devices, but then there is no way of knowing what elements of that condition would exist from one patient to another. Laboratory testing is absolutely necessary to understand the fundamental mechanisms of failure, but the retrieval specimens are the only source of information of a device's behavior *in vivo*.

3.3.4 Total Hip Replacement

To examine retrieval studies from hip replacements, we will also begin with studies that have examined specimens retrieved during both revision surgery and autopsy. Jasty et al. examined the wear rates in UHMWPE acetabular cups and noted that the median wear rate was less for the autopsy specimens than for the revision specimens [100]. They do not discuss the mean wear rates. In contrast to Lavernia et al.'s study of knees [50], Jasty et al. showed an increase in wear rate with longer implant times. Thus, the two studies show opposite behaviors in knees and hips.

Osteolysis is a direct result of wear debris generated from both the articulating and non articulating surface of total joint replacements. The examination of wear debris and its products in both revision and autopsy retrievals can indicate the role particles play in device performance. Schmalzried et al. showed that intracellular UHMWPE particles were present in both revision and autopsy retrieved specimens [207]. The overall character of the particles (number, size, etc.) was similar in both groups of retrieved specimens. Their comparison of revision and autopsy specimens indicates that particles can migrate between the prosthesis and bone even when the device is well fixed. Urban et al. examined corrosion products and showed that the corrosion products can also migrate to the bearing surface and the bone–implant interface in both revision and autopsy retrieved specimens [208]. A specific compound, chromium orthophosphate hydrate-rich corrosion product, exists in both types of specimens and appears to be the same in each.

One of the main goals of laboratory studies is to correctly model the *in vivo* situation so that investigators can learn about processes that cannot be directly observed *in vivo*. Comparing/contrasting results from *in vitro* tests with results from retrieved specimens is an essential step in the validation of *in vitro* models. The *in vitro* model can also represent the "ideal conditions" of an implant and can be used as a reference for changes that may occur *in vivo*. Simulated surgeries in cadaver femurs have been used as a comparison to specimens retrieved at autopsy [29, 47, 209]. Jasty et al. used implants in cadaveric femurs as the reference for the initial stability of the implants to compare the stability of cemented and non-cemented femoral components [209]. Both cemented and non-cemented components were stable in simulated single leg stance, but the cemented components were more stable than the non-cemented components in simulated stair climbing. They examined only cemented specimens retrieved at autopsy and showed that they retained their stability, as we would expect of well-functioning implants; however, even the well-functioning implants showed evidence of debonding at the cement–bone interface, as well as fractures in the cement. Thus, it may be that the types of damage observed by Jasty et al. do not affect the long-term stability of the cemented implant. Ohlin et al. examined the other side of the total hip, specifically the torsional strength of cemented acetabular cups [28]. Again, the cadaver surgeries represent the initial stability of the device. They observed no difference in the torsional strength between the cadaveric and autopsy specimens. However, they note that the type of failure they were creating does not reproduce the failures reported for patients who require revision surgery. Ohlin et al. were attempting to create sudden failure by a one-time overload of the device. Thus, Ohlin et al. postulated that the observed failures in revision cases came about through a gradual process, and their testing did not reproduce that failure mechanism. Miller et al. measured the relative motion at the bone cement–bone interface in both the cadaveric and autopsy-retrieved specimens [29]. The autopsy specimens were significantly more compliant (less stiff) than the cadaveric specimens. Again, we can observe that the lab specimens were not exposed to multiple years of fatigue loading generated by patient activity. The fatigue loading can affect the integrity of the interface. In addition, bone remodels during the course of an implants *in vivo* service. Since we may not know the original configuration of the bone in the retrieved specimens, it is difficult to determine how much remodeling has affected the stability of the prosthesis.

Some variables, like component geometry, can be measured directly. Berzins et al. measured the thickness of both control (never-implanted) and autopsy (well-functioning) acetabular liners using an ultrasonic method [210]. The retrieved specimens showed considerable variations in thickness in contrast to the controls. Again, because these were functioning implants, do we conclude that such variations that occur during *in vivo* service do not affect the function of the devices? Or are the changes in the device geometries inconsequential for this patient population because of patient-specific considerations?

Another potential way to create a reference for retrieved prostheses is by creating an animal model. Both dogs and humans can have stable implants with good bone ingrowth [211, 212]. For example, Bobyn et al. showed that the ingrowth-porous surface provides a barrier to UHMWPE migration in both dogs and humans [212]. Skurla and James, however, showed that wear in dogs was much lower than in humans, and questioned whether dogs are indeed a valid model for studying human implants [211]. It is clear that we have learned much from animal models, but it is important to understand that *in vivo* service conditions, and thus the natural history of the device, may not be the same in humans and animals for all mechanisms of *in vivo* change.

Acetabular components have historically had a shorter lifespan than the femoral components. Wroblewski et al. examined 59 cemented high-density polyethylene sockets obtained at revision arthroplasty [213]. Nineteen cups showed areas of wear between the outside of the socket and bone associated with a lack of cement in those areas. They raised concerns regarding impingement of the neck of the femoral stem on the edge of the cup. In a study of 26 polyethylene cups, Sychterz et al. were unable to find any evidence of delamination or fatigue cracks [30]. They did observe burnishing in all of the liners and stated that it was the most severe form of damage. Scratching and pitting also occurred in all liners, but were less severe. Of note was the finding that the rates of wear of press-fit, metal-backed liners were significantly higher than those of all-poly cemented components. While the direction of wear varied widely with the liners studied, they suggest that the direction of wear may tell us something about the design and position of the implant. Bal et al. reported that the average annual polyethylene wear was greater for modular stems than for the monoblock stems [76]. They also found that linear wear was higher for titanium stems (plasma sprayed) as compared to CoCr alloy stems (beads).

The articulation of the femoral head against the cup has been compromised by inadequate material selection, size of the femoral head, and scratching during implantation (Fig. 15.12). As the hardness of the material used for the femoral head is an important factor, cobalt–chrome alloy has been used extensively.

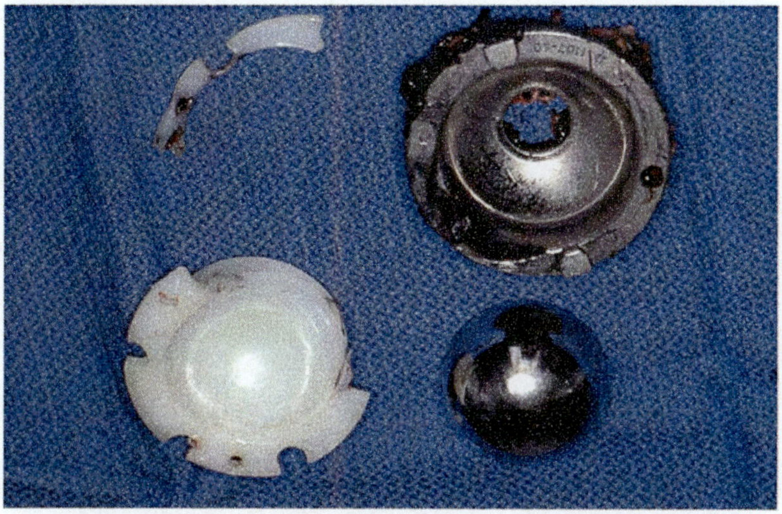

Fig. 15.12 Total hip acetabular component with complete wear through of the polyethylene component and ultimate wear of metal femoral head against metal acetabular component

However, there has been a high incidence of wear of the UHMWPE cup [76]. Cross-linked polyethylene has been introduced in an attempt to decrease the amount of wear experienced by this couple [95, 116].

Femoral head size has been found to have an impact on the wear of polyethylene cups. In a study of 128 revision and autopsy retrievals, Jasty et al. observed a significant relationship between the size of the femoral head and volumetric wear [100]. The rate was highest with 32 mm heads and lowest with 22 mm heads. Impingement was noted in 45% of the components. There was a decreased rate of wear detected with increased thickness of polyethylene with the metal-backed cups. The primary mechanism of wear was abrasion and adhesion, and not fatigue cracking or delamination. They found that "wear appeared to occur mostly at the surface and to be due to large-strain plastic deformation and orientation of the surface layers into fine, drawn out fibrils that subsequently rupture during multidirectional motion."

Other bearing couple alternatives have included ceramic-on-ceramic, ceramic-on-metal, and metal-on-metal. In THR, particulate debris with ceramic-on-polyethylene was similar to conventional bearing surface, but there was less debris with ceramic-on-ceramic, unless a mechanical problem like impingement existed [214]. Metal wear onto ceramic femoral heads has also been observed (Fig. 15.13). The taper can also be a debris generator.

As with total knee implants, several studies have attempted to find a correlation between patient, design, and surgical factors and wear. Sychterz et al. were unable to find a relationship between wear and the patient's age, weight, or gender; the duration of implantation; or the thickness. Jasty et al. also noted that age, gender, and side had no effect [30]. They did observe a relationship between duration of

Fig. 15.13 Examples of retrieved ceramic femoral heads that have worn-through the polyethylene acetabular components. There is a transfer layer of titanium from the titanium backing of the acetabular components. Retrieved specimens courtesy of the Anderson Orthopaedic Research Institute

implantation and rate of wear, with the rate being the highest initially after operation and decreasing with increasing duration *in situ*. Berzins et al. reported that thickness of the cup was not related to age, gender, Harris Hip score, size, inclination of cup, or type of femoral fixation [210]. In agreement with Jasty et al., they found that the cups with areas of reduced thickness had a longer time of service [100]. They suggested that the dimensional changes associated with the variations in thickness suggest different modes of loading.

Backside wear is also an issue with acetabular components, with metal-backed components, and with screw-enhanced fixation. Della Valle observed that backside wear was notably reduced in acetabular components with an improved locking mechanism, maximized shell liner conformity, and limited or no screw holes [215]. In a laboratory analysis of rotational and axial motion, Williams et al. also found differences in the rotational micromotion, interface slippage, and backside wear of six different designs, leading them to recommend the use of sound peripheral locking mechanism and a snug fit between the polyethylene and metal backing [216].

As in the knee, studies of autopsy specimens only for hip joints can provide valuable information on the behavior of the well-functioning prosthesis. Indeed, the postmortem retrievals may provide the best source of data for changes, e.g., wear, material properties, etc., in well-functioning devices [30]. Indeed, many studies of autopsy-retrieved implants agree that the devices appear to be well fixed, radiographically and mechanically, and do not show evidence of extensive wear [12, 25, 44–46, 121, 166, 217, 218]. Thus, the question again is, can we assume that any biomechanical or biochemical measurements in the retrieved specimens that differ from the initial or ideal implant are not related to implant failure? Perhaps, as previously mentioned, the selected group of specimens/patients was not sensitive to the changes. Alternatively, we cannot tell if any of the specimens retrieved at autopsy would have failed within a day, month, or year of retrieval had the patient continued to live. Ongoing careful study of retrieved specimens will continue to provide valuable information not available from any other source. The knowledge gained from analysis of retrieved specimens will enable researchers to understand *in vivo* failure mechanisms, to create valid models to perform controlled experiments, and to improve the longevity of implanted devices.

3.3.5 Spine

The field of spine surgery has seen an explosion of procedures requiring devices over the past decade [219]. As with total joint replacements, much of the focus has been on implants retrieved at revision surgery.

Corrosion continues to be a significant degradative process affecting spinal devices [220, 221]. Vieweg et al. examined 13 internal fixator systems (TLS-Kluger, Endotech, Germany) [221]. These implants are manufactured from chromium–nickel–molybdenum (CrNiMo) alloy and fixed with pedicle screws. Fretting corrosion at the cleft between the rod and the shell and abrasion on the surface of the rod were observed, while there was no evidence of intragranular corrosion. In a recent report, Majid et al. evaluated 118 thoracolumbar plates (made of 316 L stainless steel) [220]. They noted that corrosion was observed on 70% of the plates—including both fretting and crevice corrosion. Energy-dispersive X-ray (EDX) spectroscopy revealed a reduced nickel content of the devices, suggesting metal ion release into the host tissues. McPhee and Swanson evaluated the urinary and serum metal ions levels of patients that were implanted with stainless steel spinal instrumentation of various types [222]. There were no differences between the implant groups and the controls regarding the plasma nickel and blood chromium groups. They indicate that urinary chromium (normalized by creatine levels) was elevated in the implant groups. These findings taken together indicate that the metal ions may be present in the tissues surrounding the implant. While this is supported by the histological findings of metal particles and an inflammatory response [221], further analysis of the metal content within these tissues is warranted.

Mechanical failure of Harrington rods—an internal fixation system—has also been documented [223, 224]. Stürz et al. published case reports of two patients with evidence of fatigue fracture of their Harrington rod [224]. In one case, the fracture was noted 11 months after the removal of the EDF (elongation–derotation–flexion) cast (23 months after surgery). The fracture occurred at the first notch of the ratchet, and the rod demonstrated a lateral displacement and shortening. In a second case, a fracture of the rod was noted 6 months after the cast was removed (18 months postoperatively). Again, the fracture was at the first notch of the ratchet. Cook et al. evaluated eight Harrington rods that had mechanically failed [223]. Scanning electron micrographs demonstrated that the primary mode of failure for these devices was fatigue failure. In one case that had failed at a midshaft location, evidence of intergranular corrosion was found. The primary concern regarding the biological consequences was whether pseudoarthrosis had occurred. In these two studies, pseudoarthrosis occurred in 6 of 10 cases.

As noted, the use of lumbar intervertebral body fusion cases has increased dramatically over the past decade [219]. It is of significant interest to determine whether bony ingrowth occurs—especially with the use of different types of graft materials. Togowa et al. have evaluated the incorporation of bone graft as well as the biologic ingrowth into various spinal fusion cages [225–227]. In the most recent study, they reported on 78 cages that were removed for failed fusion (31), cage malposition or migration (16), low-back pain (5), progressive spondylosis (2), nerve-root impingement (2), compression fracture at the fusion site (1), and/or infection (2). Particles of either carbon or metal debris, to varying degrees, were noted in the tissues surrounding 53 of the 78 cages. While neither bony resorption nor inflammation was associated with the particles, the cages contained varying amounts of fibrocartilage (0–50%) and hyaline cartilage (>5%).

Dynamic stabilization systems were introduced as a treatment for fractured spines [228]. They are now utilized as an alternative to fusion or artificial discs [229]. Ianuzzi et al. evaluated 17 Dynesys systems (Zimmer Spine) [229]. While bending deformation was observed in most of the spacer components, this deformation did not correlate with implantation time. Abrasive wear was also noted in 68% of the spacers. Short surface cracks were noted on two spacers, and another spacer was fractured at the molded end. These spacers are manufactured from polycarbonate urethane (PCU). The authors suggest that prolonged exposure of this material to the inflammatory milieu may cause increased biodegradation.

One of the more recent advances in spine surgery has been the development of artificial disc replacement devices for both the cervical and lumbar spine. Jensen et al. evaluated cervical devices retrieved from an animal study as well as from two patients [230]. In the primate study, they noted relatively few wear particles and variable biologic ingrowth into the porous surface; no radiolucencies were detected. The two patients had revision of their devices to treat their symptoms. They reported that the human explants were well fixed and had an average percent ingrowth of 30.1%. Pitzen et al. evaluated a postmortem specimen retrieved from one patient [231]. They observed polyethylene and metal wear debris in the soft tissue surrounding the disc space. While there was evidence of bony ingrowth between the vertebral body (C5) and the lateral aspects of the fixation petal, the tissue surrounding the metal endplate was primarily fibrous tissue. Kurtz and his colleagues have examined retrieved lumbar total disc replacements [232–234]. In one study of 21 implants [233], they found evidence of adhesive, abrasive, fatigue, as well as fracture wear of the polyethylene components. Transverse and radial cracks were observed. There were also wire marker fractures (9 of 21 implants), which also led to debris and third-body wear in two of the cases. In a separate study, this group reported a positive correlation between dome penetration and implantation time [234].

As one would expect, spinal implants are subjected to the same concerns regarding wear properties that occur with any articulating surfaces. Arthroplasty devices, regardless of the location, are affected by stability of the fixation, the alignment of the devices, and the motions and constraints about the articulation.

3.3.6 Fracture Fixation

Depending on the type and extent of the fracture, different surgical approaches are used to treat fractures, ranging from simple cerclage wires to total joint arthroplasty. Each approach utilizes different devices and is subjected to different complications. One advantage to the scientist is that the explants are relatively available due to the frequent removal of the hardware (for various reasons).

Orthopedic wires may be used to augment numerous orthopedic procedures. A study published by Baswell et al. examined stainless steel wires retrieved from both spine (L-rod instrumentation) and hip (trochanteric reattachment) cases [235]. All wires had fractured (12 breaks for the spinal wires and 4 breaks in the hip wires). The following mechanisms of failure were noted: one for predominately tensile fracture and 15 by cyclic loading (fatigue). There was evidence of corrosion pits on the majority of the samples.

Another approach to the reduction of fractures utilizes an intramedullary (IM) rod. After observing cracking of the proximal third of an IM rod in one case, Weinstein et al. evaluated ten stainless steel IM rods (A.O., Synthes, Wayne, PA) that had been retrieved [236]. They did not find correlations between cracking and the time of implantation or grain size. Because smearing and rubbing of the fracture surfaces had occurred, it was difficult to determine the mode of failure of many of the explants. Corrosion products were noted on the exterior and fracture surfaces of four of the rods. In one implant, there were fatigue striations and extensive branching of the crack in both a transgranular and intergranular manner, which they concluded was evidence of a corrosion-mediated mechanism.

3.4 *What Have We Learned About Biological Responses to Implants?*

Jonathan Black described four types of biomaterials: inert, interactive, viable, and replant [64]. There is a question of whether “inert” or “bioinert” materials even exist, although these terms are frequently found in the literature. Most materials in clinical use today are “interactive” or bioactive biomaterials. They are designed to elicit specific, beneficial responses. "Viable" or bioresorbable biomaterials may or may not include living tissue. They are treated by the host as self and are actively resorbed and/or remodeled. "Replanted" biomaterials encompass cultured cells obtained from the specific implant patient.

With medical implants, the major concern is whether the device is "biocompatible" and that it will not be rejected by the host. Throughout the history of orthopedic implants, there have been concerns raised as to whether the materials contained within the device are toxic, carcinogenic, or stimulate hypersensitivity reactions. Preclinical *in vitro* and *in vivo* laboratory studies are conducted to screen materials and implanted devices for safety prior to the implantation into patients. These screening tests include assays for toxicity, carcinogenicity, and hypersensitivity. However, it is not possible to completely predict how an individual patient will respond to a specific device. This is partly due to the complexity of the potential biological responses with respect to the types of cells and the cell signaling that may be involved. Furthermore, many of these tests have traditionally been performed for intact surfaces and not screened for adverse responses to degradation products. Biomechanical and bioelectrical conditions may also be radically different from preclinical testing and influence the biological response.

With this in mind, the long-term biological responses to orthopedic biomaterials range from relatively inert (a thin fibrous tissue membrane encapsulating the implant with no impact on the functionality of the implant) to malignant (aggressive fibrous histocytomas). While late stage osteolysis and tumors can be visualized on radiographic images, many of the biologic reactions are more difficult to detect using imaging techniques. For these, implant retrievals are absolutely necessary. For example, the presence of bony ingrowth can be assumed to occur when there is no evidence of a radiolucency next to a porous surface of an implant. However, bone ingrowth in a three-dimensional porous surface requires histological examination of an explanted device. It is also difficult to determine the viability of a bone graft or combination product (substrate plus cells) using radiographs as dead bone is not radiolucent.

As with all medical implants, the initial biological responses reflect wound healing and responses to the surgical procedure itself. The majority of the time this proceeds uneventfully. With orthopedic implants, the interface of stable implants is usually characterized by bony apposition or a thin fibrous tissue membrane. Charnley reported the formation of a secondary cortex within the diaphyseal canal [237]. This finding was later corroborated by Schmalzried et al. [25]. Schmalried et al. also observed bone in intimate contact with the cement mantle [124].

Bone density remodeling can be measured by surgeon observation, bone densitometry, and volumetric densities. Bone remodeling in THR is related to the remodeling parameters, but is also related to the mechanical environment of the clinical loading situation and the initial patient bone biology, i.e., how much bone stock exists as well as the reactivity of the bone.

Stress shielding with resultant bone resorption can lead to failure of the composite bone and implant [238, 239]. Unfortunately, loss of bone on clinical radiographs correlated consistently with bone loss by densitometry once 70% of bone loss has occurred [218]. Composite low modulus THR stems have been successfully retrieved and modeled for early outcomes, which may decrease this problem on midterm results, i.e., 6.2 years [240]. Stem geometry, bone ingrowth coating design (i.e., circumferential versus pads and proximal versus 4/5 stem coating) along with the rigidity will also affect initial implant stability especially in the weight bearing patient and can have a direct influence on bone ingrowth early in the history of the *in situ* process. Also the longevity can be influenced when taken in conjunction with longer term failure modalities such as wear and osteolysis. Initial patient presentation with poor bone stock such as in Dorr C leading to the need for large stem diameters in more osteoporotic flexible bone can influence modulus mismatch where consideration of the cemented stem from both a biological fixation and a mechanical status in this subset of patients can be an appropriate alternative for the dynamic situation.

Over the past few decades, the primary factors contributing to an adverse biological response include responses to infection, wear particles, and metal ions.

3.4.1 Infection

The presence of infection may significantly alter the microenvironment and dynamics of the biological response. It can be associated with significant osteolysis (Fig. 15.14). However, in an examination of seven cases in which the implant was removed secondary to hematogenous infection, Cook et al. found that infection did not preclude bone ingrowth into porous coated total knee prostheses [241]. Antibiotics and debridement may be successful in the treatment of superficial infection, but the standard of care for deep infection associated with an orthopedic implant is the removal of the implant.

3.4.2 Wear Particles and Metal Ions

As evident throughout the history of orthopedic implants, and especially total joint prostheses, wear has played a significant role in compromising the long-term outcome of the procedure. Particles from any of the materials (metal, PMMA, UHMWPE, HAp, ceramic) that make up a component have been found in the tissues surrounding devices. Metal wear has been detected in the tissues surrounding total knee and total hip implants from several different designs. While wear and metal debris was detected in several metal-on-metal designs in the past, it has also been observed around other articulating couples [242]. With the original Charnley total hip implant, the fulminant biologic response to Teflon® particles contributed to a high failure rate within very short periods of time after implantation [243, 244]. Harris and others reported extensive bone loss associated with fractured cement mantles and cement debris [134].

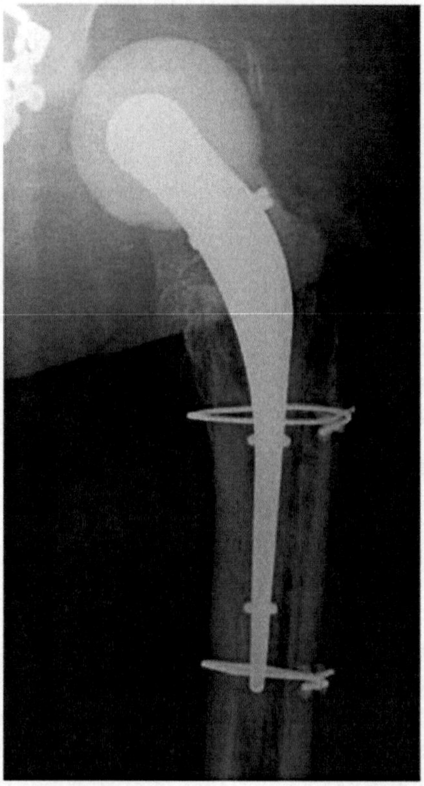

Fig. 15.14 Infected total hip revision with previous hardware placed due to pelvic fracture and *in situ* prosthetic antibiotic cement spacer for treatment of active infection

In 1987, Jones and Hungerford brought increased awareness to the biologic response to particles of PMMA that had worn away from the cement mantle surrounding total joint prostheses [139]. This was frequently associated with inadequate cementing technique. Cracks could be propagated from voids in the cement mantle caused by improper mixing or from mantles that were too thin to sustain the mechanical stresses that they were experiencing. The worst case scenario occurred with massive osteolysis around loosened and fractured cement. Two approaches to solving this problem included the introduction of porous-coated prostheses to allow for biologic fixation of the implant as well as improved cement technique utilizing techniques such as centrifugation to decrease voids and increase the homogeneity of the cement mantle. With these technological advances, osteolysis was still noted around certain implants. Although labeled as "Cement Disease," the adverse response was also found around any prosthesis with particles [139].

The periprosthetic membranes of loosened implants may also contain metal wear particles (Fig. 15.15) [32, 208]. The source of these particles may be from impingement, wear-through, or loss of surface integrity from porous-surfaced implants. Bal et al. documented the presence of corrosive metallic debris from around a CoCrMo/Ti6Al4V modular junction [76]. In addition to particles, metal

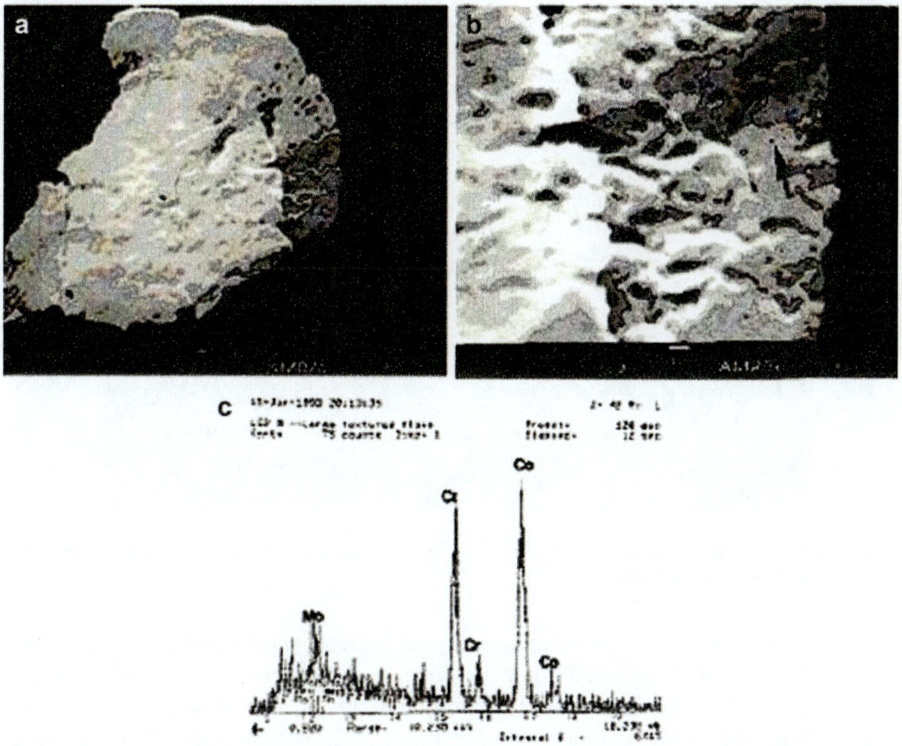

Fig. 15.15 (**a**) Low-magnification SEM micrograph of a metal particle isolated from tissue surrounding a failed TKR [34]. (**b**) Pits and cracks are evident on the surface of the metal particle. (**c**) Accompanying spectrum documents peaks of Co, Cr, and Mo. Re-published with permission of ISA.

ions may be released into the periprosthetic tissues from the surfaces themselves as well as from the particles.

With the advent of porous surfaced devices implanted without cement, it could now be seen that the wear of metal-on UHMWPE surfaces produced UHMWPE particles that could be found within the tissues of the bone–implant interface [18, 207, 245]. These particles were characterized as encompassing a variety of shapes (e.g., rounded, elongated, shards) (Fig. 15.16). While different sizes of particles were identified (from submicron to larger, >50 μm), most of the particles were in the submicron range.

The histology around mechanically compromised femoral implants reveals a significant histiocytic response associated with particles of wear debris. If left to progress, the interpositional membrane will continue to develop until there is a resultant loss of mechanical stability of the implant and significant osteolysis [246]. It is associated with a histiocytic cellular response even in well-fixed and well-functioning implants of THR.

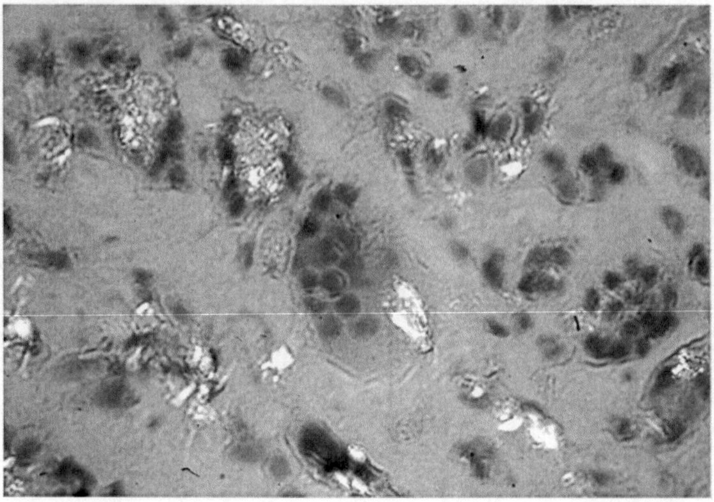

Fig. 15.16 Polyethylene debris within the interfacial membrane. Abundant macrophages and multinucleated giant cells are present

Schmalzried et al. introduced the concept of "effective joint space" [207]. These are areas which are "unsealed"; that is, areas with no cement or circumferential bone ingrowth that will allow fluid to penetrate to the interface between the implant and the bone pathways and particulate debris [212]. The fluid may contain wear debris from the articulating surfaces which become lodged in this space and is associated with osteolysis [124]. There is evidence from histological analysis of retrieved implants that the cancellous bone may be part of the effective joint space despite a well-fixed implant [247, 248]. Polyethylene has also been observed in the most proximal regions of cemented hips in the bone–cement interface [25].

3.4.3 Biological Responses to Particles and Metal Ions

Previous studies indicate a relatively benign reaction to intact materials [25, 249, 250], the body does not appear to like small particles of materials—regardless of the material itself. Size and shape of particles has been documented to be important. If the materials are small enough to be ingested by phagocytic cells, there does appear to be different responses to different materials. However, particles that are too large to be phagocytosized may still incite an adverse tissue response. This is evident by the increased numbers of giant cells that can be found surrounding large particles of UHMWPE. These cells are capable of releasing mediators into the surrounding tissues that can cause inflammation and subsequently osteolysis.

For years, there has been considerable debate as to whether the development of the periprosthetic membrane and associated osteolysis were a response of the innate or acquired immune system. Retrieval studies have sought to identify the types of cells and cytokines and other cell-signaling molecules are present. Chronic inflammation is characterized by granulation tissue containing macrophages, fibroblasts, and capillaries in varying amounts [251]. Adaptive immunity is characterized by the presence of large numbers of lymphocytes and/or plasma cells and specific cytokines. In some patient samples, the membranes contain large numbers of lymphocytes and histiocytes, while other membranes are devoid of these cells. With respect to antigens, metal ions have long been the primary culprit; polymers have long been thought to be relatively inert. Regardless of the type of immune response, a chronic reaction will lead to the secretion of substances into the microenvironment that are capable of inciting macrophages and osteoclasts to erode the host bone [27, 246]. In fact, Fornasier et al. noted that the density of the histiocytes (tissue resident macrophages) correlates with the thickness of the membrane, the density of polyethylene particles, and the time after implantation [246]. They further describe a process of cyclic reingestion of the nondigestable particles—thereby continuously stimulating the cells to produce enzymes which attract additional histiocytes while also acting to stimulate osteoclast activity.

Confirmed metal hypersensitivity associated with orthopedic implants is relatively uncommon [252–256]. Thyssen et al. reported that, based on a review of the Danish Hip Arthroplasty Registry, the risk of surgical revision was not higher in patients that had evidence of metal allergy [257]. They found that 16% of the patients had a positive patch test for metal allergy. Schuh et al. reported similar results [256]. Kosukegawa et al. described a case in which a patient underwent revision of a CoCr-on-polyethylene total hip arthroplasty for pain and a suspected abscess [255]. The joint space was filled with a white-colored fluid accompanied by necrotic soft tissue. Histology demonstrated neovascularization, lymphocytes, plasma cells, and giant cells. There was no evidence of infection; no neutrophils were present in the removed tissue. There were no metallosis or PE particles. However, the skin patch test with CoCr was positive, and the reaction abated with the removal of the implant.

Recent reports of massive pseudotumors associated with metal-on-metal implants have once again brought attention to this concern [258–262]. Metal-on-metal articulations in total hip and resurfacing implants offer an alternative to the more traditional couple of metal-on-polyethylene. Campbell et al. described their findings of an autopsy-retrieved McKee-Farrar metal-on-metal total hip replacement that was implanted in 1970 and retrieved 19 years later [263]. The bearing surfaces appeared to be in good condition, with small areas of burnishing. The appearance of the stem was consistent with motion between the metal and the cement mantle, while the acetabulum was well fixed to bone. They describe an interpositional membrane, ranging from <0.5 to 1 mm thick, that contained macrophages, giant cells, blood vessels, and wear particles of Co–Cr alloy and PMMA. They noted that lymphocytes and plasma cells were rare.

Metal-on-metal implants were reintroduced in the early 1990s. These implants were used in younger, more active adults in part due to the concerns of wear of PE in metal-on-PE implants. Soon after the reintroduction of metal-on-metal arthroplasty prostheses in the 1990s, reports appeared in the medical literature of pseudotumor-like cysts [264–266]. Willert et al. observed perivascular lymphocytes and plasma cells, macrophages, and eosinophils within these tissues [267]. Korovessis et al. also described perivascular lymphoplasmacytic infiltrates associated with metal particles [268]. The lymphocytes are indicative of an immunological response. Recently, this has been characterized as aseptic lymphocytic vasculitis-associated lesion or ALVAL [258, 269–271]. Pandit et al. have estimated that 1% of patients with metal-on-metal resurfacing implants will develop a pseudotumor within 5 years [272]. While there is increased awareness of this complication within the orthopedic community, questions remain. Is this response device specific? Fujishiro et al. examined periprosthetic tissues from 130 patients undergoing revision arthroplasty of non-metal-on-metal implants [273]. They detected perivascular lymphocytes in 50% of the patients and diffusely distributed lymphocytes in 66% of the patients. There was a direct correlation between the presence of lymphocytes and the increasing extent of metal particles. Based on these findings, it appears that there is an increased likelihood of an ALVAL-type response in any implant that is associated with the generation of metal wear debris. Campbell et al. determined that the development of pseudotumor-like tissues may be related to high levels of wear debris, but can also be detected around implants with low wear [269]. Therefore, inflammatory responses around orthopedic implants appear to be more aggressive with hypersensitivity reactions such as ALVAL and may be more host dependent, while inflammatory reactions in non-hypersensitivity reactions are likely to be correlated to numbers of wear particles. This is consistent with the findings associated with metal-on-metal implants, which are heavily technique dependent.

Hallab and Jacobs have described several mechanistic pathways that may be involved in the response to implant debris [274]. They outline a process where "both soluble and particulate debris derived from Co-Cr-Mo alloy implants can induce monocyte/macrophage activation and secretion of pro-inflammatory cytokines…via up-regulation of transcription factor NFκB, and activation of the inflammasome danger signaling in human macrophages." Whether this is the primary pathway for all types of implant debris is not known; other pathways may also operate under specific circumstances.

4 Summary

What have we learned from the analysis of orthopedic implants retrieved at revision surgery or at autopsy? While orthopedic procedures experience a high degree of success, it is important to recognize that there are limitations to their use. Host factors, particularly the quality of the bone at the implantation site, can affect the outcome. Most orthopedic implants are designed to either bear or transmit load to bone. Therefore, the alignment of the implant within the mechanical constraints of the implant site is important. There are many issues, including both mechanical and biological, involved in the selection of materials for orthopedic implants. It is important to appreciate that every effort must be undertaken to minimize the wear and corrosion of the implant. We must continue to improve implant fixation. While survivorship of orthopedic devices has improved, the ultimate outcome is still likely to be affected by wear and fixation. However, by understanding the mechanisms of failure of past implants, we can develop new strategies to address these challenges. However, it is important that, in our attempts to make improvements in the materials and designs, we do not increase the small, but significant, risk of serious adverse biological responses.

References

1. Barrack RL (2000) J Arthroplasty 15:1036–1050
2. Schroder HM, Berthelsen A, Hassani G et al (2001) J Arthroplasty 16:559–567
3. Charbord P, Livne E, Gross G et al (2011) Stem Cell Rev 7:32–42
4. Mezey E, Mayer B, Nemeth K (2010) Oral Dis 16:129–135
5. Grove JE, Bruscia E, Krause DS (2004) Stem Cells 22:487–500
6. Keaveny TM, Hayes WC (1993) J Biomech Eng 115:534–542
7. Hazelwood SJ, Martin RB (2005) In: Dorf RC (ed) The engineering handbook. CRC Press, Boca Raton, FL, pp 216–1 to –8
8. Migliaresi C, Alexander H (2004) In: Ratner BD, Hoffman AS, Schoen FJ, Lemons JE (eds) Biomaterials science. An introduction to materials in medicine. Elsevier, San Diego, CA, pp 181–197
9. FDA (2010) Code of Federal Regulations
10. Cook SD, Weinstein AM (1981) Comput Programs Biomed 13:217–224
11. Jones LC, Hungerford MW, Khanuja HS et al (2006) J Histotechnol 29:277–285
12. Collier JP, Bauer TW, Bloebaum RD et al (1992) Clin Orthop Relat Res (274):97–112
13. Hirakawa K, Jacobs JJ, Urban R et al (2004) Clin Orthop Relat Res (420):10–17
14. Noble PC, Conditt MA, Thompson MT et al (2003) Clin Orthop Relat Res (416):120–128
15. Duffy GP, Muratoglu OK, Biggs SA et al (2001) J Arthroplasty 16:42–48
16. Ranawat CS, Johanson NA, Rimnac CM et al (1986) Clin Orthop Relat Res (209):244–248
17. Kurtz SM, Rimnac CM, Pruitt L et al (2000) Biomaterials 21:283–291
18. Campbell P, Ma S, Yeom B et al (1995) J Biomed Mater Res 29:127–131
19. Collier JP (2008) Spine (Phila Pa 1976) 33:457
20. Reckling FW, Asher MA, Mantz FA et al (1975) J Bone Joint Surg Am 57:108–112
21. Bloebaum RD, Nelson K, Dorr LD et al (1991) Clin Orthop Relat Res (269):120–127
22. Mintz L, Tsao AK, McCrae CR et al (1991) Clin Orthop Relat Res (273):215–222
23. Goldberg JR, Gilbert JL, Jacobs JJ et al (2002) Clin Orthop Relat Res (401):149–161
24. Thomas KA, Cook SD, Harding AF et al (1988) Orthopaedics 11:441–451
25. Schmalzried T, Maloney WJ, Jasty M et al (1993) J Arthroplasty 8:179–188
26. Jasty M, Burke D, Harris WH (1992) Chir Organi Mov 77:349–358
27. Maloney WJ, Jasty M, Burke DW et al (1989) Clin Orthop Relat Res (249):129–140
28. Ohlin A, Balkfors B (1992) J Arthroplasty 7:87–92
29. Miller MA, Eberhardt AW, Cleary RJ et al (2010) J Orthop Res 28:170–177
30. Sychterz CJ, Moon KH, Hashimoto Y et al (1996) J Bone Joint Surg Am 78:1193–1200
31. Tsao A, Jones L, Lewallen D (2008) J Am Acad Orthop Surg 16:S7–S13
32. Schmiedberg SK, Jones LC, Chang DH et al (2007) Biomed Sci Instrum 43:104–109
33. Fox JA, Pierce M, Bojchuk J et al (2004) Arthroscopy 20:787–794
34. Que L, Topoleski L, Parks N (2000) J Biomed Mater Res 53:111–118
35. Zardiackas LD, Black RJ, Hughes JLR, R.B. et al (1989) Orthopaedics 12:85–92

36. Santaguida PL, Hawker GA, Hudak PL et al (2008) Can J Surg 51:428–436
37. Mancuso CA, Ranawat CS, Esdaile JM et al (1996) J Arthroplasty 11:34–46
38. Wright TM, Goodman SB (eds) (2001) Implant wear in total joint replacement: clinical and biologic issues, materials and design considerations. American Academy of Orthopaedic Surgeons, Rosemont, IL, 228 pp
39. Parvizi J, Kim KI, Goldberg G et al (2006) Clin Orthop Relat Res 447:60–65
40. Krackow K (1983) Adv Orthop Surg 7:69–88
41. Lakstein D, Zarrabian M, Kosashvili Y et al (2010) J Arthroplasty 25:1047–1052
42. Moskal JT, Capps SG (2010) J Am Acad Orthop Surg 18:286–296
43. Haydon CM, Mehin R, Burnett S et al (2004) J Bone Joint Surg Am 86–A:1179–1185
44. Swank M, Romanowski JR, Korbee LL et al (2007) Proc Inst Mech Eng H 221:755–761
45. Lewallen DG, Bryan RS, Peterson LF (1984) J Bone Joint Surg Am 66:1211–1218
46. Matsuno T, Katayama N (1997) Int Orthop 21:409–411
47. LaPorte DM, Mont MA, Hungerford DS (1999) Orthopaedics 22:1154–1160
48. Peterson LF, Fitzgerald RH, Jr., Johnson EW, Jr. (1979) Mayo Clin Proc 54:564–569
49. Low FH, Khoo LP, Chua CK et al (2000) Crit Rev Biomed Eng 28:33–40
50. Lavernia CJ, Sierra RJ, Hungerford DS et al (2001) J Arthroplasty 16:446–453
51. Berend KR, Sporer SM, Sierra RJ et al (2010) J Bone Joint Surg Am 92:2737–2752
52. Robbins GM, Masri BA, Garbuz DS et al (2001) Orthop Clin North Am 32:611–626, viii
53. Lavernia CJ, Sierra RJ, Gomez–Marin O (1999) Clin Orthop Relat Res (367):172–180
54. Sloan A, Hussain I, Maqsood M et al (2010) Surgeon 8:111–116
55. Richman MH, Weltman JK, Cole A (1976) Injury 8:13–19
56. Hungerford DS, Hedley A, Hanbermann E et al (1984) Total hip arthroplasty. A new approach. University Park Press, Baltimore, MD, 184 pp
57. Woolson ST, Milbauer JP, Bobyn JD et al (1997) J Bone Joint Surg Am 79:1842–1848
58. Lakstein D, Eliaz N, Levi O et al (2011) J Bone Joint Surg Am 93:57–65
59. Banovetz JM, Sharp R, Probe RA et al (1996) J Orthop Trauma 10:389–394
60. Gruen TA, McNeice GM, Amstutz HC (1979) Clin Orthop Relat Res (141):17–27
61. Morrey BF, Chao EY (1988) Clin Orthop Relat Res (228):182–189
62. Flivik G, Ljung P, Rydholm U (1990) Acta Orthop Scand 61:26–28
63. Abernethy PJ, Robinson CM, Fowler RM (1996) J Bone Joint Surg Br 78:220–225
64. Black J (1992) Biological performance of materials. Marcel Dekker, New York, 390 pp
65. Grupp TM, Weik T, Bloemer W et al (2010) BMC Musculoskelet Disord 11:3
66. McAuley JP, Gow KV, Covert A et al (1987) Can J Surg 30:424–427
67. Collier JP, Mayor MB, Jensen RE et al (1992) Clin Orthop Relat Res (285):129–139
68. Collier JP, Surprenant VA, Jensen RE et al (1992) J Bone Joint Surg Br 74:511–517
69. Collier JP, Surprenant VA, Jensen RE et al (1991) Clin Orthop Relat Res (271):305–312
70. Cook SD, Barrack RL, Clemow AJT (1994) J Bone Joint Surg Br 76:68–72
71. Kop AM, Swartz E (2009) J Arthroplasty 24:1019–1023
72. Akazawa T, Minami S, Takahashi K et al (2005) J Orthop Sci 10:200–205
73. McKellop H, Park SH, Chiesa R et al (1996) Clin Orthop Relat Res (329 Suppl):S128–S140
74. McKellop HA, Sarmiento A, Brien W et al (1992) J Arthroplasty 7:291–294
75. Jacobs JJ, Urban RM, Gilbert JL et al (1995) Clin Orthop Relat Res (319):94–105
76. Bal BS, Vandelune D, Gurba DM et al (1998) J Arthroplasty 13:492–499
77. Harding AF, Cook SD, Thomas KA et al (1985) Clin Orthop Relat Res (195):261–269
78. Cook SD, Renz EA, Barrack RL et al (1985) Clin Orthop Relat Res (194):236–247
79. Collier JP, Mayor MB, Williams IR et al (1995) Clin Orthop Relat Res (311):91–101
80. Viceconti M, Ruggeri O, Toni A et al (1996) J Biomed Mater Res 30:181–186
81. Que L, Topoleski L (1999) J Biomed Mater Res 48:705–711
82. Anderson C, Topoleski L, Engh G (1995) Trans 21st annual meeting of the society for biomaterials, San Francisco, CA, p 166
83. Wroblewski BM, Purbach B, Siney PD et al (2010) J Bone Joint Surg Br 92:486–488
84. Berzins A, Jacobs JJ, Berger R et al (2002) J Bone Joint Surg Am 84–A:1534–1540

85. Alexander N, Hungerford D, Jones LC et al (1996) J Bone Joint Surg Orthop Transactions 19:316
86. Mont MA, Shafer BL, Alexander N et al (1998) Transactions of the 24th Annual Meeting of The Society for Biomaterials, vol 21, San Diego, CA, p 94
87. Rentfrow ED, James SP, Beauregard GP et al (1996) Biomed Sci Instrum 32:135–141
88. Endo MM, Barbour PS, Barton DC et al (1999) Biomed Mater Eng 9:113–124
89. Faris PM, Ritter MA, Pierce AL et al (2006) Clin Orthop Relat Res 453:305–308
90. Bargmann LS, Bargmann BC, Collier JP et al (1999) Clin Orthop Relat Res (369):49–58
91. Currier BH, Currier JH, Mayor MB et al (2007) J Arthroplasty 22:721–731
92. Medel FJ, Kurtz SM, Parvizi J et al (2010) J Arthroplasty
93. Watanabe E, Suzuki M, Nagata K et al (2002) J Biomed Mater Res 62:540–549
94. Medel FJ, Kurtz SM, Hozack WJ et al (2009) J Bone Joint Surg Am 91:839–849
95. Wang A, Yau SS, Essner A et al (2008) J Arthroplasty 23:559–566
96. Dumbleton JH, D'Antonio JA, Manley MT et al (2006) Clin Orthop Relat Res 453:265–271
97. Oral E, Ghali BW, Rowell SL et al (2010) Biomaterials 31:7051–7060
98. Crowninshield RD, Muratoglu OK (2008) J Am Acad Orthop Surg 16(Suppl 1):S80–S85
99. Yamamoto K, Imakiire A, Masaoka T et al (2003) Int Orthop 27:286–290
100. Jasty M, Goetz DD, Bragdon CR et al (1997) J Bone Joint Surg Am 79:349–359
101. Engh GA, Dwyer KA, Hanes CK (1992) J Bone Joint Surg Br 74:9–17
102. Walker PS, Blunn GW, Lilley PA (1996) J Biomed Mater Res 33:159–175
103. Wroblewski BM, Siney PD, Dowson D et al (1996) J Bone Joint Surg Br 78:280–285
104. Glyn–Jones S, Isaac S, Hauptfleisch J et al (2008) J Arthroplasty 23:337–343
105. Currier JH, Bill MA, Mayor MB (2005) J Biomech 38:367–375
106. Tsao A, Mintz L, McRae CR et al (1993) J Bone Joint Surg Am 75:19–26
107. Weber AB, Morris HG (1996) J Arthroplasty 11:856–858
108. Kiely PD, Harty JA, McElwain JP (2005) Acta Orthop Belg 71:429–434
109. Chmell MJ, Poss R, Thomas WH et al (1996) J Arthroplasty 11:351–353
110. Nenshi R, Takata J, Stegienko S et al (2010) Surg Innov 17:291–294
111. Muratoglu OK, Bragdon CR, O'Connor DO et al (2002) Clin Orthop Relat Res (404):89–95
112. Pruitt LA (2005) Biomaterials 26:905–915
113. McKellop H, Shen FW, DiMaio W et al (1999) Clin Orthop Relat Res (369):73–82
114. Muratoglu OK, Bragdon CR, O'Connor DO et al (1999) Biomaterials 20:1463–1470
115. Sobieraj MC, Rimnac CM (2009) J Mech Behav Biomed Mater 2:433–443
116. McKellop H, Shen FW, Lu B et al (1999) J Orthop Res 17:157–167
117. Popoola OO, Yao JQ, Johnson TS et al (2010) J Orthop Res 28:1120–1126
118. Baxter RM, Freeman TA, Kurtz SM et al (2011) Clin Orthop Relat Res 469:2308–2317
119. MacDonald D, Sakona A, Ianuzzi A et al (2011) Clin Orthop Relat Res 469:2278–2285
120. Topoleski L, Duchyene P, Cuckler J (1990) J Biomed Mater Res 24:135–154
121. Maloney WJ, Schmalzried T, Harris WH (2002) Clin Orthop Relat Res (405):70–78
122. Jasty M, Jiranek W, Harris WH (1992) Clin Orthop Relat Res (285):116–128
123. Bishop NE, Schoenwald M, Schultz P et al (2009) Hip Int 19:87–95
124. Schmalzried TP, Kwong LM, Jasty M et al (1992) Clin Orthop Relat Res (274):60–78
125. Hernigou P, Le Mouel S (1999) J Arthroplasty 14:1005–1010
126. Evans SL (2006) Proc Inst Mech Eng H 220:1–10
127. Kim YH, Oh SH, Kim JS et al (2003) J Bone Joint Surg Am 85–A:675–681
128. Smith SW, Estok DM, 2nd, Harris WH (1998) J Bone Joint Surg Am 80:1632–1640
129. Hanson PB, Walker RH (1995) J Arthroplasty 10:683–688
130. Cristofolini L, Erani P, Savigni P et al (2007) Clin Biomech (Bristol, Avon) 22:410–421
131. Cabanela ME, Campbell DC, 2nd, Henderson ED (1979) Mayo Clin Proc 54:559–563
132. Koenig JH, Maheshwari AV, Ranawat AS et al (2009) Orthopaedics 32(9):pii
133. Kurtz S, Mowat F, Ong K et al (2005) J Bone Joint Surg Am 87:1487–1497
134. Harris WH, Schiller AL, Scholler JM et al (1976) J Bone Joint Surg Am 58:612–618

135. Amstutz HC, Markolf KL, McNeice GM et al (1976) In: The hip. Proceedings of the fourth open scientific meeting of the hip society, St Louis, CV Mosby, pp 192–216
136. Beckenbaugh RD, Ilstrup DM (1978) J Bone Joint Surg Am 60:306–313
137. Ries MD, Guiney W, Jr., Lynch F (1994) J Arthroplasty 9:555–558
138. Jones LC, Frondoza CG, Hungerfrod DS (1999) J Biomed Mater Res 48:889–898
139. Jones LC, Hungerford DS (1987) Clin Orthop Relat Res 225:192–206
140. Goldring SR, Schiller AL, Roelke M et al (1983) J Bone Joint Surg Am 65:575–584
141. Lennox DW, Schofield BH, McDonald DF et al (1987) Clin Orthop Relat Res (225):171–191
142. Schmalzried TP, Harris WH (1993) J Bone Joint Surg Br 75:608–615
143. Okamoto T, Inao S, Gotoh E et al (1997) J Bone Joint Surg Br 79:83–86
144. Buckwalter AE, Callaghan JJ, Liu SS et al (2006) J Bone Joint Surg Am 88:1481–1485
145. Engh CA, Hooten JP, Jr., Zettl–Schaffer KF et al (1994) Clin Orthop Relat Res (225):89–96
146. Coultrup OJ, Hunt C, Wroblewski BM et al (2010) J Orthop Res 28:565–570
147. Jasty M, Maloney WJ, Bragdon CR et al (1990) J Bone Joint Surg Am 72:1220–1229
148. Kawate K, Maloney WJ, Bragdon CR et al (1998) Clin Orthop Relat Res 355:70–76
149. Jasty M, Maloney WJ, Bragdon CR et al (1991) J Bone Joint Surg Br 73:551–558
150. Hedley AK, Gruen TA, Borden LS et al (1987) Hip:225–250
151. Woolson ST, Maloney WJ (1992) J Arthroplasty 7(Suppl):381–388
152. Cook SD, Thomas KA, Dalton JE et al (1992) J Biomed Mater Res 26:989–1001
153. Sumner DR, Kienapfel H, Jacobs JJ et al (1995) J Arthroplasty 10:157–167
154. Bloebaum RD, Rubman MH, Hofmann AA (1992) J Arthroplasty 7:483–493
155. Cook SD, Thomas KA (1991) J Bone Joint Surg Br 73:20–24
156. Vigorita VJ, Minkowitz B, Dichiara JF et al (1993) Clin Orthop Relat Res 293:211–218
157. Collier JP, Mayor MB, Chae JC et al (1988) Clin Orthop Relat Res 235:173–180
158. Brooker AF Jr, Collier JP (1984) J Bone Joint Surg Am 66:619–621
159. Kang JS, Dorr LD, Wan Z (2000) J Arthroplasty 15:730–735
160. Coventry MB (1992) Clin Orthop Relat Res 274:22–29
161. Agins HJ, Alcock NW, Bansal M et al (1988) J Bone Joint Surg Am 70:347–356
162. Jacobs MA, Bhargava T, Lathroum JM et al (2009) J Bone Joint Surg Am 91:961–964
163. Kleinhans JA, Jakubowitz E, Seeger JB et al (2009) Orthopaedics 32:364
164. Wasielewski RC, Parks N, Williams I et al (1997) Clin Orthop Relat Res 345:53–59
165. Akizuki S, Takizawa T, Horiuchi H (2003) J Bone Joint Surg Br 85:1123–1127
166. Tonino A, Oosterbos C, Rahmy A et al (2001) J Bone Joint Surg Am 83–A:817–825
167. Aebli N, Krebs J, Schwenke D et al (2003) J Bone Joint Surg Br 85:499–503
168. Geesink RGT (1993) Acta Orthopaedica Belgica 59:160–164
169. Watanabe H, Akizuki S, Takizawa T (2004) J Bone Joint Surg Br 86:824–829
170. Hardeman F, Vandenneucker H, Van Lauwe J et al (2006) Knee 13:419–421
171. Ritter MA (2009) J Bone Joint Surg Br 91:745–749
172. Rodriguez JA, Bhende H, Ranawat CS (2001) Clin Orthop Relat Res (388):10–17
173. Parks NL, Engh GA, Topoleski LD et al (1998) Clin Orthop Relat Res 356:10–15
174. Engh GA, Lounici S, Rao AR et al (2001) J Bone Joint Surg Am 83–A:1660–1665
175. Harman MK, S A, Banks, Hodge WA (2001) Clin Orthop Relat Res 392:383–393
176. Rohrbach M, Luem M, Ochsner PE (2008) J Orthop Surg Res 3:8
177. Wright TM, Rimnac CM, Stulberg SD et al (1992) Clin Orthop Relat Res (276):126–134
178. Garcia RM, Kraay MJ, Messerschmitt PJ et al (2009) J Arthroplasty 24:131–138
179. Cook SD, Thomas KA, Haddad RJ, Jr. (1988) Clin Orthop Relat Res 234:90–101
180. Hood RW, Wright TM, Burstein AH (1983) J Biomed Mater Res 17:829–842
181. Landy MM, Walker PS (1988) J Arthroplasty 3:S73–S85
182. Sumner DR, Kienapfel H, Jacobs JJ et al (1995) J Arthroplasty 10:157–167
183. Feng EL, Stulberg SD, Wixson RL (1994) Clin Orthop Relat Res (299):60–71
184. Blunn GW, Joshi AB, Minns RJ et al (1997) J Arthroplasty 12:281–290
185. Mow VC, Hayes WC (1991) Basic orthopaedic biomechanics. Raven Press, New York

186. Hungerford DS, Kenna RV, Haynes DW (1984) In: Hungerford DS, Krackow KA, Kenna RV (eds) Total knee arthroplasty: a comprehensive approach. Williams and Wilkins. Baltimore, pp 20–34
187. Williams A, Logan M (2004) Knee 11:81–88
188. Goodfellow J, O'Connor J (1978) J Bone Joint Surg Br 60–B:358–369
189. Mariconda M, Silvestro A, Mansueto G et al (2010) Arch Orthop Trauma Surg 130:61–64
190. Kilgus DJ, Moreland JR, Finerman GA et al (1991) Clin Orthop Relat Res 273:223–231
191. Stulberg SD, Stulberg BN, Hamati Y et al (1988) Clin Orthop Relat Res (236):88–105
192. Small SR, Berend ME, Ritter MA et al (2010) J Arthroplasty
193. Harris WH (1984) Clin Orthop Relat Res (183):4–11
194. Rodriguez JA, Baez N, Rasquinha V et al (2001) Clin Orthop Relat Res 392:174–183
195. Surace MF, Berzins A, Urban RM et al (2002) Clin Orthop Relat Res 404:14–23
196. Lee HH, Lo YC, Lin LC et al (2008) J Formos Med Assoc 107:84–88
197. Chen CE, Juhn RJ, Ko JY (2011) J Arthroplasty 26:339.e11–339.e13
198. Chen CY, Huang KY, Chang JK et al (2000) Kaohsiung J Med Sci 16:368–374
199. Rapuri VR, Clarke HD, Spangehl MJ et al (2010) J Arthroplasty
200. Atwood SA, Currier JH, Mayor MB et al (2008) J Arthroplasty 23:431–440
201. Fisher J, Jennings LM, Galvin AL et al (2010) Clin Orthop Relat Res 468:12–18
202. Matsuda S, Ishinishi T, White SE et al (1997) J Arthroplasty 12:790–797
203. Matsuda S, White SE, Williams VG, 2nd et al (1998) J Arthroplasty 13:699–706
204. Rosenberg AG, Andriacchi TP, Barden R et al (1988) Clin Orthop Relat Res (236):106–114
205. Kraay MJ, Darr OJ, Salata MJ et al (2001) Clin Orthop Relat Res (392):239–244
206. Hashiguchi T, Hirano T, Shindo H et al (1999) Arch Orthop Trauma Surg 119:30–34
207. Schmalzried TP, Jasty M, Harris WH (1992) J Bone Joint Surg Am 74:849–863
208. Urban RM, Jacobs JJ, Gilbert JL et al (1994) J Bone Joint Surg Am 76:1345–1359
209. Jasty M, Burke D, Harris WH (1992) Chir Organi Mov 77:349–358
210. Berzins A, Sumner DR, Galante JO (1998) J Biomed Mater Sci 39:120–129
211. Skurla CT, S.P.James (2001) Biomed Sci Instrum 37:245–250
212. Bobyn JD, Jacobs JJ, Tanzer M et al (1995) Clin Orthop Relat Res (311):21–39
213. Wroblewski BM, Lynch M, Atkinson JR et al (1987) J Bone Joint Surg Br 69:61–63
214. Mochida Y, Boehler M, Salzer M et al (2001) Clin Orthop Relat Res (389):113–125
215. Della Valle AG, Rana A, Furman B et al (2005) Clin Orthop Relat Res 440:184–191
216. Williams VG, 2nd, Whiteside LA, White SE et al (1997) J Arthroplasty 12:25–31
217. Creighton RA, Romeo AA, Brown FM, Jr. et al (2007) Arthroscopy 23:703–709
218. Engh CA, Jr., McAuley JP, Sychterz CJ et al (2000) J Bone Joint Surg Am 82–A:1414–1420
219. Deyo RA, Gray DT, Kreuter W et al (2005) Spine (Phila Pa 1976) 30:1441–1445; discussion 6–7
220. Majid K, Crowder T, Baker E et al (2011) J Spinal Disord Tech
221. Vieweg U, van Roost D, Wolf HK et al (1999) Spine (Phila Pa 1976) 24:946–951
222. McPhee IB, Swanson CE (2007) Spine (Phila Pa 1976) 32:1963–1968
223. Cook SD, Barrack RL, Georgette FS et al (1985) Spine (Phila Pa 1976) 10:313–316
224. Sturz H, Hinterberger J, Matzen K et al (1979) Arch Orthop Trauma Surg 95:113–122
225. Togawa D, Bauer TW, Lieberman IH et al (2004) J Bone Joint Surg Am 86–A:70–79
226. Togawa D, Bauer TW, Brantigan JW et al (2001) Spine (Phila Pa 1976) 26:2744–2750
227. Togawa D, Bauer TW, Lieberman IH et al (2003) Spine (Phila Pa 1976) 28:246–253; discussion 54
228. Weiss M (1975) Clin Orthop Relat Res (112):150–158
229. Ianuzzi A, Kurtz SM, Kane W et al (2010) Spine (Phila Pa 1976) 35:E1310–E1316
230. Jensen WK, Anderson PA, Nel L et al (2005) Spine (Phila Pa 1976) 30:2497–2502
231. Pitzen T, Kettler A, Drumm J et al (2007) Eur Spine J 16:1015–1020
232. Shkolnikov YP, Bowden A, MacDonald D et al (2010) J Biomed Mater Res B Appl Biomater 94:312–317
233. Kurtz SM, van Ooij A, Ross R et al (2007) Spine J 7:12–21

234. Kurtz SM, Patwardhan A, MacDonald D et al (2008) Spine (Phila Pa 1976) 33:481–489
235. Baswell IL, Sander T, Allen B, Jr. et al (1986) J Biomed Mater Res 20:887–894
236. Weinstein AM, Clemow AJ, Starkebaum W et al (1981) J Bone Joint Surg Am 63:1443–1448
237. Charnley J (1979) Low friction arthroplasty of the hip. Springer: New York
238. Kerner J, Huiskes R, van Lenthe GH et al (1999) J Biomech 32:695–703
239. McGovern TF, Engh CA, Zettl–Schaffer K et al (1994) Clin Orthop Relat Res (306):145–154
240. Akhavan S, Matthiesen MM, Schulte L et al (2006) J Bone Joint Surg Am 88:1308–1314
241. Cook SD, Barrack RL, Thomas KA et al (1989) J Arthroplasty 4(Suppl):S33–S43
242. Mirra JM, Amstutz HC, Matos M et al (1976) Clin Orthop Relat Res (117):221–240
243. Gheorghiu D, Peter V, Lynch M (2010) Acta Orthop Belg 76:129–131
244. Charnley J (1963) Lancet 7322:1379
245. Tsao A, Jones LC, Choung S et al (1993) Quantitative characterization of particulate polyethylene wear debris from failed total joint arthroplasties. In: Proceedings of the 19th Annual Meeting of the Society for biomaterials, pp 239
246. Fornasier V, Wright J, Seligman J (1991) Clin Orthop Relat Res (271):272–282
247. Massin P, Chappard D, Flautre B et al (2004) J Biomed Mater Res B Appl Biomater 69:205–215
248. Massin P, Viguier E, Flautre B et al (2004) J Biomed Mater Res B Appl Biomater 68:140–148
249. Bauer TW, Geesink RC, Zimmerman R et al (1991) J Bone Joint Surg Am 73:1439–1452
250. Vernon–Roberts B, Freeman MAR (1977) In: Swanson SAV, Freeman MAR (eds) The scientific basis of joint replacement. Wiley, New York, pp 86–129
251. Anderson JM, Cook G, Costerton B et al (2004) In: Ratner BD, Hoffman AS, Schoen FJ, Lemons JE (eds) Biomaterials science. An introduction to materials in medicine. Elsevier Academic Press, San Diego, CA, pp 293–354
252. Bergschmidt P, Bader R, Mittelmeier W (2011) Knee
253. Nater JP, Brain RG, Deutman R et al (1976) Contact Dermatitis 2:259–261
254. Merritt K, Rodrigo JJ (1996) Clin Orthop Relat Res (326):71–79
255. Kosukegawa I, Nagoya S, Kaya M et al (2010) J Arthroplasty
256. Schuh A, Lill C, Honle W et al (2008) Zentralbl Chir 133:292–296
257. Thyssen JP, Jakobsen SS, Engkilde K et al (2009) Acta Orthop 80:646–652
258. Watters TS, Cardona DM, Menon KS et al (2010) Am J Clin Pathol 134:886–893
259. Kwon YM, Ostlere SJ, McLardy–Smith P et al (2010) J Arthroplasty
260. Kwon YM, Thomas P, Summer B et al (2010) J Orthop Res 28:444–450
261. Singh C, Kaplan A, Pambuccian SE (2010) Diagn Cytopathol
262. Guyer RD, Shellock J, Maclennan B et al (2011) Spine (Phila Pa 1976)
263. Campbell P, Urban RM, Catelas I et al (2003) J Bone Joint Surg Am 85–A:2218–2222
264. Parvizi J, Picinic E, Sharkey PF (2009) Instr Course Lect 58:183–191
265. Svensson O, Mathiesen EB, Reinholt FP et al (1988) J Bone Joint Surg Am 70:1238–1242
266. Counsell A, Heasley R, Arumilli B et al (2008) Acta Orthop Belg 74:870–874
267. Willert HG, Buchhorn GH, Fayyazi A et al (2005) J Bone Joint Surg Am 87:28–36
268. Korovessis P, Petsinis G, Repanti M et al (2006) J Bone Joint Surg Am 88:1183–1191
269. Campbell P, Ebramzadeh E, Nelson S et al (2010) Clin Orthop Relat Res 468:2321–2327
270. Hutt JR, Busch C, Hughes RA (2010) Rheumatol Int
271. Crawford R, Ranawat CS, Rothman RH (2010) J Arthroplasty 25:1–2
272. Pandit H, Glyn–Jones S, McLardy–Smith P et al (2008) J Bone Joint Surg Br 90:847–851
273. Fujishiro T, Moojen DJ, Kobayashi N et al (2010) Clin Orthop Relat Res
274. Hallab NJ, Jacobs JJ (2009) Bull NYU Hosp Jt Dis 67:182–188

Biography

Dr. Jones is an Associate Professor of Orthopaedic Surgery, with an adjunct faculty appointment in Materials Science, at the Johns Hopkins University (JHU). She is the Director of JHU Orthopaedic Research at the Good Samaritan Hospital, the Director of the Center for Osteonecrosis Research and Education, and the Technical Director of the JHU Arthritis Surgery Bone Bank. Jones received her bachelor's and master's degrees in biology/biological sciences from Ursinus College and Towson State University. She received her Ph.D. degree in Molecular Biology and Immunology from the JHU. She is a past President of the Society for Biomaterials (SFB) as well as an active member of the Orthopaedic Research Society (ORS), the American Academy of Orthopaedic Surgeons, ASBMR, ARCO International, Rocky Mountain Bioengineering Symposium, and NONF. Jones is a member of the College of Fellows of the American Institute for Medical & Biological Engineering (AIMBE) and a Fellow of the International Union of Societies for Biomaterials Science and Engineering (IUSBSE). Her research has focused on the pathogenesis and treatment of musculoskeletal diseases. She has published numerous articles and lectures around the country on the biological response to implant materials, osteonecrosis, and bone graft materials.

Dr. Tsao received her engineering degrees in Biomedical Engineering and Materials Science & Engineering from Johns Hopkins University (JHU). She attended medical school at Cornell University and completed her orthopedic residency at Northwestern University in Chicago. She returned to JHU to complete a fellowship in Arthritis and Adult Reconstruction Surgery. Tsao's previous position was as a Professor in the Department of Orthopaedic Surgery & Rehabilitation with the University of Mississippi Medical Center. She currently practices in the western Phoenix metropolitan area in private practice. She specializes in total joint reconstruction of the hip and knee. Her specialty interests include muscle sparing approaches and customization of surgical procedures and implants for individual patients. She has published numerous articles and lectures around the country to practicing orthopedic surgeons and is involved in research for osteonecrosis, primary and revision total hip, and knee replacement surgery. Currently, Tsao is active as an AO Faculty Member and a consultant for several surgical companies where she is also active in teaching surgical techniques, the use of new implant technologies, and difficult primary and revision joint replacements. Tsao is a member of the American Academy of Orthopaedic Surgeons, Orthopaedic Research Society, American Association of Hip and Knee Surgeons, and the National Osteonecrosis Foundation. She has been a Diplomat of the American Board of Orthopaedic Surgery since 1994.

Dr. Topoleski received his B.S., M. Eng. and M.S. degrees from Cornell University, with an emphasis on biomechanics. He received his Ph.D. in Bioengineering from the University of Pennsylvania, where he focused on biomaterials. He joined the faculty in the Department of Mechanical Engineering at UMBC (the University of Maryland, Baltimore County) in 1990, where he is currently a Professor. He has served as the Graduate Program Director for the Department, is currently the President of the UMBC Faculty Senate, and has also been named a UMBC Presidential Teaching Professor for excellence in teaching. Topoleski's research interests include mechanics of biomaterials, both human-made and natural, with an emphasis on orthopedic and cardiovascular materials. He remains an active member of the Society for Biomaterials.

Chapter 16
The Use of Finite Element Analysis in Design, Life Prediction, and Failure Analysis of Biomaterials and Medical Devices

Ming Wu and Paul Briant

Abstract This chapter describes the adoption and application of finite element analysis (FEA) during the design and failure analysis of medical devices. FEA has been an important product development tool in the medical device industry, by allowing designers to explore various design options quickly and quantitatively. The chapter discusses some of the key steps and important considerations in applying the finite element method to the design and failure analysis of devices, and provides an example on how these steps are implemented in a nitinol *inferior vena cava filter* (IVCF) wire frame design and validation.

1 Introduction

The medical device industry is continually pushed to develop new devices that provide new treatments, improve the quality of medical care, and improve the way treatment is delivered. This growth is primarily driven by the quality of life demands of an aging population and the escalating treatment costs. *Finite element analysis* (FEA) has increasingly become a critical development tool for new devices due to its ability to help device designers quickly explore various design options, optimize designs, and gain a deeper insight into how a device is performing. With the advancement of FEA software and computer hardware technology, FEA is being more readily adopted by the medical device industry. However, FEA cannot be treated as a black box that generates results; understanding how to perform practical FEA that is appropriate to the problem, and how to validate the results from the FEA, is something all designers/engineers need to learn and appreciate.

M. Wu (✉) • P. Briant
Exponent, Inc., 149 Commonwealth Drive, Menlo Park, CA 94025, USA
e-mail: mwu@exponent.com

N. Eliaz (ed.), *Degradation of Implant Materials*, DOI 10.1007/978-1-4614-3942-4_16,

2 Finite Element Analysis and Its Application

FEA is a numerical analysis method for obtaining approximate solutions to a wide range of engineering problems. FEA can be dated back to the work by Richard Courant and Alexander Hrennikoff first developed in the 1940s, who utilized the Ritz method of numerical analysis and minimization of variational calculus to obtain approximate solutions to structural systems [1, 2]. The development of the finite element method continued its momentum in the 1950s and 1960s for the application of airframe structural applications and civil structural applications. In the years since 1960, FEA has received widespread acceptance in engineering. Thousands of papers and books have been published on the subject [3–7].

Early on, the application of FEA to solve structural mechanics problems was limited due to the limitation of computer hardware capabilities. With the rapid increase of computing power and reduction of the cost of computers starting in 1990s, FEA has evolved into a powerful design tool that has significantly improved both the engineering design standard and the design process methodology in various industrial applications including the medical device industry. FEA allows clear visualization of how the structures behave and deform and calculates the distribution of stresses and displacements. FEA provides the option for entire designs to be constructed at any design complexity and optimized before any prototyping and manufacturing of structures. The introduction of FEA has substantially decreased the time to take products from concept to production, as well as provided a powerful tool for failure analysis and life assessment.

FEA can be performed in two- or three-dimensional analyses. A two-dimensional (2D) analysis is simpler, runs faster, and requires less computer power and cost. A three-dimensional (3D) analysis generally produces more accurate results, but requires more powerful computers and computational cost. An engineer can choose to perform either a 2D or 3D analysis based on his or her understanding of the problem and the method that can provide the most appropriate results with the highest efficiency. In addition, an engineer must determine whether the structure or device behaves linearly or non-linearly. Linear systems are far less complex and generally do not take into account plastic deformation, large component rotation, or contact between components. Non-linear systems can account for plastic deformation, large component rotation, and component–component contact. The material constitutive laws used in a FE-based analysis are another very important part of the analysis input that require a thorough understanding by the engineer who performs the analysis. The material constitutive models in a FE-based analysis need to capture the realistic behavior of the device material from rigorous testing.

3 Finite Element Analysis Applied to Medical Devices

3.1 Uses of FEA in Medical Devices

FEA is well suited to be used in the medical device industry to analyze the *structural integrity* of devices, especially given the small, intricate, and complex nature of medical devices. Cardiovascular devices, such as stents and valve replacement devices, are often too small for experimental methods to assess the mechanical state of stress in a device under load. In addition, traditional experimental techniques for measuring strain, such as installing strain gages on a device, may add too much stiffness to the system, thereby altering the stress state of the device. *Numerical modeling* can therefore be used as an accurate means of quantitatively obtaining an estimate of the mechanical stresses in the device.

The two primary mechanisms by which medical devices fail due to mechanical stress are overload and fatigue. An *overload failure* is the application of a one-time load that is beyond the strength of the structure, causing the formation and rapid propagation of a crack. During *fatigue failures*, however, small loads that the structure can initially withstand are applied repeatedly, which slowly grow a crack over time. To protect against overload failure, the stresses calculated from FEA should be compared to the known mechanical strength of the material, such as the *yield strength* or *ultimate tensile strength*. This is usually of concern in regions where tensile loads are dominant, but *compressive failure* can also occur, e.g., during the formation and shape setting of nitinol devices. For fatigue failures, a *Goodman diagram* or *S–N diagram* analyzing the cyclic stress or strain amplitude and mean stress or strain is often constructed to help assess the fatigue load in the device and compare to known fatigue strength properties. It should be noted that FEA is typically used to calculate the mechanical stresses in a device assuming a defect-free material. Since all materials have some inherent defects present in them as well as other unforeseeable uncertainties in the design, an appropriate *safety factor* should be used when comparing the calculated stresses/strains to experimentally determined ultimate tensile strengths or fatigue strengths. In fatigue failures especially, the cracks that eventually lead to failure tend to initiate at defects in the material or at a location where an abrupt geometry discontinuity acts as a *stress concentration*.

For implantable medical devices, such as cardiovascular stents, fatigue failure is one of the primary issues of concerns, although overload failures can occur, e.g., when a device is crimped down onto a catheter during implantation. Stents and other cardiovascular devices are currently expected to last 15 years *in vivo*, which results in approximately 600 million loading cycles due to the cardiac rhythm. For short-term use devices, such as drug delivery patches, overload of the device is a primary concern. An example is overloading the adhesive that bonds the drug delivery patch to the skin as a person moves. Other handheld devices, such as blood glucose meters, should be designed to be able to withstand expected loads during operation by the user or rapid events, such as being dropped.

3.2 Device Analysis: Overview

The first steps in performing any design assessment from a mechanical perspective are (1) characterization of the raw material properties for all of the components in the device and (2) identification of the potential mechanical loads that may be applied to the device. The primary material properties required for all analyses are the elastic response of the material, which is typically obtained from tensile testing, as well as the strength of the material (either ultimate or fatigue). In addition, the plastic response of the material may be required (if yielding is expected), or the *coefficient of thermal expansion (CTE)*—if the device goes through a large temperature range and there are mating components with different CTEs, or if the device is undergoing a large temperature gradient. It should be noted that all material testing should be performed on a material that has undergone all of the treatment processes that a finished device would undergo, and the testing should be performed at the operating temperature and conditions.

After the necessary material properties of the device have been obtained, the loads applied to the device need to be determined. Although the interaction between a device and a biologic tissue is always difficult to assess, assessment of the mechanical loads is generally easier for medical devices that operate outside of the body. For implanted devices, assessing the mechanical loads applied to the device is much more difficult. For cardiovascular devices, the *in vivo* mechanical environment is frequently assessed by using an imaging technique, such as CT, MRI, or real-time X-ray, to track the deformation of the device. The measured displacements are then applied directly to the device in the model to measure the stresses and strains, which are compared to published data or test results for strength and/or fatigue life properties.

The necessity of requiring experimental imaging to obtain the applied loads inhibits one of the advantages of using FEA early in the design process, namely the ability to quickly evaluate different design modifications in a quantitative manner. Since many implanted devices operate under load-control situations, a change in the device design will result in a change in the device displacements. Therefore, FEA models that directly model the tissue and device–tissue interactions in load control are currently being developed and evaluated. However, for many medical devices, conservative estimates of the mechanical loads and/or device displacements can be derived, which can be applied directly to the device. In addition, FEA can also be used to perform sensitivity studies to determine how sensitive different design parameters are to a particular type of loading.

Once a (conservative) model of the *in vivo* mechanical environment has been created, and the resulting stresses and strains in the device have been analyzed, testing of the device at material, component, and assembly levels is performed. FEA is an integral part of this step in the design process because almost always, a simplified loading scheme is implemented in laboratory testing, as compared to the more complicated *in vivo* loading environment. Therefore, an FEA model of the laboratory test conditions can be made to ensure the stresses and strains developed during the testing condition are of the same magnitude and in the same location as during *in vivo* loading. S–N testing at a material or component level should be performed in the laboratory to be able to estimate the fatigue strength of the material at the expected number of loading cycles. The peak cyclic stresses or strains calculated from the *in vivo* FEA analysis can then be compared with the fatigue strengths determined from fatigue testing to determine the safety factor for the device.

When failure of medical devices occurs, FEA can again be used to provide insight into the mechanical stresses that may have led to the failure. FEA allows the device to be quickly assessed under a range of loading conditions that may have been applied and caused the failure. In addition, FEA again plays a crucial role by being able to calculate stresses for both variations of the *in vivo* condition, as well as under any laboratory test conditions that may be performed.

3.3 Device Analysis: Summary

Below are some of the key aspects of performing FEA for medical devices:

1. Material properties
   (a) Unless the properties of a given material are well established, tensile testing is typically used to obtain the elastic and/or plastic response of the material.
   (b) All testing should be performed at the operating temperature and on a material that has undergone all treatment processes as a finished device.
   (c) Material property input into FEA is commonly in true stress–true strain.
2. Applied loads and boundary conditions
   (a) Conservative estimates of applied loads or displacements should be used and based on experimental measurements.
3. Mesh density
   (a) The mesh density of the finite element model has a direct impact on the accuracy of the results. To accurately characterize the response of a structure, appropriate mesh density has to be used in the structure analyzed, especially at the locations of interest and where a high stress gradient is expected. In addition to the mesh density, the shape of the elements used to construct the structure needs to be well controlled to minimize excessively distorted and warped elements.
   (b) To confirm the final results from the model are not sensitive to the mesh used, a mesh convergence study should be performed where the analysis is iterated with different mesh densities.
   (c) Often reduced integration elements are used in the analysis to help reduce the computational cost. Fully integrated or high order elements are used when the reduced integration elements are not able to capture the desired

accuracy. Carefully planning the mesh density of the model can provide added benefit in reducing computational cost and time without sacrificing accuracy. Engineers should also know the formulations of the different types of elements in an FEA software and what they are suited for, so that they can make the proper selection of the element types, along with the mesh density, to achieve correct analysis results.

4. Tolerance assessment
   (a) Often, analysis iterations need to be performed to assess the effect of the device geometric tolerance and to identify the worst–case stresses and strains in the device design. For example, a finite element model of the device is constructed and analyzed for the device geometry at both the maximum tolerance and minimum tolerance to ensure the capturing of the worst–case condition.
5. Model validation
   (a) An overall model validation should be performed to confirm that the response of the model is similar to the response of an actual device. The model validation is often done by comparing the FEA results of the device or a subdevice to the bench-top test results.

4 Example of Using FEA in Medical Device Design Fatigue Analysis

The following is an example of using FEA for the design analysis of a medical device. This illustration is only intended to show some of the key aspects of using FEA in the medical device design process, and the results used for the illustration have been altered to protect the confidentiality of the design company. Figure 16.1 shows the 3D geometry of the device, which is a *nitinol* wire frame based *inferior vena cava filter (IVCF).*[1] IVFCs are designed to prevent an often fatal condition known as pulmonary embolism. Pulmonary embolism is caused by the migration of a large blood clot arising from deep veins of the legs (a condition known as deep vein thrombosis) to the lungs.

Before the analysis, tensile testing was performed at 37 °C (body temperature) on samples of the nitinol wire used for the IVCF frame. The superelastic material properties of the nitinol wire were obtained from tensile testing for the construction of the material constitutive model used in the FEA, as shown in Fig. 16.2. A finite element model of the IVCF wire frame structure was then created based on the design specifications. The loads applied to the device consist of crimping the wire

[1] Courtesy of Crux Biomedical, Inc.

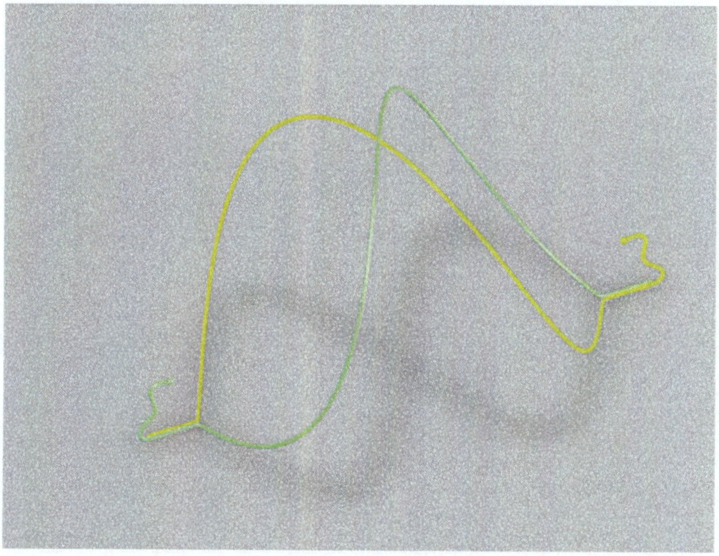

Fig. 16.1 Three-dimensional geometry of the IVCF device wire frame

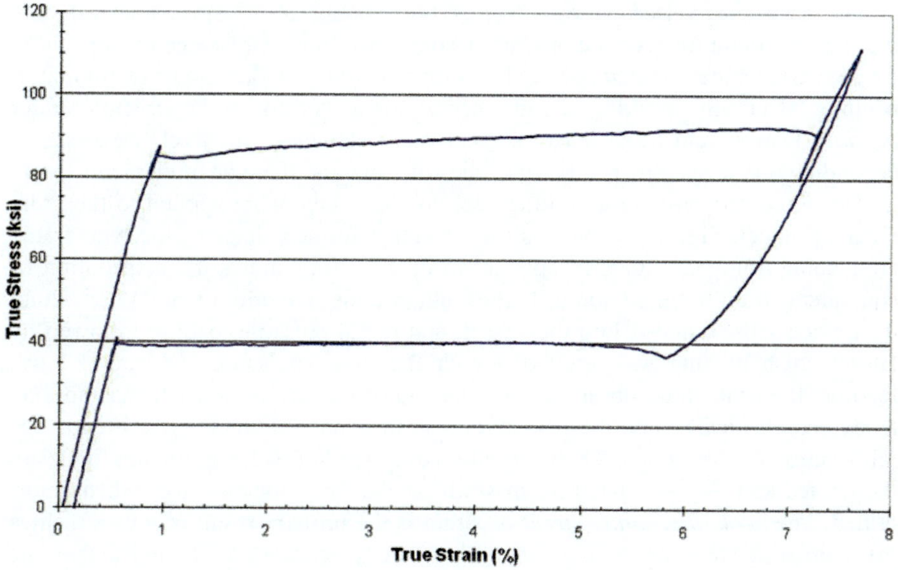

Fig. 16.2 True stress-strain curve from tensile testing of the nitinol wire

frame onto a catheter, self-expansion into a vessel, and finally *in vivo* pulsatile loading including radial pulsatile loading and lateral crush loading. In order to be able to accurately capture the stress state of the device, all of these loading stages were simulated in the analysis.

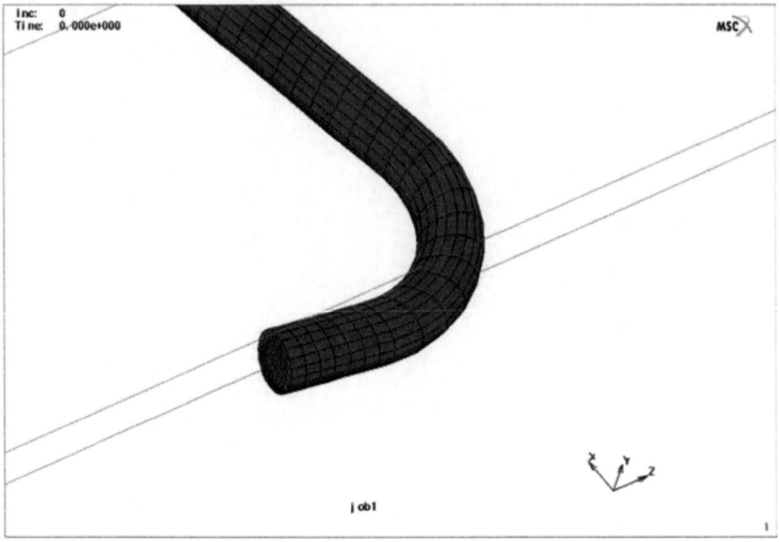

Fig. 16.3 Mesh density of the model

The mesh density used in the analysis is shown in Fig. 16.3. Using a relatively fine mesh is important for accurate predictions of stresses and displacements, especially for devices that are crimped onto catheters and therefore undergo a large amount of bending. Generally speaking, the number of elements through the structure under *bending* load should be at least six, and more elements will likely be required depending on the specific geometry of the structure and the type of elements used.

The simulated loads representing each loading stage were applied to the finite element model. Figure 16.4 shows the overall deformed shape of the wire frame after being crimped onto the catheter, while Fig. 16.5 shows the distribution of equivalent strain in the region of highest strain under the crimp load. After crimp, the device self-expanded into the vessel, and radial pulsatile loading followed by lateral crush loading was applied. Under the radial pulsatile and lateral crush loading, the peak strain observed near the end of the device under the crimp load recovered elastically, and the peak cyclic strain amplitude location shifted to the mid-section of the device. The peak alternating strain (cyclic strain amplitude) in the device and the associated mean strain at the same location were then determined. The *peak alternating stress* or strain is the primary result of interest, since the alternating stress or strain generally has a larger impact on fatigue life than the mean stress or strain. Therefore, the *peak alternating strain* and mean strain calculated in this portion of the analysis represent the highest expected *in vivo* cyclic loads and will be compared to the fatigue strength from testing.

Once the predicted *in vivo* peak strain amplitude and mean strain were determined, experimental S–N testing was performed to determine the safety factor at the required lifetime for the device of 80 million cycles. S–N testing was performed on wire apexes that were designed and fabricated for the testing, and that had

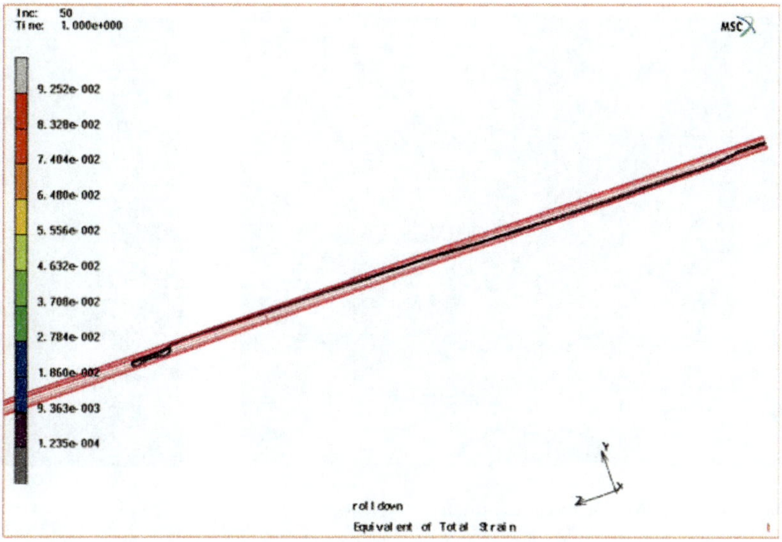

Fig. 16.4 Shape of the device after being crimped on to the catheter

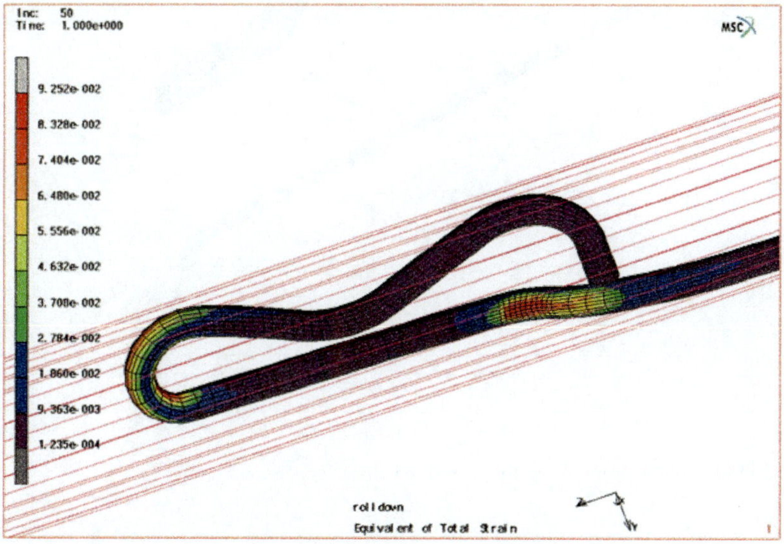

Fig. 16.5 Close-up view of one end of the device showing the strain distributions

undergone the same processing as the finished device wire frame. A finite element model of the apex test specimen was used to estimate the required testing displacements to achieve equivalent target strains for each loading stage the device wire frame would experience (crimping, self-expansion, and radial pulsatile loading and lateral crush loading).

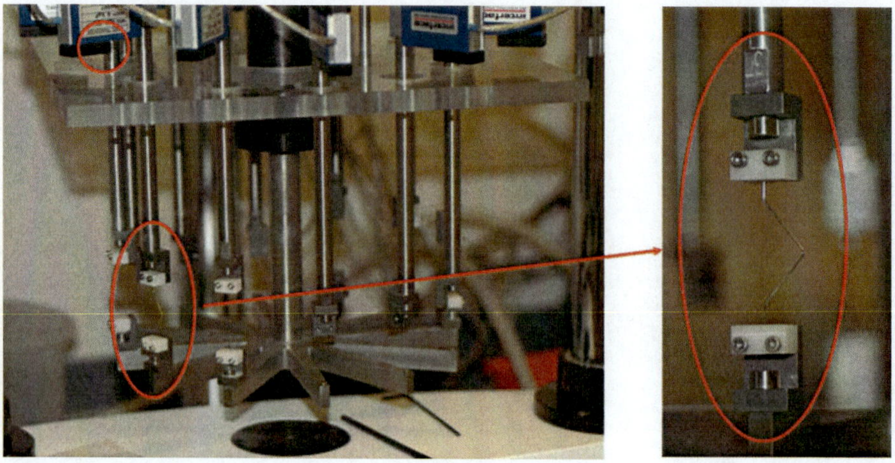

Fig. 16.6 Apex fatigue test specimen mounted in tester

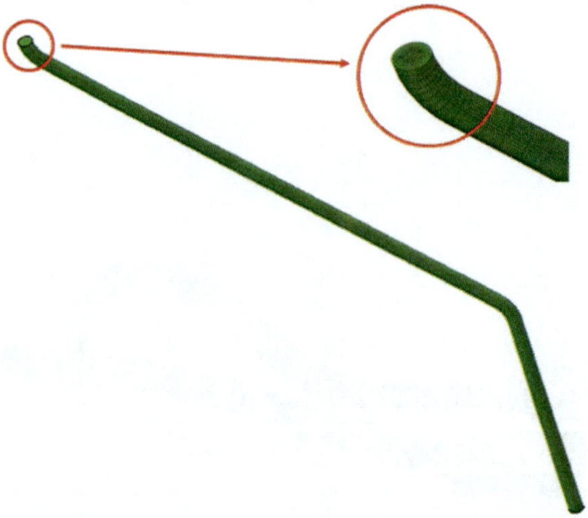

Fig. 16.7 Finite element model of the apex fatigue test specimen

The test conditions were selected so that the apex test specimens were subjected to the same mean strain predicted at the location of the device wire frame where the peak alternating strain occurred. Various strain amplitudes were selected for the S–N testing, to determine the cycles to failure as a function of strain amplitude. Figure 16.6 shows the apex specimen S–N testing setup.

Figure 16.7 shows the finite element model created for the apex test specimen, while Fig. 16.8 shows the maximum principal strain distribution in the apex test specimen model during cycle loading. Figure 16.9 shows the fatigue tested apex

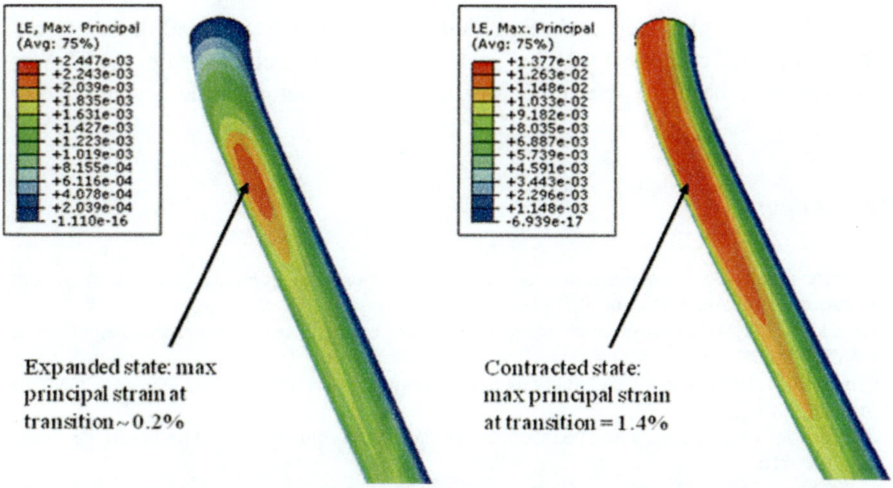

Fig. 16.8 Contour plots of maximum principal strains for one of the test conditions, under the minimum cyclic load (*left*) and maximum cyclic load (*right*)

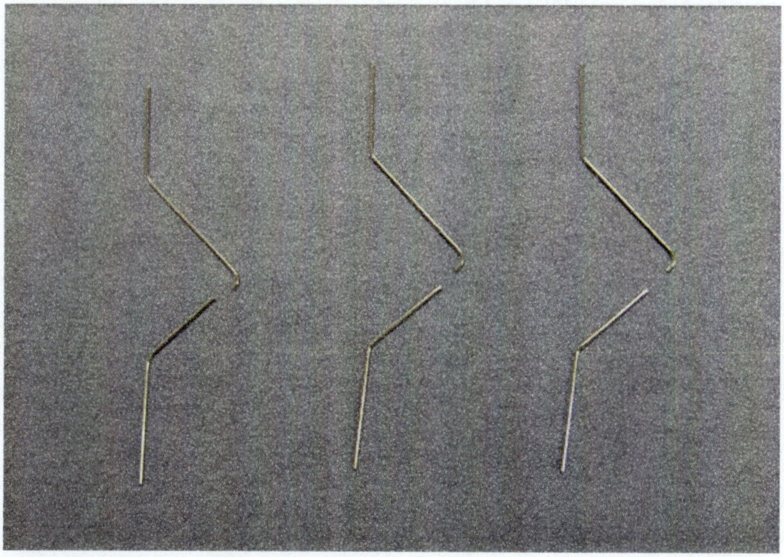

Fig. 16.9 Failed apex specimens from fatigue testing

specimens with the fracture location consistent with the peak strain amplitude location predicted from the apex specimen FEA.

After all laboratory fatigue testing was performed, the resulting cycles to failure were graphed on an S–N plot, and a statistical analysis was performed to conservatively extrapolate the S–N test data to 80 million cycles (the required life span of the device). This projected strain amplitude corresponding to the 80 million cycle lifetime was then compared with the peak cyclic strains predicted from the FEA of the device wire frame to determine the fatigue safety factor for the device.

References

1. Courant R (1943) Variational methods for the solutions of problems of equilibrium and vibrations. Bull Am Math Soc 49:1–23
2. Hrenikoff A (1941) Solution of problems in elasticity by the framework method. J Appl Mech 8:169–175
3. Clough RW (1980) The finite element method after twenty-five years: a personal view. Comput Struct 12:361–370
4. Zienkiewicz OC, Cheung YK (1965) Finite elements in the solution of field problems. Engineer 220:507–510
5. Noor AK (1991) Bibliography of books and monographs on finite element technology. Appl Mech Rev 44(8):307–317
6. Hughes TJR (1987) Finite element method—linear static and dynamic finite element analysis. Prentice-Hall, Englewood Cliffs
7. Zienkiewicz OC, Taylor RL (1989) The finite element method, vol I, Basic formulations and linear problems. McGraw-Hill, London

Biography

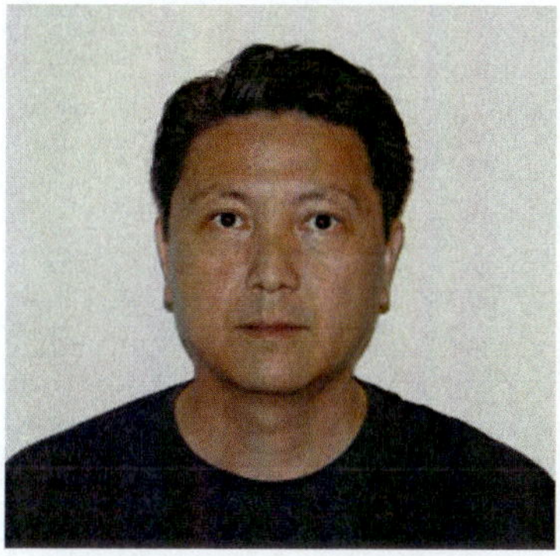

Dr. Ming Wu is a Principal Engineer in Exponent's Mechanical Engineering practice. Dr. Wu's areas of specialization include mechanical system design; experimental, analytical, and numerical stress analyses; mechanical testing; instrumentation and data acquisition; fracture mechanics and mathematical modeling of fatigue processes; macroscopic and microscopic failure analyses; and mechanical metallurgy, microstructure, and compositional analyses.

Dr. Paul Briant is a Managing Engineer in Exponent's Mechanical Engineering practice. He specializes in solid mechanics, finite element analysis, mechanical engineering, biomechanics, and digital image processing.

Chapter 17
Biological Safety Evaluation of Polymers

Bong Joo Park and Jong-Chul Park

Abstract During the last several years, a variety of degradable polymeric materials and medical devices have been developed, and their biological safety evaluation has evolved from screening assays. However, the clinical application of these materials and medical devices is still a critical issue due to the host responses resulting from inflammatory reactions and foreign-body response in their presence or their degraded toxic components in the body. Therefore, every degradable polymeric material and medical device used in biomedical applications needs to be screened for its biological safety.

This chapter focuses on the biological response following implantation of degradable biomaterials and some methods for biological safety evaluation of degradable biomaterials recommended by the International Organization for Standardization (ISO) 10993 standard and US Food and Drug Administration (FDA) agency.

1 Introduction

A variety of biodegradable polymers as biomaterials and medical devices that include cardiovascular, orthopaedic, and soft tissue implants have been widely used in short-term and long-term applications, because it is generally believed

B.J. Park
Yonsei Medical Technology and Quality Evaluation Center, Yonsei University Health System, Seoul, South Korea

J.-C. Park (✉)
Yonsei Medical Technology and Quality Evaluation Center, Yonsei University Health System, Seoul, South Korea

Department of Medical Engineering, Yonsei University College of Medicine, 134 Shinchon-dong, Seodaemun-ku, Seoul 120-752, South Korea
e-mail: parkjc@yuhs.ac

N. Eliaz (ed.), *Degradation of Implant Materials*, DOI 10.1007/978-1-4614-3942-4_17,

Table 17.1 ISO 10993 and FDA biological response tests

Cytotoxicity
Sensitization
Irritation
Intracutaneous reactivity
System toxicity (acute toxicity)
System toxicity (subacute toxicity)
Genotoxicity
Implantation
Hemocompatibility
Chronic toxicity
Carcinogenicity
Reproductive and developmental toxicity
Biodegradation

that biodegradable polymers have good biocompatibility. However, biodegradable implants made of them often trigger fibrotic and inflammatory responses in the body by leaching of toxic contaminants from the degradable implanted polymeric materials, such as emulsifiers, residual monomers, and many other types of additives. These responses are critical problems to the development of the biodegradable polymeric materials, because the final goal for the development of new polymeric biomaterials is to serve human health by their clinical use as commercially available medical devices. Therefore, the evaluation of the biological response to the materials should follow procedures that allow for objective evaluation of the medical devices safety and biocompatibility [1, 2].

The most important requirement for a biodegradable polymer is its biocompatibility, not only in terms of physical and chemical properties, but also in those that define their behaviour at the time they contact the body [3]. Nevertheless, the biocompatibility of implantable biomaterials remains a critical issue in limiting material longevity and functionality, and the foreign body reaction (FBR) to implantable biomaterials also presents a significant risk to patients [4].

Various natural, synthetic, and semi-synthetic materials are currently utilized in the fabrication of implantable materials [5–26]. A number of studies have shown that biomaterials and medical devices made from these materials have biocompatibility [27]. In addition, it has been shown that biological reactions that are adverse for a material in an application may not be adverse for the same material in a different application. Similarly, the material found to be safe may not be safe in other application [28]. Therefore, every biomaterial that targets to be used in medical applications needs to be screened for its biological safety.

This chapter focuses on the biological response following implantation of degradable polymeric materials and some methods for biological safety evaluation of degradable biomaterials. Recommended methods by ISO 10993 and the FDA are shown in Table 17.1 [29–32], and summarized afterwards.

2 Biological Response to Implantable Biomaterials

2.1 Inflammatory Response Following Material Implantation

Most biodegradable polymers, such as poly(glycolic acid) and poly(lactic acid), are composed of biological residues normally present in the human body, but some implants made of these biodegradable polymers often trigger acute or chronic inflammatory responses [33–41]. The chronic inflammatory responses to the degradable implants have been associated with the pathogenesis of fibrosis [38], amyloidosis [42, 43], and carcinogenesis [44, 45]. The mechanisms causing adverse tissue responses mediated by the degradable polymeric biomaterials are not fully understood, and the insufficiency of such knowledge hinders in the development of biomaterials with improved bio- and tissue compatibility [46].

Generally, host responses following implantation of degradable polymeric biomaterials include injury, blood–material interactions, provisional matrix formation, acute inflammation, chronic inflammation, granulation tissue development, FBR, and fibrosis/fibrous capsule development [47–51].

Inflammation is generally defined as the reaction of vascularized living tissue to local injury [52, 53]. The inflammatory response is caused by the tissue injury that results from the implantation of biomaterial as well as the continual presence of biomaterial in the body. When a tissue is injured by implantation, a wound healing response is initiated through a series of complex events. The main stages in this process include acute inflammation, chronic inflammation, and the formation of granulomatous tissue [4, 47, 54].

The acute inflammation lasts from hours to days and is marked by fluid and protein exudation as well as a neutrophilic reaction. This phase is mostly responsible for the provisional matrix formation and cleaning of the wound site. Vessels dilate and excess blood flows into the injury site [55–59]. Numerous blood and tissue proteins such as cytokines and growth factors are released, and leukocytes adhere to the endothelium of the blood vessels and infiltrate the injury site. Monocytes are then called into the site, and these differentiate into macrophages that are involved in immune response and inflammation [60, 61]. The acute inflammatory response with biomaterials usually resolves in less than 1 week, depending on the extent of the injury at the implant site [51].

Chronic inflammation induced by the continual presence or degradable components of the biomaterial and medical device in the body is histologically less uniform compared to acute inflammation, and the wound healing response is generally dependent on the size and/or degree of injury [4]. This phase is generally characterized by the presence of monocytes, macrophages, and lymphocytes, as well as the proliferation of blood vessels and connective tissue to restructure the affected area [29, 57–63].

With biocompatible materials, early resolution of the acute and chronic inflammatory responses occurs with the chronic inflammatory response composed of mononuclear cells, usually lasting no longer than 2 weeks. Following the acute and chronic inflammatory responses, granulation tissue identified by the presence of macrophages, the infiltration of fibroblasts, and neovascularization in the new healing tissue is identified. The granulation tissue is the precursor to fibrous capsule formation, and it is separated from the implant or biomaterial by the cellular components of the FBR [51].

2.2 *Foreign-Body Response*

When the degradable polymeric material reaches the stages of its degradation process, the materials elicit a local FBR [51, 64]. The FBR is composed of foreign-body giant cells and the components of granulation tissue, which consist of macrophages, fibroblasts, and capillaries in varying amounts, depending on the form and topography of the implanted material [47].

According to several articles, the degree of this reaction depends on the properties of the material, such as shape, size, surface chemistry, roughness, morphology and porosity, composition, sterility, contact duration, and degradation [64–68]. Following implantation, an interface between biomaterial and blood is immediately created, and nonspecific absorption of blood and tissue fluid proteins onto the surface of the biomaterial is induced [69]. In fact, the extent of nonspecific protein absorption can be used to evaluate the degree of biocompatibility of the biomaterial. Following nonspecific protein absorption, immune and inflammatory cells, such as monocytes, leukocytes and platelets, intervene in order to protect the body from the foreign object. The end stage of the FBR involves walling off the biomaterial by a vascular, collagenous fibrous capsule that is typically 50–200 nm thick. This fibrous wall confines the implanted biomaterial and prevents it from interacting with the surrounding tissues [4, 29].

3 Biological Safety Evaluation Methods

Biological safety of degradable biomaterials and medical devices is evaluated through biological testing and clinical investigation by several national and international standards. Among these standards [70–81], ISO 10993 has recently expedited the harmonization of the testing and provides a series of standard methods for evaluating the biocompatibility of a biomaterial and medical device prior to clinical testing [74–81]. The choices of biological safety tests shall be selected according to the recommended methods by ISO 10993 and the FDA. Initial evaluation tests, such as cytotoxicity, sensitization, and irritation or intracutaneous reactivity, are requisite, while the supplementary tests, such as chronic toxicity and carcinogenicity, may be considered logically necessary in some cases, but there are no standard methods for these yet [31, 71, 73].

3.1 *Cytotoxicity Test*

The main goal of cytotoxicity testing is to assess the biocompatibility of polymeric materials and elastomer portions of biomaterials and medical devices using mammalian cell culture system. Cytotoxicity test is useful in evaluating the toxicity of biomaterials and medical device components; it provides a good way to screen biomaterials prior to *in vivo* tests. All biomaterials and medical devices, including degradable and non-degradable materials, are required to be evaluated for cytotoxicity. Unlike the other studies utilized in biological safety testing, cytotoxicity is not a pass/fail test [71, 73].

For cytotoxicity test, three cell culture assays are commonly used for biomaterials and medical devices; the three main types of assays are the extract dilution, agar diffusion, and direct contact assays. These assays are morphological assays, in that the outcome is measured by observations of changes in cell morphology, and measure the initial reaction of cells to biomaterials and its components. At least one type of cytotoxicity test should be performed on each device or material [70–73]. Basic methodologies, as described in the ISO 10993-5, are described in the following paragraph [76].

3.1.1 Extract Dilution Assay

The extract dilution assay uses several extracting media and extraction conditions to test materials with actual use conditions or to exaggerate those conditions. ISO standard methods for preparing samples are contained in Part 12 of ISO 10993, *Sample preparation and reference materials* [80]. Extracts can be titrated to yield a semi-quantitative measurement of cytotoxicity. After sample extraction, the extract is placed on a layer of L-929 fibroblast cells and incubated for 48 h at 37 ± 1°C. Following incubation, live and dead cells are examined using microscopical and histochemical or vital stains, such as neutral red, for malformation, degeneration, and lysis of the cells [30, 71, 73, 76].

3.1.2 Agar Diffusion Assay

This assay is recommended for high-density materials, such as elastomeric closures. In this method, a thin layer with a culture medium containing 2 % agar is placed over the cultured L-929 fibroblast cells. After the agar has solidified, test material (or an extract of the test material dried on filter paper) and specimens of negative and positive controls are placed on the surface of the agar layer, and the cells are incubated for at least 24 h at 37 ± 1°C. Cytotoxicity is evaluated by a zone of malformed and lysed cells, or by the loss of the vital stain under and around the periphery of the specimens [30, 71, 73, 76].

3.1.3 Direct Contact Assay

The procedure of this assay is appropriate for low-density materials, such as soft contact lens polymers, hydroxyethyl methacrylate (HEMA). In this method, a piece of test material and specimen of negative and positive controls are carefully placed onto the surface of the cell layer. The cells are then incubated for 24 h at 37 ± 1°C. After incubation, the culture medium and the test materials are removed, and the cells are fixed and stained with a cytochemical stain such as hematoxylin blue. Cytotoxicity is indicated by malformation, degeneration, and lysis of cells with leachable chemicals under and around the periphery of the test material [30, 71, 73, 76].

3.2 Sensitization Test

Exposure to even minute amounts of potential leachable chemicals in biomaterials and medical devices can result in allergic or sensitization reactions [1, 30]. Sensitization test is done to evaluate whether a material contains hazardous and leachable chemicals causing adverse local or systemic effects after repeated or prolonged exposure [30, 73]. Sensitization is the result of immunologically medicated reactions resulting in erythema and edema. Because most such reactions are of the dermal cell-mediated type, rather than the humoral or antigen type, the skin of animals (usually guinea pigs, because they exhibit dermal sensitivity similar to humans) is used in this testing [2–4, 70, 72]. Most biomaterials and medical devices require evaluation by a sensitization test. There are three assays commonly used to evaluate biomaterials and medical devices for sensitization test: the guinea pig maximization test (GPMT), the closed-patch test, and the murine local lymph node assay (LLNA).

The GPMT and closed-patch assays for evaluating the delayed contact sensitization potential of extracts of biomaterials and medical devices are contained in Part 10 of ISO 10993, *Tests for irritation and sensitization*. The LLNA accepted for testing single chemicals as a stand-alone alternative to the GPMT. All classes of biomaterials and medical devices require assessment by a delayed contact sensitization assay with guinea pigs [71, 78].

GPMT is the most sensitive method; it is preferred for single chemicals, and is recommended for biomaterials and medical devices that will have externally communicating or internal contact with the body or body fluids. In GPMT, the complete Freund's adjuvant (CFA) to enhance the skin sensitization response is used with the test material [70–73, 78]. There are three phases in GPMT: intradermal induction, topical induction, and challenge phases. The procedure is described in the following paragraph.

In the intradermal induction phase, pairs of injections of the test material in its solvent, CFA mixed with the solvent, and CFA mixed with the test material are injected to the skin of ten guinea pigs. Then, 1 week after exposure, the test materials are reapplied for topical induction phase. Two weeks later, all guinea pigs are challenged with a topical application of the test material. Control animals are injected with the solvent alone for the induction and challenge phases. The degree of erythema and edema is rated at 24, 48, and 72 h after the challenge dose. The results are considered positive if more animals in the test group have reaction scores of 1 or greater than in the control [70–72, 78].

3.3 *Irritation Test*

Irritation test evaluates a localized irritation potential causing a generally localized inflammatory response to single, repeated, or continuous exposure to a biomaterial and medical device without the involvement of an immunological mechanism [70–73]. Methods for evaluating the irritation potential of biomaterials and medical devices are contained in Part 10 of ISO 10993, *Tests for irritation and sensitization*. Irritation tests also require to evaluate most biomaterials and medical devices, and the methods in Part 10 are intracutaneous (intradermal), skin, and ocular irritation assays [78].

The intracutaneous irritation assay with rabbit evaluates biomaterials and medical devices for possible contact irritation. This assay is similar to the sensitization test. For this assay, extracts of the test material are injected intradermally. The injection sites are scored for erythema and edema. This procedure is recommended for devices that will have externally communicating or internal contact with the body or body fluids. It reliably detects the potential for local irritation due to the chemicals which may be extracted from a biomaterial [70–73].

The skin irritation test with rabbit recommended for topical devices that have external contact with intact or breached skin. The test materials or extracts are directly placed to intact and shaved sites on the back skin of a rabbit. After a 24-h exposure, the test material is removed, and the sites are scored for erythema and edema [70–73].

The ocular irritation assay in rabbits describes a single or repeated application of the test sample, but is significantly different from the other local tissue irritation assays. This assay is performed only when the device is intended to be put into the eye. For this assay, a small sample of devices is placed directly into the lower eyelid of a rabbit. After 72 h, the eye is evaluated for redness and swelling of the conjunctiva, cornea and iris, conjunctivia, and presence of discharge. The ISO standard for ocular irritation does not require calculation of a Draize eye irritation score [70–72, 78].

3.4 *Systemic and Subchronic Toxicity*

Systemic toxicity test evaluates the potential harmful effects on target and organs with single or multiple exposure to biomaterials and medical devices [30, 71, 72]. Systemic effects are categorized on the basis of time to initiate adverse effects: acute (within 24 h), subacute (in 24 h to 28 days), subchronic (in 90 days, but not exceeding 10% of an animal's lifespan), and chronic (longer than 10% of an animal's lifespan) [72]. ISO standard methods for evaluating the systemic toxicity are contained in Part 11 of ISO 10993, *Tests for systemic toxicity* [79]. This standard contains methods for administration of a single-dose or multiple-doses by the oral, dermal, inhalation, intravenous, and intraperitoneal routes of administration. In systemic toxicity tests, single-dose toxicity tests are recommended for all biomaterials or medical devices that contact directly or indirectly the blood path. Multiple-doses toxicity tests are also required for any other materials or devices that contact internal tissues, such as breached or compromised surfaces, mucosal membranes, and blood path for prolonged or permanent. Following the injection of single dose or multiple doses, the mice are observed for toxic signs, just after injection, and at four other time points [71–73, 79].

For assessing biocompatibility, rabbit pyrogenicity test is preferred and is also included in the systemic toxicity category to evaluate material-mediated fever-causing reactions to the extracts of biomaterials or medical devices. Rabbits are intravenously injected with a saline extract of the test material, and the temperatures of rectal are measured periodically. Although the rabbit pyrogenicity test has been the standard, the Limulus amebocyte lysate (LAL) reagent test has been used increasingly in the recent years [30, 71–73, 79].

3.5 Genotoxicity

Genotoxicity includes all potential means with serious consequences in genetic damage. Genotoxicity test applies mammalian or non-mammalian cells to evaluate whether gene mutations, chromosomal aberrations, or other DNA or gene changes are caused by the test materials. In genotoxicity tests, a set of *in vitro* and *in vivo* assays should be used for assessing substances or materials that can directly or indirectly induce genetic damage through a variety of mechanisms [71–73].

Part 3 of ISO 10993, *Tests for genotoxicity, carcinogenicity, and reproductive toxicity*, requires a series of three genotoxicity tests. The major genotoxic effects are: DNA effects, gene mutations, and chromosomal aberrations [74]. Three assays, the Ames test for bacterial reverse mutations using *Salmonella typhimurium*, a chromosome aberration assay in mammalian cells, and sister chromatid exchange in mammalian cells, are commonly used for *in vitro* genotoxicity tests. Among the assays, the Ames test is only used for biomaterials or medical devices with less critical body contact. *In vivo* genotoxicity assays, including the micronucleus test, bone marrow cytogenic test, or dominant lethal test using rodent, may be suitable for selected biomaterials and medical devices [71–74].

3.6 Implantation Test

ISO 10993-6, *Tests for local effects after implantation*, describes test methods for assessing the local effects after implantation of biomaterials and medical devices intended for use in medical applications [77].

Biomaterials or medical devices involving permanent or prolonged contact with bone, tissues, and blood require assessment by implantation. Implantation tests are also used to assess the biological safety of degradable polymeric materials that directly contact the living tissue. These tests can evaluate if the implantable materials have absorbable or nonabsorbable properties using clinical trials for either short-term or long-term periods, and can assess the dynamics of biochemical exchange and cellular and immunologic responses in implantation sites using histopathology [71–73, 77].

3.7 Hemocompatibility

Hematocompatibility is the most complex concerns of the standard safety for devices to be evaluated. The tests for hematocompatibility evaluate the effects of blood-contacting materials, such as blood transfusion sets, haemodialysis sets, intravenous catheters, and vascular prostheses, on blood coagulation, hemolysis, and thrombus formation. The blood-contacting materials must be evaluated for biocompatibility to establish their safety. In practice, most of the biomaterials and medical devices can be either haemolysis of the blood cells or thrombogenicity [71–73, 75]. Part 4 of ISO 10993, *Selection of tests for interactions with blood*, describes five test categories for evaluating blood–material interactions (between biomaterials or medical devices and the blood path or blood). The testing categories are thrombosis, coagulation, platelets, haematology, and complement system [75].

The haemolysis assay is designed to evaluate the haemolytic properties of all biomaterials or medical devices, especially those intended for use in contact with blood, excluding some materials which contact only intact skin or mucous membranes. This assay assesses the acute haemolytic properties of test materials by detecting the damaged red blood cells when they are brought in contact with test materials or their extracts. Blood coagulation is measured by the prothrombin time assay and the partial thromboplastin time assay that detect coagulation abnormalities in the extrinsic and intrinsic pathways. Platelet testing includes quantification of platelet numbers as well as analysis of their structure and function. The testing can include analysis of platelet factors, or components on the platelet surface which are released from platelets or adherent to the surface of the device. Testing for thrombogenicity is normally done by detecting aggregated platelet, fibrin, and other cellular elements, and by observing thrombus formation. But, these tests are usually difficult, controversial, and expensive. To decrease unwanted tissue damages by inappropriate or excess complement activation in circulating blood, complement activation testing is recommended for implantable biomaterials or medical devices that contact the circulatory blood. This test detects complement activation in plasma as a result of exposure of the plasma to the test materials or their extracts. The measure of complement activation indicates whether a test article involves a complement-induced inflammatory immune response in humans [30, 71–73, 75].

3.8 Carcinogenicity

Most degradable biomaterials, medical devices, or their components, including harmful chemicals, have tumorigenic potential in the body over the total lifespan of the test animal. Carcinogenicity tests are therefore needed and are used to determine the tumorigenic potential of biomaterials, medical devices, or their extracts after single or multiple exposures over the total lifespan of the test animal (e.g. 2 years for rat, 18 months for mouse). But, these tests are infrequently required to apply the biomaterials or medical devices with a permanent or cumulative contact for 30 days or longer, because the testing is expensive, highly problematic, and controversial [71–74]. According to ISO 10993-3, *Tests for genotoxicity, carcinogenicity and reproductive toxicity*, situations suggesting the need for carcinogenicity tests may include resorbable materials, unless there are significant and adequate data on human use. Biomaterials or medical devices having positive results in genotoxicity on mammalian cells would be required to be tested for carcinogenicity before clinical testing [74].

3.9 Reproductive and Developmental Toxicity

Reproductive toxicity tests evaluate the potential effects of biomaterials, medical devices, or their extracts on reproductive function, teratogenicity, and prenatal and early postnatal development. They are often required for biomaterials or medical devices with permanent contact with internal tissues and organs [71–73]. Standard test methods for reproductive toxicity are found in ISO 10993-3, *Tests for genotoxicity, carcinogenicity and reproductive toxicity* [74].

3.10 Biodegradation

Biodegradation tests evaluate the effects of biodegradable biomaterials and medical devices and their degradable products on the body in the biological environment. Most biodegradable polymeric biomaterials or medical devices can degrade, crumble, become rubbery, or become rigid with time, and the released products from biomaterials may be toxic to some tissues in the body. A variety of these products from materials are identified and quantified by appropriate methods, which are performed on the extractable chemicals and elements to establish the safety of biomaterials or medical devices [70–73]. ISO standard methods for identification and quantification of degradation products, such as additives, corrosion products, impurities, monomers, etc., from polymeric materials are contained in Part 13 of ISO 10993, *Identification and quantification of degradation products from polymeric medical devices* [81].

4 Conclusions

The number of degradable polymeric materials has increased enormously over the past decade, and the biological safety evaluations of these materials have also become a routine in most laboratories working on the development of degradable biomaterials for biomedical applications. Some evaluation methods and regulations by FDA and ISO are now routinely established; they can be performed at any process for developing the materials. However, standardization of test methods for screening biodegradable new polymeric materials is still a difficulty at present, because they have some problems and several phenomena, such as water uptake and degradation with leaching of compounds to the medium when testing the materials.

Therefore, new biodegradable polymeric materials that aim to be used in biomedical applications need to be screened for their biological safety by means of a combination of evaluation methods that provide different data, thus allowing reliable conclusions. Moreover, collecting data through several types of test methods, such as morphological, biochemical, and biological levels, will allow for biocompatibility determination of the biodegradable system before further *in vivo* tests are carried out.

References

1. Pizzoferrato A, Vespucci A, Ciapetti G et al (1985) Biocompatibility testing of prosthetic implant materials by cell-cultures. Biomaterials 6:346–351
2. Williams D (1991) Objectivity in the evaluation of biological safety of medical devices and biomaterials. Med Device Technol 2:44–48
3. Silva GA, Marques AP, Gomes ME et al (2005) Cytotoxicity screening of biodegradable polymeric systems. In: Reis RL (ed) Biodegradable systems in tissue engineering and regenerative medicine. CRC Press, New York
4. Onuki Y, Bhardwaj U, Papadimitrakopoulos F et al (2008) A review of the biocompatibility of implantable devices: current challengs to overcome foreign body response. J Diabetes Sci Technol 2:1003–1015
5. Sano A, Hojo T, Maeda M et al (1998) Protein release from collagen matrices. Adv Drug Deliv Rev 31:247–266
6. Geiger M, Li RH, Friess W (2003) Collagen sponges for bone regeneration with rhBMP-2. Adv Drug Deliv Rev 55:1613–1629
7. Uchegbu IF, Schätzlein AG, Tetley L et al (1998) Polymeric chitosan-based vesicles for drug delivery. J Pharm Pharmacol 50:453–458
8. Khor E, Lim LY (2003) Implantable applications of chitin and chitosan. Biomaterials 24:2339–2349
9. Borchard G, Junginger HE (2001) Modern drug delivery applications of chitosan. Adv Drug Deliv Rev 52:103
10. de Vos P, Hoogmoed CG, Busscher HJ (2002) Chemistry and biocompatibility of alginate-PLL capsules for immunoprotection of mammalian cells. J Biomed Mater Res 60:252–259
11. Vercruysse KP, Prestwich GD (1998) Hyaluronate derivatives in drug delivery. Crit Rev Ther Drug Carrier Syst 15:513–555

12. Cadée JA, Brouwer LA, den Otter W et al (2001) A comparative biocompatibility study of microspheres based on crosslinked dextran or poly(lactic-co-glycolic)acid after subcutaneous injection in rats. J Biomed Mater Res 56:600–609
13. Draye JP, Delaey B, Van de Voorde A et al (1998) *In vitro* and *in vivo* biocompatibility of dextran dialdehyde cross-linked gelatin hydrogel films. Biomaterials 19:1677–1687
14. Shive MS, Anderson JM (1997) Biodegradation and biocompatibility of PLA and PLGA microspheres. Adv Drug Deliv Rev 28:5–24
15. Athanasiou KA, Niederauer GG, Agrawal CM (1996) Sterilization, toxicity, biocompatibility and clinical applications of polylactic acid/polyglycolic acid copolymers. Biomaterials 17:93–102
16. Daugherty AL, Cleland JL, Duenas EM et al (1997) Pharmacological modulation of the tissue response to implanted polylactic-co-glycolic acid microspheres. Eur J Pharm Biopharm 44:89–102
17. Lunsford L, McKeever U, Eckstein V et al (2000) Tissue distribution and persistence in mice of plasmid DNA encapsulated in a PLGA-based microsphere delivery vehicle. J Drug Target 8:39–50
18. Ronneberger B, Kissel T, Anderson JM (1997) Biocompatibility of ABA triblock copolymer microparticles consisting of poly(L-lactic-co-glycolic-acid) A-blocks attached to central poly (oxyethylene) B-blocks in rats after intramuscular injection. Eur J Pharm Biopharm 43:19–28
19. Ronneberger B, Kao WJ, Anderson JM et al (1996) *In vivo* biocompatibility study of ABA triblock copolymers consisting of poly(L-lactic-co-glycolic acid) A blocks attached to central poly(oxyethylene) B blocks. J Biomed Mater Res 30:31–40
20. Royals MA, Fujita SM, Yewey GL et al (1999) Biocompatibility of a biodegradable in situ forming implant system in rhesus monkeys. J Biomed Mater Res 45:231–239
21. Dalsin JL, Hu BH, Lee BP et al (2003) Mussel adhesive protein mimetic polymers for the preparation of nonfouling surfaces. J Am Chem Soc 125:4253–4258
22. Tsai W-B, Grunkemeier JM, McFarland CD et al (2002) Platelet adhesion to polystyrene-based surfaces preadsorbed with plasmas selectively depleted in fibrinogen, fibronectin, vitronectin, or von Willebrand's factor. J Biomed Mater Res 60:348–359
23. Shen M, Horbett TA (2001) The effects of surface chemistry and adsorbed proteins on monocyte/macrophage adhesion to chemically modified polystyrene surfaces. J Biomed Mater Res 57:336–345
24. Paradossi G, Cavalieri F, Chiessi E et al (2003) Poly(vinyl alcohol) as versatile biomaterial for potential biomedical applications. J Mater Sci Mater Med 14:687–691
25. Oka M, Ushio K, Kumar P et al (2000) Development of artificial articular cartilage. Proc Inst Mech Eng H 214:59–68
26. Maruoka S, Matsuura T, Kawasaki K et al (2006) Biocompatibility of polyvinylalcohol gel as a vitreous substitute. Curr Eye Res 31:599–606
27. Mendes SC, Reis RL, Bovell YP et al (2001) Biocompatibility testing of novel starch-based materials with potential application in orthopaedic surgery: a preliminary study. Biomaterials 22:2057–2064
28. Anderson JM, Langone JJ (1999) Issues and perspectives on the biocompatibility and immunotoxicity evaluation of implanted controlled release systems. J Control Release 57:107–113
29. Morais JM, Papadimitrakopoulos F, Burgess DJ (2010) Biomaterials/tissue interactions: possible solutions to overcome foreign body response. AAPS J 12:188–196
30. Ratner BD, Bianco RW, Grehan JF et al (2004) Biological testing of biomaterials. In: Ratner BD, Hoffman AS, Schoen FJ et al (eds) Biomaterials science: an introduction to materials in medicine, 2nd edn. Elsevier Academic Press, San Diego
31. International Standard ISO 10993 (2003) Biological evaluation of medical devices-parts 1: evaluation and testing. International Organization for Standardization, Geneva, Switzerland

32. Abramson S, Alexander H, Best S et al (2004) Classes of materials used in medicine. In: Ratner BD, Hoffman AS, Schoen FJ et al (eds) Biomaterials science: an introduction to materials in medicine, 2nd edn. Elsevier Academic Press, San Diego
33. Holland SJ, Tighe BJ, Gould PL (1986) Polymers for biodegradable medical devices. Part 1. The potential of polyesters as controlled macromolecular release systems. J Controlled Release 4:155–180
34. Leenslag JW, Pennings AJ, Bos RR et al (1987) Resorbable materials of poly(L-lactide). VI. Plates and screws for internal fracture fixation. Biomaterials 8:70–73
35. Kulkarni RK, Moore EG, Hegyeli AF et al (1971) Biodegradable poly(lactic acid) polymers. J Biomed Mater Res 5:169–181
36. Lam KH, Schakenraad JM, Esselbrugge H et al (1993) The effect of phagocytosis of poly(L-lactic acid) fragments on cellular morphology and viability. J Biomed Mater Res 27:1569–1577
37. Lam KH, Schakenraad JM, Groen H et al (1995) The influence of surface morphology and wettability on the inflammatory response against poly (L-lactic acid): A semi-quantitative study with monoclonal antibodies. J Biomed Mater Res 29:929–942
38. Schakenraad JM, Oosterbaan JA, Nieuwenhuis P et al (1988) Biodegradable hollow fibers for the controlled release of drugs. Biomaterials 9:116–120
39. Schakenraad JM, Nieuwenhuis P, Molenaar I et al (1989) *In vivo* and *in vitro* degradation of glycine/DL-lactic acid copolymers. J Biomed Mater Res 23:1271–1288
40. Bos RR, Rozema FB, Boering G et al (1991) Degradation of and tissue responses to biodegradable poly(L-lactide) for use as internal fixation of fractures: a study in rats. Biomaterials 12:32–36
41. Bergsma EJ, Rozema FR, Bos RR et al (1993) Foreign-body reactions to resorbable poly(L-lactide) bone plates and screws used for the fixation of unstable zygomatic fractures. J Oral Maxillofac Surg 51:666–670
42. Picken MM, Gallo GR, Frangione B (1990) Amyloid enhancing factor and inflammatory reaction. Lab Invest 63:586–587
43. Amyloidosis KR (1999) In: Rubin E, Farber JL (eds) Pathology. Philadelphia, Lippincott
44. Nakamura T, Shimizu Y, Okumura N et al (1994) Tumorigenicity of poly-L-lactide (PLLA) plates compared with medical grade polyethylene. J Biomed Mater Res 28:17–25
45. Weitzman SA, Gordon LI (1990) Inflammation and cancer: Role of phagocyte-generated oxidants in carcinogenesis. Blood 76:655–663
46. Jiang WW, Su SH, Eberhart RC et al (2007) Phagocyte responses to degradable polymers. J Biomed Mater Res 82:492–497
47. Anderson JM (2001) Biological responses to materials. Annu Rev Mater Res 31:81–110
48. Anderson JM (2000) Multinucleated giant cells. Curr Opin Hematol 7:40–47
49. Gretzer C, Emanuelsson L, Liljensten E et al (2006) The inflammatory cell influx and cytokines changes during transition from acute inflammation to fibrous repair around implanted materials. J Biomater Sci Polym Ed 17:669–687
50. Luttikhuizen DT, Harmsen MC, Van Luyn MJ (2006) Cellular and molecular dynamics in the foreign-body reaction. Tissue Eng 12:1955–1970
51. Anderson JM, Rodriguez A, Chang DT (2008) Foreign-body reaction to biomaterials. Semin Immunol 20:86–100
52. Cotran RS, Kumar V, Collins T (1999) Robbins Pathologic Basis of Disease, 6th edn. W.B. Saunders, Philadelphia
53. Synderman R, Gallin JI, Goldstein IM (1992) Inflammation: Basic Principles and Clinical Correlates, 2nd edn. Raven, New York
54. Ratner BD, Bryant SJ (2004) Biomaterials: where we have been and where we are going. Annu Rev Biomed Eng 6:41–75
55. Anderson JM (1988) Inflammatory response to implants. ASAIO Trans 34:101–107
56. Jutila MA (1992) Leukocyte traffic to sites of inflammation. APMIS 100:191–201

57. Pober JS, Cotran RS (1990) The role of endothelial cells in inflammation. Transplantation 50:537–544
58. Williams GT, Williams WJ (1983) Granulomatous inflammation-a review. J Clin Pathol 36:723–733
59. Wahl SM, Wong H, McCartney-Francis N (1989) Role of growth factors in inflammation and repair. J Cell Biochem 40:193–199
60. Johnston RB (1988) Current concepts: immunology, monocytes and macrophages. N Engl J Med 318:747–752
61. Collins T (2002) Acute and chronic inflammation. In: Cotran RS, Kumar V, Collins T (eds) Robbins Pathologic Basis of Disease, 6th edn. W.B.Saunders, Philadelphia
62. Kovacs EJ (1991) Fibrogenic cytokines: the role of immune mediators in the development of scar tissue. Immunol Today 12:17–23
63. Xia Z, Triffitt JT (2006) A review on macrophage response to biomaterials. Biomed Mater 1:R1–R9
64. Fournier E, Passirani C, Montero-Menei CN et al (2003) Biocompatibility of implantable synthetic polymeric drug carriers: focus on brain biocompatibility. Biomaterials 24:3311–3331
65. Laurencin CT, Elgendy H (1994) The biocompatibility and toxicity of degradable polymeric materials: implication for drug delivery. In: Domb A, Maniar M (eds) Site specific drug delivery. Wiley, New York
66. Sieminski AL, Gooch KJ (2000) Biomaterial-microvasculature interactions. Biomaterials 21:2232–2241
67. Rihova B (2000) Immunocompatibility and biocompatibility of cell delivery systems. Adv Drug Deliv Rev 42:65–80
68. Babensee J, Anderson JM, Mclntire LV et al (1998) Host response to tissue engineered devices. Adv Drug Deliv Rev 33:111–139
69. Williams DF (1987) Tissue-biomateial interactions. J Mater Sci 22:3421–3445
70. ASTM F720-81 (2002) Testing guinea pig for contact allergens: guinea pig maximization test. In: Annual Book of ASTM standards vol 13.01 Medical devices. Emergency medical services ASTM international: West Conshohocken, PA
71. Northup SJ (1999) Safety evaluation of medical devices: US food and drug administration and international standards organization guidelines. Int J Toxicol 18:275–283
72. Joel DB et al (2004) Biocompatibility testing. In: Gary EW, Gary LB (eds) Encyclopedia of biomaterials and biomedical engineering. Marcel Dekker, New York
73. Pacific boilabs, Assessing biocompatibility: a guide for medical device manufactures. http://www.pacificbiolabs.com/biocomp_intro.asp
74. International Standard ISO 10993 (2003) Biological evaluation of medical devices-parts 3: tests for genotoxicity and reproductive toxicity. International Organization for Standardization, Geneva, Switzerland
75. International Standard ISO 10993 (2002) Biological evaluation of medical devices-parts 4: selection of tests for interactions with blood. International Organization for Standardization, Geneva, Switzerland
76. International Standard ISO 10993 (2009) Biological evaluation of medical devices-parts 5: tests for in vitro cytotoxicity. International Organization for Standardization, Geneva, Switzerland
77. International Standard ISO 10993 (2007) Biological evaluation of medical devices-parts 6: tests for local effects after implantation. International Organization for Standardization, Geneva, Switzerland
78. International Standard ISO 10993 (2002) Biological evaluation of medical devices-parts 10: tests for irritation and delayed-type hypersensitivity. International Organization for Standardization, Geneva, Switzerland

79. International Standard ISO 10993 (2006) Biological evaluation of medical devices-parts 11: tests for systemic toxicity. International Organization for Standardization, Geneva, Switzerland
80. International Standard ISO 10993 (2006) Biological evaluation of medical devices-parts 12: sample preparation and reference materials. International Organization for Standardization, Geneva, Switzerland
81. International Standard ISO 10993 (2009) Biological evaluation of medical devices-parts 13: identification and quantification of degradation products from polymeric medical devices. International Organization for Standardization, Geneva, Switzerland

Biography

Dr. Park is professor at Department of Medical Engineering, Yonsei University College of Medicine, Seoul, Korea. He is also Director at Yonsei Medical Technology and Quality Evaluation Center, Yonsei University Health System. Park graduated in 1987 from Yonsei University with a B.S. degree in Biochemistry, received his M.S. degree in Public Health from Seoul National University in 1990, and obtained his Ph.D. degree in Biotechnology and Life Science from Tokyo University of Agriculture and Technology, Tokyo, Japan, in 1995. After doing postdoctoral work at National Institute of Health and Sciences (NIHS), Tokyo, Japan, he served as Assistant Professor and Associate Professor at Yonsei University from 1996 to 2009. He was appointed Professor in 2009. Dr. Park is with the editorial board of Biomedical Materials as well as Biomaterials Research Journal. He is also a member of the Korean Society for Biomaterials, Korea Society of Medical & Biological Engineering, The Japanese Society for Medical Mycology, Society for Biomaterials, USA, and New York Academy of Science. Park has published more than 150 research papers. His specialty is biocompatibility, bio-sensors, biological activity evaluation of bioactive compounds, and control of microorganism. His current major research interests are cell viability and migration, application of bioactive compounds in tissue engineering, tissue preservation at physiological conditions, biological evaluation of artificial organs and medical devices, polymeric scaffolds for tissue engineering, and sterilization of microorganisms by electric current.

Dr. Park is a Senior Research Scientist at Yonsei Medical Technology and Quality Evaluation Center (MTEC), Yonsei University Health System, Seoul, Korea. He graduated in 1998 from Korea National Open University with a B.S. degree in Hygienic Science, received his M.S. degree in Biomedical Engineering from Yonsei University in 2003, and obtained his Ph.D. degree in Veterinary Sciences from Gifu University, Gifu, Japan, in 2006. Park served as a Research Scientist at MTEC from 1999 to 2009. He also worked as a Research Scientist and a postdoctoral researcher at National Institute of Health and Sciences (NIHS), Tokyo, from 2003 to 2006, and served as a Technical Official for Ministry of Health, Labour and Welfare, Japan, at NIHS from 2007 to 2008. Park is a member of the Korean Society for Biomaterials, The Japanese Society for Medical Mycology, and The Mycological Society of Japan. He is the author of more than 40 research papers. His specialty is biocompatibility, biomedical engineering, and identification of microorganism. His current major research interests include biological safety evaluation for medical devices, application of bioactive compounds in tissue engineering, and sterilization of microorganisms by plasma technique.

Chapter 18
Biological Responses to and Toxicity of Nanoscale Implant Materials

Lei Yang and Thomas J. Webster

Abstract Nanomaterial safety and toxicity are of great importance for nanomaterial-based medical implants. A better understanding of the fate of nanomaterials after production and after implantation is clearly necessary. In terms of implant degradation, nanoscale materials can be generated and released into peripheral host tissues regardless of their constituent grain sizes (or other characteristic features, such as particle size). Unfortunately, the biological responses to and toxicity of nanoscale implant materials have not been sufficiently studied to date, partially due to the complexity of such studies and the lack of well-established methods to do so. In this chapter, the advances and progression of biological responses (especially concerning the toxicity of nanoscale implant materials either after production or implantation) are summarized. Prior to that discussion, host responses to implant materials and properties of nanomaterials pertinent to their altered biological responses are introduced.

L. Yang
School of Engineering, Brown University,
Providence, RI 02912, USA

Department of Orthopaedic Surgery, The First Affiliated Hospital of Soochow University,
Suzhou 215006, P.R. China
e-mail: yleibrown@gmail.com

T.J. Webster (✉)
School of Engineering, Brown University,
Providence, RI 02912, USA

Institute for Molecular and Nanoscale Innovation (IMNI), Brown University,
Providence, RI 02912, USA

Department of Orthopaedics Brown University, Providence, RI 02912, USA
e-mail: Thomas_Webster@brown.edu

N. Eliaz (ed.), *Degradation of Implant Materials*, DOI 10.1007/978-1-4614-3942-4_18,

1 Introduction

The degradation and/or wear of implant materials are inevitable *in vivo*, and the fate of the degradation or wear products from such devices is not only a scientific question but also a safety issue pivotal to patients. The prevalent use of artificial implants such as orthopedic, cardiovascular, dental, and ophthalmological prostheses has brought increasing concerns for the safety of implant materials. These concerns have become even more urgent and imperative as nanoscale materials (materials with feature sizes less than 100 nm) have quickly emerged as novel implant materials over the past decade. Such concerns largely exist because these materials are composed of sizes small enough to penetrate cells and tissues unnoticeably; nanomaterial toxicity is especially a concern when used for long-term applications inside patients.

Generally, nanoscale implant debris *in vivo* can be traced back to three sources: nanoscale materials implanted or administrated (e.g., calcium phosphate nanoparticles incorporated into and onto tissue engineering scaffolds or used as implant coatings), degradation products of implanted materials, and wear debris from articulating components of implants. For the latter two sources, the parent implant materials can possess any grain size (or characteristic feature size). In other words, *in vivo* degradation or the wear of conventional, micrometer-grained implant materials can also generate nanoscale debris, which has been discovered in patient periprosthetic tissues or other organs [1–3]. Therefore, the exploration and full understanding of the toxicity and safety risks associated with nanoscale materials are of great importance. Here, the term nanoscale implant material refers to materials yielded from any of the sources mentioned earlier.

In the first part of this chapter, host responses to biomaterials and properties of nanomaterials that are pertinent to biological responses are reviewed. In the second part, studies aimed at understanding the toxicity of nanoscale implant materials are summarized. Due to the scope of the book and limited space, many popular nanomaterials of particular interest for applications other than medical implants (e.g., nanoparticles for drug delivery and bioimaging) are not extensively covered here.

2 Overview of Host Responses to Biomaterials

While degradation of implant materials *in vivo* is a type of material response to the host, host responses towards biomaterials include a series of complicated physiological activities triggered at the biomaterial–tissue interface at a time scale from milliseconds (e.g., protein adsorption) to even years (e.g., foreign body reactions (FBRs), chronic inflammation, and fibrous encapsulation), as shown in Fig. 18.1. Several important host responses or mechanisms will be briefly reviewed here, including inflammatory reactions, immunological responses, toxicity, blood–material interactions, thrombosis, and infection. To a great extent, these host responses comprise aberrations of common host defense processes [4].

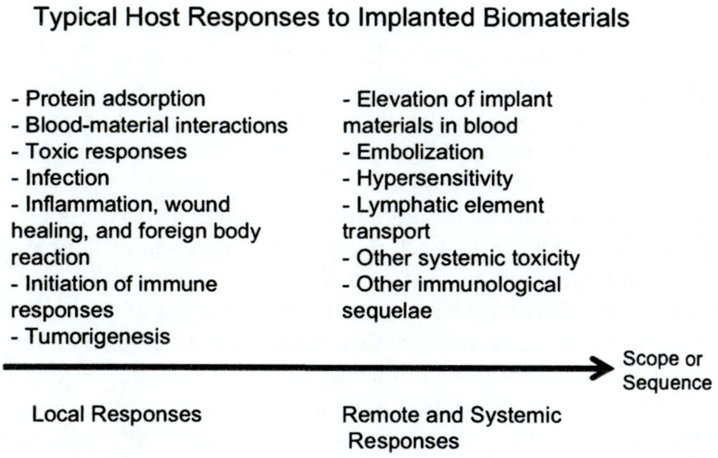

Fig. 18.1 Typical host responses to implanted biomaterials

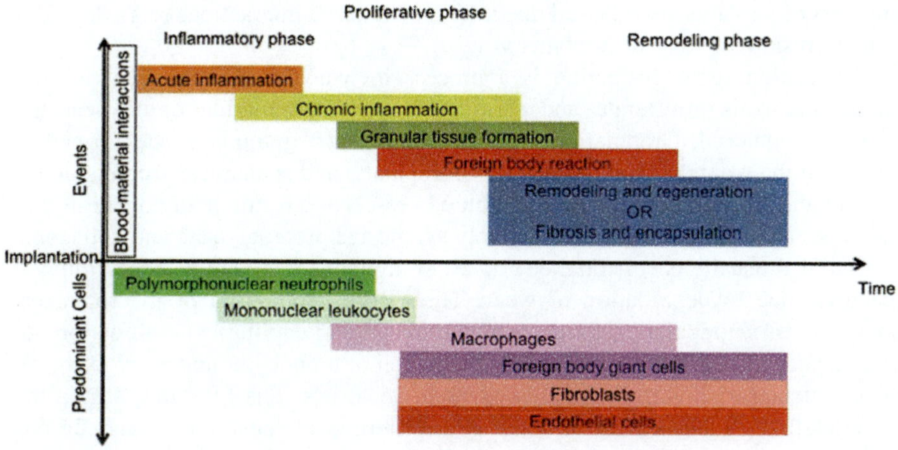

Fig. 18.2 Inflammation and subsequent responses to biomaterials after implantation

2.1 Inflammatory Reactions and Subsequent Responses

When placing an implant in the *in vivo* environment, injury of tissues or organs is inevitable and the injury triggers a sequence of events including acute and chronic inflammation, granulation tissue generation, FBR, and fibrous encapsulation (or fibrosis), as illustrated in Fig. 18.2. In the context of implant material degradation after implantation, degradation products can also cause local inflammatory reactions like chronic inflammation and FBRs, and the inflammatory responses to degradation products do not necessarily include all the sequential events mentioned earlier. The progression of inflammatory reactions is mediated by a number of chemical mediators released from plasma, cells, and tissues. These chemical mediators (such as cytokines, growth factors, platelet activating factors, plasma proteases, vasoactive amines, lysosomal proteases, etc.) also play an important role in the degradation of implant materials, and in turn, degradation products may initiate the release of chemical mediators to prolong the inflammatory responses, which ultimately can cause implant failure.

Acute inflammation can last for a relatively short time period, from minutes to days after the injury, and it is characterized by the exudation of fluid and plasma proteins and the migration of leukocytes (predominantly neutrophils). Next, activation of neutrophils and macrophages lead to phagocytosis of foreign materials, which may occur in both acute and chronic inflammation. Characteristics of chronic inflammation include the presence of macrophages, monocytes, and lymphocytes, and the proliferation of blood vessels and connective tissues. Macrophages are of great importance in chronic inflammation, not only because of numerous biologically active products or chemical mediators (e.g., neutral proteases, oxygen free radicals, cytokines, and growth factors) they secret, but also because of their possible role in the systemic effects on tissues or organs when the components or products of implants are released due to tissue–material interactions (e.g., degradation, corrosion, or wear of implants *in vivo*).

Granulation tissue formation is a process in which fibroblasts and vascular endothelial cells proliferate and start to generate granulation tissue near the implanted material. Granulation tissue is a pink, soft granular tissue formed at the site of the wound (or on the surface of implants) and is characterized histologically by the proliferation of new small blood vessels as a result of angiogenesis and the presence of fibroblasts which actively synthesize proteoglycans and collagen.

The FBR usually occurs on the surfaces of implants or around detached implant materials due to degradation or wear. The FBR is composed of macrophages, components of granulation tissue (e.g., fibroblast and capillaries), and/or foreign body giant cells (large cells formed by the fusion of monocytes and macrophages), which attempt to engulf and phagocytose the materials. The FBR may last at the material–tissue interface for as long as the lifetime of the implant, and fibrous encapsulation generally covers the implant material to separate the implant and its FBR from the local tissue environment.

Generally, surface or bulk chemistry, surface roughness, porosity, or surface-to-volume ratios, and other surface properties of implants are very important towards mediating the type of FBR; such implant surface properties are responsible for macrophage activation and the composition (e.g., ratio of macrophages and foreign body giant cells) of the FBR.

Fibrous encapsulation (or fibrosis) is often considered as the end-stage of the duration of the inflammatory responses. It is a formation of a fibrous capsule on implant materials by connective tissue. Fibrosis is controlled by the proliferative capacity of cells and is largely influenced by the properties of the implant materials (such as roughness, size of implant–tissue interface, biodegradability, etc.). For example, the size or porosity of implants or implanted materials can affect the extent of fibrosis at the implant site. Larger implants which may cause increased

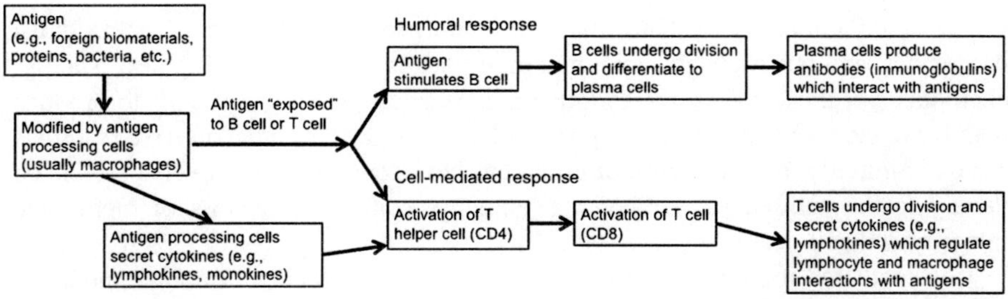

Fig. 18.3 The pathways of immune responses to antigens such as implant degradation products (foreign biomaterials)

injury or defects can lead to significant fibrosis around the materials, while highly porous materials implanted into bone may result in less fibrosis [4].

2.2 *Immunological Responses*

Pathogens like bacteria, fungi, virus, and foreign materials will trigger immune system response, which recognizes and differentiates these "foreigners" from the host and protects the host from exposure to them. Obviously, degradation product from implants *in vivo* is a type of pathogen, which can almost inevitably cause immunological responses. The immune system is a complex network of proteins, cells, and specific organs, which has been viewed historically as having humoral and cellular branches, though both branches are closely related (Fig. 18.3). In humoral immunity, there are a variety of antibodies (e.g., immunoglobulins (Ig)), complement proteins, and cytokines facilitating many aspects of the immunological response. The antibodies are produced in cells of the B lymphocyte lineage (B cell). In fact, these humoral elements and complement activation processes (as described below) have been of particular interest in biomaterial-induced immune responses.

Complement is an immune response that can nonspecifically recognize and eliminate foreign elements (including biomaterials) by coating the materials with complement components that permit and facilitate phagocytosis. The complement system is composed of more than 20 different plasma proteins as it directly and indirectly contributes to acute inflammation and immune responses. Activation of the complement cascade can follow two separate but interacting pathways: classical and alternative. These two pathways are distinct in the immune complexes (antigens and specific antibodies) involved. The alternative pathway is usually considered to be responsible for most of the complement activation triggered by biomaterials. Either of the pathways leads to the formation of specific complement proteins (C4b and/or C3b), which can be recognized by granulocytes and can activate these cells to produce degradative enzymes and destructive oxygen species. In addition, both pathways end with the formation of a complement receptor (C5a) and a protein complex known as the membrane attack complex (MAC, initiated by C5b).

C5a is an inflammatory mediator that can bind to specific receptors on neutrophils, monocytes, and macrophages, and the binding leads to numerous responses such as chemotaxis of these cells towards inflammatory sites, activation of cells to produce reactive oxygen species and subsequent cell death, as well as other adverse reactions *in vivo*. Similarly, the formation of MAC has been found to have adverse effects on damaging the cellular lipid bilayer, eventually leading to the loss of membrane integrity and cell death.

Cell-mediated immune responses do not involve antibodies or complement but rather involve the activation of T lymphocytes (T cells), macrophages (or natural killer cells), and the release of various cytokines in response to an antigen (or foreign substance). These cells originate from stem cells in the bone marrow and differ in morphology, function, and the expression of cell surface antigens, but they all maintain cell surface receptors that can recognize and/or eliminate foreign materials.

2.3 Toxicity

From a biochemistry viewpoint, toxicity is ubiquitously associated with reactive oxygen species such as superoxide anions, superoxide dismutase, and hydrogen peroxide, just to name a few. Although much work is needed to understand the mechanisms of how reactive oxygen species lead to toxicity, it is generally believed that the reactive oxygen species can denature proteins, disrupt lipid bilayers, or destroy the nuclei of cells, which will initiate further adverse effects and/or responses at larger scales. Therefore, toxicity is often viewed at the cellular and systemic levels. As described earlier, inflammatory and immunological responses can result in toxic effects at the cellular level, which is usually known as cytotoxicity. However, excessive or severe inflammatory and immunological responses, together with direct chemical toxicity of wear, corrosion and degradation products, may cause systemic toxicity. This is defined as toxicity remote from the initial insult. Systemic toxicity may be readily detected due to the damage to target organs, which reveals apparent signs and symptoms; however, this is not always the case. In the particular interest of this book, degradation products of implant materials are one of the typical factors that cause both nonimmune and immune systemic toxicity.

Usually, systemic toxicity caused by degradation products or, generally, biomaterials, is dose dependent. For nonimmune systemic toxicity, a threshold below which the material reveals little toxicity can be determined by careful *in vitro* studies and appropriate animal studies for each degradation product or biomaterial. However, this threshold is usually higher for nonimmune systemic toxicity than immune systemic toxicity, because immune systemic toxicity is also individually dependent and antigen- (or foreign material-) dependent. In fact, systemic toxicity due to immune responses to degradation products or foreign biomaterials is extremely difficult to predict, because these immune responses depend on the genetics of the individual and also the properties, dosage, and location of the released materials. Therefore, release of a large quantity of

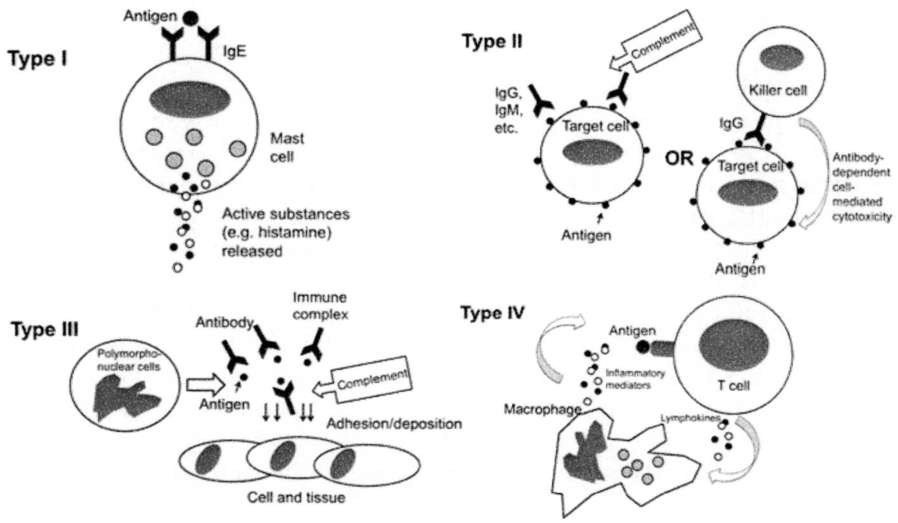

Fig. 18.4 Illustration of four types of hypersensitivity

degradation products may or may not cause severe reactions in the individual. This fact also causes difficulty in using animal models to evaluate the potential risks of degradation products or biomaterials, differing from the success in evaluating nonimmune systemic toxicity.

Hypersensitivity is considered as a type of systemic toxicity resulting from excessive and unusual immune responses which may damage the body's own tissue. Hypersensitivity usually has four types, as illustrated in Fig. 18.4. In type I hypersensitivity, IgE antibodies attach to cells and interact with exogenous antigens (e.g., implant degradation products), while in type II hypersensitivity, antigens attach to cells (e.g., platelets) and interact with antibodies like IgG or IgM. Type III hypersensitivity involves immune complexes (antigen–antibody IgG and IgM complexes) formed in the blood, which become lodged in various tissues where they may trigger an immune response. Type IV hypersensitivity (or delayed-type hypersensitivity) is cell mediated, antibody independent, and involves the reaction of T cells with antigenic substances. For hypersensitivity caused by degradation products of implant materials or biomaterials, Type I and II hypersensitivity are rare (but have been reported) while Type IV hypersensitivity is the most common. Type III hypersensitivity has been reported in material systems which slowly release antigenic substances into the body.

2.4 Blood–Material Interactions

When biomaterials or their degradation products are exposed to blood, blood–material interactions like hemostatic reactions may kick in. A main function of hemostatic mechanisms is to prevent and control bleeding by the formation of a clot, and this process involves interactions between material surfaces, platelets, and coagulation proteins. Briefly, the hemostatic reaction to biomaterial surfaces may follow primary hemostasis and secondary hemostasis. In primary hemostasis; (1) protein mediators adsorb onto the material surface, (2) platelets adhere onto the material surfaces and release granule contents, and (3) a large number of platelets aggregate to form a platelet plug, which is stimulated by the granule contents. Simultaneously, secondary hemostasis, a coagulation cascade initiated by clotting factors (proteins from blood plasma), occurs to form a fibrin gel that strengthens the platelet plug (platelet thrombus). However, there are also mechanisms to control platelet activation and the coagulation cascade, inhibiting excessive thrombus formation. These mechanisms involve a number of proteins, inhibitors, or anticoagulants which can terminate the pathway of platelet activation and the coagulation cascade. For the degradation products of implants that are of particular interest to this book, adverse consequences may occur when excessive thrombi coagulate on the material surface due to imbalances between these activation and inhibition processes.

2.5 *Infection*

Implant-associated infection has received exceptional attention because infection may lead to re-operation, amputation, or even death of patients. Studies have revealed that artificial hearts reach an infection rate of about 100% after implantation for 90 days [4], and that 14.8% of revised total hip arthroplasties were due to infection between 2005 and 2006 in the United States [5]. *S. epidermidis* and *S. aureus* are responsible for most implant-associated infection, but in many cases, there are multiple bacteria leading to infection of implants. *S. epidermidis* exists on the human skin and primarily causes infection on polymeric implant materials. In contrast, *S. aureus* is a natural tissue pathogen which may cause infection due to tissue damage around implants; it is frequently associated with metallic-related materials. The mechanism of implant-associated infection includes a complex process of bacterial attachment and adhesion to material surfaces, aggregation or dispersion of bacterial colonies, as well as biofilm (a thick surface composite consisting of microorganisms, extracellular materials, environmental adsorbents and debris) formation. Details of these mechanisms are not reviewed here due to space constraints.

Degradation of implant materials also plays an important role in implant-associated infection, because the degradation establishes environmental conditions that microbes can exploit. Specifically, degradation products provide surfaces for bacterial adhesion as well as nutrients (e.g., iron ions) and stimuli (e.g., Mg^{2+} and Ca^{2+}) for bacterial adhesion, growth, and propagation.

3 Properties of Nanomaterials Which Influence Biological Responses

The previous overview demonstrates that properties of biomaterials are critical to mediating host responses, and altered structural or surface properties of biomaterials may result drastically in different host responses. Nanoscale implants may amplify the effects of these material properties on biological responses, because many material properties change when the size of constituent material structures or features decrease to the nanometer scale (usually, less than 100 nm). Although more studies are needed to further understand the properties of nanomaterials that are pertinent to mediating biological responses, a number of material properties have been correlated to cellular and host responses (such as single cell functions, inflammatory and immune responses, toxicity, etc.) and, thus, have demonstrated great potential to mediate these biological responses.

In this section, several important properties of nanomaterials that cells and tissues recognize (subsequent to initial protein interactions) *in vivo* and *in vitro* are summarized. These properties include the size, mechanical properties (e.g., stiffness), surface chemistry (which can be closely related to bulk chemistry), topography, surface energy, and surface electrical properties. Figure 18.5 illustrates the interface between cells and nanoscale implant materials, showing possible interactions between cell membranes, proteins, and nanomaterial properties. It is

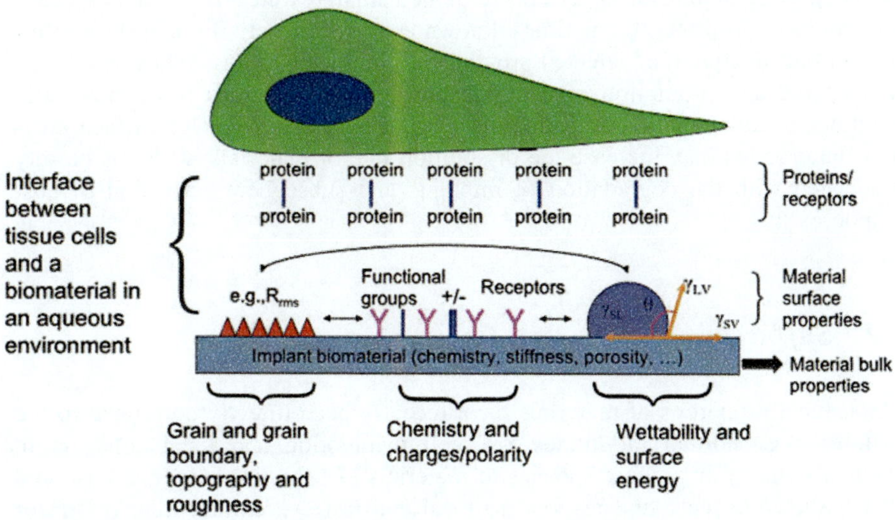

Fig. 18.5 Schematic illustrating potential interfacial interactions between a cell (or tissue) and nano biomaterials. *Double arrows* indicate interactions among surface properties of nanobiomaterials. In the schematic, examples of nanomaterial surface properties include: root mean square roughness (R_{rms}), surface electrical charges (+/−), contact angles (θ), interfacial tension between solids and vapors (γ_{SV}), interfacial tension between solids and liquids (γ_{SL}), and interfacial tension between liquids and vapors (γ_{LV})

worth emphasizing that a great number of different proteins (such as adhesion proteins, cytokines, growth factors, etc.) are involved in the interactions between materials and biological systems in an aqueous environment (e.g., bodily fluids, culture media, etc.), which has been briefly discussed in the previous section. Therefore, studying the fate of proteins in response to nanomaterial (surface) properties is an effective approach to understanding subsequent responses of biological systems.

3.1 Size and Surface Area

Compared to conventional micrometer-scale materials, extremely small sizes of the features on, or components in, nanomaterials can result in numerous distinct biological or host responses. First, the small size of nanomaterials favors uptake by cells through several distinguished pathways. The type of pathways and the efficiency of cellular uptake for implant materials are also strongly dependent on the particle size of the materials, which will be discussed later. Second, the small size of nanomaterials also greatly affects their circulation time if the materials have been released or transported to the bloodstream. For example, nanoparticles of various chemistries with sizes from 70 to 200 nm revealed greater prolonged circulation times compared to smaller or larger particles [6]. Third, the increased surface area of nanomaterials as a result of their small feature sizes confers specific conditions or properties (sometimes, known as surface reactivity or surface affinity) for the adsorption of selected proteins or biological entities, adhesion of cells or bacteria, and promotion of catalytic biochemical reactions (e.g., enzymatic reactions in inflammation and immune responses). Lastly, greater surface area-to-volume ratios may increase the dissolution rate of materials, which is closely associated with the degradation of implants and subsequent activation of host responses.

3.2 Stiffness

Mechanical properties of materials have recently been directly correlated to the cellular (e.g., normal rat kidney-52E cell locomotion and focal adhesion in Pelham's study) or tissue responses to materials [7]. Studies have demonstrated that tissue cells sense and respond to local matrix (e.g., extracellular matrix or synthetic material) stiffness by the formation of molecular adhesion complexes and the actin–myosin cytoskeleton, and feedback from the matrix stiffness has important implications for cellular development, motility, and differentiation [7, 8]. Moreover, an increasing number of studies have indicated that cellular behavior is mediated by matrix stiffness, which perhaps may be pivotal for understanding disease processes, morphogenesis, and tissue regeneration [9].

The role of material stiffness on mediating cellular or tissue responses is only partially known, and the role of nanoscale dimensional changes in mechanical properties is also unclear. However, the impact of nanomaterial stiffness should not be neglected.

3.3 Surface Chemistry

Biological responses are very sensitive to the changes in material surface chemistry as a result of the modification of the bulk chemistry and/or surface molecules on the materials. The bulk chemistry of materials usually refers to the chemical or phase composition, and crystallinity of materials. However, surface chemistry is not necessarily the same as the bulk chemistry of implant materials. For example, anodized nanotubular titanium developed for orthopedic implant applications has nanometer titanium oxides (mostly titania, TiO_2) on the surface while the bulk remains metallic titanium. Implant materials different in bulk chemistry can be categorized as bioinert or bioactive materials, which differentiate in the manner of fixation to the host tissue [10]. For example, bone tissue forms a morphological encapsulation to bioinert materials (e.g., most metals and ceramics), but forms a bioactive interfacial bond with bioactive materials (e.g., hydroxyapatite).

On the other hand, surface chemistry can be readily modified by conjugating different functional molecules onto nanomaterial surfaces regardless of the underlying chemistry of nanomaterials, leading to altered biological responses. For instance, a wide range of peptides [most notably, arginine–glycine–aspartic acid (RGD) and bone morphogenetic protein-2 (BMP-2)] have been conjugated on various nanomaterial surfaces (titanium and its alloys, hydrogels, polymers, etc.) to improve bone cell functions from adhesion to osteogenic functions *in vitro* and *in vivo* [11–13]. In fact, nanomaterials provide much larger specific surface areas, more substructures (such as grain boundaries), and more active surfaces for chemical modification than conventional micrometer-scale materials.

3.4 Topography and Roughness

Surface landscapes composed of undulations and even steep gradients and pores constitute the topography of material surface [14]. Topography is another predominant surface property pertinent to material-mediating biological responses. Topography is difficult to define by a few simple parameters, thus roughness is a widely used parameter (usually it is a statistical variation in the height of the surface, e.g. root mean square height of the surface around some mean value), whose effects on tissue/cellular responses have been studied [15]. An important lesson from natural biological systems is that many systems (e.g., bone, tooth, and blood vessels) have nanoscale topography and roughness due to their hierarchical micro-to-nano structures. Biologically inspired by this information, enhanced tissue/cellular responses leading to greater efficacy of implants have been achieved by creating nanoscale roughness on traditional material surfaces. This strategy has been applied to a variety of nanoscale materials (including metals [16], carbon nanofibers (CNFs)/nanotubes [17], polymers [18], ceramics [19], and polymer/ceramic composites [20]) to increase normal cell functions while decreasing inflammatory reactions (and inhibiting cancer cell growth).

The mechanisms behind such nanotopography-mediated tissue responses are still not clear. Increased adsorption of cell adhesive proteins (such as fibronectin, vitronectin, etc.) on nanoscale-rough surfaces, due to either uneven surface landscapes or increased specific surface area, is a plausible mechanism. However, differential adsorption of proteins (by means of selectivity, quantity, etc.) on the same nanorough surfaces has been observed, indicating that the current understanding of promoted tissue growth on nanomaterials is far from being sufficient. In addition, due to the limitation of describing topography by roughness, the correlation between nanoscale roughness and biological responses at the bone–implant interface could be inadequate and difficult to explain. For example, there is a lack of evidence to explain the fact that many different types of cells "experience" the same roughness, but their responses are different from one cell type to another.

3.5 Surface Wettability and Surface Energy

It has also been widely observed that cell adhesion and subsequent activities are dependent on implant surface wettability [21]. Surface wettability (i.e., whether the surface is hydrophilic or hydrophobic) has been traditionally determined by water contact angles (θ), as Vogler suggested a definition of hydrophilic surfaces having a θ less than 65° [22]. θ is closely related to surface energy, and a classic definition of θ has been actually given by the Young–Dupré equation:

$$\cos\theta = (\gamma_{SV} - \gamma_{SL})/\gamma_{LV}, \tag{18.1}$$

where γ_{SV} is the surface energy (or surface tension, in the unit of J/m^2) between the solid and vapor, γ_{SL} is the surface energy between the solid and liquid, and γ_{LV} is the surface energy between the liquid and vapor (refer to Figure 18.5). This equation indicates that ways to control surface wettability essentially include changing the three surface energy terms mentioned earlier (i.e., γ_{SV}, γ_{SL}, and γ_{VL}). Changing nanomaterial surface chemistry is a common approach taken to alter their surface energy and resulting wettability. Moreover, changing nanometer surface topography may also alter surface energy and wettability [23].

Because surface energy measurements demonstrate a close relationship between the high surface energy of hydrophilic nanomaterials (compared to hydrophobic materials) and the adsorption of hydrophilic cell adhesive proteins (e.g., fibronectin and vitronectin), it has been speculated that hydrophilic (i.e., high surface energy) nanomaterials have a higher affinity for such cell adhesive proteins and, subsequently, promote cell functions and tissue responses better than hydrophobic nanomaterials. For example, maximum vitronectin adsorption was noted on hydrophilic surfaces [24]. However, inconsistent experimental results have been reported on this speculation, and further investigations on the correlation between surface energy (or wettability) and biological responses are needed.

3.6 Surface Electrical Properties

Electrostatic interactions also play a major role in mediating biological responses to biomaterials, since cell membranes are charged and virtually all material surfaces are charged in aqueous solutions. Surface charges on nanomaterials can be induced or modified by altering surface chemistry or charging the nanomaterials in an electrical field (not common). However, the effects of surface charge on cell functions are not clear and recent findings vary between cell type and cell functions [25]. For example, a study revealed that positively charged indium tin oxide (ITO) enhanced the adhesion of rat marrow stromal cells but impaired subsequent cell spreading and differentiation [26]. Because many tissue cells and bacteria are negatively charged on the surface, positively charged nanomaterial surfaces may exhibit a strong affinity to these cells or bacteria. For example, a study has shown that positively charged polymer nanoparticles (specifically, polystyrene nanobeads (diameter 200 nm) modified by poly-L-lysine, chitosan, and heparin sulfate) were phagocytosed by mouse macrophages to a greater extent than negatively charged ones [27]. However, contradictory results have also been reported for this argument based on surface electrostatics. A few studies on hydrophilic polystyrene particles revealed that phagocytosis of the particles increased using negatively charged particles, which promoted an inflammatory response [28, 29]. In addition, there has also been some early evidence showing the influence of piezoelectric and/or conductive materials on osteoblast responses. For instance, Itoh *et al.* [30] created electrically polarized hydroxyapatite with pores to increase bone growth and decrease osteoclast activity, and they attributed these effects to the electrical polarity of material surfaces with pores.

In summary, surface properties of nanomaterials greatly influence and mediate biological responses (e.g., cell or bacterial responses, inflammatory reactions, etc.) on nanomaterial surfaces. However, it is difficult to predict the permeation in cellular uptake and other biological responses to nanomaterials by such generalized concepts as surface wettability, partially because all the surface properties described earlier exert synergetic effects between the interactions of nanomaterials and biological systems.

4 Toxicity of Nanoscale Implant Materials

4.1 Cellular Uptake of Nanoscale Implant Materials

Cellular uptake is an important biological process by which cells internalize substances such as nutrients, proteins, foreign materials, and biological entities. Although cellular uptake of materials does not necessarily implicate adverse host responses, it is of great importance to understand the biological responses to implant materials, especially nanoscale implant materials because their exceptional small size favors cellular uptake compared to the conventional materials with larger sizes.

Cellular uptake of nanoscale implant materials is size dependent. Clearly, extremely small size particles may cross cell membranes readily. Most of the studies so far have observed endocytosis of nanoparticles with sizes less than 200 nm [6, 31]. The endocytic pathways for internalizing nanoscale implant materials include phagocytosis, macropinocytosis, clathrin-mediated endocytosis, and non-clathrin-mediated endocytosis [6, 31], all of which are closely related to the host responses addressed in the previous section. The type of endocytic pathways along which nanoparticles progress is strongly affected by the size of nanoscale implant materials. For example, latex nanospheres of less than 200 nm are prevailingly endocytosed by nonphagocytic cells via clathrin-coated pits (i.e., clathrin-mediated endocytosis), while the endocytosis pathway is dominated by caveolae internalization (i.e., non-clathrin-mediated) when the sphere size increases to 500 nm [32]. Furthermore, endocytic efficiency is also a function of the nanoparticle size. Rejman et al. also reported that nanoparticles with sizes of 50–100 nm were internalized rapidly by cells, whereas 200 nm nanoparticles revealed a substantially slower cellular processing.

Surface physicochemical properties of nanoscale implant materials also strongly influence their cellular uptake. As mentioned earlier, these surface properties include surface area, surface charge, surface composition and chemical functionality, surface wettability, and surface roughness (topography). On one hand, these surface properties determine the aggregate or agglomeration state, secondary geometry and architecture (e.g., structure of self-assembled nanoparticles), and dissolution kinetics of the nanomaterial itself *in vitro* and *in vivo*, which are key factors affecting cellular uptake. On the other hand, the synergy of surface physicochemical properties directly mediates cellular internalization processes by interacting differently with endocytic receptors on cell membranes and/or controlling diffusion kinetics of the particles. As a good example of nanomaterial cellular uptake, many studies have demonstrated that the physicochemical characteristics of carbon nanotubes (CNTs) are crucial to their cellular uptake. CNTs are also promising nanomaterials for orthopedic implant applications because of their extraordinary electrical and mechanical properties. Raffa et al. [33] reviewed recent CNT studies and proposed a unified explanation to how CNTs are internalized depending on their surface properties (specifically, length and diameter, surface charge, and hydrophilicity). The internalization pathway of CNTs that form into micrometer-size aggregates has been due to surface charge or surface wettability interactions leading to phagocytosis (a process of engulfment by which macrophages ingest cellular fragments or foreign substances, see Sect. 2). Phagocytosis also occurs for nonagglomerated CNTs with lengths more than 1 μm. However, for more hydrophobic CNTs, or CNTs that formed supramolecular complexes via binding proteins, cellular components (e.g., cytokines and growth factors) and other ligands (e.g., antibodies), the receptor-mediated endocytosis pathway dominates. For CNTs whose surface properties prevent the formation of supramolecular complexes, a possible diffusion pathway has been speculated from imaging, theoretical and simulation studies. It is also worth mentioning that the receptor-mediated endocytosis of nanoscale materials is well established, and this pathway is an important internalization mechanism for not only CNTs but also many other biological macromolecules and nanomaterials.

However, knowledge of cellular uptake of nanoscale materials is far from complete. Fundamental issues such as internalization pathways and intracellular traffic at the molecular level remain unclear. Information concerning manipulation of nanomaterial properties to enhance or suppress cellular uptake of nanoparticles is still scarce. A better understanding of these issues will render the design of safe and efficacious artificial implants possible.

4.2 Toxicity of Metallic Nanoscale Implant Materials

Metals are widely used for orthopedic, cardiovascular, and dental implants, and corresponding metal oxides exist on implant surfaces or in the debris generated by mechanochemical processes. The toxicological effects of nanoscale metallic or metal oxide particles have been studied, both *in vitro* and *in vivo*, and in most cases, these toxicological effects have demonstrated a strong dependence on the particle size and on other surface properties. An *in vivo* study has reported that carcinogenicity was found for Ni particles implanted in the soft tissue of rats when their size decreased to 500 nm, whereas micrometer-sized Ni particles caused just an allergic reaction instead [34]. Cellular internalization and tissue inflammation have been indentified with Ti and TiO_2 particles less than 3 μm when exposed to human neutrophils and rat soft tissue [34]. If nanoparticle size decreases below an even lower threshold (10–50 nm reported by different studies), the responses of the body's defense system (e.g., immune system) to nanoparticles are reduced, but the nanoparticles are more likely to travel through the circulatory system and/or to target organs [34, 35]. For example, an *in vivo* study implanted Ti wires into rat femurs for 18 months and investigated Ti levels in the blood or other organs (such as the heart, lung, spleen, liver, kidney, etc.) due to nonwear physiological release of Ti [3]. The Ti content in all of the selected organ tissues and blood was higher than the predetermined basal Ti level measured in the rats without Ti implantation, indicating both corrosion of the Ti implant (in the absence of wear) and systemic Ti accumulation in target organ tissues. Nanosized debris or organometallic complexes were proposed as one of the mechanisms causing systemic distribution of Ti after implant degradation. Kadar *et al.* [36] investigated the uptake and biological responses to nanosized Fe particles (reduced from iron oxides) and soluble Fe^{3+} (from $FeCl_3$) in gills and blood cells of blue mussels; the impairment of lysosomal stability in blood cells and increased lipid peroxidation in the gills were reported. Other studies have also demonstrated DNA and chromosome damage and cytotoxicity caused by CoCr micro- and nano-debris from surgical implants; more DNA damage, chromosomal damage, and toxic effects were identified with CoCr nanoparticles compared to micrometer-size particles [37, 38].

The toxicological effects of noble metallic nanoparticles (such as gold and silver nanoparticles) have been investigated in a great number of studies. Although these metallic nanomaterials may not be commonly used as medical implants, several important findings are of great interest towards understanding the biological effects of metallic nanomaterials. The studies indicate that the location of nanoparticles after cellular uptake is unclear, and nanoparticles entering the nucleus may not cause cell death. For example, plain gold nanoparticles were readily internalized by cells through receptor-mediated endocytosis but were unable to enter the nucleus; however, surface modified gold nanoparticles (conjugated with peptide) had the ability to reach the nucleus, and 95% of HepG2 cells under this nuclear exposure remained viable after 12 hours [39]. The toxicological effects of nanoparticles on cells may also differ per cell lines and/or particle surface charges. For example, Tkachenko *et al.* [40] reported different cell viability to the same peptide–BSA–gold nanoparticles in HeLa and 3T3/NIH cell lines, whereas Goodman *et al.* [41] reported similar cytotoxicity of the same gold nanoparticles among COS-1 cells, red blood cells, and *E. coli*, but with a more toxic effect of cationic gold nanoparticles (ammonium-modified) than anionic gold nanoparticles (carboxylate-modified). These studies also suggest that surface coatings, which alter surface properties of nanoparticles, play an important role in toxicity of (metallic) nanomaterials.

4.3 Toxicity of Ceramic Nanoscale Implant Materials

To date, studies on the toxicity of nanoscale ceramic particles have focused on titanium oxides (TiO_2 or TiO), iron oxides (Fe_2O_3, Fe_3O_4 or compounds of these two), aluminum oxides, chromium oxides, silicon oxides, and zinc oxides, some of which are common degradation products from metallic and ceramic implants. On the other end, bioactive ceramic nanoparticles, like calcium phosphates ($Ca_3(PO_4)_2$, hydroxyapatite $Ca_{10}(PO_4)_6(OH)_2$, etc.), or bioglasses, which are widely used in orthopedic applications, are considered to possess good cyto- and bio-compatibility properties.

In vitro cell studies have shown that titanium oxide nanoparticles produce free radicals with a strong oxidizing ability, which could further catalyze DNA damage [42].

A recent study also showed that TiO_2 nanoparticles (nano-anatase TiO_2, crystallite size ~5 nm) could insert into DNA base pairs and cause DNA cleavage in mouse liver cells [43]. However, Landsiedel *et al.* [44] reported contradictory results in which no genotoxicity was observed *in vitro* and *in vivo* on TiO_2 nanoparticles dispersed in test media. In addition, Zhu *et al.* [22] reported that TiO_2 nanoparticles had no developmental toxic effect on zebrafish embryos and larvae.

Similar contradictions have been found for iron oxide and silicon oxide nanoparticles. A substantial number of opposing results have been reported for the cytotoxicity of both bare and coated (coatings such as pullulan, poly(ethylene glycol) (PEG), polyvinyl acetate (PVA), and amino- and thiol-terminated polymers, etc.) iron oxide nanoparticles, and the well-known debate on whether iron oxide nanoparticles are biocompatible or toxic at a variety of concentrations still remains unclear [45]. Recently, different results for SiO_2 nanoparticle cytotoxicity have also been reported. On one hand, SiO_2 nanoparticles (sizes from the nanometer range to submicron range) have been considered nontoxic by many researchers and have been used as a negative control in cytotoxicity studies with other nanomaterials [46]. On the other hand, oxidative stress, cell damage, and apoptosis (to hepatic and myocardial cells) were induced by SiO_2 nanoparticles (sizes of 21 and 48 nm) in several recent studies [47, 48].

There are many possible reasons behind the inconsistency in these reported studies. First, the cytotoxicity of titanium oxide and iron oxide nanoparticles is probably strongly dose dependent [47–50]. Second, cytotoxicity of these nanoparticles may be cell specific, resulting in different cell responses to the same nanomaterial [46]. Third, agglomeration of the nanoparticles due to interactions with proteins or cellular elements may also add complexity to the results [45].

On the contrary, comparatively consistent results for the toxicity of chromium oxide, aluminum oxide, and zinc oxide nanoparticles have been reported. For example, a study has observed that CrO_3 nanoparticles had no toxic effect on Neuro-2A cells after a 72-h exposure at concentrations up to 200 μg/mL [50]. Similarly, most studies have reported little or no adverse effects of Al_2O_3 nanoparticles, either *in vitro* or *in vivo*, on tissue cells, zebrafish embryos, and fruit flies [22, 51]. A recent study also reported that Al_2O_3 nanoparticles had no effect on human fibroblast viability and little genotoxicity after 5 days at doses from 0.1 to 10 mg/T-75 flask [51]. However, there is a case that Al_2O_3 nanoparticles elicited proinflammatory responses (as demonstrated by increased cytokine release and adhesion of activated monocytes) [52], indicating that more investigation is needed to determine Al_2O_3 nanotoxicity.

ZnO nanoparticles have apparent toxicological effects, as indicated by most studies so far. For example, a comprehensive study on the cytotoxicity of ZnO nanoparticles (sizes of 70 and 420 nm) demonstrated significantly reduced cell viability in a dose- and time-dependent manner, resulting from the elevated levels of reactive oxygen species (ROS) in human bronchoalveolar carcinoma-derived cells (A549) [53]. Apoptosis, cell deformation, and significantly decreased mitochondrial function and increased membrane leakage of lactate dehydrogenase

(LDH) have been observed for Neuro-2A cells exposed to ZnO nanoparticles (50–100 μg/mL) [50]. ZnO nanoparticles have also demonstrated a very toxic effect to zebrafish embryo and larva development *in vivo* [22].

4.4 Toxicity of Polymeric Nanoscale Implant Materials

Extraordinary recent developments in prosthetic implant coatings and implantable scaffolds for tissue engineering applications have stimulated the growth of numerous new polymeric nanomaterials. These new emerging polymeric nanomaterials include biodegradable nanopolymers (poly(lactic acid) (PLA), poly(glycolic acid) (PGA), poly(lactic-co-glycolic acid) (PLGA), poly(ε-caprolactone)(PCL), poly (ethylene glycol) (PEG), etc.), nanofeatured hydrogel systems (poly(hydroxyethyl methacrylate)(PHEMA), poly(*N*-isopropylacrylamide) (PNIPAAm), poly(methyl methacrylate)(PMMA), PEG, etc.), and nanocomposite blends with other nanomaterials. Toxicity studies of many of these nanopolymers have indicated acceptable biocompatibility or low toxicity properties. For example, PEG revealed little toxic effects and reduced immunogenicity, and many PEG-coated nanoparticles (the coating process is called PEGylation) have not shown PEG-specific toxicity or immune responses [27, 54]. PHEMA and PNIPAAm hydrogels have also been shown to be highly biocompatible and nontoxic in most of the studies reported to date [55].

However, similar to their micrometer counterparts, polymeric nanoscale implant materials degrade through processes like hydrolysis, resulting in acidic products which induce protein damage and inflammatory reactions in the host. As mentioned before, this degradation process may accelerate dramatically due to the extremely high surface area of nanomaterials. This occurs mostly to biodegradable nanopolymers like PLA, PGA, and PLGA, therefore surface modification of these nanopolymers (e.g., PEGylation) is usually necessary to reduce possible toxic effects [56]. Toxic effects of nanopolymers (like hydrogel systems) may also emerge from unreacted monomers and initiators that can leach out during the intended application [57]. For instance, it is known that the NIPAAm monomer which forms PNIPAAm may be carcinogenic or teratogenic [58]. In addition, small-sized debris may also directly activate chronic inflammation, FBRs, and immunological responses (see previous sections).

4.5 Toxicity of Carbon Nanostructures

Carbon nanostructures (including CNTs, CNFs, fullerene (C_{60}) and derivatives, and nanodiamonds) are attractive new material forms for medical implant applications due to their extraordinary electrical, antiwear, and mechanical properties, and the capacity to promote the regeneration of various tissues and reduce immunological responses [59, 60]. In contrast to the well-studied material properties which enhance desirable biological functions, health risks and toxicological data of these newly emerging carbon nanomaterials are not completely known.

To date, research efforts have focused on C_{60} and CNT (or CNF). Mixed results of C_{60} cytotoxicity have been reported, and the cell type appears to be a significant factor affecting test results. For example, at concentrations less than 60 μg/mL, C_{60} was reported to have a low cellular uptake, no apoptotic effect, and generally be nontoxic to macrophages [45, 61, 62]. In contrast, the dose-dependent cytotoxicity of C_{60} and its derivatives were observed in human dermal fibroblasts and liver carcinoma cells at concentrations from 0.24 to 2,400 ppb [63]. In the same study, C_{60} appeared to be more toxic than surface modified derivatives which possess water solubility (hydroxylated C_{60}). Another study on amino acid-modified C_{60} reported dose-dependent proinflammatory responses and cell death on human epidermal keratinocytes (HEK) at concentrations between 40 μg/mL and 400 μg/mL [64].

Compared to the mixed results with C_{60}, CNTs have shown more consistent results concerning toxicological effects at high concentrations, above tens of micrograms per milliliter. Two types of CNTs, single-walled CNT (SWCNT) and multi-walled CNT (MWCNT), have been well studied, both yielding similar results [45, 65]. Different cell lines (including human embryo kidney cells, human keratinocytes, HeLa cells, and lung carcinoma cells (A549 and H1299)) exposed to SWCNT all demonstrated dose- and time-dependent apoptosis and the inhibition of cell proliferation [66, 67]. Similar dose- and time-dependent increases in cell viability have been observed with MWCNT [68]. High concentrations (0.6 μg/mL) of MWCNT have also been reported to induce immune and inflammatory gene over-expression [69]. Surface modification can reduce such high toxicity. For instance, carboxyl-, biotin-, and fluorescein-coated SWCNTs and water-soluble SWCNTs demonstrated reduced or little cytotoxicity compared to unmodified SWCNTs [70, 71]. A few groups reported that surface modified, hydrophobic MWCNTs were less toxic than hydroxyl- or carboxyl-coated MWCNTs in the concentration range of 0.002–0.2 μg/mL [72].

Due to their excellent mechanical and biological properties, nanodiamond (including nanodiamond particles and nanocrystalline coatings) have emerged as promising materials for bioanalytical, drug delivery, and orthopedic implant applications [73–75]. At the same time, the toxicity of nanodiamond has also been of interest. Early studies since the 1960s on micrometer-sized diamond particles suggest its inertness, low chemotactic activity, and biocompatibility to a variety of cell types [76]. Most of the recent studies on nanodiamond have found consistent results, showing low toxicity to various cell types (osteoblasts, fibroblasts, human kidney cells, neuroblastoma, macrophage, keratinocyte, and PC-12 cells) and little production of reactive oxygen species [77, 78]. Direct transmission electron microscopy (TEM) evidence has shown that nanodiamond particles were internalized by neuroblastoma cells, but no significant cytotoxicity (in cell mitochondrial functions) was observed after the internalization [78].

4.6 Mechanisms Behind the Toxicity of Nanomaterials

Although the mechanisms behind the toxicity of nanomaterials have not been clearly outlined to date and need further investigation, several molecular and biochemical mechanisms have been proposed based on available results. These mechanisms mainly include cellular uptake of materials, ROS production, nanoparticle aggregation, and leaching of toxic species.

Cellular uptake of nanoparticles has been discussed in the previous section and can be directly observed though techniques like TEM. Many studies have attributed the cytotoxicity of CNTs and iron oxides to the internalization of nanoparticles [45]. However, the intracellular fate of numerous nanoparticles remains unknown, yet cellular internalization of nanoparticles is closely related to other molecular or biochemical activities leading to toxicity.

Nanoparticle-induced ROS generation has been widely accepted as the molecular mechanism of nanomaterial toxicity. Nanoparticles such as Fe, ZnO, SiO_2, CNT, and C_{60} have been shown to produce free radicals or cause oxidative stress, which further results in lipid peroxidation, DNA damage, cell membrane and cytoskeleton disruption, protein oxidation, and eventually, apoptosis or cell injury [36, 45, 48, 53, 63, 79]. Nevertheless, detailed mechanisms of ROS generation by exposure to nanoparticles remain unclear.

Particle aggregation or agglomeration in the aqueous environment is almost inevitable for all kinds of nanomaterials due to their extremely high surface area. Studies have suggested that nanoparticle aggregation also influences, or even dictates, nanomaterial toxicity. Wick *et al.* [80] hypothesized that SWCNT agglomeration causes *in vitro* cytotoxic effects on human MSTO-211H cells based on their experimental results, whereas contradictory results have demonstrated lower cytotoxicity with agglomerated SWCNTs in human fibroblasts [81]. This contradiction indicates that the particle agglomeration effect on nanomaterial toxicity is important and needs further understanding.

A few groups have attributed CNT cytotoxicity to trace amounts of catalysts (e.g., Fe, Pt, and Y) remnant from manufacturing of these nanomaterials [45, 82]. As mentioned before, the toxicity of nanopolymers (like hydrogels) may come from the unreacted monomers or initiators in the polymer structures. These results indicate that leaching of toxic species can also be a decisive factor in nanomaterial toxicity.

Lastly, it is important to emphasize that the mechanisms described earlier probably work collectively towards toxicity, and that all mechanisms are tightly related to the bulk and surface properties of the nanomaterial, as previously discussed. A better understanding of nanomaterials toxicity and successful solutions to reduce their toxic effects will largely rely on the discovery, design, and fabrication of appropriate nanomaterials. As a summary, several desirable properties of nanoscale implant materials which may reduce toxicity are summarized in Table 18.1; however, the information may not apply to all material formulations and may vary depending on specific conditions *in vivo* and *in vitro*.

Table 18.1 Desirable material properties that may reduce toxicity of nano implant materials

Material properties	Possible effect or consequence
Proper particle sizes (e.g., >200~500 nm)	Reduced endocytosis and carcinogenicity
PEGylated surfaces	Reduced toxic effects and immunogenicity
Hydroxylated or hydrophilic surfaces; water-soluble surfaces	Decreased cytotoxicity and macrophage functions; reduced complement activation and inflammatory reactions
Proper surface charges	Negatively charged: may reduce phagocytosis and platelet activation Positively charged: may reduce inflammatory reactions in some cases
Increased nanoscale roughness and/or decreased micron-scale roughness	Reduced macrophage functions and inflammatory reactions; increased osteoblast and endothelial viability
Low impurity (catalysts, unreacted materials, etc.)	Decreased toxic effects due to the impurities

5 Concluding Remarks

Understanding biological responses, especially, toxicity of nanoscale implant materials is an imperative, complex, and challenging journey which has exceptional impact on using nanomaterials to improve human health. This topic interfaces biology, toxicology, medicine, and materials science, and will not have easy answers. A collective consideration combining possible host responses, toxicological effects, material degradation consequences, and the synergy of material surface and bulk science will be necessary for designing and manufacturing the next generation of safe medical implants based on nanomaterials. Prior to this, establishing molecular, biophysical, and biochemical mechanisms behind nanomaterial-induced toxicity and other biological responses is extremely urgent, and timely efforts on this establishment will benefit and facilitate the development of improved medical implants.

References

1. MacDonald SJ, McCalden RW, Chess DG, Bourne RB, Rorabeck CH, Cleland A, Leung F (2003) Metal-on-metal versus polyethylene in hip arthroplasty: a randomized clinical trial. Clin Orthop Relat Res 406:282–296
2. Campbell P, Urban RM, Catelas I, Skipor AK, Schmalzried TP (2003) Autopsy analysis thirty years after metal-on-metal total hip replacement: a case report. J Bone Joint Surg Am 85-A (11):2218–2222
3. Sarmiento-Gonzalez A, Encinar J, Marchante-Gayon JM, Sanz-Medel A (2009) Titanium levels in the organs and blood of rats with a titanium implant, in the absence of wear, as determined by double-focusing ICP-MS. Anal Bioanal Chem 393(1):335–343

4. Anderson JM, Gristina AG, Hanson SR, Harker LA, Johnson RJ, Merritt K, Naylor PT, Schoen FJ (1996) Host reactions to biomaterials and their evaluation. In: Ratner BD, Hoffman AS, Schoen FJ, Lemons JE (eds) Biomaterials science: an introduction to materials in medicine. Academic, San Diego, pp 165–213
5. Bozic KJ (2009) The increasing number of THA revisions in the United States: why is it happening? http://www.orthosupersite.com/view.asp?rID=44153. Accessed 10/5 2009
6. Goldberg M, Langer R, Jia X (2007) Nanostructured materials for applications in drug delivery and tissue engineering. J Biomater Sci Polym Ed 18(3):241–268
7. Pelham RJ Jr, Wang Y (1997) Cell locomotion and focal adhesions are regulated by substrate flexibility. Proc Natl Acad Sci USA 94(25):13661–13665
8. Chan CE, Odde DJ (2008) Traction dynamics of filopodia on compliant substrates. Science 322(5908):1687–1691. doi:322/5908/1687 [pii] 10.1126/science.1163595
9. Discher DE, Janmey P, Wang YL (2005) Tissue cells feel and respond to the stiffness of their substrate. Science 310(5751):1139–1143. doi:310/5751/1139 [pii] 10.1126/science.1116995
10. Hench LL (1998) Bioactive materials: the potential for tissue regeneration. J Biomed Mater Res 41(4):511–518. doi:10.1002/(SICI)1097-4636(19980915)41:4<511::AID-JBM1>3.0.CO;2-F [pii]
11. Schuler M, Owen GR, Hamilton DW, De Wilde M, Textor M, Brunette DM, Tosatti SGP (2006) Biomimetic modification of titanium dental implant model surfaces using the RGDSP-peptide sequence: a cell morphology study. Biomaterials 27(21):4003–4015
12. Kroese-Deutman HC, Van Den Dolder J, Spauwen PHM, Jansen JA (2005) Influence of RGD-loaded titanium implants on bone formation in vivo. Tissue Eng 11(11–12):1867–1875
13. Balasundaram G, Yao C, Webster TJ (2008) TiO_2 nanotubes functionalized with regions of bone morphogenetic protein-2 increases osteoblast adhesion. J Biomed Mater Res A 84(2):447–453. doi:10.1002/jbm.a.31388
14. Assender H, Bliznyuk V, Porfyrakis K (2002) How surface topography relates to materials properties. Science 297(5583):973–976
15. Thomas KA, Cook SD (1985) An evaluation of variables influencing implant fixation by direct bone apposition. J Biomed Mater Res 19(8):875–901. doi:10.1002/jbm.820190802
16. Webster TJ, Ejiofor JU (2004) Increased osteoblast adhesion on nanophase metals: Ti, Ti6Al4V, and CoCrMo. Biomaterials 25(19):4731–4739
17. Price RL, Waid MC, Haberstroh KM, Webster TJ (2003) Selective bone cell adhesion on formulations containing carbon nanofibers. Biomaterials 24(11):1877–1887. doi: S0142961202006099 [pii]
18. Washburn NR, Yamada KM, Simon CG Jr, Kennedy SB, Amis EJ (2004) High-throughput investigation of osteoblast response to polymer crystallinity: influence of nanometer-scale roughness on proliferation. Biomaterials 25(7–8):1215–1224. doi:S0142961203006689 [pii]
19. Webster TJ, Hellenmeyer EL, Price RL (2005) Increased osteoblast functions on theta plus delta nanofiber alumina. Biomaterials 26(9):953–960. doi:10.1016/J.Biomaterials.0204.03.040
20. Liu H, Slamovich EB, Webster TJ (2006) Increased osteoblast functions among nanophase titania/poly(lactide-co-glycolide) composites of the highest nanometer surface roughness. J Biomed Mater Res A 78A(4):798–807
21. Balasundaram G (2007) Nanomaterials for Better Orthopedics. In: Webster TJ (ed) Nanotechnology for the regeneration of hard and soft tissues. World Scientific, Hackensack, NJ; London, pp 53–78
22. Zhu XS, Zhu L, Duan ZH, Qi RQ, Li Y, Lang YP (2008) Comparative toxicity of several metal oxide nanoparticle aqueous suspensions to Zebrafish (*Danio rerio*) early developmental stage. J Environ Sci Health A 43(3):278–284
23. McHale G, Shirtcliffe NJ, Aqil S, Perry CC, Newton MI (2004) Topography driven spreading. Phys Rev Lett 93(3):036102
24. Liu H, Webster TJ (2006) Nanomedicine for implants: a review of studies and necessary experimental tools. Biomaterials 28(2):354–369. doi:S0142-9612(06)00763-0 [pii] 10.1016/j.biomaterials.2006.08.049

25. Wilson CJ, Clegg RE, Leavesley DI, Pearcy MJ (2005) Mediation of biomaterial–cell interactions by adsorbed proteins: a review. Tissue Eng 11(1–2):1–18
26. Qiu Q, Sayer M, Kawaja M, Shen X, Davies JE (1998) Attachment, morphology, and protein expression of rat marrow stromal cells cultured on charged substrate surfaces. J Biomed Mater Res 42(1):117–127. doi:10.1002/(SICI)1097-4636(199810)42:1<117::AID-JBM15>3.0.CO;2-I [pii]
27. Zahr AS, Davis CA, Pishko MV (2006) Macrophage uptake of core-shell nanoparticles surface modified with poly(ethylene glycol). Langmuir 22(19):8178–8185
28. Gbadamosi JK, Hunter AC, Moghimi SM (2002) PEGylation of microspheres generates a heterogeneous population of particles with differential surface characteristics and biological performance. FEBS Lett 532(3):338–344
29. Cui ZR, Mumper RJ (2001) Chitosan-based nanoparticles for topical genetic immunization. J Control Release 75(3):409–419
30. Itoh S, Nakamura S, Nakamura M, Shinomiya K, Yamashita K (2006) Enhanced bone ingrowth into hydroxyapatite with interconnected pores by electrical polarization. Biomaterials 27(32):5572–5579
31. Lanone S, Boczkowski J (2006) Biomedical applications and potential health risks of nanomaterials: molecular mechanisms. Curr Mol Med 6(6):651–663
32. Rejman J, Oberle V, Zuhorn IS, Hoekstra D (2004) Size-dependent internalization of particles via the pathways of clathrin-and caveolae-mediated endocytosis. Biochem J 377:159–169
33. Raffa V, Ciofani G, Vittorio O, Riggio C, Cuschieri A (2010) Physicochemical properties affecting cellular uptake of carbon nanotubes. Nanomedicine-UK 5(1):89–97
34. Watari F, Abe S, Koyama C, Yokoyama A, Akasaka T, Uo M, Matsuoka M, Totsuka Y, Esaki M, Morita M, Yonezawa T (2008) Behavior of in vitro, in vivo and internal motion of micro/nano particles of titanium, titanium oxides and others. J Ceram Soc Jpn 116(1349):1–5
35. Vinogradov SV, Bronich TK, Kabanov AV (2002) Nanosized cationic hydrogels for drug delivery: preparation, properties and interactions with cells. Adv Drug Deliv Rev 54(1):135–147
36. Kadar E, Lowe DM, Sole M, Fisher AS, Jha AN, Readman JW, Hutchinson TH (2010) Uptake and biological responses to nano-Fe versus soluble $FeCl_3$ in excised mussel gills. Anal Bioanal Chem 396(2):657–666
37. Papageorgiou I, Yin ZR, Ladon D, Baird D, Lewis AC, Sood A, Newson R, Learmonth ID, Case CP (2007) Genotoxic effects of particles of surgical cobalt chrome alloy on human cells of different age in vitro. Mutat Res 619(1–2):45–58. doi:Doi 10.1016/J.Mrfmmm.2007.01.008
38. Papageorgiou I, Brown C, Schins R, Singh S, Newson R, Davis S, Fisher J, Ingham E, Case CP (2007) The effect of nano- and micron-sized particles of cobalt-chromium alloy on human fibroblasts in vitro. Biomaterials 28(19):2946–2958. doi:S0142-9612(07)00190-1 [pii] 10.1016/j.biomaterials.2007.02.034
39. Tkachenko AG, Xie H, Coleman D, Glomm W, Ryan J, Anderson MF, Franzen S, Feldheim DL (2003) Multifunctional gold nanoparticle-peptide complexes for nuclear targeting. J Am Chem Soc 125(16):4700–4701. doi:Doi 10.1021/Ja0296935
40. Tkachenko AG, Xie H, Liu YL, Coleman D, Ryan J, Glomm WR, Shipton MK, Franzen S, Feldheim DL (2004) Cellular trajectories of peptide-modified gold particle complexes: comparison of nuclear localization signals and peptide transduction domains. Bioconj Chem 15(3):482–490. doi:Doi 10.1021/Bc034189q
41. Goodman CM, McCusker CD, Yilmaz T, Rotello VM (2004) Toxicity of gold nanoparticles functionalized with cationic and anionic side chains. Bioconj Chem 15(4):897–900. doi:Doi 10.1021/Bc049951i
42. Dunford R, Salinaro A, Cai LZ, Serpone N, Horikoshi S, Hidaka H, Knowland J (1997) Chemical oxidation and DNA damage catalysed by inorganic sunscreen ingredients. FEBS Lett 418(1–2):87–90

43. Li N, Ma LL, Wang J, Zheng L, Liu J, Duan YM, Liu HT, Zhao XY, Wang SS, Wang H, Hong FS, Xie YN (2010) Interaction between nano-anatase TiO_2 and liver DNA from mice in vivo. Nanoscale Res Lett 5(1):108–115. doi:Doi 10.1007/S11671-009-9451-2
44. Landsiedel R, Ma-Hock L, Van Ravenzwaay B, Schulz M, Wiench K, Champ S, Schulte S, Wohlleben W, Oesch F (2010) Gene toxicity studies on titanium dioxide and zinc oxide nanomaterials used for UV-protection in cosmetic formulations. Nanotoxicology 4(4):364–381. doi:Doi 10.3109/17435390.2010.506694
45. Lewinski N, Colvin V, Drezek R (2008) Cytotoxicity of nanoparticles. Small 4(1):26–49. doi: Doi 10.1002/Smll.200700595
46. Brunner TJ, Wick P, Manser P, Spohn P, Grass RN, Limbach LK, Bruinink A, Stark WJ (2006) In vitro cytotoxicity of oxide nanoparticles: comparison to asbestos, silica, and the effect of particle solubility. Environ Sci Technol 40(14):4374–4381
47. Ye YY, Liu JW, Chen MC, Sun LJ, Lan MB (2010) In vitro toxicity of silica nanoparticles in myocardial cells. Environ Toxicol Pharmacol 29(2):131–137
48. Ye Y, Liu J, Xu J, Sun L, Chen M, Lan M (2010) Nano-SiO_2 induces apoptosis via activation of p53 and Bax mediated by oxidative stress in human hepatic cell line. Toxicol In Vitro 24(3):751–758. doi:S0887-2333(10)00002-0 [pii] 10.1016/j.tiv.2010.01.001
49. Gupta AK, Gupta M (2005) Cytotoxicity suppression and cellular uptake enhancement of surface modified magnetic nanoparticles. Biomaterials 26(13):1565–1573. doi: S0142961204004983 [pii] 10.1016/j.biomaterials.2004.05.022
50. Jeng HA, Swanson J (2006) Toxicity of metal oxide nanoparticles in mammalian cells. J Environ Sci Health A 41(12):2699–2711. doi:Doi 10.1080/10934520600966177
51. Tsaousi A, Jones E, Case CP (2010) The in vitro genotoxicity of orthopaedic ceramic (Al_2O_3) and metal (CoCr alloy) particles. Mutat Res 697(1–2):1–9. doi:Doi 10.1016/J. Mrgentox.2010.01.012
52. Oesterling E, Chopra N, Gavalas V, Arzuaga X, Lim EJ, Sultana R, Butterfield DA, Bachas L, Hennig B (2008) Alumina nanoparticles induce expression of endothelial cell adhesion molecules. Toxicol Lett 178(3):160–166. doi:S0378-4274(08)00079-9 [pii] 10.1016/j. toxlet.2008.03.011
53. Lin WS, Xu Y, Huang CC, Ma YF, Shannon KB, Chen DR, Huang YW (2009) Toxicity of nano- and micro-sized ZnO particles in human lung epithelial cells. J Nanopart Res 11(1):25–39. doi:Doi 10.1007/S11051-008-9419-7
54. Wang AZ, Gu FX, Farokhzad OC (2009) Nanoparticles for Cancer Diagnosis and Therapy. In: safety of Nanoparticles. Nanostructure Science and Technology. Springer New York, pp 1–27. doi:10.1007/978-0-387-78608-7_10
55. Meenach SA, Anderson KW, Hilt JZ (2009) Hydrogel Nanocomposites: biomedical Applications, Biocompatibility, and Toxicity Analysis. In: Safety of Nanoparticles. Nanostructure Science and Technology. Springer New York, pp 1–27. doi:10.1007/978-0-387-78608-7_7
56. Wei X, Lee Y-k, Huh KM, Kim S, Park K (2009) Safety and Efficacy of Nano/Micro Materials. In: Safety of nanoparticles. Nanostructure science and technology. Springer New York, pp 1–26. doi:10.1007/978-0-387-78608-7_4
57. Peppas NA, Bures P, Leobandung W, Ichikawa H (2000) Hydrogels in pharmaceutical formulations. Eur J Pharm Biopharm 50(1):27–46
58. Weng H, Zhou J, Tang LP, Hu ZB (2004) Tissue responses to thermally-responsive hydrogel nanoparticles. J Biomater Sci Polym Ed 15(9):1167–1180
59. Chun YW, Webster TJ (2009) The role of nanomedicine in growing tissues. Ann Biomed Eng 37(10):2034–2047
60. Vandrovcova M, Vacik J, Svorcik V, Slepicka P, Kasalkova N, Vorlicek V, Lavrentiev V, Vosecek V, Grausova L, Lisa V, Bacakova L (2008) Fullerene C-60 and hybrid C-60/Ti films as substrates for adhesion and growth of bone cells. Phys Status Solidi A 205(9):2252–2261. doi:Doi 10.1002/Pssa.200879730

61. Fiorito S, Serafino A, Andreola F, Bernier P (2006) Effects of fullerenes and single-wall carbon nanotubes on murine and human macrophages. Carbon 44(6):1100–1105. doi:Doi 10.1016/J.Carbon.2005.11.009
62. Jia G, Wang HF, Yan L, Wang X, Pei RJ, Yan T, Zhao YL, Guo XB (2005) Cytotoxicity of carbon nanomaterials: single-wall nanotube, multi-wall nanotube, and fullerene. Environ Sci Technol 39(5):1378–1383
63. Sayes CM, Fortner JD, Guo W, Lyon D, Boyd AM, Ausman KD, Tao YJ, Sitharaman B, Wilson LJ, Hughes JB, West JL, Colvin VL (2004) The differential cytotoxicity of water-soluble fullerenes. Nano Lett 4(10):1881–1887
64. Rouse JG, Yang JZ, Barron AR, Monteiro-Riviere NA (2006) Fullerene-based amino acid nanoparticle interactions with human epidermal keratinocytes. Toxicol In Vitro 20(8):1313–1320
65. Webster TJ (2009) Safety of nanoparticles: from manufacturing to medical applications. Nanostructure science and technology, Springer, New York
66. Cui D, Tian F, Ozkan CS, Wang M, Gao H (2005) Effect of single wall carbon nanotubes on human HEK293 cells. Toxicol Lett 155(1):73–85. doi:S0378-4274(04)00410-2 [pii] 10.1016/j.toxlet.2004.08.015
67. Manna SK, Sarkar S, Barr J, Wise K, Barrera EV, Jejelowo O, Rice-Ficht AC, Ramesh GT (2005) Single-walled carbon nanotube induces oxidative stress and activates nuclear transcription factor-kappa B in human keratinocytes. Nano Lett 5(9):1676–1684
68. Monteiro-Riviere NA, Inman AO (2006) Challenges for assessing carbon nanomaterial toxicity to the skin. Carbon 44(6):1070–1078
69. Ding LH, Stilwell J, Zhang TT, Elboudwarej O, Jiang HJ, Selegue JP, Cooke PA, Gray JW, Chen FQF (2005) Molecular characterization of the cytotoxic mechanism of multiwall carbon nanotubes and nano-onions on human skin fibroblast. Nano Lett 5(12):2448–2464
70. Kam NWS, Jessop TC, Wender PA, Dai HJ (2004) Nanotube molecular transporters: internalization of carbon nanotube-protein conjugates into mammalian cells. J Am Chem Soc 126(22):6850–6851
71. Sayes CM, Liang F, Hudson JL, Mendez J, Guo WH, Beach JM, Moore VC, Doyle CD, West JL, Billups WE, Ausman KD, Colvin VL (2006) Functionalization density dependence of single-walled carbon nanotubes cytotoxicity in vitro. Toxicol Lett 161(2):135–142
72. Magrez A, Kasas S, Salicio V, Pasquier N, Seo JW, Celio M, Catsicas S, Schwaller B, Forro L (2006) Cellular toxicity of carbon-based nanomaterials. Nano Lett 6(6):1121–1125
73. Yang L, Sheldon BW, Webster TJ (2009) The impact of diamond nanocrystallinity on osteoblast functions. Biomaterials 30(20):3458–3465
74. Yang L, Sheldon BW, Webster TJ (2009) Orthopedic nano diamond coatings: control of surface properties and their impact on osteoblast adhesion and proliferation. J Biomed Mater Res A 91A(2):548–556
75. Schrand AM, Hens SAC, Shenderova OA (2009) Nanodiamond particles: properties and perspectives for bioapplications. Crit Rev Solid State 34(1–2):18–74. doi:Doi 10.1080/10408430902831987 Pii 910822117
76. Schrand AM, Johnson J, Dai L, Hussain SM, Schlager JJ, Zhu L, Hong Y, Ōsawa E (2009) Cytotoxicity and genotoxicity of carbon nanomaterials. In: Safety of Nanoparticles. Nanostructure Science and Technology. Springer New York, pp 1–29. doi:10.1007/978-0-387-78608-7_8
77. Yu SJ, Kang MW, Chang HC, Chen KM, Yu YC (2005) Bright fluorescent nanodiamonds: no photobleaching and low cytotoxicity. J Am Chem Soc 127(50):17604–17605
78. Schrand AM, Huang HJ, Carlson C, Schlager JJ, Osawa E, Hussain SM, Dai LM (2007) Are diamond nanoparticles cytotoxic? J Phys Chem B 111(1):2–7
79. Sayes CM, Gobin AM, Ausman KD, Mendez J, West JL, Colvin VL (2005) Nano-C-60 cytotoxicity is due to lipid peroxidation. Biomaterials 26(36):7587–7595

80. Wick P, Manser P, Limbach LK, Dettlaff-Weglikowska U, Krumeich F, Roth S, Stark WJ, Bruinink A (2007) The degree and kind of agglomeration affect carbon nanotube cytotoxicity. Toxicol Lett 168(2):121–131
81. Tian FR, Cui DX, Schwarz H, Estrada GG, Kobayashi H (2006) Cytotoxicity of single-wall carbon nanotubes on human fibroblasts. Toxicol In Vitro 20(7):1202–1212
82. Shvedova AA, Castranova V, Kisin ER, Schwegler-Berry D, Murray AR, Gandelsman VZ, Maynard A, Baron P (2003) Exposure to carbon nanotube material: assessment of nanotube cytotoxicity using human keratinocyte cells. J Toxicol Environ Health A 66(20):1909–1926

Biography

Thomas J. Webster is an associate professor in the School of Engineering and Department of Orthopaedics at Brown University. His degrees are in chemical engineering from the University of Pittsburgh (B.S., 1995), and in biomedical engineering from Rensselaer Polytechnic Institute (M.S., 1997; Ph.D., 2000). To date, his lab group has generated 8 textbooks, 48 book chapters, 233 invited presentations, at least 343 peer-reviewed literature articles and/or conference proceedings, at least 504 conference presentations, and 24 provisional or full patents. Some of these patents led to the formation of three companies. He is the founding editor-in-chief of the International Journal of Nanomedicine (the first international journal in nanomedicine), has organized 13 conferences emphasizing nanotechnology in medicine, and has organized over 49 symposia at numerous conferences emphasizing biological interactions with nanomaterials. Webster has received numerous honors including: 2002, Biomedical Engineering Society Rita Schaffer Young Investigator Award; 2003, Outstanding Young Investigator Award Purdue University College of Engineering; 2005, American Association of Nanomedicine Young Investigator Award Finalist; 2005, Coulter Foundation Young Investigator Award; 2006, Fellow, American Association of Nanomedicine; and 2010, Distinguished Lecturer in Nanomedicine, University of South Florida. He was recently appointed co-director of the Indo-US Center for Biomaterials for Healthcare.

Dr. Lei Yang is currently a postdoctoral research associate in the School of Engineering at Brown University. He received a Ph.D. (2011) in Engineering from Brown University, and a B.S. (2004) and a M.S. (2006) in Materials Science and Engineering from Tsinghua University, Beijing. His research interests focus on biomaterials for orthopedic applications and advanced carbon nanostructures. He has published over 16 peer-reviewed journal papers, 7 conference proceedings, and 4 patents. His work on nanocrystalline diamond (NCD) orthopedic implant coatings and mathematical modeling of cell behaviors on nanotopography has been presented at 18 conferences. He received the 2011 Society for Biomaterials (SFB) Student Award for Outstanding Research, 2010 Materials Research Society (MRS) Graduate Student Silver Award, 2010 SFB Student Travel Achievement Recognition (STAR), and Undergraduate and Graduate Student Poster Award (first place) at the 2010 Sigma Xi Northeast Regional Conference.

Index

N. Eliaz (ed.), *Degradation of Implant Materials*, DOI 10.1007/978-1-4614-3942-4,

S

T